AF823656

BIOLOGICAL AND MEDICAL PHYSICS, BIOMEDICAL ENGINEERING

For further volumes:
http://www.springer.com/series/3740

BIOLOGICAL AND MEDICAL PHYSICS, BIOMEDICAL ENGINEERING

The fields of biological and medical physics and biomedical engineering are broad, multidisciplinary and dynamic. They lie at the crossroads of frontier research in physics, biology, chemistry, and medicine. The Biological and Medical Physics, Biomedical Engineering Series is intended to be comprehensive, covering a broad range of topics important to the study of the physical, chemical and biological sciences. Its goal is to provide scientists and engineers with textbooks, monographs, and reference works to address the growing need for information.

Books in the series emphasize established and emergent areas of science including molecular, membrane, and mathematical biophysics; photosynthetic energy harvesting and conversion; information processing; physical principles of genetics; sensory communications; automata networks, neural networks, and cellular automata. Equally important will be coverage of applied aspects of biological and medical physics and biomedical engineering such as molecular electronic components and devices, biosensors, medicine, imaging, physical principles of renewable energy production, advanced prostheses, and environmental control and engineering.

Editor-in-Chief:

Elias Greenbaum, Oak Ridge National Laboratory, Oak Ridge, Tennessee, USA

Editorial Board:

Masuo Aizawa, Department of Bioengineering, Tokyo Institute of Technology, Yokohama, Japan

Olaf S. Andersen, Department of Physiology, Biophysics & Molecular Medicine, Cornell University, New York, USA

Robert H. Austin, Department of Physics, Princeton University, Princeton, New Jersey, USA

James Barber, Department of Biochemistry, Imperial College of Science, Technology and Medicine, London, England

Howard C. Berg, Department of Molecular and Cellular Biology, Harvard University, Cambridge, Massachusetts, USA

Victor Bloomfield, Department of Biochemistry, University of Minnesota, St. Paul, Minnesota, USA

Robert Callender, Department of Biochemistry, Albert Einstein College of Medicine, Bronx, New York, USA

Steven Chu, Lawrence Berkeley National Laboratory, Berkeley, California, USA

Louis J. DeFelice, Department of Pharmacology, Vanderbilt University, Nashville, Tennessee, USA

Johann Deisenhofer, Howard Hughes Medical Institute, The University of Texas, Dallas, Texas, USA

George Feher, Department of Physics, University of California, San Diego, La Jolla, California, USA

Hans Frauenfelder, Los Alamos National Laboratory, Los Alamos, New Mexico, USA

Ivar Giaever, Rensselaer Polytechnic Institute, Troy, New York, USA

Sol M. Gruner, Cornell University, Ithaca, New York, USA

Judith Herzfeld, Department of Chemistry, Brandeis University, Waltham, Massachusetts, USA

Mark S. Humayun, Doheny Eye Institute, Los Angeles, California, USA

Pierre Joliot, Institute de Biologie Physico-Chimique, Fondation Edmond de Rothschild, Paris, France

Lajos Keszthelyi, Institute of Biophysics, Hungarian Academy of Sciences, Szeged, Hungary

Robert S. Knox, Department of Physics and Astronomy, University of Rochester, Rochester, New York, USA

Aaron Lewis, Department of Applied Physics, Hebrew University, Jerusalem, Israel

Stuart M. Lindsay, Department of Physics and Astronomy, Arizona State University, Tempe, Arizona, USA

David Mauzerall, Rockefeller University, New York, New York, USA

Eugenie V. Mielczarek, Department of Physics and Astronomy, George Mason University, Fairfax, Virginia, USA

Markolf Niemz, Medical Faculty Mannheim, University of Heidelberg, Mannheim, Germany

V. Adrian Parsegian, Physical Science Laboratory, National Institutes of Health, Bethesda, Maryland, USA

Linda S. Powers, University of Arizona, Tucson, Arizona, USA

Earl W. Prohofsky, Department of Physics, Purdue University, West Lafayette, Indiana, USA

Andrew Rubin, Department of Biophysics, Moscow State University, Moscow, Russia

Michael Seibert, National Renewable Energy Laboratory, Golden, Colorado, USA

David Thomas, Department of Biochemistry, University of Minnesota Medical School, Minneapolis, Minnesota, USA

Brendan Z. Allison • Stephen Dunne
Robert Leeb • José del R. Millán
Anton Nijholt
Editors

Towards Practical Brain-Computer Interfaces

Bridging the Gap from Research to Real-World Applications

With 107 Figures

Springer

Editors
Brendan Z. Allison
Graz University of Technology
Austria

Anton Nijholt
University of Twente
Enschede, The Netherlands

Stephen Dunne
StarLab Barcelona
Spain

Robert Leeb
José del R. Millán
Swiss Federal Institute of Technology Lausanne
Switzerland

Biological and Medical Physics, Biomedical Engineering ISSN 1618-7210
ISBN 978-3-642-29745-8 ISBN 978-3-642-29746-5 (eBook)
DOI 10.1007/978-3-642-29746-5
Springer Heidelberg New York Dordrecht London

Library of Congress Control Number: 2012944560

© Springer-Verlag Berlin Heidelberg 2012
This work is subject to copyright. All rights are reserved by the Publisher, whether the whole or part of the material is concerned, specifically the rights of translation, reprinting, reuse of illustrations, recitation, broadcasting, reproduction on microfilms or in any other physical way, and transmission or information storage and retrieval, electronic adaptation, computer software, or by similar or dissimilar methodology now known or hereafter developed. Exempted from this legal reservation are brief excerpts in connection with reviews or scholarly analysis or material supplied specifically for the purpose of being entered and executed on a computer system, for exclusive use by the purchaser of the work. Duplication of this publication or parts thereof is permitted only under the provisions of the Copyright Law of the Publisher's location, in its current version, and permission for use must always be obtained from Springer. Permissions for use may be obtained through RightsLink at the Copyright Clearance Center. Violations are liable to prosecution under the respective Copyright Law.
The use of general descriptive names, registered names, trademarks, service marks, etc. in this publication does not imply, even in the absence of a specific statement, that such names are exempt from the relevant protective laws and regulations and therefore free for general use.
While the advice and information in this book are believed to be true and accurate at the date of publication, neither the authors nor the editors nor the publisher can accept any legal responsibility for any errors or omissions that may be made. The publisher makes no warranty, express or implied, with respect to the material contained herein.

Printed on acid-free paper

Springer is part of Springer Science+Business Media (www.springer.com)

Preface

Brain–computer interface (BCI) research is advancing rapidly. The last few years have seen a dramatic rise in journal publications, academic workshops and conferences, books, new products aimed at both healthy and disabled users, research funding from different sources, and media attention. This media attention has included both BCI fi (BCI-based science fiction) and stories in mainstream magazines and television news programs.

Despite this progress and attention, most people still do not use BCIs, or even know what they are. While the authors of this book generally have access to the best BCI equipment, and they know how to use it, the chapters are written in the old-fashioned way, with keyboards and mice instead of BCIs. This may be surprising because BCIs are generally presented inaccurately in the popular media, where undeserved hype and sloppy reporting often create a gap between expectations and reality.

This book aims to bridge that gap by educating readers about BCIs, with emphasis on making BCIs practical in real-world settings. Experts in BCI research widely agree that one of the major challenges in the field is moving BCIs from laboratory gadgets that work with some healthy users to tools that are reliable, straightforward, and useful in field settings for whoever needs them. Many of these experts discuss the state of the art and major challenges across four sections. Three of the sections address the three main components of BCIs: sensors, signals, and signal processing; devices and applications; and interfaces and environments. The last section summarizes other challenges that relate to complete BCI systems instead of one component.

BCI research is inherently interdisciplinary, requiring contributions from neuroscience, psychology, medicine, human–computer interaction (HCI), many facets of engineering, and other disciplines. Similarly, many sectors are involved in BCI research, including academia, small and large businesses, government, medicine, and different types of nonprofit institutions. The authors who contributed to this book represent an eclectic mix of these disciplines and sectors. This breadth of contributors provides different perspectives and should make this book relevant to a wide variety of readers.

However, while this book could be useful for different specialists in the BCI community, we also made a strong effort to keep the chapters practical and readable for people who do not have a background in BCI research or any related discipline. Chapters are written in plain English, without unnecessary technical detail, and acronyms and special terms are defined within chapters and in our glossary. Ample references are provided in case readers want more information. Hence, many readers outside of the conventional BCI community may enjoy this book for different reasons. Nurses, doctors, therapists, caretakers, and assistive technology practitioners may want to learn more about what real-world BCIs can (and cannot) do, which may help them decide whether a BCI is viable as an assistive technology. Other readers may instead be curious about BCIs for other user groups, including healthy users. Students might use this book to learn about BCIs, and teachers might assign chapters in relevant classes. Business experts and policy makers may want to learn more about whether BCIs are promising enough to merit additional funding through commercial investment or grants. Journalists, writers, or other people interested in developing articles, documentaries, or other shows might find helpful background information or inspiration here. Finally, we hope our book appeals to people who are just curious about a technology that has long captured the human imagination and could revolutionize how people interact with each other and their environments.

Acknowledgements: The editors gratefully acknowledge the help of the following chapter reviewers: Tom Carlson, Günter Edlinger, Jan van Erp, Shangkai Gao, Gary Garcia Molina, Gangadhar Garipelli, Cuntai Guan, David Ibañez, Andrea Kübler, Bram Van de Laar, Fabien Lotte, Massimiliano Malavasi, Behnam Molaee, Roderick Murray-Smith, Tim Mullen, Femke Nijboer, Dani Perez Marcos, Mannes Poel, Aureli Soria-Frisch, Olga Sourina, Michael Tangermann, Aleksander Väljamäe, Yijun Wang, Tomas Ward, and Thorsten Zander. Their often extensive and always careful comments certainly helped the authors in improving the chapters. The editors also want to express their gratitude to their "technical editor," Hendri Hondorp from the HMI group of the University of Twente, for improving uniformity, consistency, and completeness of the book. Finally, preparation of many chapters in this book has benefited from funding from the *European Union Seventh Framework Programme (FP7/2007-2013)*. In particular the editors gratefully acknowledge the support of the *Future BNCI project (Project number ICT-248320)*.

Enschede
November 2011

Brendan Z. Allison
(Graz University of Technology, Austria)
Stephen Dunne
(StarLab Barcelona, Spain)
Robert Leeb
(École Polytechnique Fédérale de Lausanne, Switserland)
José del R. Millán
(École Polytechnique Fédérale de Lausanne, Switserland)
Anton Nijholt
(University of Twente, The Netherlands)

Contents

List of Contributors

Brendan Z. Allison Institute for Knowledge Discovery, Laboratory of Brain–Computer Interfaces, Graz University of Technology, Graz, Austria

Giuseppe Andreoni INDACO, Politecnico di Milano, Milan, Italy

Günther Bauernfeind Institute for Knowledge Discovery, Laboratory of Brain–Computer Interfaces, Graz University of Technology, Graz, Austria

Lugi Bianchi Neuroscience Department, Tor Vergata University of Rome, Rome, Italy

Martin Billinger Institute for Knowledge Discovery, Laboratory of Brain–Computer Interfaces, Graz University of Technology, Graz, Austria

Benjamin Blankertz Machine Learning Laboratory, Berlin Institute of Technology, Berlin, Germany

Christian Breitwieser Institute for Knowledge Discovery, Laboratory of Brain–Computer Interfaces, Graz University of Technology, Graz, Austria

Anne-Marie Brouwer TNO, Soesterberg, The Netherlands

Clemens Brunner Institute for Knowledge Discovery, Laboratory of Brain–Computer Interfaces, Graz University of Technology, Graz, Austria Swartz Center for Computational Neuroscience, UC San Diego, La Jolla, CA, USA

Ian Daly Institute for Knowledge Discovery, Laboratory for Brain–Computer Interfaces, Graz University of Technology, Graz, Austria

Lorenzo Desideri AIAS Bologna onlus, Ausilioteca AT Centre, Corte Roncati, Bologna, Italy

Stephen Dunne StarLab Teodor Roviralta, Barcelona, Spain

Günter Edlinger g.tec Medical Engineering GmbH, Schiedlberg, Austria Guger Technologies OG, Graz, Austria

Jan B.F. van Erp TNO, Soesterberg, The Netherlands

Josef Faller Institute for Knowledge Discovery, Laboratory of Brain–Computer Interfaces, Graz University of Technology, Graz, Austria

Gary Garcia-Molina Philips Research Europe, Eindhoven, The Netherlands

Christoph Guger g.tec Medical Engineering GmbH, Schiedlberg, Austria Guger Technologies OG, Graz, Austria

Hayrettin Gürkök Human Media Interaction, University of Twente, Enschede, The Netherlands

Christian Henle Laboratory for Biomedical Microtechnology, Department of Microsystems Engineering – IMTEK, University of Freiburg, Freiburg, Germany Cortec GmbH, Freiburg, Germany

Johannes Höhne Berlin Institute of Technology, Department of Machine Learning, Berlin Brain–Computer Interface (BBCI) group, Berlin, Germany

Elisa Holz Department of Psychology I, University of Würzburg, Würzburg, Germany

Evert-Jan Hoogerwerf AIAS Bologna onlus, Ausilioteca AT Centre, Corte Roncati, Bologna, Italy

Petar Horki Graz University of Technology, Institute for Knowledge Discovery, BCI-Lab, Graz, Austria

Jing Jin Key Laboratory of Advanced Control and Optimization for Chemical Processes, East China University of Science and Technology, Shanghai, China

Tzyy-Ping Jung Swartz Center for Computational Neuroscience, Institute for Neural Computation Institute of Engineering in Medicine, University of California San Diego, La Jolla, CA, USA

Vera Kaiser Institute for Knowledge Discovery, Laboratory of Brain–Computer Interfaces, Graz University of Technology, Graz, Austria

Shin'ichiro Kanoh Department of Electronics and Intelligent Systems, Tohoku Institute of Technology, Taihaku-ku, Sendai, Japan

Tobias Kaufmann Department of Psychology I, University of Würzburg, Würzburg, Germany

Christian A. Kothe Swartz Center for Computational Neuroscience, Institute for Neural Computation, University of California San Diego, La Jolla, CA, USA,

Alex Kreilinger Graz University of Technology, Institute for Knowledge Discovery, BCI-Lab, Graz, Austria

Andrea Kübler Department of Psychology I, University of Würzburg, Würzburg, Germany

Bram van de Laar Human Media Interaction, University of Twente, AE Enschede, The Netherlands

Anatole Lécuyer INRIA Rennes Bretagne-Atlantique, Campus Universitaire de Beaulieu, Rennes Cedex, France

Robert Leeb Chair in Non-Invasive Brain-Machine Interface, École Polytechnique, Fédérale de Lausanne, Lausanne, Switzerland

Yisi Liu Nanyang Technological University, Nanyang Ave, Singapore

Fabien Lotte INRIA Bordeaux Sud-Ouest, Talence, France

Scott Makeig Swartz Center for Computational Neuroscience, Institute for Neural Computation, University of California San Diego, La Jolla, CA, USA

Massimiliano Malavasi AIAS Bologna onlus, Ausilioteca AT Centre, Corte Roncati, Bologna, Italy

Steven G. Mason Left Coast Biometrics Ltd., Vancouver, BC, Canada

Donatella Mattia Clinical Neurophysiology, Neuroelectrical Imaging and BCI Lab, Fondazione Santa Lucia, IRCCS, Rome, Italy

Jürgen Mellinger Institute of Medical Psychology and Behavioral Neurobiology, University of Tübingen, Tübingen, Germany

José d. R. Millán Chair in Non-Invasive Brain-Machine Interface, École Polytechnique, Fédérale de Lausanne, Lausanne, Switzerland

Marco Molinari Spinal Cord Injury Unit, Fondazione Santa Lucia, IRCCS, Rome, Italy

Melody M. Moore Jackson Georgia Institute of Technology, College of Computing, NW Atlanta, GA, USA

Gernot R. Müller-Putz Institute for Knowledge Discovery, Laboratory of Brain–Computer Interfaces, Graz University of Technology, Graz, Austria

Roderick Murray-Smith School of Computing Science, University of Glasgow, Glasgow, Scotland

Minh Khoa Nguyen Nanyang Technological University, Nanyang Ave, Singapore

Femke Nijboer Human Media Interaction, University of Twente, AE Enschede, The Netherlands

Anton Nijholt Human Media Interaction, University of Twente, AE Enschede, The Netherlands

Paolo Perego INDACO, Politecnico di Milano, Milan, Italy

Gert Pfurtscheller Institute for Knowledge Discovery, Laboratory of Brain–Computer Interfaces, Graz University of Technology, Graz, Austria

Floriana Pichiorri Neuroelectrical Imaging and BCI Lab, Fondazione Santa Lucia, IRCCS, Rome, Italy

Danny Plass-Oude Bos Human Media Interaction, University of Twente, AE Enschede, The Netherlands

Melissa Quek School of Computing Science, University of Glasgow, Glasgow, Scotland

Adriane B. Randolph Kennesaw State University, Information Systems, Kennesaw, GA, USA

Yann Renard Independent Brain–Computer Interfaces & OpenViBE Consultant, Rennes, France

Jörn Rickert Bernstein Center Freiburg, University Freiburg, Freiburg, Germany Cortec GmbH, Freiburg, Germany

Rüdiger Rupp Spinal Cord Injury Center, Heidelberg University Hospital, Heidelberg, Germany

Gerwin Schalk Laboratory of Nervous System Disorders, Division of Genetic Disorders, Wadsworth Center, New York State Department of Health, Albany, NY, USA

Reinhold Scherer Graz University of Technology, Institute for Knowledge Discovery, BCI-Lab, Graz, Austria

Martin Schuettler Laboratory for Biomedical Microtechnology, Department of Microsystems Engineering – IMTEK, University of Freiburg, Freiburg, Germany Cortec GmbH, Freiburg, Germany

Aureli Soria-Frisch Starlab Barcelona SL, Barcelona, Spain

Olga Sourina Nanyang Technological University, Nanyang Ave, Singapore

Thomas Stieglitz Laboratory for Biomedical Microtechnology, Department of Microsystems Engineering - IMTEK, University of Freiburg, Freiburg, Germany Bernstein Center Freiburg, University Freiburg, Freiburg, Germany Cortec GmbH, Freiburg, Germany

I. Putu Susila Nuclear Equipment Engineering Center, National Atomic Energy Agency of Indonesia (BATAN), Tangerang Selatan, Indonesia

Michael Tangermann Berlin Institute of Technology, Department of Machine Learning, Berlin Brain–Computer Interface (BBCI) Group, Berlin, Germany

Marieke E. Thurlings TNO, Soesterberg, The Netherlands

Aleksander Väljamäe Institute for Knowledge Discovery, Laboratory for Brain–Computer Interfaces, Graz University of Technology, Graz, Austria

Bastian Venthur Machine Learning Laboratory, Berlin Institute of Technology, Berlin, Germany

Isabella C. Wagner Donders Institute for Brain, Cognition and Behaviour, Centre for Cognitive Neuroimaging, Radboud University Nijmegen, Nijmegen, The Netherlands

Qiang Wang Nanyang Technological University, Nanyang Ave, Singapore

Yijun Wang Swartz Center for Computational Neuroscience, Institute for Neural Computation, University of California San Diego, La Jolla, CA, USA

Tomas E. Ward Department of Electronic Engineering, National University of Ireland Maynooth, Maynooth, Co. Kildare, Ireland

Peter J. Werkhoven TNO, Soesterberg, The Netherlands

Danhua Zhu College of Biomedical Engineering and Instrument Science, Zhejiang University, Hangzhou, China

Acronyms

AD	Assistive device
ANFIS	Adaptive neuro-fuzzy inference systems
ANOVA	ANalysis Of VAriance
AR	Augmented reality
ASSR	Auditory steady-state responses
AT	Assistive technology
BCI	Brain computer interface
BMI	Brain-machine interface
BNCI	Brain/neuronal computer interface
BSS	Blind source separation
CAD	Computer aided design
CLIS	Complete locked-in syndrome
CSP	Common spatial patterns
ECG	ElectroCardioGram
ECoG	ElectroCortic**o**Gram
EDA	ElectroDermal Activity
EEG	ElectroEncephaloGraphy
EM	Expectation maximization
EMG	ElectroMyoGram
EOG	ElectroOculoGraphy
ERD	Event related de-/synchronisation
ERP	Event-related potential
ERS	Event related de-/synchronisation
FES	Functional electrical stimulation
fNIRS	**f**unctional Near infrared spectroscopy
GMM	Gaussian mixture models
GSR	Galvanic skin response
hBCI	**h**ybrid BCI
HMM	Hidden Markov models
HR	Heart rate
ICA	Independent component analysis

ITR	Information transfer rate
KNN	K-nearest neighbors
LDA	Linear discriminant analysis
LED	Light emitting diode
LiS	Locked-in syndrome
LVQ	Linear vector quantization
MEG	MagnetoEncephaloGram
ME	Motor execution
MI	Motor imagery
MLP	Multi-layer perceptron
NIRS	Near InfraRed Spectroscopy
NN	Neural network
PCA	Principal component analysis
RESE	Random electrode selection ensemble
RLDA	Regularized linear discriminant analysis
SCI	Spinal cord injury
SFFS	Sequential floating forward search
SSSEP	Steady-state somatosensory evoked potential
SSVEP	Steady-state visual evoked potential
SVM	Support vector machine
UCD	User-centred design
VE	Virtual environment
VR	Virtual reality

Chapter 1
Recent and Upcoming BCI Progress: Overview, Analysis, and Recommendations

Brendan Z. Allison, Stephen Dunne, Robert Leeb, José del R. Millán, and Anton Nijholt

1.1 Introduction

Brain–computer interfaces (BCIs) let people communicate without using muscular activity. BCIs have been developed primarily as communication devices for people who cannot move because of conditions like Lou Gehrig's disease. However, recent advancements like practical electrodes, usable and adaptive software, and reduced cost have made BCIs appealing to new user groups. People with mild to moderate disabilities might benefit from BCIs, which were previously so cumbersome and technically demanding that other assistive communication technologies were preferable. Simple and cheap BCIs have gained attention among a much larger market: healthy users.

Right now, healthy people who use BCIs generally do so for fun. These types of BCIs will gain wider adoption, but not as much as the next generation of field BCIs and similar systems, which healthy people will use because they consider them useful. These systems could provide useful communication in situations

B.Z. Allison (✉)
Department of Cognitive Science, University of California at San Diego, USA
e-mail: bci2k2@yahoo.com

S. Dunne
Starlab Teodor Roviralta 45, 08022 Barcelona, Spain
e-mail: stephen.dunne@starlab.es

R. Leeb · J.d.R. Millán
Chair in Non-Invasive Brain-Machine Interface, École Polytechnique Fédérale de Lausanne, Station 11, CH-1015 Lausanne, Switzerland
e-mail: robert.leeb@epfl.ch; jose.millan@epfl.ch

A. Nijholt
Human Media Interaction, University of Twente, PO Box 217, 7500 AE Enschede, The Netherlands
e-mail: a.nijholt@utwente.nl

B. Z. Allison et al. (eds.), *Towards Practical Brain-Computer Interfaces*, Biological and Medical Physics, Biomedical Engineering, DOI 10.1007/978-3-642-29746-5_1,
© Springer-Verlag Berlin Heidelberg 2012

when conventional means such as keyboards or game controllers are unavailable or inadequate. Future BCIs will go beyond communication in different ways, such as monitoring error, alertness, frustration, or other cognitive and emotive states to facilitate human–computer interaction (HCI). The hardware, software, and functionality afforded by BCIs will be more effectively integrated with any devices that the user already wears or carries. BCIs that contribute to rehabilitation or functional improvement could go further beyond communication and make BCIs appealing to far more users, such as persons with stroke, autism, or attentional disorders. The next 5 years will help resolve which of these areas are promising.

The BCI community also faces growing challenges. Because BCIs are generally not well known or understood, many end users and others may have unrealistic expectations or fears. Groups might unnecessarily conduct research that was already done, or miss opportunities from other disciplines or research projects. In addition to developing and sharing knowledge about BCIs, we also need practical infrastructural issues like terms, definitions, standards, and ethical and reporting guidelines. The appeal of the brand "BCI" could encourage unjustified boasting, unscrupulous reporting in the media or scientific literature, products that are not safe or effective, or other unethical practices. The acronym is already used much more broadly than it was just 5 years ago, such as to refer to devices that write to the brain or literally read minds [8, 23].

On the other hand, several key advances cannot be ignored. With improved flexibility and reliability, new applications, dry electrodes that rely on gold and composites rather than gel, practical software, and growing public appeal, we could be on the verge of a Golden Age of BCI research. Key performance indicators like sales, cost, and dependence on support should reflect substantial progress in the next 5 years. While the spirit of camaraderie and enthusiasm should remain strong within the BCI community, the BCIs in 5 years will be significantly better in many ways. This sentimental elan was captured best by Jacques Vidal, the inventor of BCIs, who gave a lecture after many years of retirement at a workshop that we authors hosted in Graz, Austria in September 2011. "It still feels like yesterday," he said, "but it isn't."

1.2 Overview of This Book

This book is divided into four sections. These sections are structured around the four components of a BCI (Fig. 1.1). Articles about BCIs generally describe four components, which are responsible for:

1. Directly measuring brain activity
2. Identifying useful information from that activity
3. Implementing messages or comments through devices or applications
4. Providing an application interface or operating environment.

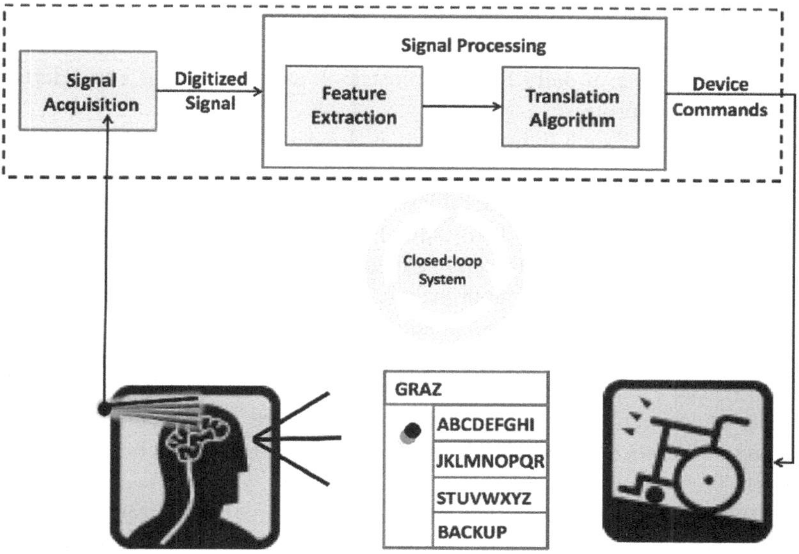

Fig. 1.1 The components of any BCI system from [2]. The different sections of this book are structured around these different components

In this book, the first two components are jointly addressed in the first section. The second section discusses the devices and applications that implement user commands, and the third covers interfaces and environments. The last section addresses practical issues that span all the components of a BCI.

1.2.1 Overview of Section One

In this first part of the book we start at the beginning, with the signals, the sensors used to capture those signals and the signal processing techniques used to extract information. The majority of recent BCI research and development, particularly in Europe and Asia, has been based on electroencephalogram (EEG) activity, recorded using resistive electrodes with conductive gel. This is the BCI standard, and is sufficient for many purposes. However, many researchers, including those involved in writing this book, feel that much more can be done in terms of usability, robustness and performance if we look beyond the standard platform.

The term hybrid BCI is used in various ways, as discussed in Chap. 18 of this book and some recent articles [3, 21]. Chap. 2 discusses hybrid sensor systems that combine different technologies that measure brain activity. Here we see an example of a hybrid optical–electrical sensor system providing functional near-infrared spectroscopy (fNIRS) and EEG in a single system. The resulting "compound" signal provides information on neural activity and haemodynamic response in coincident brain areas. There are many possible hybrid systems but for practical and useful BCI systems, for use in daily life, we must look at mobility and cost. Here, too, such systems show promise.

A consequence of such hybrid systems is the need for some sort of data fusion to make sense of these compound signals in a coherent way. In Chap. 3, we have a critical review of classifier ensembles and their use in BCI applications. This Machine Learning approach is ideally suited to hybrid systems and to BCI in general as it copes particularly well with variable data sources such as physiological signals.

For many EEG based BCI approaches, the focus has moved to performance enhancement in recent years. Independent component analysis (ICA) continues to provide improvements in three important and practical aspects, as discussed in Chap. 4. The chapter discusses artifact removal, improved SNR and optimal electrode selection, and how these techniques might be implemented in real-time. Such improvements are essential if we are to move from the lab into real world scenarios.

Finally we look at the world of invasive sensors, where chronic BCI makes sense for some applications [17]. While there are many different points of view on whether the perceived advantages justify the procedures necessary to implant such electrodes, and on whether this is as risky or invasive as often perceived, there can be no doubt that some groups are making significant steps towards wholly and long term implantable Electrocorticogram (ECoG) BCIs. Chap. 5 talks about the short term possibilities for such systems and what they might look like.

1.2.2 Overview of Section Two

Recording the brain signals, applying sophisticated signal processing and machine learning methods to classify different brain patterns is only the beginning of establishing a new communication channel between the human brain and a machine. This Part focuses on how to provide new devices and applications for different users, a challenge that goes beyond simple control tasks.

The first chapter in this section (Chap. 6) by Leeb and Millán gives an overview on current devices and application scenarios for various user groups [18]. Up to now, typical BCI applications require a very good and precise control channel to achieve performances comparable to users without a BCI. However, modern BCIs offer low throughput information and are insufficient for the full dexterous control of such complex applications. Techniques like shared control can enhance the interaction, yielding performance comparable to systems without a BCI [9, 26]. With shared control the user sends high-level commands at a fairly slow pace (e.g., directions of a wheelchair) and the system executes fast and precise low-level interactions (e.g., obstacle avoidance) [7, 27]. Chapter 6 also includes examples of how the performance of such applications can be improved by novel hybrid BCI architectures [3,22], which are a synergetic combination of a BCI with other residual input channels.

The impact and usage of BCI for neurological rehabilitation to lessen motor impairment and for restoration and recovery of hand motor functions is discussed by Mattia and colleagues in Chap. 7. On the one hand, BCI systems can be utilized to bypass central nervous system injury by controlling neuroprosthetics for patients' arms to manage reach and grasp functional activities in peripersonal space [20]. On the other, BCI technology can encourage motor training and practice by offering an on-line feedback about brain signals associated with mental practice, motor intention and other neural recruitment strategies, and thus helping to guide neuroplasticity associated with post-stroke motor impairment and its recovery [6].

Brain–Computer Interfaces are no longer only used by healthy subjects under controlled conditions in laboratory environments, but also by patients, controlling applications in their homes under real-world settings [18]. But which types of applications are useful for them, and how much can BCIs influence other applications that may already be in development with other communication technologies? Holz and co-authors discuss the different aspects of user involvement and the roles that users could or should have in the design and development of BCI driven assistive applications. Their focus is on BCI applications in the field of communication, access to ICT and environmental control, typical areas where assistive technology solutions can make the difference between participation and exclusion. User-centered design is an important principle gaining attention within BCI research, and this issue is addressed from an application interface perspective in Chap. 11.

The next chapter by Quek and colleagues addresses similar issues. Here, the focus is on how new BCI applications have to be designed to go beyond basic BCI control and isolated intention detection events. Such a design process for the overall system comprises finding a suitable control metaphor, respecting neuro-ergonomic principles, designing visually aesthetic feedback, dealing with the learnability of the system, creating an effective application structure (navigation), and exploring the power of social aspects of an interactive BCI system. Designing a human-machine system also involves eliciting a user's knowledge, preferences, requirements and priorities. To avoid overloading end users with evaluation tasks and to take into account issues specific to BCI, techniques and processes from other fields that aim to acquire these must be adapted for applications that use BCI [29].

The last chapter of this section is focused on an emerging application field. Recently BCIs have gained interest among the virtual reality (VR) community, since they have appeared as promising interaction devices for virtual environments [12]. These implicit interaction techniques are of great interest for the VR community. For example, users might imagine movement of their hands to control a virtual hand, or navigate through houses or museums by thoughts alone or just by looking at some highlighted objects [13, 16]. Furthermore, VR can provide an excellent testing ground for procedures that could be adapted to real world scenarios. Patients with disabilities can learn to control their movements or perform specific tasks in a virtual environment (VE). Lotte and co-authors provide several studies which highlight these interactions.

1.2.3 Overview of Section Three

While the term "BCI" has three words, the "interface" part has not received enough attention. Sensors to detect brain activity are making great strides, with dry electrodes that are increasingly cheap and effective. Pattern classification has long been an active research area, with numerous articles and data analysis competitions. But, especially in the early days of BCI research, relatively few BCI articles focused on improved usability, immersive and natural environments, evaluating user experience, user-centered interface design, accounting for the needs of special user populations, and other issues relating to the human–computer interaction (HCI) side of BCIs [1, 2, 10, 11, 19].

Section three summarizes progress and issues in application interfaces and operating environments for BCIs. The first chapter reviews how to evaluate users' experiences, including case studies. The second considers multimodal interfaces and how to integrate them seamlessly and effectively in a multimodal environment. This issue is further explored in Chap. 17. The third chapter of Section three describes newer, broader applications of BCI technology to improve human–computer interaction. The next two chapters show how phase detection and dry sensors could improve performance and usability.

In Chap. 11, van de Laar and colleagues discuss some issues that are emerging as BCI research draws on issues from the broader HCI community. They note that usability is a critical factor in adopting new technologies, which underscores the importance of evaluating user experience (UX). They review work showing that UX and BCIs both affect each other, including the methods used to evaluate UX such as observation, physiological measurement, interviews, and questionnaires. The authors use two different case studies as exercises in identifying and applying the correct UX evaluation methods. The chapter provides a strong argument that UX evaluation should be more common in BCI research.

As BCIs are put into service in real world, high-end applications, they will become one element in a multi-modal, multi-task environment. This brings with it new issues and problems that have not been prevalent in single task controlled environment BCI applications. In Chap. 12, we see what these possible problems may be and are presented with guidelines on how to manage this in a multi-modal environment. These issues are later explored in the fourth section of this book.

Another consequence of advanced BCI applications is the potential for enhanced user interfaces based on brain state. In this scenario, the current state of the user provides context to system in order to improve the user experience. These states may include alertness, concentration, emotion or stress. Chap. 13 introduces two application areas, medical and entertainment, based on recognition of emotion and concentration.

Steady-state-visual-evoked potentials (SSVEPs; [24]) are frequently used as control signals for BCIs. However, there is a practical limitation in the high frequency range (>30 Hz), because only a few frequencies can be used for BCI purposes. Garcia-Molina and co-authors show in Chap. 14 how repetitive visual stimuli, with the same frequency but different phases, can be used as control signals.

The last chapter of this section addresses a recurrent problem in the area of BCI research, which is practical EEG recording. A limiting factor in the widespread application is the usage of abrasive gel and conductive paste to mount EEG electrodes, which is a technology that is finally beginning to change after 20 years of relatively poor progress. Therefore, many research groups are now working on the practical usability of dry electrodes to completely avoid the usage of electrode gel. In Chap. 15, Edlinger and colleagues compare dry versus wet electrodes. Raw EEG data, power spectra, the time course of evoked potentials, ERD/ERS values and BCI accuracy are compared for three BCI setups based on P300, SMR and SSVEP BCIs.

1.2.4 Overview of Section Four

The previous sections each discussed different BCI components. This concluding section takes a step back by broadening the focus to complete BCI systems. Which software platforms are available to integrate different BCI components? What are the best ways to evaluate BCIs? What are the best ways to combine BCIs with other systems? Are any non-visual BCIs available? These important questions cannot be easily addressed without considering all the components of a BCI holistically.

The development of flexible, usable software that works for non-experts has often been underappreciated in BCI research, and is a critical element of a working BCI infrastructure [1, 2, 10]. In Chap. 16, Brunner and numerous co-authors describe the major software platforms that are used in BCI research. The lead developers of seven different publicly available platforms were asked to contribute a summary of their platform. The summaries describe technical issues such as supported devices and programming languages as well as general issues such as licensing and the intended user groups. The authors conclude that each platform has unique benefits, and therefore, tools that could help combine specific features of different programs (such as the TOBI Common Implementation Platform) should be further developed.

As BCIs gain attention, the pressure to report new records increases. In 2011 alone, three different journal papers, each from different institutions, claimed to have the fastest BCI [4, 5, 28]. Similarly, the influx of new groups includes some people who are not familiar with the methods used by established researchers to measure BCI performance and avoid errors. These two factors underscore the importance of developing, disseminating, and using guidelines. Chap. 17 reviews different methods to measure performance, account for errors, test significance and hypotheses, etc. Billinger and colleagues identify specific mistakes to avoid, such as estimating accuracy based on insufficient data, using the wrong statistical test in certain situations, or reporting the speed of a BCI without considering the delays between trials. We note that accuracy and information transfer rate are not at all the only ways to evaluate BCIs, and authors should report other factors too.

This book, like many emerging BCI publications [3, 14, 15, 21, 22, 25], has many references to hybrid BCIs. In Chap. 18, Müller-Putz and colleagues review the different types of hybrid BCIs. Hybrid BCIs combine different ways to send information, and so they are often categorized according to the types of signal combinations they use. While one signal must be a BCI, the other signal could also involve EEG, or heart rate, eye movement, a keyboard or joystick, etc. Different sections discuss the different types of BCIs, including technical details and examples of relevant papers. We conclude that BCIs could help people in different ways, and that most BCIs will be hybrid BCIs.

Most BCIs require vision. BCIs based on the brain's response to flashing or oscillating lights require lights, and even BCIs based on imagined movement usually require visual cues, such as observing a robot or cursor movement. But what if the user has trouble seeing, or wants to look somewhere else? Chap. 19 reviews non-visual and multisensory BCIs that could work for users with visual deficits. In addition, non-visual BCIs allow alternative communication pathways for healthy people who prefer to keep their vision focused elsewhere, such as drivers or gamers. Finally, emerging research shows the benefits of multisensory over unisensory cues in BCI systems. Wagner and colleagues review four categories of noninvasive BCI paradigms that have employed non-visual stimuli: P300 evoked potentials, steady-state evoked potentials, slow cortical potentials, and other mental tasks. After comparing visual and non-visual BCIs, different pros and cons for existing and future multisensory BCI are discussed. Next, they describe multimodal BCIs that combine different modalities. The authors expect that more multisensory BCI systems will emerge, and hence effective integration of different sensory cues is important in hybrid BCI design.

Chap. 20 returns to the general issue of evaluating BCIs, but from a different perspective. Randolph and colleagues first review major factors in BCI adoption. They then present the BioGauges method and toolkit, which has been developed and validated extensively over the years. Drawing on their earlier experience categorizing different facets of BCIs and other assistive technologies, they parametrically address which factors are important and how they are addressed through BioGauges. They review how these principles have been used to characterize control with different transducers—not just conventional EEG BCIs but also fNIRS BCI and communication systems based on skin conductance. The authors' overall goal is to help match the right BCI to each user, and BioGauges could make this process much faster and more effective.

1.3 Predictions and Recommendations

BCI research does have an air of mystery about it. Indeed, BCI research and development depends on a wide variety of factors that can make predictions and recommendations difficult. Nonetheless, we recently completed a roadmap that includes our expectations and recommendations for BCI research over the next

5 years. This roadmap, like this book, entailed extensive collaboration with other stakeholders in the BCI community and surrounding fields. Over more than 2 years, we hosted workshops, gave talks, scheduled meetings, sent emails, and otherwise engaged people to learn their views about what is, and should be, next.

This roadmap was developed during the same time period as this book, and involves many of the same people. However, the book and roadmap were separate projects, addressing different topics and goals, without any effort to synchronize them. Thus, it is somewhat gratifying to note that the major issues that our chapter authors addressed generally aligned with the issues we considered important in the roadmap. This roadmap is publicly available from http://www.future-bnci.org/. Our predictions for the next 5 years are summarized across the top ten challenges that we identified within BCI research. The first two of these challenges, reliability and proficiency, are presented jointly because our expectation is that these issues will increasingly overlap in the near future.

Reliability and Proficiency: "BCI illiteracy" will not be completely solved in the near future. However, matching the right BCI to each user will become easier thanks to basic research that identifies personality factors or neuroimaging data to predict which BCI approach will be best for each user. Hybrid BCIs will make it much easier to switch between different types of inputs, which will considerably improve reliability and reduce illiteracy.

Bandwidth: There will be substantial but not groundbreaking improvements in noninvasive BCIs within the next 5 years. Invasive BCIs show more potential for breakthroughs, although translating major improvements to new invasive BCIs for human use will take more time. Matching the right BCI to each user will also improve the mean bandwidth. Tools to increase the effective bandwidth, such as ambient intelligence, error correction and context awareness, will progress considerably.

Convenience: BCIs will become moderately more convenient. New headwear will more seamlessly integrate sensors with other head-mounted devices and clothing. However, BCIs will not at all become transparent devices within 5 years.

Support: Expectations are mixed. Various developments will reduce the need for expert help. In 5 years, there will be a lot more material available online and through other sources to support both experts and end users. Simple games are already emerging that require no expert help. On the other hand, support will remain a problem for many serious applications, especially with patients. In 5 years, most end users who want to use a BCI, particularly for demanding communication and control tasks, will still need help.

Training: Two trends will continue. First, BCI flexibility will improve, making it easier to choose a BCI that requires no training. Second, due to improved signal processing and experimentation, BCIs that do require training will require less training.

Utility: This is an area of considerable uncertainty. It will be easier to switch between BCI applications and adapt to new applications. However, it is too early to say whether BCIs for rehabilitation will gain traction, which would greatly increase utility.

Image: Unfortunately, many people will either not know about BCIs or have unrealistic and overly negative opinions about them. Inaccurate and negative portrayals in science fiction and news media will continue unchecked. We are concerned that the "bubble will burst," meaning that excess hype and misrepresentation could lead to a backlash against BCI research, similar to the neurofeedback backlash that began in the late 1970s. This could hamstring public funding, sales, and research.

Standards: We anticipate modest progress in the next 5 years. At least, numerous technical standards will be established, including reporting guidelines. Ethical guidelines will probably also proceed well. We think the disagreement over the exact definition of a BCI will only grow, and cannot be stopped with any reasonable amount of funding. We are helping to form a BCI Society, which could help encourage and disseminate standard terms, guidelines, methods, and events.

Infrastructure: We also anticipate modest progress. Many software tools will improve. Improved online support will advise people on the best systems and walk people through setup and troubleshooting. Infrastructure development depends heavily on outside funding.

In addition to our 5 year view, we also developed recommendations for the next 5 years. These are directed mainly at decision-makers who will decide on funding BCI research and development, such as government officials or corporate decision-makers. However, they also can and should also influence individual developers and groups trying to decide where to focus their time and energy in the near future. Our recommendations are:

- Encourage new sensors that are comfortable and easy to set up, provide good signal quality, work in real-world settings, look good, and are integrated with other components.
- Pursue invasive and noninvasive BCIs, recognizing that they do not represent competing fields but different options that each may be better suited to specific users and needs.
- Signal processing research should focus not only on speed and accuracy but also reliability and flexibility, especially automated tools that do not require expert help.
- New BCI software platforms are not recommended. Rather, existing platforms should be extended, emphasizing support for different inputs, flexibility, usability, and convenience.
- Hybrid BCIs, which combine different BCI and BNCI inputs, are extremely promising and entail many new questions and opportunities.
- Passive BCIs and monitoring systems could improve human–computer interaction in many ways, although some directions (such as realtime emotion detection) remain elusive.

- BCI technology can be applied to related fields in scientific and diagnostic research. This tech transfer should be strongly encouraged and could lead to improved treatment.
- Many aspects of BCI and BNCI research are hampered by poor infrastructure. We recommend numerous directions to improve BCI infrastructure, including a BCI Society.
- Ethical, legal, and social issues (ELSI) should be explicitly addressed within each project, and the next cluster should include at least one Work Package (WP) to explore broader issues.
- Support BCI competitions, videos, expositions, and other dissemination efforts that present BCIs in a fair and positive light to patients, carers, the public, and other groups.
- Grant contracts should include all expected work, including clustering events, expositions, and unwritten expectations. Streamlining administration would help.
- Research projects should specify target user groups and address any specific needs or expectations they have. Testing with target users in field settings should be emphasized.
- Interaction with other research groups and fields needs improvement. Opportunities to share data, results, experience, software, and people should be identified sooner.

1.4 Summary

All BCIs require different components. This book discusses these components, as well as issues relating to complete BCI systems. In the last few years, BCIs have gained attention for new user groups, including healthy users. Thus, developing practical BCIs that work in the real-world is gaining importance. The next 5 years should see at least modest progress across different challenges for BCI research.

One of the most prevalent themes in BCI research is practicality. Perhaps 10 years ago, simply getting any BCI to work in a laboratory was an impressive feat. Today, the focus is much more on developing practical, reliable, usable systems that provide each user with the desired functionality in any environment with minimal inconvenience. While there was always some interest in making BCIs practical, this has become much more prevalent in recent years.

However, as BCI research and development gains attention, it also develops new challenges. Newcomers to BCI research may bring promising ideas and technologies, but may also bring different expectations and methods that might not be well suited to BCI research. The influx of new people also broadens the definition of "BCI" and may create new possibilities that are difficult to analyze and predict.

These factors underscore why the future is both promising and unpredictable. Some predictions seem reasonably safe. For example, we think that BCIs will be combined with new systems more often, leading to hybrid BCIs and intelligent systems that incorporate context and ambient intelligence. We are also optimistic about dry electrodes and improved usability. On the other hand, some emerging BCI systems, such as neuromodulation systems, could go in many different directions. Perhaps the safest prediction of all is that the next 5 years will be exciting and dynamic, with significant changes in BCIs and especially in how they are marketed, perceived, and used.

References

1. Allison, B.Z.: Toward ubiquitous BCIs. Brain–computer interfaces. The Frontiers Collection, pp. 357–387 (2010)
2. Allison, B.Z.: Trends in BCI research: progress today, backlash tomorrow? XRDS: Crossroads. The ACM Magazine for Students **18**(1), 18–22 (2011). doi:10.1145/2000775.2000784
3. Allison, B.Z., Leeb, R., Brunner, C., Müller-Putz, G.R., Bauernfeind, G., Kelly, J.W., and Neuper, C. (2012). Toward smarter BCIs: Extending BCIs through hybridization and intelligent control. *Journal of Neural Engineering, 013001.*
4. Bin, G., Gao, X., Wang, Y., Li, Y., Hong, B., Gao, S.: A high-speed BCI based on code modulation VEP. J. Neural Eng. **8**, 025,015, (2011). doi: 10.1088/1741–2560/8/2/025015
5. Brunner, P., Ritaccio, A.L., Emrich, J.F., Bischof, H., Schalk, G.: Rapid communication with a "P300" matrix speller using electrocorticographic signals (ECoG). Front. Neurosci. **5** (2011)
6. Buch, E., Weber, C., Cohen, L.G., Braun, C., Dimyan, M.A., Ard, T., Mellinger, J., Caria, A., Soekadar, S., Fourkas, A., Birbaumer, N.: Think to move: a neuromagnetic brain–computer interface (BCI) system for chronic stroke. Stroke **39**, 910–917 (2008)
7. Carlson, T., Monnard, G., Leeb, R., Millán, J.: Evaluation of Proportional and Discrete Shared Control Paradigms for Low Resolution User Inputs. Proceedings of the 2011 IEEE International Conference on Systems, Man, and Cybernetics, pp. 1044–1049 (2011)
8. Demetriades, A.K., Demetriades, C.K., Watts, C., Ashkan, K.: Brain-machine interface: The challenge of neuroethics. Surgeon **8**, 267–269 (2010)
9. Flemisch, O., Adams, A., Conway, S., Goodrich, K., Palmer, M., Schutte, P.: The H-Metaphor as a Guideline for Vehicle Automation and Interaction. (NASA/TM–2003–212672) (2003)
10. Gürkök, H., Nijholt, A.: Brain–computer interfaces for multimodal interaction: a survey and principles. International Journal of Human–Computer Interaction, ISSN 1532–7590 (electronic) 1044–7318 (paper), Taylor & Francis, Oxford, United Kingdom (2011)
11. Kübler, A., Kotchoubey, B., Kaiser, J., Wolpaw, J.R., Birbaumer, N.: Brain–computer communication: unlocking the locked in. Psychol. Bull. **127**(3), 358–375 (2001)
12. Lecuyer, A., Lotte, F., Reilly, R., Leeb, R., Hirose, M., Slater, M.: Brain–computer interfaces, virtual reality, and videogames. Computer **41**(10), 66–72 (2008)
13. Leeb, R., Keinrath, C., Friedman, D., Guger, C., Scherer, R., Neuper, C., Garau, M., Antley, A., Steed, A., Slater, M., Pfurtscheller, G.: Walking by thinking: the brainwaves are crucial, not the muscles! Presence (Camb.) **15**, 500–514 (2006)
14. Leeb, R., Sagha, H., Chavarriaga, R., Millán, J.: A hybrid brain–computer interface based on the fusion of electroencephalographic and electromyographic activities. J. Neural Eng. **8**(2), 025,011, (2011). doi:10.1088/1741–2560/8/2/025011, http://dx.doi.org/10.1088/1741-2560/8/2/025011
15. Long, J., Li, Y., Yu, T., Gu, Z.: Target selection with hybrid feature for BCI-based 2-D cursor control. IEEE Trans. Biomed. Eng. **59**(1), 132–140 (2012)
16. F. Lotte, "*Brain-Computer Interfaces for 3D Games: Hype or Hope?*", Foundations of Digital Games (FDG'2011), pp. 325-327, 2011. ACM, New York, USA
17. Millán, J., Carmena, J.M.: Invasive or noninvasive: understanding brain-machine interface technology. IEEE Eng. Med. Biol. Mag. **29**(1), 16–22 (2010)

18. Millán, J., Rupp, R., Müller-Putz, G., Murray-Smith, R., Giugliemma, C., Tangermann, M., Vidaurre, C., Cincotti, F., Kübler, A., Leeb, R., Neuper, C., Müller, K., Mattia, D.: Combining brain–computer interfaces and assistive technologies: State-of-the-art and challenges. Front. Neurosci. **4**, 161 (2010). doi:10.3389/fnins.2010.00161
19. Moore, M.M.: Real-world applications for brain–computer interface technology. IEEE Trans. Neural Syst. Rehabil. Eng. **11**(2), 162–165 (2003)
20. Müller-Putz, G.R., Scherer, R., Pfurtscheller, G., Rupp, R.: Brain–computer interfaces for control of neuroprostheses: From synchronous to asynchronous mode of operation. Biomedizinische Technik **51**, 57–63 (2006)
21. Müller-Putz, G.R., Breitwieser, C., Cincotti, F., Leeb, R., Schreuder, M., Leotta, F., Tavella, M., Bianchi, L., Kreilinger, A., Ramsay, A., Rohm, M., Sagebaum, M., Tonin, L., Neuper, C., Millán, J.: Tools for brain–computer interaction: A general concept for a hybrid BCI. Front. Neuroinform. **5**, 30 (2011)
22. Pfurtscheller, G., Allison, B., Bauernfeind, G., Brunner, C., Solis Escalante, T., Scherer, R., Zander, T., Müller-Putz, G., Neuper, C., Birbaumer, N.: The hybrid BCI. Front. Neurosci. **4**, 42 (2010)
23. Racine, E., Waldman, S., Rosenberg, J., Illes, J.: Contemporary neuroscience in the media. Soc. Sci. Med. **71**(4), 725–733 (2010)
24. Regan, D.: Human brain electrophysiology: evoked potentials and evoked magnetic fields in science and medicine. Elsevier, New York (1989)
25. Su, Y., Qi, Y., Luo, J.X., Wu, B., Yang, F., Li, Y., Zhuang, Y.T., Zheng, X.X., Chen, W.D.: A hybrid brain–computer interface control strategy in a virtual environment. J. Zhejiang Univ. Sci. C **12**, 351–361, (2011). doi:10.1631/jzus.C1000208
26. Tonin, L., Leeb, R., Tavella, M., Perdikis, S., Millán, J.: The role of shared-control in BCI-based telepresence. Proceedings of 2010 IEEE International Conference on Systems, Man and Cybernetics, pp. 1462–1466 (2010)
27. Vanhooydonck, D., Demeester, E., Nuttin, M., Van Brussel, H.: Shared control for intelligent wheelchairs: An implicit estimation of the user intention. Proc. 1st Int. Workshop Advances in Service Robot, pp. 176–182 (2003)
28. Volosyak, I.: SSVEP-based Bremen-BCI interface – boosting information transfer rates. J. Neural Eng. **8**, 036,020 (2011). doi: 10.1088/1741–2560/8/3/036020
29. Williamson, J., Murray-Smith, R., Blankertz, B., Krauledat, M., Müller, K.: Designing for uncertain, asymmetric control: Interaction design for brain–computer interfaces. Int J. Hum. Comput. Stud. **67**(10), 827–841 (2009)

Part I
Sensors, Signals and Signal Processing

Chapter 2
Hybrid Optical–Electrical Brain Computer Interfaces, Practices and Possibilities

Tomas E. Ward

2.1 Introduction

In this chapter we present an overview of the area of electroencephalography-functional near infrared spectroscopy (EEG-fNIRS) measurement as an activity monitoring technology for brain computer interfacing applications. Our interest in this compound neural interfacing technology is motivated by a need for a motor cortical conditioning technology suitable for use in a neurorehabilitation setting [15, 50]. Specifically we seek BCI technology that allows a patient with a paretic limb (as a consequence of stroke) to engage in movement-based rehabilitation exercises which will, we hope, encourage neuroplastic processes in the brain so that recovery and function is ultimately restored [38]. As we are interested in rehabilitation following stroke haemodynamic signatures of motor cortical activity coupled with the corresponding direct measures of the electrical activity of the neurons involved could be a rich source of new information on the recovering brain areas. While most neural engineers will be familiar with the concepts underpinning the electroencephalogram (EEG), the same cannot be said for fNIRS. Consequently this chapter will discuss much of the foundational concepts underlying this measurement before describing an EEG-fNIRS probe and early experiments which illustrate the concept and highlight aspects of the utility of this hybrid BCI approach.

2.2 The Underlying Physiological Origins of EEG and fNIRS

It is appropriate at this juncture to consider the physical basis of the measurements generated during both electroencephalography and fNIRS. While both measurement

T.E. Ward (✉)
Department of Electronic Engineering, National University of Ireland Maynooth, Maynooth, Co. Kildare, Ireland
e-mail: tomas.ward@eeng.nuim.ie

B. Z. Allison et al. (eds.), *Towards Practical Brain-Computer Interfaces*, Biological and Medical Physics, Biomedical Engineering, DOI 10.1007/978-3-642-29746-5_2,
© Springer-Verlag Berlin Heidelberg 2012

modalities produce signals which correlate with neural activation the precise relationships between neural activity and the measured responses are very different [45].

As stated previously many researchers in the BCI community are familiar with the EEG and its underlying neurophysiological origins but the corresponding background to fNIRS is not as widely known. The concept that there is a single vasoactive agent associated with neural activation which in turn causes dilation of the vasculature to increase blood flow is a common misconception for example and a gross oversimplification. The true picture is still being revealed through active research however it is already clear that the process is a complex one [3, 7, 23]. In this section, we briefly summarise the origin of the EEG as commonly measured in a BCI context before presenting a more comprehensive exposition of the agents and events surrounding the haemodynamic response which drives fNIRS.

2.2.1 Origin of the EEG

The EEG represents the electrical potential, usually a difference in potential measured between various points on the scalp. This potential on the scalp arises as a result of neural activity whose action can be considered as a set of distributed current sources embedded in a volume conductor (the head). When the brain is active, patterns of communication are altered across large numbers of neurons, primarily in the form of synaptic state changes. At the cellular level such synaptic activity leads to local changes in the membrane potential which are electrotonically conducted in the form of post-synaptic potentials (PSP). These can be either excitatory (depolarizing) or inhibitory (hyperpolarising) in nature altering the propensity of the neural membrane to generate an action potential [44]. The resultant changes in ionic currents acting through a localised volume conductor constitute what is termed a *local field potential* (LFP). The LFP associated with single cell synaptic activity is very small, however, because of synchronised activation of large numbers of specific sets of neurons in the cortex sharing similar orientation during brain activity, these LFP sum together with their aggregated volume conductor to constitute a substantial current source. Different brain states give rise to different sets of current sources which are unfortunately mixed and filtered as they manifest as biopotentials on the scalp. This makes the reconstruction of the position and geometry of such sources (and hence volumetric localisation of neural activity) an ill-defined inverse problem. The temporal localisation of neural activity is unaffected however and therefore the EEG contains accurate information regarding timing of neural activation patterns.

EEG instrumentation is conceptually simple to understand comprising a sensor and a biopotential difference amplifier. The sensors are called electrodes which converts ionic current flow in the body to electron-based current flow in the amplifier circuitry. As the biopotentials generated on the scalp as a result of neural activity are typically of very low amplitude (10^{-6} V) and extraneous sources of noise sometimes many orders of magnitude greater, good amplifier and electrode system design is key to producing reliable responses. Consequently the technical development of EEG systems is an active and important field of endeavour [51].

2.2.2 *Origin of fNIRS Responses*

fNIRS is based on the optical measurement of the haemodynamic response to neural activation [30]. One aspect of this response—the blood oxygen level dependent (BOLD) signal is the basis for functional magnetic resonance imaging (fMRI), a brain imaging modality closely related to fNIRS in terms of the underlying measurand. The responses measuring during fNIRS are usually interpreted in terms of changes in oxy- and deoxyhaemoglobin concentration changes—a somewhat richer set of variables than those available from basic fMRI. As in fMRI, an interpretation of the haemodynamic responses in terms of neural activation is often considered on the simple basis that significant changes in haemodynamics corresponds to increases in neural activation [10]. However, the picture is more complicated that this—indeed much more so and in order to equip neural engineers appropriately for experimentation with this modality and interpretation of data a background on the underlying cellular and even molecular signalling dynamics involved will be presented.

Haemodynamic changes associated with brain activity or more precisely the relationship between local neural activity and cerebral blood flow is termed *neurovascular coupling* [18]. Understanding neurovascular coupling (NVC) is important in terms of interpreting the responses acquired during fNIRS so as to avoid naïve interpretation of the signal. This is especially true in the case of measurement in damaged brain such as following a stroke when pathological conditions of neurovascular mechanisms may exist [29]. Just as it is important to have a basic appreciation of neuronal anatomy and physiology to understand the origin of the EEG a basic understanding of the anatomy of the neurovasculature is useful in understanding the origin of fNIRS responses.

2.2.2.1 Anatomy of the Neurovasculature

The blood supply to the brain (Fig. 2.1a) is carried by extracerebral and intracerebral arteries and arterioles. The main supply to the brain comes from two pairs of cranial arteries; the *internal carotid arteries* (which are bifurcations of the common carotid arteries at the neck) and the *vertebral arteries*. The internal carotid arteries branch at the base of the brain to form two major cerebral arteries; the *anterior cerebral artery* (ACA) and the *middle cerebral arteries* (MCA). The ACA and MCA form the anterior circulation which supplies the *forebrain*. The vertebral arteries consist of a right and left branch which come together to form the *Basilar artery* at the level of the *pons*. This artery then joins up with the internal carotid arteries to form an arterial ring at the base of the brain called the *Circle of Willis*.

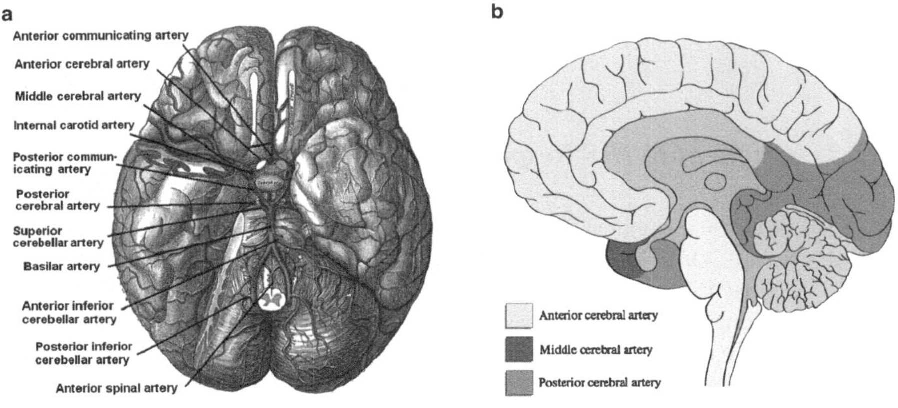

Fig. 2.1 (**a**) Major vessels of the cerebral circulation and (**b**) associated vascular territories

The *posterior cerebral arteries* and two other small arteries called the anterior and posterior communicating arteries also arise at this juncture. The Circle of Willis offers redundancy of circulation between posterior and anterior vasculature systems in the event of blockage in any of the feeding arteries. The posterior cerebral, basilar and vertebral arteries together give rise to the posterior circulation which supplies the posterior cortex. It comprises a number of arterial branches two of which in particular are significant in terms of their vascular territory; the *posterior inferior cerebral artery* and the *anterior inferior cerebral artery*. These arteries supply the *medulla* and pons and their occlusion during a stroke, which is relatively common, leads to specific deficits in somato, sensory and motor function. The vascular territories associated with the various arterial processes are illustrated in Fig. 2.1b.

The posterior and anterior circulation branch into smaller pial arteries and arterioles that branch out over the surface of the brain. These give rise to arterioles which penetrate orthogonally into the brain parenchyma (the functional part of the brain, i.e., neurons and glial cells). These parenchymal arterioles subdivide further into an extensive and distributed capillary network which reflects the metabolic requirements of the underlying neuronal system (Fig. 2.2).

The cerebral vasculature is equipped with neurovascular control mechanisms which match cerebral blood flow (CBF) with local cellular energy needs. These neurovascular coupling mechanisms are distributed and vary in type according to their location along the blood vessel, however the basic regulatory processes arise through interactions between neurons, glia and vascular cells. Neurons and glia in particular produce vaso-dilation or -constriction signals which are in turn transformed into neural activation-matched changes in CBF through the intricately choreographed action of endothelial, pericytes and smooth muscle cells constituting the cerebral vessel walls. The intimate structural and functional relationship between

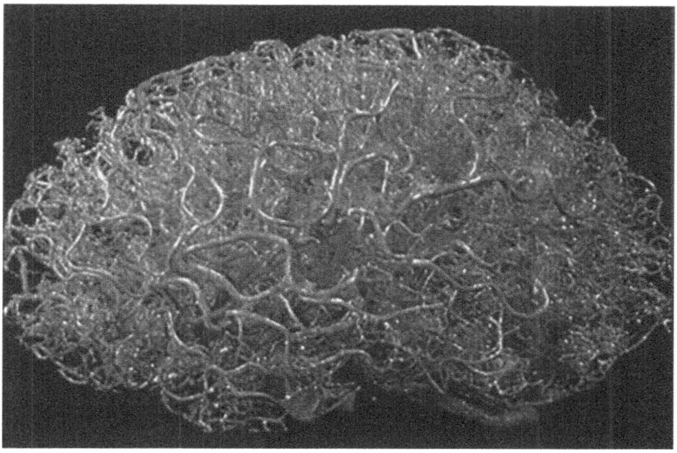

Fig. 2.2 The cerebral vascular system as revealed through plastic emulsion injection and dissolution of brain parenchyma [53]

cerebral vessels and neural/glial processes involved is significantly important to warrant description with a unique term—the neurovascular unit (NVU). Figure 2.3 illustrates important anatomical aspects of the NVU. At the pial artery stage the tissue consists of an endothelial layer surrounded by smooth muscle cells which in turn is contained in an outer layer (termed adventitia) comprising collagen, fibroblasts and perivascular nerves. Changes in vascular tone at this extracerebral stage are communicated through extrinsic innervation by peripheral nerves originating from cranial autonomic ganglia. As the vessel progresses as a parenchymal arteriole (an intracerebral microvessel) they become progressively smaller and changes in tone are communicated increasingly by local interneurons, glial cells and more centralised forms of intrinsic innervation. Finally as the vessel further penetrates deeper into the parenchyma it loses the smooth muscle layers and branches into cerebral capillaries. These capillaries comprise endothelial cells, contractile cells called pericytes and basal lamina upon which astrocytes (the most common type of glia cell) are attached via specialised processes called "feet." The interface between the walls of capillaries and the surrounding tissue is a very important one as it keeps vascular and extravascular concentrations of ions and molecules at appropriate levels in their respective regions. In the brain, this interface is especially tight and is termed *the blood-brain barrier*.

2.2.2.2 Physiology of the Neurovasculature

NVC dynamics drive the responses measured during fNIRS and act through the anatomical structures identified in the previous section. The role of NVC mechanisms are to provide autoregulation and functional hyperaemia in the brain.

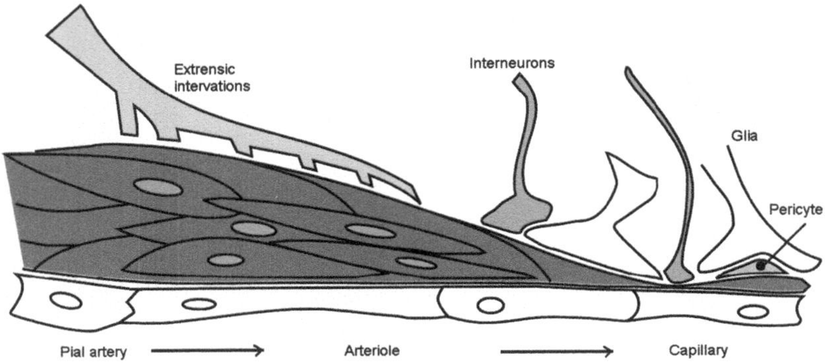

Fig. 2.3 Anatomy of the neurovascular unit (NVU)

Functional hyperaemia is a term which describes increased blood flow associated with increased activation of tissue while autoregulation in this context describes the ability of the cerebral vasculature to maintain the necessary homeostatic blood pressure during periods of changing blood flow. The science of NVC [39] is an active area with a long history [42] and in this article we can only summarise the current understanding of the anatomical and physiological processes involved. Regardless, what is known is that NVC can be understood as a dynamical system comprising sensing apparatus and actuating apparatus. The actuating apparatus act either to relax the vessel (which, all other factors being equal, should increase CBF) in a process termed vasodilation or to constrict the vessel (reducing CBF) in a process called vasoconstriction. These agonistic and antagonistic forces are responsible for generating the appropriate blood flow conditions required for optimal metabolic functioning of the neuronal networks served. Taking the view of Kleinfeld and his colleagues [23], we can group the mechanisms involved as one global and two local pathways. The global pathway is essentially driven through the vasodilatory substance acetylcholine (ACh) and the vasoconstrictor serotonin (5HT) released from various subcortical nuclei in response to blood oxygenation level, cortical state and perhaps even breathing patterns. Of the local pathways one comprises the action of different local interneurons which directly act on the smooth muscle wall of the vessel to produce dilation (through the production of nitric oxide or vasoactive intestinal peptide) or constriction effects (through the production of somatostatin and neuropeptide Y). To complicate matters further some interneurons appear to be capable of releasing both vasodilationary and vasoconstrictive substances. The second local pathway is provided through astrocytes in response to the activity of excitatory neurons [41]. The excitatory neurotransmitter involved is glutamate, the volume conduction of which causes an increase in intracellular levels of Ca^{2+} in astrocytes [35]. During periods of high synaptic activity waves of Ca^{2+} are propagated to nearby blood vessels which are critical in astrocyte-induced vasodilation. It also appears that these changes in levels of Ca^{2+} trigger the conversion of arachidonic acid to the vasoconstrictor 20-hydroxyeicosatetraenoic (20-HETE) and the dilatory substances prostaglandin $E(PG_E)$ and epoxyeicosatrienoic acid (EET). The relative balance in levels of these substances is a function of the partial pressure of oxygen. Low pO_2 leads to predominately vasodilatory conditions [19].

Even the picture above is a simplification of the processes at work and many new observations and hypothesis are being generated at the time of writing. For example, distinctions are now being highlighted between remote and local vasodilatory mechanisms [36]. The pial arteries which are the primary source of vascular resistance must also undergo dilation/constriction activity in response to downstream activity. Furthermore these adjustments result in increased flow to active areas and reduced flow to nearby inactive areas. The mechanisms driving these upstream activations are still being researched although it appears that signalling along the vessel wall via smooth muscle cells and endothelium is responsible in part. Finally it would appear that under certain conditions stimulus-induced *vasoconstriction* can occur at the site of neural activity [6]. The physiological significance of such behaviour has not yet been elucidated. And finally, results are emerging to suggest that bi-directional information flow is occurring at the vessel-parachyma boundary which has led to speculation that vessel-to-neuron and vessel-to-glial signalling may have a role for information processing in the brain [33].

In summary, what is certain is that there is no single vasoactive agent which simply diffuses through to the capillary beds feeding active neurons to produce the required functional hyperaemia. The emerging understanding is that there is a whole host of vasoactive substances released during neural activation via neurons and glial cells [52] which act on cerebral endothelial cells, pericytes and smooth muscle cells at different levels of the vasculature to produce a coordinated haemodynamic response which results in the appropriate increase in CBF for the active brain area. The interpretation of haemodynamic responses then, in this context of neural activation should therefore be considered carefully as the processes involved are revealing themselves to be increasingly complex and elaborate.

2.2.2.3 The fNIRS Signal

fNIRS measures the haemodynamic response associated with neural activation acting through the mechanisms above. Neural activation essentially causes an increase in glucose and oxygen consumption which in turn through NVC processes cause an increase in cerebral blood flow (CBF) [34]. While the increases in the cerebral metabolic rate of glucose (CMRglu) are matched by the increases in CBF the $CMRO_2$ is much less [14] leading to a net increase in the concentration of oxyhaemoglobin [HbO] with a corresponding change in concentration of deoxyhaemoglobin [Hb]. Figure 2.4 illustrates these changes in the relative states of haemoglobin along with the time course of the concentration changes which can be termed the *haemodynamic response (HR)*. The HR to specific

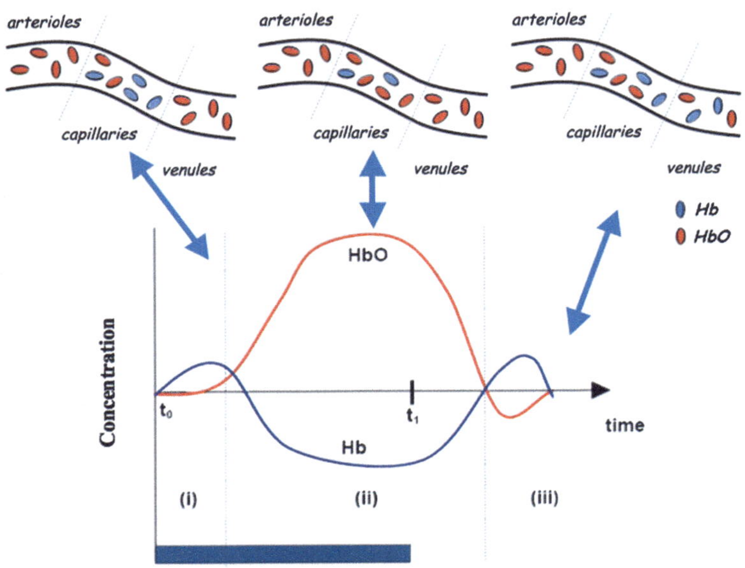

Fig. 2.4 Changes in [HbO][Hb] associated with neural (**a**) early increase in [Hb] (**b**) increase in CBF resulting in increase in [HbO], decrease in [Hb] and (**c**) return to basal state

endogenous/exogenous stimuli is the signal of interest measured with fNIRS. fNIRS in fact measures the optical properties of tissue and its changes in the spectral range of 700–900 nm [21]. This is the near infrared band. Photons of these wavelengths can penetrate the scalp, skull and meninges surrounding the brain to interrogate the superficial layers of the cerebral cortex. Such tissue constitutes a highly scattering medium and consequently backscattered photons can be collected from a detector positioned appropriately on the surface of the scalp to yield information on optical property changes at the cortical level.

The principal optical absorbers (termed *chromophores*) which undergo changes in concentration during neural activation are, conveniently, oxy- and deoxyhaemoglobin. Figure 2.5 illustrates the various types of photon–tissue interaction events which are all highly scattering. For a given photonic flux I_0 only a very small set will arrive at a detector situated a distance L from the optical source. The set of paths taken by photons which are collected at the detector have a geometry which has been described as an "optical banana" [32]. Such a set of photon paths is illustrated in the inset of Fig. 2.5. It is clear from this image which is the output of a Monte Carlo simulation of photon–tissue interaction that the mean path length is L' where $L' > L$. A factor called the differential pathlength factor B, which has been derived experimentally for different tissues, is used to account for this extended pathlength. Therefore $L' = B \cdot L$ is used for subsequent spectroscopic calculations which rely on the mean pathlength. One such calculation which is of great importance in near infrared spectroscopy is the modified Beer–Lambert Law

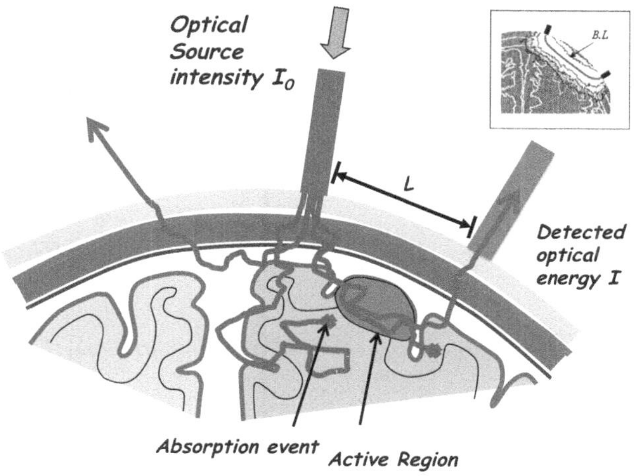

Fig. 2.5 Photon interactions in the head as measured non-invasively from scalp mounted source and detectors

(MBLL). The MBLL relates chromophore concentration levels to optical absorption values. The MBLL can be expressed in terms of the parameters pertinent to fNIRS as follows

$$A_\lambda = \log\left(\frac{1}{T}\right) = (\alpha_{\lambda,Hb} \cdot [Hb] + \alpha_{\lambda,HbO} \cdot [HbO]) \cdot B_\lambda \cdot L + G \qquad (2.1)$$

where T is the *transmittance,* which is the ratio of incident power to transmitted power, i.e. $\frac{I_\lambda}{I_{0,\lambda}}$.

The term A_λ is termed the *optical density* and is wavelength specific. The wavelength dependency comes from the wavelength-specific absorption tendencies (represented by the specific extinction coefficients $\alpha_{\lambda,Hb}, \alpha_{\lambda,HbO}$.) of oxy- and deoxyhaemoglobin. These values have been experimentally derived and tabulated elsewhere [9]. The term G is used to account for optical losses dues to scattering and is assumed constant over the measurement period. Usually a differencing operation is used to eliminate the effect of scattering to yield,

$$\Delta A_\lambda = (\alpha_{\lambda,Hb} \cdot .\Delta[Hb] + \alpha_{\lambda,HbO} \cdot \Delta[HbO]) \cdot B_\lambda \cdot L \qquad (2.2)$$

and therefore *changes* in chromophore concentrations are a common measurement made during fNIRS studies. In order to resolve the separate contributions of $\Delta[Hb]$ and $\Delta[HbO]$ a number of wavelengths of light are used to yield a set of simultaneous equations which are solved to yield the individual chromophore concentration changes.

The conversion of raw optical density signals to measures of [HbO] and [Hb] proceeds as follows. In this example the optical brain computer interfacing technology described by [11] is the source of the optical density measurements. This basic continuous wave system uses light emitting diodes rather than lasers as an optical source. The detectors are avalanche photodiodes (APD)-Hammamatsu C5460–01 which are used in many commercial systems. The wavelengths used in this case were 760 nm and 880 nm. Consequently Eq. (2.3) expresses the relationships between the optical variables as follows:

$$\frac{\Delta A_{760\,\mathrm{nm}}}{B_{760\,\mathrm{nm}} \cdot L} = (\alpha_{760\,\mathrm{nm},Hb} \cdot \Delta[Hb] + \alpha_{760\,\mathrm{nm},HbO} \cdot \Delta[HbO]) \tag{2.3}$$

$$\frac{\Delta A_{880\,\mathrm{nm}}}{B_{880\,\mathrm{nm}} \cdot L} = (\alpha_{880\,\mathrm{nm},Hb} \cdot \Delta[Hb] + \alpha_{880\,\mathrm{nm},HbO} \cdot \Delta[HbO]) \tag{2.4}$$

In matrix form these can be expressed as

$$\mathbf{A}/BL = \boldsymbol{\alpha}\mathbf{C} \tag{2.5}$$

where

$$\mathbf{A} = \begin{pmatrix} \Delta A_{760\,\mathrm{nm}} \\ \Delta A_{880\,\mathrm{nm}} \end{pmatrix}, \boldsymbol{\alpha} = \begin{pmatrix} \alpha_{760\,\mathrm{nm},Hb} & \alpha_{760\,\mathrm{nm},HbO} \\ \alpha_{880\,\mathrm{nm},Hb} & \alpha_{880\,\mathrm{nm},HbO} \end{pmatrix} \text{ and } \mathbf{C} = (\Delta[Hb] \quad \Delta[HbO])$$

Equation (2.5) is solved to extract **C** for each time sample as

$$\mathbf{C} = \boldsymbol{\alpha}^{-1} \cdot \mathbf{A}/B \cdot L. \tag{2.6}$$

The differential path length factor is age-dependent as well as altering with wavelength and has been described as follows:

$$B_{780} = 5.13 + 0.07 A_y^{0.81} \tag{2.7}$$

A_y is the age of the subject in years and B_{780} is the differential path length factor for 780 nm [9]. Values for other wavelengths can be derived from this measure through a tabulated scaling parameter B_N [9]. Equation (2.6) can be applied at each time step to yield the temporal dynamics of [HbO] and [Hb]. Figure 2.6 shows the results of this calculation for a simple finger tapping exercise lasting 20 s per trial using the system described above. A single source and detector were used measuring over the C3 position (using the 10–20 EEG electrode placement standard). The parameters values used are summarised in Table 2.1.

Figure 2.6 shows the response of 6 trials. The detected light signals were linearly de-trended and low-pass filtered using a fourth order Butterworth filter with a cut-off frequency of 0.5 Hz to remove the cardiac pulsations. A clear elevation in [HbO] levels is apparent as well as a reduction in [Hb]. There is a considerable lag in the

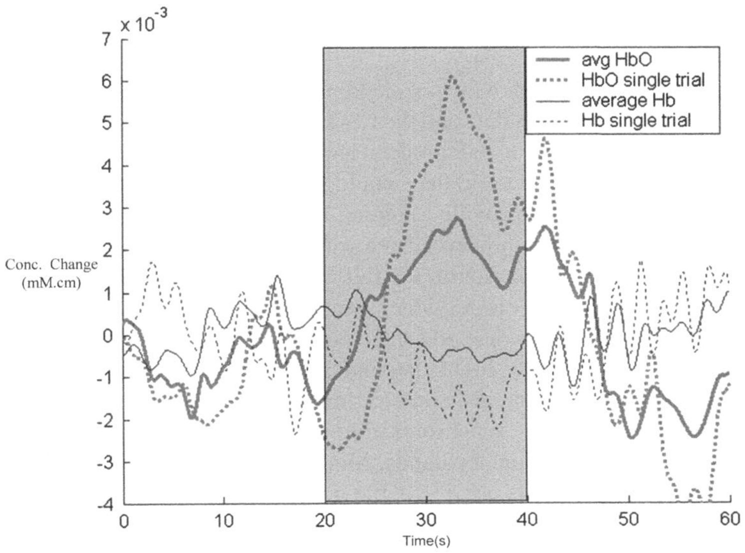

Fig. 2.6 [HbO] and [Hb] changes during a motor task (*shaded area*) illustrating average and single trial responses over the *left hand side* primary motor cortex

Table 2.1 Values of parameters used to calculate NIRS response in Fig. 2.6

Wavelength (nm)	HbO–Extinction Co-efficient ($mM^{-1}cm^{-1}$)	Hb–Extinction Co-efficient ($mM^{-1}cm^{-1}$)	B_N
760	0.6096	1.6745	1.12
880	1.2846	0.3199	0.84

haemodynamic response of about 6–8 s which of course has implications for use in brain computer interfacing applications.

The fNIRS responses in Fig. 2.6 are stereotypical for the modality and clearly demonstrate the evolution of the temporal responses. When measured using a number of source-detector pairs over the scalp and processed through an appropriate tomographic algorithm images of brain activation over cortical areas can be derived [2]. Such images provide the spatial localisation capability which are of utility for increasing the bit rate when used as a BCI [46] or for more general studies of brain function [1, 28]. For the purposes of clarity and simplicity, however, the remainder of this chapter considers only the temporal aspects of fNIRS, as it is only that aspect which has been incorporated in the fNIRS-EEG systems discussed later. In the next section time domain models for fNIRS responses are given which both helps reveal the signal processing problems inherent in the modality and provides a basis for understanding the response vis-a-vis the EEG.

2.3 Signal Models

Recently data has emerged which is leading to a better understanding of the relationships between fNIRS data and the EEG. Blankertz in particular has proposed that EEG activity leads to a damping in haemodynamics. This line of research is important and will ultimately, one would hope, lead to the required bridging subcomponents which will provide a unified fNIRS-EEG signal model. Currently such observations have not however been transformed into a model which lends itself to improving signal extraction for fNIRS-EEG systems. In this section then some of the basic building blocks which might be useful for the development of such signal-oriented models are presented with an emphasis on fNIRS systems. Signal models for the EEG are highly dependent on the neurophysiological origin of the underlying active components, however for many BCI applications the EEG can be interpreted as a series of synchronisation and desynchronisation events across a relatively small set of frequency bands. Models based on phase modulation (and even shifting) of these bands are a suitable conceptual model in many cases. The optical haemodynamic response is more straightforward and brain activation can be considered simply as the aggregation of active "voxels" not unlike fMRI. Simple as it may be, signal models which may be of practical utility to neural engineers interested in the processing of these responses are only beginning to emerge. Full dynamical system representations of the underlying haemodynamic responses have been developed, however, these models are not easy to work with and a simpler model is presented here which captures many aspects of the signal including all relevant components of extrinsic and physiological origin. This model is used to illustrate the effect of various parameters on signal characteristics and is a useful explanatory tool. This model comprises both a tractable physiological model of the neurovascular coupling events and a spectrophotometric model which captures the effect of the sensor apparatus in converting these neurovascular dynamics to changes in signal levels in the optical detectors based on Eq. (2.1). The model is expanded to include a number of extraneous noise sources normally presented during real fNIRS studies such as cardiac pulsations, respiration and other fluctuations in blood pressure which given rise to haemodynamic changes.

2.3.1 *Modelling the Vascular Response*

Several models have been proposed to account for the changes in blood volume, flow, oxy- and deoxyhaemoglobin concentrations which characterise the haemodynamic signal associated with neural activation. Of these, the best known and most widely invoked are the Windkessel-based models of Mandeville and his colleagues [31] and the balloon model of Buxton [5]. Both of these biomechanical models attempt to capture the dynamic changes in the post arteriole vasculature as a function of neural stimulus. Here, we utilise the balloon model which makes the assumption that the vascular bed which constitutes cerebral blood volume (CBV)

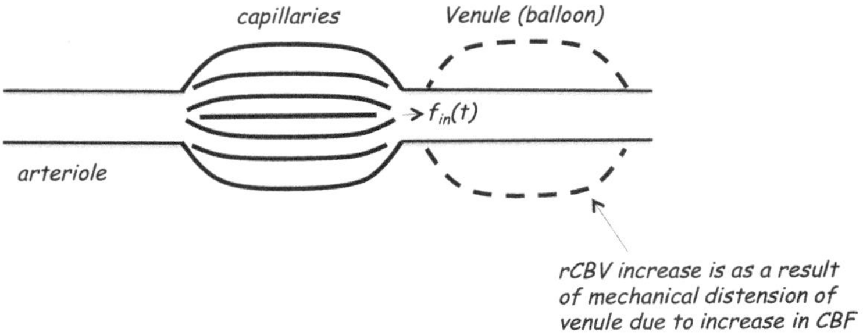

Fig. 2.7 The balloon model of the neurovascular response relating neural activation to changes in cerebral blood volume

can be modelled to some extent as an expandable venous compartment (hence the term "balloon"). The expansion of this balloon is driven by CBF which is assumed to be directly proportional to neural activation (Fig. 2.7).

A key part of the model then is the outflow from this vascular compartment which is a function of the compartmental volume. The precise form of this blood volume function governs the haemodynamics generated, and in the original exposition of this work Buxton and his colleagues show a number of different examples of this dependency. Generally the model is an appropriate balance between simplicity of representation and explanatory physiological power. It captures many of the well known aspects of the HRF as measured during fMRI such as transient changes in deoxyhaemoglobin and oxy-haemoglobin concentrations as well as the initial dip and other peculiarities of the BOLD response as measured experimentally.

The differential equation form of the model is as follows:

$$E(t) = 1 - (1 - E_0)^{\frac{1}{f_{in}(t)}} \tag{2.8}$$

$$\dot{q}(t) = \frac{f_{in}(t)}{\tau_0}\left[\frac{E(t)}{E_0} - \frac{q(t)}{v(t)}\right] + \frac{1}{\tau_v}\left[f_{in}(t) - v^{\frac{1}{\alpha}}\right]\frac{q(t)}{v(t)} \tag{2.9}$$

$$\dot{v}(t) = \frac{1}{\tau_v}\left[f_{in}(t) - v^{\frac{1}{\alpha}}\right] \tag{2.10}$$

$$\dot{p}(t) = \frac{1}{\tau_v}\left[f_{in}(t) - v^{\frac{1}{\alpha}}\right]\frac{p(t)}{v(t)} \tag{2.11}$$

E, q, v and p represent oxygen extraction rate, normalised [Hb], normalised blood volume and normalised total haemoglobin concentration respectively. [HbO] is obtained by subtracting q from p. Neural activation is represented by the CBF function $f_{in}(t)$ and is usually modelled as a binary function representing the stimulus input.

Figure 2.8 shows a solution for the Eqs. (2.8)–(2.11) for a trapezoidal binary input for f_{in}. This corresponds to a binary neural activation event. The output of the model displays the canonical form for the haemodynamic response to neural

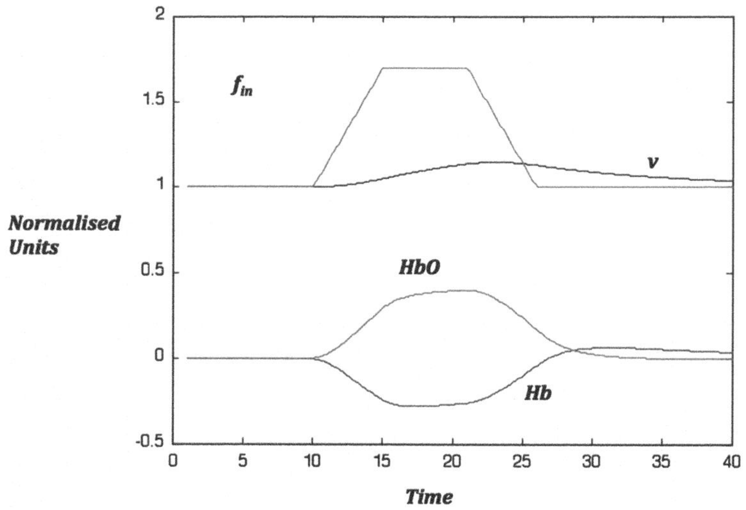

Fig. 2.8 The balloon model of the neurovascular response relating neural activation to changes in CBV. From *top* to *bottom* the variables plotted are CBF, normalised blood volume, [HbO] and [Hb]

activation. The model above has been used previously to generate a priori estimate for the BOLD response to specific stimuli patterns. A similar approach can be taken with fNIRS using such a model for predicting [HbO] and [Hb] changes for a given stimulus [13].

2.3.2 Spectrophotometric Translation

In fMRI the measured BOLD response is derived from the balloon model through known relationships between the relative contributions of the magnetic susceptibilities involved for oxy- and deoxyhaemoglobin and the signal measured at the detection apparatus. In contrast, fNIRS is an optical measurement and therefore it is changes in the optical properties of the tissue volume during neural activation that induces signal changes. Consequently a spectrophotometric modelling component is required to capture this aspect of the signal. The signal acquired at the detector can be approximated as a linear mixture of a number of components [27]. The basic equation involved is the MBLL in Eq. (2.1) however this equation must be altered to account for additional sources of optical density changes due to other sources of physiological origin especially low frequency blood pressure oscillations (Mayer wave), scattering and absorption changes due to the cardiac cycle and respiration [12]. A basic representation is as follows

$$S(\lambda, t) = \varphi_b(\lambda, t) + \varphi_c(\lambda, t) + \varphi_m(\lambda, t) + \varphi_n(\lambda, t) \quad (2.12)$$

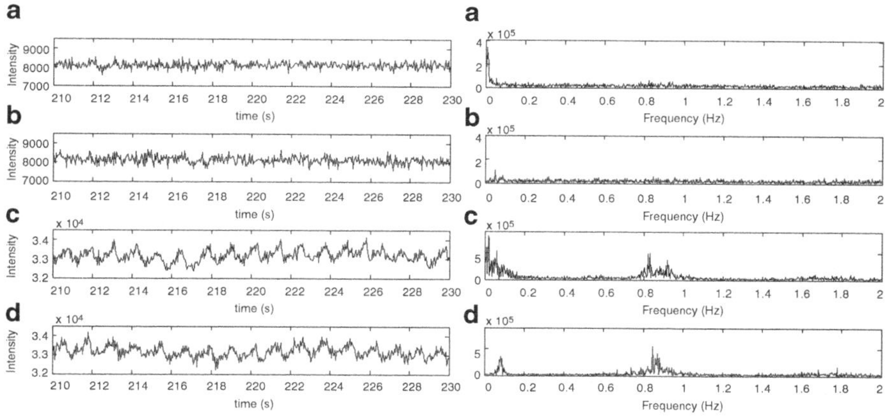

Fig. 2.9 Visual comparison of synthetic fNIRS model with real optical density signals. The *left hand side* is the time domain while the corresponding spectra are on the *right hand side*. (**a**) Actual measurement at 690 nm, (**b**) synthetic output for 690 nm, (**c**) actual measurement at 830 nm and (**d**) synthetic output at 830 nm

where

$$\varphi_b(\lambda, t) = e^{\Delta A_\lambda} \quad (2.13)$$

$$\varphi_c(\lambda, t) = K(\lambda) \cdot f(k(t), R(t)) \quad (2.14)$$

$$\varphi_m(\lambda, t) = M(\lambda) \cdot \sin(2 \cdot \pi \cdot f_m \cdot t + \theta) \quad (2.15)$$

Equation (2.13) represents the transmittance as given by Eq. (2.2) driven by the [HbO] and [Hb] values predicted by Eqs. (2.8) through (2.11). Equation (2.14) represents transmittance changes associated with the cardiac cycle and in this instance is a wavelength dependent scaling of a piecewise linear cardiac pulse $k(t)$ which is temporally scaled according to the rate function $R(t)$. The Mayer wave is represented in Eq. (2.15) as a sinusoidal component at frequency f_m whose amplitude is a function of wavelength. The term $\varphi_n(\lambda, t)$ accounts for optical environmental noise sources and can be adequately represented by a normally distributed noise signal.

2.3.3 Synthetic Signal Generation

Simulation of (2.12) with appropriately tuned parameters can yield realistic optical signals. The neural activation is incorporated through the cerebral blood flow signal as described in Sect. 2.2.2.3. Figure 2.9 shows sample output from this model along with real fNIRS signals both in the time and frequency domain. It is apparent that the model captures many of the characteristic features of the signal in both domains.

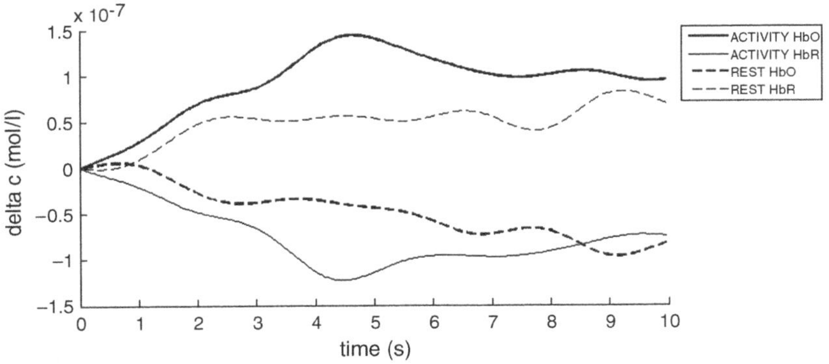

Fig. 2.10 Averaged derived changes in [HbO] and [Hb] responses both at rest and in response to activation using the synthetic signal model (compare with Fig. 2.6)

The output of such a model can be put through the same signal processing pipeline as actual fNIRS data to yield responses such as $\Delta[Hb]$ and $\Delta[HbO]$. Figure 2.10 illustrates the results of this processing for the synthetic data above in which $f_{in}(t)$ was modelled as a trapezoid with a rise time of 5 s, plateau time of 5 s, fall time of 5 s, and a rest time of 5 s. The amplitude of $f_{in}(t)$ was 1.7 units for a 10 s active period followed by a 10 s rest period. This was then repeated to match the number of active and rest periods of the real fNIRS data (there were 20 such periods in this case) and the responses averaged [27].

The most useful feature of this model for synthetic fNIRS data generation is the flexibility afforded when constructing the data. All parameters can be fine-tuned to replicate a real fNIRS signal. Such a signal model is useful for signal processing research as it allows a systematic investigation of the effect of various parameters on the fNIRS recording process. For example, the effect of movement artefact can be examined through varying the L parameter in Eq. (2.6) during the measurement simulation. Although fNIRS systems are a lot less sensitive to movement than fMRI, this type of artefact can cause problems and methods for motion artefact removal are an active research area [16, 47, 48].

The above section provides a sketch for a number of practical approaches to modelling the fNIRS signal. Such ideas can be combined with basic EEG models to yield a compound model encompassing both modalities. The neural activation function $f_{in}(t)$ is clearly the nexus between the two models and emerging research especially in the EEG-fMRI domain will elucidate more precisely the coupling mechanisms involved [22, 24, 43]. Even in the absence of comprehensive models it is clear that fNIRS and EEG are each measuring some aspects of neural activation [25, 40] and in the next section we summarise some early results we have obtained through combining the modalities for brain computer interfacing-like applications.

2.4 Combined EEG-fNIRS Measurements in Overt and Imagined Movement Tasks

To demonstrate the utility of combining the fNIRS and EEG modalities a pair of experiments are described. Both experiments involve the monitoring of motor cortex however in the first experiment the task involves overt finger tapping while the second experiment involves imagined movement only. The EEG processing is based on a standard motor rhythm paradigm—Event Related Synchronisation/Desynchronisation (ERS/ERD), which is a relative increase/decrease in the band power of a chosen frequency range that coincides with some event [37]. In ERS/ERD analysis, a baseline "reference" period of EEG data is recorded before the event and then compared to an "activity" period of EEG data, recorded during or following the event. ERD is known to occur in the μ frequency range (8–12 Hz) on movement onset and ERS is known to occur in the β frequency range (12–30 Hz) following movement offset.

2.4.1 *fNIRS/EEG Sensor*

A hybrid probe was designed to hold three fNIRS light sources (laser diodes), three fNIRS light detectors (APDs) and seven EEG electrodes in the array shown in Fig. 2.11 [26]. There are seven fNIRS channels with the corresponding EEG electrodes located directly above the centre point of each fNIRS channel. The centre point of an fNIRS channel is the interrogated area of cortex (as in Fig. 2.5), so with this set-up, we are recording electrical and haemodynamic activity from approximately the same area of cortex. An alternate combination probe is to include the electrode as part of the optical fibre housing however such a design is more complex to fabricate [8, 49]. Thus, we have seven co-locational, dual-modality recording sites. fNIRS data was recorded using a TechEn CW6 system (TechEn Inc., USA). Wavelengths used were 690 nm and 830 nm, sampled at 25 Hz. EEG data was recorded using a BioSemi Active-Two system (BioSemi Inc., The Netherlands) at 2,048 samples/s.

2.4.2 *Experimental Description*

In this simple proof-of-concept experiment data was collected from two healthy individuals. Both subjects gave voluntary consent. Subject A was male, 37 years old and left-handed (self-reported). Subject B was male, 26 years old and right-handed (self reported). During the experiment, the subjects were seated in a comfortable chair viewing a computer screen which presented instructions. Subjects were instructed to tap each of their fingers to their thumb on both hands. Tapping was self-paced. Individual trials lasted for 20 s, during which time the on-screen instruction read either "TAP" (an "active" trial) or "RELAX" (a "rest" trial). Twenty trials were

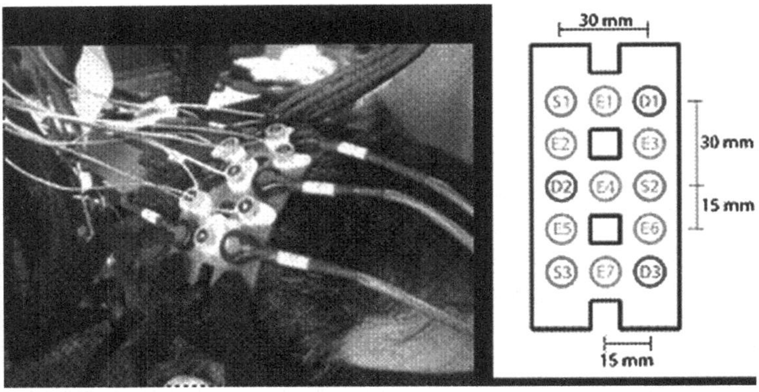

Fig. 2.11 fNIRS/EEG probe used in the experiments. SX, DX and EX denote source, detector and electrode positions respectively

carried out per experimental run, which alternated between active and rest, lasting 400 s in total. Two experimental runs were recorded for each subject with a short break between runs. The central electrode of the fNIRS/EEG recording module was located at C3 for Subject A (left-handed) and C4 for Subject B (right-handed). In a second series of experiments the above was repeated however rather than overt movement subjects were asked to perform imagined movement tasks.

2.4.3 Signal Processing

EEG data was first analysed to identify the frequencies at which ERS and ERD occurred in the μ frequency range and β frequency range with respect to the transition events, i.e., the transition from rest to active periods and vice-versa. The frequency ranges at which ERS and ERD occurred were identified through inspection of spectral plots for the reference and activity periods during both events. Raw EEG data was bandpass filtered with a sixth-order Butterworth filter to the identified ERS/ERD ranges, squared to obtain a power signal and then smoothed using a lowpass sixth-order Butterworth filter at 5 Hz. For ERS/ERD analysis, the reference window was chosen to be between 4.5 and 3.5 s before both types of event. For a transition from a rest trial to an active trial, the activity window was selected to be from 0 to 1 s after the transition. For a transition from active to rest, the activity window was selected to be from 0.5 to 1.5 s after the transition. These windows were chosen to capture the expected timing of pre-movement μ-rhythm ERD and β post-movement ERS. These changes in μ and β power were used as features for EEG classification. For fNIRS, the 690 nm and 830 nm raw intensity measurements were processed according to the techniques described in Sect. 2.2. The amplitude of these responses were used as features for motor cortical activation detection.

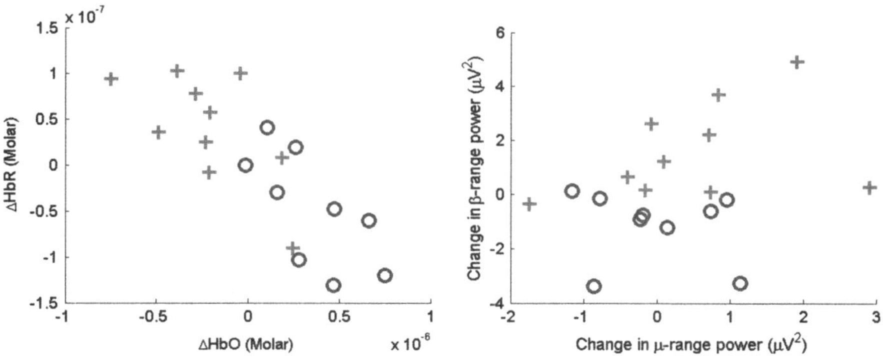

Fig. 2.12 *Left hand side*, 2D fNIRS feature space for Channel 2 of Subject A, Trial 1. *Crosses* indicate feature locations when subject is in a rest period. *Circles* indicate feature locations when subject is in a finger-tapping period. *Right hand side*, 2D EEG feature space for Channel 2 of Subject A, Trial 1

Classification was performed on the fNIRS and EEG signals to classify the activity into one of two classes: "active" and "rest." We employed the Linear Discriminant Analysis (LDA) classifier and calculated classification accuracy via leave-one-out cross-validation (LOOCV). In particular, for N trials, $N - 1$ trials were used for training the classifier and the remaining trial was used for testing. This was repeated N times with each trial used for testing once. Classification accuracy was calculated as the number of correct classifications over N.

For EEG, the feature extracted was change in μ-rhythm and β-rhythm power from the reference period to the activity period at the beginning of a trial. This resulted in a two-dimensional EEG feature space (Fig. 2.12). For fNIRS, the average change in amplitude of the $\Delta[HbO]$ and $\Delta[Hb]$ signals over a trial were used to define a two-dimensional fNIRS feature space. By combining the fNIRS and EEG feature spaces, an fNIRS/EEG four-dimensional feature space was also created for classification.

2.4.4 Results

A table of classification results are presented in Tables 2.2 and 2.3. Shown is the classification accuracy of the classifier when operating on fNIRS features only, EEG features only and the combined feature space. The results show that utilising both fNIRS and EEG features for classification yields an improvement on classification accuracy.

The classification accuracy particularly for imagined movement is not particularly impressive, however, there was no subject training for such imagined movement. More importantly the goal of the above experiments is not to demonstrate

Table 2.2 Classification accuracy for overt movement tasks

	Subject A						Subject B					
	Trial 1			Trial 2			Trial 1			Trial 2		
Channel	fNIRS (%)	EEG (%)	Comb. (%)	fNIRS (%)	EEG (%)	Comb. (%)	fNIRS (%)	EEG (%)	Comb. (%)	fNIRS (%)	EEG (%)	Comb. (%)
1	84	79	90	100	84	95	79	84	95	79	84	84
2	79	79	84	95	79	95	47	79	63	53	84	84
3	100	74	95	100	84	100	79	74	79	84	84	84
4	95	84	95	84	84	79	74	74	63	53	90	79
5	74	74	84	42	74	58	47	68	47	74	79	95
6	100	90	100	95	74	90	58	74	68	79	84	79
7	58	84	84	68	68	79	68	63	68	58	74	63
Average	84	80	90	83	78	85	65	74	69	68	83	81

Table 2.3 Classification accuracy for imagined movement tasks

	Subject A			Subject B		
Channel	fNIRS (%)	EEG (%)	Dual (%)	fNIRS (%)	EEG (%)	Dual (%)
1	59	51	64	64	46	62
2	56	59	67	51	54	59
3	56	54	64	61	41	56
4	69	67	72	64	59	67
5	61	51	72	41	36	46
6	56	77	64	74	59	69
7	56	59	62	15	43	49
Average	59	60	66	53	48	58

a high performance imagined movement task-based BCI operation but to illustrate how a combined EEG/fNIRS probe can work to yield both electrical and haemodynamic signatures of motor cortical activation which results in a higher information carrying compound signal. The features used are only of a rudimentary nature and it is likely that there is significant additional information regarding cortical function residing in this combined signal space. Indeed, the hybrid signal produced is a measure of neurovascular coupling and could have enhanced value when measured for subjects who have suffered stroke or similar cerebro-vascular damage. Fusion approaches such as the above are part of the next phase of BCI development [4, 17, 20] which will require enhanced, robust performance outside the laboratory setting.

2.5 Conclusion

In this chapter we have provided a description of EEG-fNIRS hybrid neural-haemodynamic technology for basic brain computer interfacing applications. The primary contributions has been an introduction to the neurovascular response which bridges neural activity and the haemodynamic response, the presentation of a synthetic signal model which facilitates a better understanding of the fNIRS signal especially for signal processing engineers, and the presentation of a set of results demonstrating how a compound EEG-fNIRS probe can yield a neural interfacing technology with a superior capacity to accurately monitor motor cortical activations.

It is clear that hybrid measurement modalities have great potential in creating the next generation of BCI. The fNIRS-EEG technology discussed here is but one example of such an approach although it is one that might be quite fruitful particularly when used with the damaged brain where measurements of neurovascular coupling may have great diagnostic and ongoing cortical status monitoring capability. The field of compound fNIRS-EEG interface technology is only beginning to develop and we believe that as fNIRS technology improves and becomes less expensive more ingenious methods for extracting neural activation signatures will be developed leading to more powerful and useful BCI applications.

Acknowledgements This work was supported by Science Foundation Ireland: Research Frontiers Program 2009, Grant No. 09/RFP/ECE2376.

References

1. Arenth, P.M., Ricker, J.H., Schultheis, M.T.: Applications of functional near-infrared spectroscopy (fNIRS) to Neurorehabilitation of cognitive disabilities. Clin. Neuropsychol. **21**(1), 38–57 (2007)
2. Arridge, S.R.: Optical tomography in medical imaging. Inverse Probl. **15**(2), R41–R93 (1999)
3. Bernardinelli, Y., Salmon, C., Jones, E.V., Farmer, W.T., Stellwagen, D., Murai, K.K.: Astrocytes display complex and localized calcium responses to single-neuron stimulation in the hippocampus. J. Neurosci. **31**(24), 8905–8919 (2011)
4. Brunner, P., Bianchi, L., Guger, C., Cincotti, F., Schalk, G.: Current trends in hardware and software for brain–computer interfaces (BCIs). J. Neural Eng. **8**(2), 025001 (2011)
5. Buxton, R.B., Wong, E.C., Frank, L.R.: Dynamics of blood flow and oxygenation changes during brain activation: the balloon model. Magn. Reson. Med. **39**(6), 855–864 (1998)
6. Cauli, B., Tong, X.K., Rancillac, A., Serluca, N., Lambolez, B., Rossier, J., Hamel, E.: Cortical GABA interneurons in neurovascular coupling: relays for subcortical vasoactive pathways. J. Neurosci. **24**(41), 8940–8949 (2004)
7. Cloutier, M., Bolger, F.B., Lowry, J.P., Wellstead, P.: An integrative dynamic model of brain energy metabolism using in vivo neurochemical measurements. J. Comput. Neurosci. **27**(3), 391–414 (2009)
8. Cooper, R.J., Everdell, N.L., Enfield, L.C., Gibson, A.P., Worley, A., Hebden, J.C.: Design and evaluation of a probe for simultaneous EEG and near-infrared imaging of cortical activation. Phys. Med. Biol. **54**(7), 2093–2102 (2009)
9. Cope, M.: The application of near-infrared spectroscopy to non-invasive monitoring of cerebral oxygenation in the newborn infant. PhD thesis, University of London (1991)
10. Coyle, S., Ward, T., Markham, C., McDarby, G.: On the suitability of near-infrared (NIR) systems for next-generation brain–computer interfaces. Physiol. Meas. **25**(4), 815–822 (2004)
11. Coyle, S.M., Ward, T.E., Markham, C.M.: Brain–computer interface using a simplified functional near-infrared spectroscopy system. J. Neural Eng. **4**(3), 219–226 (2007)
12. Coyle, S., Ward, T., Markham, C.: Physiological noise in near-infrared spectroscopy: implications for optical brain computer interfacing. Conf. Proc. IEEE Eng. Med. Biol. Soc. **6**, 4540–4543 (2004)
13. Cui, X., Bray, S., Reiss, A.L.: Functional near infrared spectroscopy (NIRS) signal improvement based on negative correlation between oxygenated and deoxygenated hemoglobin dynamics. Neuroimage. **49**(4), 3039–3046 (2010)
14. Davis, T.L., Kwong, K.K., Weisskoff, R.M., Rosen, B.R.: Calibrated functional MRI: mapping the dynamics of oxidative metabolism. Proc. Natl. Acad. Sci. USA. **95**(4), 1834–1939 (1998)
15. Dobkin, B.H.: Brain–computer interface technology as a tool to augment plasticity and outcomes for neurological rehabilitation. J. Physiol. **579** (Pt 3), 637–642 (2007)
16. Falk, T.H., Guirgis, M., Power, S., Chau, T.T.: Taking NIRS-BCIs outside the lab: towards achieving robustness against environment noise. IEEE Trans. Neural Syst. Rehabil. Eng. **19**(2), 136–146 (2011)
17. Fazli, S., Mehnert, J., Steinbrink, J., Curio, G., Villringer, A., Müller, K.R., Blankertz, B.: Enhanced performance by a hybrid NIRS-EEG brain computer interface. Neuroimage. **59**(1), 519–29 (2011)

18. Filosa, J.A.: Vascular tone and neurovascular coupling: considerations toward an improved in vitro model. Front. Neuroenergetics. **2**(16), 1–8 (2010)
19. Gordon, G.R., Choi, H.B., Rungta, R.L., Ellis-Davies, G.C., MacVicar, B.A.: Brain metabolism dictates the polarity of astrocyte control over arterioles. Nature. **456**(7223), 745–749 (2008)
20. Green, A.M., Kalaska, J.F.: Learning to move machines with the mind. Trends Neurosci. **34**(2), 61–75 (2011)
21. Jöbsis, F.F.: Noninvasive, infrared monitoring of cerebral and myocardial oxygen sufficiency and circulatory parameters. Science **198**(4323), 1264–1267 (1977)
22. Kilner, J.M., Mattout, J., Henson, R., Friston, K.J.: Hemodynamic correlates of EEG: a heuristic. Neuroimage **28**(1), 280–286 (2005)
23. Kleinfeld, D., Blinder, P., Drew, P.J., Driscoll, J.D., Muller, A., Tsai, P.S., Shih, A.Y.: A guide to delineate the logic of neurovascular signaling in the brain. Front. Neuroenergetics. **3**, 1 (2011)
24. Laufs, H., Holt, J.L., Elfont, R., Krams, M., Paul, J.S., Krakow, K., Kleinschmidt, A.: Where the BOLD signal goes when alpha EEG leaves. Neuroimage **31**(4), 1408–1418 (2006)
25. Lauritzen, M., Gold, L.: Brain function and neurophysiological correlates of signals used in functional neuroimaging. J. Neurosci. **23**(10), 3972–3980 (2003)
26. Leamy, D.J., Ward, T.E.: A novel co-locational and concurrent fNIRS/EEG measurement system: design and initial results. Conf. Proc. IEEE Eng. Med. Biol. Soc. **2010**, 4230–4233 (2010)
27. Leamy, D.J., Ward, T.E., Sweeny, K.T.: Functional near infrared spectroscopy (fNIRS) synthetic data generation. Conf. Proc. IEEE Eng. Med. Biol. Soc. 6589–6592 (2011)
28. Leff, D.R., Orihuela-Espina, F., Elwell, C.E., Athanasiou, T., Delpy, D.T., Darzi, A.W., Yang, G.Z.: Assessment of the cerebral cortex during motor task behaviours in adults: A systematic review of functional near infrared spectroscopy (fNIRS) studies. Neuroimage **54**(4), 2922–2936 (2011)
29. Lin, W.H., Hao, Q., Rosengarten, B., Leung, W.H., Wong, K.S.: Impaired neurovascular coupling in ischaemic stroke patients with large or small vessel disease. Eur. J. Neurol. **18**(5), 731–736 (2011)
30. Lloyd-Fox, S., Blasi, A., Elwell, C.E.: Illuminating the developing brain: the past, present and future of functional near infrared spectroscopy. Neurosci. Biobehav. Rev. **34**(3), 269–284 (2010)
31. Mandeville, J.B., Marota, J.J., Ayata, C., Zaharchuk, G., Moskowitz, M.A., Rosen, B.R., Weisskoff, R.M.: Evidence of a cerebrovascular postarteriole windkessel with delayed compliance. J. Cereb. Blood Flow Metab. **19**(6), 679–689 (1999)
32. Mansouri, C., L'huillier, J.P., Kashou, N.H., Humeau, A.: Depth sensitivity analysis of functional near-infrared spectroscopy measurement using three-dimensional Monte Carlo modelling-based magnetic resonance imaging. Lasers Med. Sci. **25**(3), 431–438 (2010)
33. Moore, C.I., Cao, R.: The hemo-neural hypothesis: on the role of blood flow in information processing. J. Neurophysiol. **99**(5), 2035–2047 (2008)
34. Nair, D.G.: About being BOLD. Brain Res. Rev. **50**(2), 229–243 (2005)
35. Panatier, A., Vallée, J., Haber, M., Murai, K.K., Lacaille, J.C., Robitaille, R.: Astrocytes are endogenous regulators of Basal transmission at central synapses. Cell **146**(5), 785–798 (2011)
36. Pelligrino, D.A., Vetri, F., Xu, H.L.: Purinergic mechanisms in gliovascular coupling. Semin. Cell Dev. Biol. **22**(2), 229–236 (2011)
37. Pfurtscheller, G., Lopes da Silva, F.H.: Event-related EEG/MEG synchronization and desynchronization: basic principles. Clin. Neurophysiol. **110**(11), 1842–1857 (1999)
38. Richards, L., Hanson, C., Wellborn, M., Sethi, A.: Driving motor recovery after stroke. Top Stroke Rehabil. **15**(5), 397–411 (2008)
39. Riera, J., Sumiyoshi, A., Brain oscillations. Ideal scenery to understand the neurovascular coupling. Curr. Op. Neurobiol. **23**, 374–381 (2010)
40. Rosa, M.J., Daunizeau, J., Friston, K.J.: EEG-fMRI integration: A critical review of biophysical modeling and data analysis approaches. J. Integr. Neurosci. **9**(4), 453–476 (2010)
41. Rouach, N., Koulakoff, A., Abudara, V., Willecke, K., Giaume, C.: Astroglial metabolic networks sustain hippocampal synaptic transmission. Science **322**(5907), 1551–1555 (2008)

42. Roy, C.S., Sherrington, C.S.: On the regulation of the blood supply of the brain. J. Physiol. **11**, 85–108 (1890)
43. Scheeringa, R., Fries, P., Petersson, K.M., Oostenveld, R., Grothe, I., Norris, D.G., Hagoort, P., Bastiaansen, M.C.: Neuronal dynamics underlying high- and low-frequency EEG oscillations contribute independently to the human BOLD signal. Neuron **69**(3), 572–583 (2011)
44. Schomer, D.L., Lopes da Silva, F.H.: (eds.) Niedermeyer´s Electroencephalography: Basic Principles, Clinical Applications, and Related Fields. 6th Edition, Lippincott, Williams and Wilkins, Philadelphia (Penn.) (2011)
45. Shibasaki, H.: Human brain mapping: hemodynamic response and electrophysiology. Clin. Neurophysiol. **119**(4), 731–743 (2008)
46. Sitaram, R., Caria, A., Birbaumer, N.: Hemodynamic brain–computer interfaces for communication and rehabilitation. Neural Netw. **22**(9), 1320–1328 (2009)
47. Sweeney, K.T., Leamy, D.J., Ward, T.E., McLoone, S.: Intelligent artifact classification for ambulatory physiological signals. Conf. Proc. IEEE Eng. Med. Biol. Soc. **2010**, 6349–6352 (2010)
48. Sweeney, K.T., Ayaz, H., Ward, T.E., Izzetoglu, M., McLoone, S.F., Onaral, B.: A Methodology for Validating Artifact Removal Techniques for fNIRS. Conf. Proc. IEEE Eng. Med. Biol. Soc. 4943–4946 (2011)
49. Wallois, F., Patil, A., Héberlé, C., Grebe, R.: EEG-NIRS in epilepsy in children and neonates. Neurophysiol. Clin. **40**(5–6), 281–292 (2010)
50. Ward, T.E., Soraghan, C.J., Matthews, F., Markham, C.: A concept for extending the applicability of constraint-induced movement therapy through motor cortex activity feedback using a neural prosthesis. Comput. Intell. Neurosci. 51363 (2007)
51. Webster, J.G. (eds.): Medical Instrumentation, Application and Design, 3rd edn. Wiley, Hoboken, N.J. (1998)
52. Zonta, M., Angulo, M.C., Gobbo, S., Rosengarten, B., Hossmann, K.A., Pozzan, T., Carmignoto, G.: Neuron-to-astrocyte signaling is central to the dynamic control of brain microcirculation. Nat. Neurosci. **6**(1), 43–50 (2003)
53. Zlokovic, B., Apuzzo, M.: Strategies to circumvent vascular barriers of the central nervous system. Neurosurgery **43**, 877–878 (1998)

Chapter 3
A Critical Review on the Usage of Ensembles for BCI

Aureli Soria-Frisch

3.1 Introduction

We review in this chapter the State of the Art on ensemble classification techniques employed in Brain Computer Interfaces (BCIs). Different reviews and survey papers on general BCI systems and technologies have been already published [11, 32, 54, 56]. However no works in the analyzed literature are focused on the employment of classifier ensembles. Therefore we attain to the best of our knowledge the first complete review on the utilization of these kind of techniques in the BCI application field.

Several features have been proposed for its usage within BCI systems [32]: amplitude values of EEG signals, band powers, power spectral density values, autoregressive and adaptive autoregressive parameters, time-frequency features, spatial filters, and inverse model-based features. But it is the classification stage that currently captures the attention of the researchers in the field as exemplified by the target of [32]. In this context the increasing interest in the employment of ensemble classifiers among pattern recognition researchers in different application fields [36], has boosted the popularity of ensemble methodologies within BCI research. This is a paradigm originated in the Machine Learning community that has flowed into other research areas. Generally a group of classifiers is applied to a data set and the results are then combined through an operator in this kind of systems. Being this the basic structure the several variants existent in the literature are discussed.

The review herein particularly targets classifier ensembles, which are characterized in [32] as one of the best alternatives for the development of BCI systems. Although classifier ensembles have been used from an early stage of BCI research [38], they have not been extensively analyzed. Ensembles deserve further

---

A. Soria-Frisch (✉)
Starlab Barcelona SL, C/ Teodor Roviralta 45, 08022 Barcelona, Spain
e-mail: aureli.soria-frisch@starlab.es

B. Z. Allison et al. (eds.), *Towards Practical Brain-Computer Interfaces*, Biological and Medical Physics, Biomedical Engineering, DOI 10.1007/978-3-642-29746-5_3,
© Springer-Verlag Berlin Heidelberg 2012

attention as stated both in [32] and [52], what seems to be confirmed by the absence of the topic on recent monographs in the field like [11].

The paper in [32] discusses some theoretical aspects on classifiers including different taxonomies. Besides this the paper makes a well structured presentation of classifiers, distinguishing among the following groups: linear classifiers (Linear Discriminant Analysis—LDA, Support Vector Machines—SVM), neural networks (most focused on Multilayer Perceptron—MLP), bayesian (bayesian quadratic, Hidden Markov Models—HMM), neighbor classifiers (K-Nearest Neighbors—KNN, Mahalanobis distance based), and classifier ensembles. Furthermore it briefly describes each of these types. The analysis of the properties of each classifier group results very interesting. Furthermore it targets an unsolved question in pattern recognition research. As stated by the No-Free Lunch theorem [12] there is no way to predict the general superiority of a classifier over another one. This means that a classifier is better than another one just on a particular data set, what can only be assessed experimentally [12]. Therefore looking at general characteristics of classifiers, as is done in [32], is in my opinion the right approach for *a priori* selecting one classifier or another one. However their analysis is not grounded on the particular features of the data set, but on high level system and application features like the employed Brain/Neuronal Computer Interaction (BNCI) paradigm, the synchronous/asynchronous quality of the needed output, and the existence or not of comparative studies of techniques. All of these lay on theoretical expert knowledge on BNCI data and applications. Hence we take as inspiration the work in [32] but try to emphasize the pattern recognition related aspects.

Some experimental studies have been published on classifier ensembles. Hence [2] evaluates the performance of 13 classifiers including four classifier ensemble approaches: AdaBoost, bagging, stacking, and Random Forest all based on decision trees. Moreover boosting (AdaBoost), bagging, and random forests based on three different classifiers: KNN, C4.5 decision trees, and linear SVM are compared in [52]. They derive general guidelines in the employment of classifier ensembles but are definitely biased by the selected methodologies and the experimental results they obtain. So we try to complement these works by following a more general study, which takes also the theoretical facets into account.

Our intention is therefore: to describe different design principles that can help new users to quickly identify how to proceed when developing a new ensemble based BCI system, to give an extensive review of nomenclature in order to ease the interdisciplinary communication, to summarize best practices and construction principles in order for users to make a good use of this powerful tool, and lastly summarize results obtained for different data sets as a reference.

The paper is structured as following. Section 3.2 describes some theoretical questions, which are then used for organizing the literature review: the pattern recognition perspective for BCI implementation, integration and fusion in multimodal systems, and some theoretical motivation for the usage of ensembles. We use in Sect. 3.3 the integration and fusion level for giving a taxonomy of approaches presented in the literature. Other ways of looking at the ensembles are given then: different types of ensembles depending on the nature of the "ensembled"

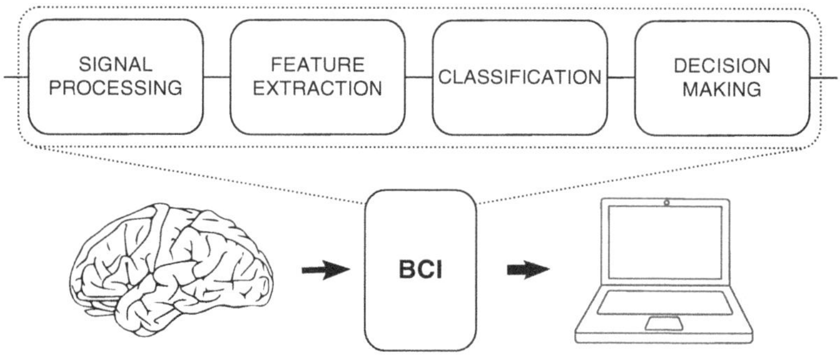

Fig. 3.1 Typical stages of a pattern recognition based BCI system. Usually *Signal Processing* includes as well *Signal Enhancement*, and *Feature Extraction* includes *Feature Selection*

methodologies (see Sect. 3.4); the various partitioning strategies for generating the data subsets whereby each ensemble member is trained (see Sect. 3.5); and different combining approaches used in the literature (see Sect. 3.6). Finally we summarize the analyzed works in tabular form in Sect. 3.7 and give some conclusions in Sect. 3.8.

3.2 Theoretical Background

3.2.1 Pattern Recognition Ensemble Definition and Context

A pattern recognition system attains mapping any kind of data into a decision by discovering structure in the data. Hence, in the case of a BNCI application the data presents a signal representation acquired through a physiological sensor and the decision concerns the issue of a command for controlling a device. The typical stages of a pattern recognition system are signal enhancement, signal processing, feature extraction, feature selection, classification, and decision making, where a decision threshold is applied over the real valued outputs of the classifier in order to generate a decision label (see Fig. 3.1). These stages apply as well for the case of BNCI systems based on Pattern Recognition [24, 35].

A pattern recognition ensemble is a methodological approach whereby the stabilization of a classification procedure is realized through the combination of several pattern recognition stages up to a particular one, usually the classification stage (see Fig. 3.2). Here we use the stability definition given by [32], which defines a stable classifier as one presenting a high bias and a low variance. On the contrary, an unstable classifier presents a low bias and a high variance, usually with respect to the training set [6]. Although the usual approach is to create a classifier ensemble, i.e. different classification approaches are combined, it is worth pointing out that in

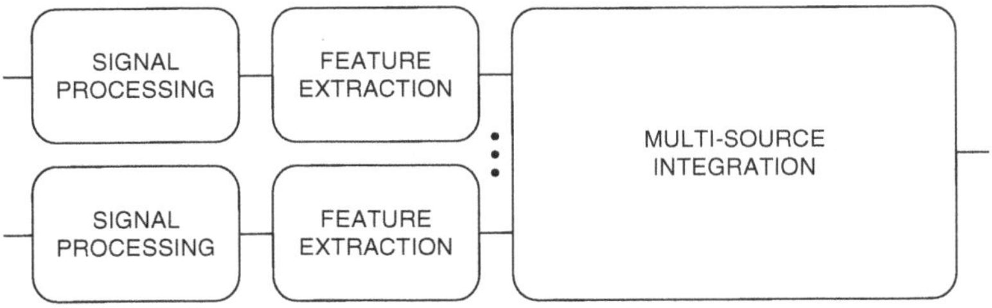

Fig. 3.2 Typical stages of a multi-source/multi-modal pattern recognition based system. The *Multi-Source Integration* usually includes several *Classification*, and *Decision Making* stages (see Fig. 3.1), whose organization depends on the type of integration (see Fig. 3.3)

some cases one combines another stage than the classification one (see Sect. 3.3). Classifier ensembles have received other names in the literature as mentioned in [28]: combination of multiple classifiers, classifier fusion, mixture of experts, consensus aggregation, voting pool of classifiers, divide-and-conquer classifiers, stacked generalization, etc (see the given reference for a list of related works). Composite classifier systems [10], and collective recognition methods [42] were the terms used in the two first works proposing the usage of classifier ensembles [28].

3.2.2 Pattern Recognition Perspective on Fusion

As formerly mentioned the basic structure of a pattern recognition system is modified when creating an ensemble procedure. This is achieved by repeating the processing chain depicted in Fig. 3.1 up to a particular stage. Once repeated, the stages must be combined in a unique system, which is realized through integration (see Fig. 3.2). This is a term often confused with fusion, but we differentiate them herein following [33, 50]. Hence integration stands for the combination of different information sources in a more complex system, whereas fusion, for an instance of integration.

Hence in general three different types of multi-sensor integration have been distinguished traditionally [33]: separate operation, cueing, and fusion. We define herein an additional class corresponding to concatenation, a term we take from [18]. This taxonomy can be applied as well for ensembles in BNCI applications:

Separate operation. In this case the multiple information sources control different aspects of the system (see Fig. 3.3a). This would apply for instance in a BNCI system where a user can control the mouse cursor position through motor imagery, and, the clicks through blinks extracted from the Electrooculography (EOG) signal.

Guiding/cueing/switching. In this case one information source can serve to guide the application of another one or to select among them (see Fig. 3.3b). This has been suggested in [34] as a methodology for the implementation of hybrid

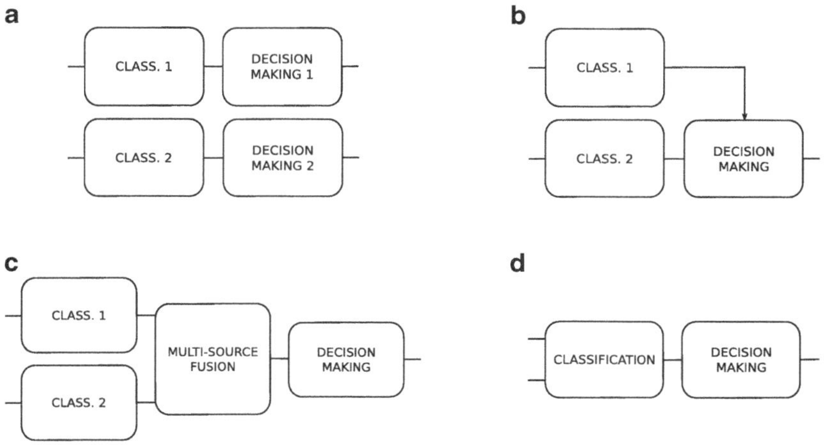

Fig. 3.3 Different types of multi-source integration, which represent particular implementation instances of the corresponding module in Fig. 3.2. (**a**) Integration through separate operation, (**b**) exemplary integration through guiding the decision making, (**c**) integration through fusion at the classification level, (**d**) integration through feature concatenation

BNCIs, whereby the user is allowed to change from one modality to another one depending on its state, e.g. tiredness.

Fusion. Data fusion attains the transformation of the data delivered by multiple information sources into one representational form [1]. This normally is realized through a fusion stage (see Fig. 3.3c). Here the information sources can range from different sensors to different classification algorithms as is usually the case of ensembles in BNCI systems. We extensively analyze this type of integration in Sect. 3.6.

Concatenation. Concatenation implies grouping the data components delivered by multiple information sources into a vector for further processing (see Fig. 3.3d). This type of integration has been particularly used in BNCI systems for the integration of both features [9,39,46] and classification results [14,15,18].

Some works dealing with classifier combination [5, 27, 47], which is closely related to classifier ensembles, distinguish only between integration through selection and through fusion of classifiers. In my opinion selection corresponds to strategies described herein by the terms *separate operation* and *guiding/cueing/switching*, whereas the term fusion in that sense groups *concatenation* and *fusion* itself.

A taxonomy of ensemble methodologies with respect to the stage in which the different processing chains are combined can be established. This property has been denoted in the literature as the fusion level [33, 50]. Hence we can find ensemble procedures with signal, feature, classification or symbol fusion, also known as decision fusion. In the case of BNCI systems the most common fusion levels are the classification and decision ones as further described in Sect. 3.3. Moreover the implementation of an ensemble system with fusion implies the utilization of a fusion operator (see Sect. 3.6). Therefore we can distinguish among different types of systems depending on the operator they use. In this context it is worth pointing out the work in [18], where the results obtained with different integration strategies, namely concatenation vs. fusion, at different levels, namely feature vs. classification level, and fusion operators are compared.

3.2.3 Grounding the Superiority of Ensembles

Classifier ensembles have been described as being particularly efficient for synchronous BCI [32]. They are capable of decreasing the error variance [16, 36]. Moreover [32] extends this main advantage to BNCI. Hence the classification error is formed by the three components noise, bias, and variance. Since the variability of signals is rather large in BNCI systems, i.e., the main component of the error function is the variance, decreasing it is of enormous interest [32, 35]. However the successful improvement in terms of error variance depends on the stability of the classifiers included in the ensemble. Therefore the combined classifiers must be unstable in the sense described in [12, 32] in order to succeed decreasing the error variance. On the contrary if the combined methodologies are stable, i.e., they present a low variance, the resulting ensemble will probably present the same error, since the combination mainly targets the variance error.

In some ensemble types the error improvement can be even described in analytical form. The application of random forests, a particular type of ensembles based on random subsampling and decision trees [7], exemplary allows such a definition. Hence the improvement of the ensemble can be quantified by giving a theoretical upper bound of the random forest generalization error PE^* [7, 52]:

$$PE^* \leq \bar{\rho}(1 - s^2)/s^2, \tag{3.1}$$

where $\bar{\rho}$ stands for the average correlation among individual decision trees, and s, for the strength of the classifier. The strength of the classifier is related to the generalization error of each individual classifier. So the larger s, the smaller the upper bound of the ensemble generalization error. This superiority of random subspaces is even analytically defined for the methodology denoted as Random Electrode Selection Ensemble (RESE) [53]. Furthermore the other term $\bar{\rho}$ formally shows the importance of the diversity among the members of the ensemble, which has to be large as described in [27, 28, 44].

The work in [27] describes a similar analytical characterization for some other ensemble methods. For instance, AdaBoost, which constitutes a particular application of boosting, allows defining the error bound on the training set ϵ for an ensemble of L classifiers with individual errors $\epsilon_i < 0.5$ as [27]:

$$\epsilon < 2^L \prod_{i=1}^{L} \sqrt{\epsilon_i(1 - \epsilon_i)}. \tag{3.2}$$

Another positive feature of ensembles is their capability to cope with small training sets of high-dimensional data [49, 52]. As is well known the larger the dimensionality of the feature space, the more samples have to be taken into account for training a classifier. This so-called curse of dimensionality is caused by the increase of complexity in high-dimensional spaces when estimating the decision surface, which is the surface in the feature space generated by training the classification procedure for discriminating among classes [12]. A rule of thumb even advises a ratio of 5–10 training samples per class and per feature component [21, 32, 43]. The abundance of high-dimensional data in BNCI fosters therefore the usage of ensembles in this application field. Their advantage is attributable to the fact that they divide the complexity of the original decision surface estimation in simpler problems. This reduction even relates in some cases to a reduction in the dimensionality of the feature space, e.g., in ensembles based on bagging, feature subsampling. However other resampling strategies like random subsampling without replacement reduce even more the training data sets. Consequently they should not be applied on small training sets. We come back to this question when discussing the partition strategies used by different approaches (see Sect. 3.5).

Finally classifier ensembles tackle the enormous time variability of EEG signals as described in [52]. This advocates for the extension in the number of classifiers to generate a particular decision surface. In this particular case the employment of a partition strategy in the time domain results particularly interesting.

3.3 Integration and Fusion Level

We discuss in this section different types of ensemble upon their integration or fusion level [33, 50]. In this context we take into account systems based on concatenation, fusion, and guiding (see Sect. 3.2.2) excluding separate operation.

3.3.1 Feature Concatenation

One usual approach in BNCI systems [32] is the concatenation of different types of features, e.g., spatial filters at different positions, autoregression, frequency band power, in a unique feature vector. The resulting feature is then passed through a classifier. This is for instance the approach followed in an early work in the field [39], where frequency features are extracted from different time segments and then concatenated. This same strategy was followed in [20], where an ensemble of Multilayer Perceptron classifiers is used after concatenating the extracted features in the different time intervals.

A more recent work takes into account the extraction of local Common Spatial Patterns (CSPs), whose center is distributed among different electrodes [46]. In this case and because of the high-dimensionality of the resulting feature space, a feature selection stage is added. The spatial distributed features are concatenated and finally classified through Linear Discriminant Analysis. This is a similar approach as the one presented in [31]. However the output of the spatial distributed filters is directly used in this case after applying a decision threshold [31], whereas the former approach [46] concatenates the features and uses the aforementioned linear discriminant classification. Accordingly we describe [31] within decision fusion approaches (see Sect. 3.3.4).

Further approaches with feature integration are described in [9, 18]. In the first case feature extraction in the temporal, spatial, and frequency domains are sequentially combined. The result is though a set of temporal sequences that are used as features for entering a final classifier. Interesting in this case is the usage of either a sample-based classification, or a temporal fusion both implemented through a Bayes classifier (see Sect. 3.6.2). On the other hand [18] concatenates features generated by setting up different configurations of a basic processing chain, which include a sequence of a decimation, a frequency filter, a normalization, a channel selection, a spatial filtering, a frequency band decomposition, and a logarithmic post-processing stages. The result of this feature concatenation approach is delivered to a final classifier, where the performance of a SVM and a logistic regression classifiers are compared. Actually the feature concatenation approach is further compared in the paper with a classification concatenation, and several classification fusion approaches.

It is lastly worth mentioning that the high-dimensionality of the feature space achieved through feature concatenation encourages the application of resampling strategies as done in [52] (see Sect. 3.5).

3.3.2 Classification Concatenation

In some cases dealing with classification of motor imagery data, the feature extraction is realized in three different domains: the time, the spatial, and the frequency domain [13–15]. This extraction is realized in sequential stages and for different frequency intervals simultaneously. Once the features are extracted they are passed to a Linear Discriminant Analysis classifier, one for the result of the extraction in each band. These classifiers form the ensemble. While [13] averages the results, other works by the same authors makes use of a second level of classification, i.e., KNN, LDA, SVM, linear programming machine, and two regression approaches classify the concatenation of the different classification results [14, 15].

One further example of classification concatenation by some other authors is given in [18]. Here the features generated as mentioned in the former section go separately to one classifier each. The classification results are then concatenated and delivered to a second classification stage.

3.3.3 Classification Fusion

The first approach we found with fusion at the classification level is applied on BCI competition III data by Gao Xiarong and colleagues [58]. Although we have not found any paper in the literature describing the proposed ensemble approach its structure has been analyzed in [8]. In this approach the feature extraction is realized through One-Versus-the-Rest, a generalization of the well known Common Spatial Patterns. Following, three different classifiers (LDA, fuzzy KNN, and SVM) are used in a bagging procedure together with an adaptive fusion stage, where an operator is selected from a group of six options on a sample by sample basis. The work in [8] proposes to change the feature stages by using the extraction of Morlet wavelet coefficients and a further selection stage. This is realized through two different procedures, Analysis of Variance (ANOVA) and Genetic Algorithms, whose performance is compared in that work.

One common alternative of ensembles with fusion at the classification level takes into account applying a classifier for several types of features [18, 19]. In [18] one classifier is applied to each of the features resulting from different configurations of the same processing chain as described in Sects. 3.3.1 and 3.3.2. Besides comparing with a feature and a classification concatenation approaches, the classification results are fused, where the performance of product, average, and majority voting is compared as well. The classification fusion outperforms in the reported results both feature and classification concatenation. The methodology proposed in [19] by the same prime author, which is used to classify actual movement data, lightly differs. In this case up to eight different types of feature extraction procedures are used. The following features are extracted in all cases for each signal channel: third order autoregressive coefficients; power estimates in five spectral bands based on a filter bank; EEG signals after artifact-removal and downsampling; a downsampled wavelet decomposition of three levels based on a symlet function; and three different feature sets based on Independent Component Analysis (ICA). A classifier stage is then applied on the eight extracted feature sets. So we have eight classifiers, one per feature set, whose results are lastly combined. Average is used as fusion operator.

An already mentioned work for mental imagery employs classification fusion based on an initial random partition of the electrodes [53]. Following they apply the same multi-stage classification methodology to each of the resulting signal subset: Principal Component Analysis (PCA) for dimensionality reduction, Fisher LDA for feature extraction, and Gaussian Mixture Models (GMM) with Expectation Maximization (EM) for classification. Once performed the classification results are fused through an average operator.

The main purpose of [2] is the comparison of different approaches for the classification of motor imagery data. Hence it evaluates the performance of 13 classifiers including four different ensemble methodologies: boosting, bagging, stacking, and random forest. These four approaches are based on decision trees. The classifiers deal with simple statistical features extracted from three different frequency bands. While boosting, stacking and random forest use a fusion at the classification stage, Alzoubi et al. [2] further proposes to use decision fusion for bagging.

Three different works for p300 detection use similar approaches based on fusion at the classification level [23, 41, 45]. All of them make use of classification ensembles, but differ in the partition strategy. Thus each ensemble is trained by using a specific partition of the training data set. This question is further analyzed in Sect. 3.5. In [41] SVM classifiers are used for the ensemble, where each of them classifies a group of channels selected through accuracy analysis and is tuned with a particular parameter set. On the other hand LDA is used in [45]. In this case the feature extraction succeeds in form of wavelet coefficient computation for different types of wavelets plus a stage for the automate selection of channels denoted as Sequential Floating Forward Search (SFFS). Lastly [23] makes use of stepwise Linear Discriminant Analysis. In this case the ensemble outperforms a single classifier of the same type. Different fusion operators are used in each approach, with [23] comparing the performance of several ones.

A further work on p300 is based on fusion at the classification level [37]. In this case the application of p300 is not the typical spelling one, but the detection of targets in images. Different linear discriminant functions are learned from data windowed at different time intervals. The results of this time-dependent classification is then spatially weighted, so that the fusion stage is defined in the spatial domain. The time domain windowing and the spatially defined fusion make the system robust in front of signal drifts and sample-by-sample fluctuations.

Lastly we comment on a classification fusion work [3] where five different so-called mental states are detected: two motor imagery states, one mental rotation state, one arithmetic operation, and a relaxation state. For this purpose data is filtered in the delta band and wavelet coefficients from four electrodes out of ten selected. These features undergo a classifier stage, where MLP and an ensemble of Adaptive Neuro-Fuzzy Inference Systems (ANFIS) are compared. A rule system is applied in the output of the ensemble for final decision making.

3.3.4 Decision Fusion

To the best of our knowledge the first publication describing an ensemble classifier for BCI [38] made already use of a fusion at the decision level. Here different Linear Vector Quantization (LVQ) classifiers were applied to features related to the Bereitschftspotential. In the last stage voting logic was applied for fusing the decisions of each classifier.

A slight different approach is presented in [55], where simulated neuron spike signals are used in a BNCI system. The work tries to use these signals for controlling a robotic arm. The data go through three different so-called neural decoders that map the spike signals into motor control signals. The result of these three neural decoders go then through the decision fusion stage, which is implemented either with a Kalman filter or a Multilayer Perceptron. As it can be observed this approach differs from the other classifier ensemble approaches described herein, both in terms of the used type of signals and of the employed methodology. However we mention it here for the sake of completeness.

In [48] the authors combine different feature extraction methods and apply a selection algorithm for selecting the best extractor–classifier pairs, which are called experts. Since the proposed methodology uses a majority voting fusion stage, the fusion is done at the decision level. Also in [31] the majority voting operator is used. In this case though Common Spatial Patterns constitute the employed classifiers. In contrast to [46] (see 3.3.1) they do not use the output of this supervised feature extraction methodology for feeding an additional classification stage, but to apply a decision threshold on it. Lastly the obtained decisions are fused.

3.4 Ensemble Type

In the following section we group the different approaches in the literature upon the type of the employed classifier ensemble. They differ from each other in the nature of the information source that deliver the piece of evidence for its posterior combination in the overall methodology. In this context the different sources range from different acquisition devices through different acquisition units up to different methodologies.

3.4.1 Classifier Ensembles

Classifier ensembles are characterized by the fact that each ensemble component is implemented through the same methodological approach. Hence in [19,48] the same classifier is applied on several feature sets. While multinomial logistic regression is used in [19, 48] employs SVMs over a feature subset previously selected through statistical analysis and genetic algorithms. Linear SVMs are as well used in [41,52]. In this last work a classifier ensemble based on SVM is compared with ensembles based on KNN, and C4.5 decision trees.

The method mostly used in classifier ensembles is the Linear Discriminant Analysis. It is sometimes used out of the box like in [45] or with some variants [23]. Salvaris and Sepulveda [45] evaluates the optimal number of ensemble components, which is four. The ensemble performance decreases when augmenting this number, because the data sets become each time smaller.

A further specific BNCI employment of LDAs is this in combination with CSPs and temporal windowing [13, 14, 37]. Rare cases of classifier ensembles are represented by the usage of LVQ [38], decision trees [2], MLPs [20], or ANFIS [3] classifiers.

3.4.2 Stacked Ensemble

Stacked ensembles were introduced in [57]. They receive as well the name of multi-classifier systems [8, 28]. A stacked ensemble is formed by different classification methodologies, whose results are fused in order to increase the generalization capability of the overall system.

The first approach we found in this context was submitted by Xiarong Gao and colleagues in a BCI competition [58] and described in [8]. A LDA, a fuzzy KNN, and a SVM form the ensemble. In [55] the stacked ensemble is realized through the application of a Kalman filter, a population vector algorithm, and an optimal linear decoder. Lastly the stacked procedure compared with three classifier ensembles in [2] results from the combination of a decision tree and a neural network. This same neural network implements then the fusion stage.

3.4.3 Multi-Channel Ensemble

This is a type of ensembles very specific of BNCI systems. The extensive usage of CSP in the literature makes natural its application to selected subsets of channels. Some examples of this ensemble type can be found in [31, 46, 53]. In [31] the multichannel ensemble outperforms the single classifiers LDA, RLDA and SVM. Four strategies for selecting the channels are compared: no selection, sensorimotor, heuristic, and bank selected by cross-fold validation for each subject. On the other hand [53] proposes a random selection of channels for generating the subsets. LDAs are then used for the realization of the feature extraction stage, and Gaussian Mixture Models, as classifier of each subset. An interesting aspect of this last work is the performance evaluation w.r.t. the number of classifiers in the ensemble. Here convergence is obtained on every analyzed data set from 15 to 20 on.

3.4.4 Multimodal Ensemble

We denote herein a multimodal ensemble this formed by different input modalities. This can be generated by different BNCI paradigms as the examples described in [34] for the implementation of so-called hybrid BCIs. Moreover different sensors can be as well applied for gathering the BNCI response. This is the case of the emotion recognition system presented in [30].

3.5 Resampling Strategies

The term resampling strategies, which we take from [12] and first appeared in [6], stands for a set of procedures whereby ensembles can be constructed. Here the objective is to create different data subsets for training each of the members of the ensemble. This is achieved by partitioning the original training set. Resampling can be denoted as well as meta-learning. However this last term is not to be confused with the term meta-classifier, which in [19] stands indistinguishably for the ensemble combination (see Sect. 3.4.1) and the fusion stage (see Sect. 3.6).

The original objective of resampling as described in [12] is to improve the classification performance by generating classifiers for different subsets of the data set. Although resampling refers to sample grouping, the concept can be extended to include other types of data partitioning, what has been definitely done within BNCI research. Hence each component of the ensemble will be trained on a particular data partition. In this more general context the objective is to fulfill the motto "divide and conquer." Taking this fact into account we can then distinguish between procedures originally developed for any pattern recognition application field, i.e., bagging, boosting, and random feature partitioning, and those specially developed within the BNCI application field. In this case the main motivation of the partition is to reduce the variance error due to the different types of BCI variability [32, 35]: time, session-to-session, and subject-to-subject.

It is worth mentioning that some ensemble types, like stacked ones (see Sect. 3.4.2), do not explicitly use any partitioning strategy but train each component in the complete training set. On the other hand some partitioning strategies were designed to work with particular ensemble types, most of them, with classifier ensembles, e.g. bagging. So there is a close relationship between the partition strategy and the type of applied ensemble.

In the following sections we organize the different resampling strategies with respect to the partitioning target. Hence we can distinguish among procedures that divide the feature space, the data set, or which are based on the signal nature of physiological signals.

3.5.1 Data Set Partitioning

3.5.1.1 Bagging

Bagging or bootstrap aggregation is one of the simpler but still effective way of constructing training subsets. Bagging improves its performance when being used to train component classifiers that are unstable [12, 32]. It is based on the random selection with replacement of a data subset of lower cardinality than the original training data set (see Fig. 3.4a). Both the number of subsets and its cardinality are parameters of the procedure. Bagging has been taken into account in two comparative performance evaluations on ensembles [2, 52].

One particular type of bagging is realized when an ensemble is based on different parameterization of the classifier that forms the ensemble [27]. This type of bagging has been used in [20], where all of five MLPs in the ensemble are trained with the complete training data set but with different initial training conditions.

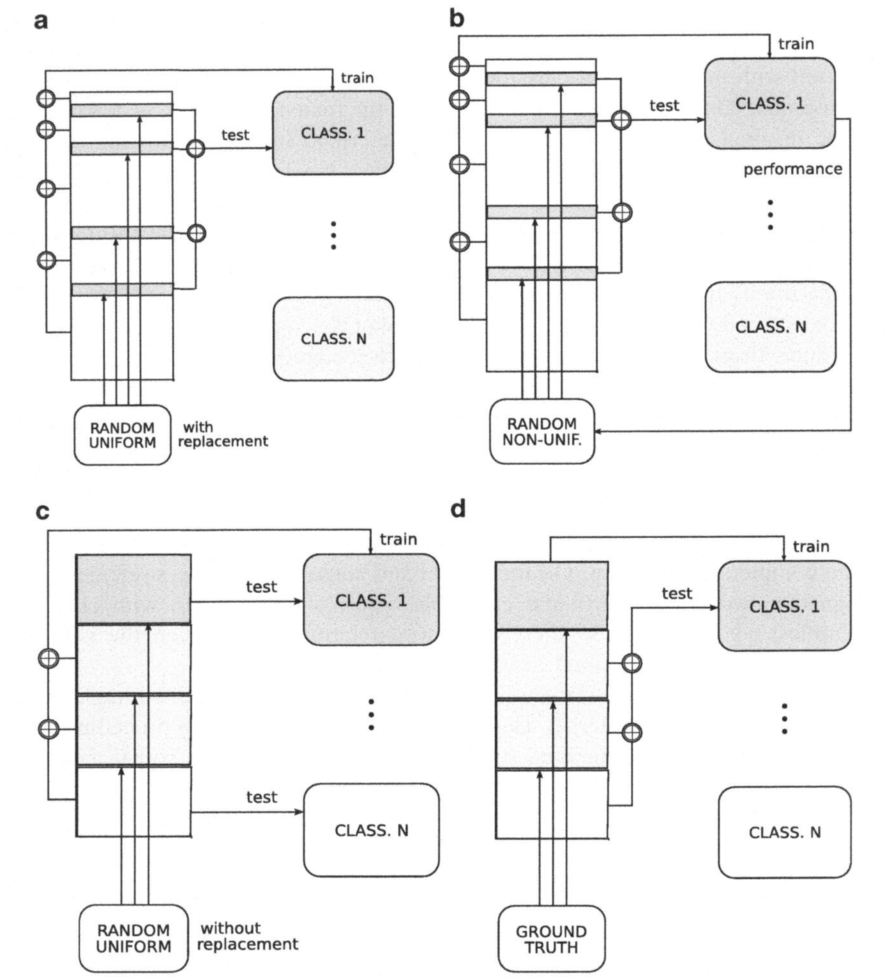

Fig. 3.4 Different types of resampling strategies based on data-set partitioning. (**a**) Bagging, (**b**) Boosting, (**c**) Random without replacement, (**d**) Class partitioning

3.5.1.2 Boosting

Another way of generating training subsets for each member of the ensemble is denoted as boosting. This case presents several sequential stages where a subset is randomly generated. However the probability distribution of the selection process in each stage does depend on the performance evaluation of the former stage. Thus those patterns that are misclassified by former stages present a larger probability of being in the training subset of the subsequent stages (see Fig. 3.4b). The minimal requirement for a boosting procedure to be effective is that the involved classifiers are weak [44], i.e., light over random guessing. Boosting has been compared within [52].

Moreover [2] uses a variant of the basic procedure formerly described denoted as AdaBoost, which stands for adaptive boosting. In this case the initial probability distribution of the samples in the training set is uniform. As we add components in the ensemble, and thus we have to generate new subsets, the samples modify its probability of being selected proportionally to the error produced by it in the former stage. The result of the ensemble components is finally fused through a weighted sum, where the weights of each classifier result depend on that error. AdaBoost's main problem is the overfitting, which can be mitigated by reducing the number of iterations [44]. One further problem that can appear are bad classifications attributable to falsely assigned ground truth [22, 32]. This is not a problem in benchmark data sets, but its importance might increase in real applications.

3.5.1.3 Random Selection Without Replacement

In this case we refer to the partition of the data set in disjoint subsets. This can be done for instance based on the subsets generated for the cross-fold validation as done in [23] (see Fig. 3.4c). One of the possible problems of such an approach is that training data sets become smaller for each member of the ensemble. Hence data partitioning strategies like this one are better applied on problems with very large data sets.

Another factor that can hinder the performance improvement when using random selection without replacement refers to the instability of the ensemble classifiers. This partition strategy might not be effective when some components of the ensemble present already performance near the maximum level. Such an effect has been reported in [23]. Being this procedure similar to bagging, it needs as well unstable classifiers in the ensemble. This feature is difficult to fulfill when most classifiers in the ensemble are in the largest performance range.

3.5.1.4 Class Partitioning

In [3, 41] a procedure is used whereby the data set is partitioned taking its class as coded in the ground truth into account. The associated idea is to train a classifier with groups of similar elements, where similarity is based on the class membership (see Fig. 3.4d). This can be understood as well as having each classifier devoted to detect each class. One Versus the Rest [58] and other multi-class variant of CSPs present a similar strategy. In this case however the classifiers are applied for feature extraction, but they adapt to distinguish samples from one class w.r.t. the rest of them.

It is worth mentioning that this is a type of partitioning strategy we have not found exactly reflected in general ensemble research [27]. A somehow similar strategy is to add a previous clustering algorithm, whereby similarity is based on cluster membership. Hence each member of the ensemble is trained on data belonging to each cluster [44].

3.5.2 Feature Space Partitioning

This strategy can be useful when taking small data sets into account, since it divides the general classification in other problems dealing with smaller feature spaces. For this to be effective, fusion operators with few parameters might be preferred. This somehow contradicts the statement in [32].

3.5.2.1 Feature Set Subsampling

The straightest strategy in this category is the employment of different types of features with separate classifiers. In this way each classifier in the ensemble have to work in a feature space of smaller dimensionality. This has been done for instance in [19,48]. An interesting point in [48] is the inclusion of a further selection procedure, that selects the optimal feature-classifier pairs denoted as experts. The experts are selected from two pools of respectively feature sets and parameter sets of a SVM basis classifier.

A more standard procedure results from randomly subsampling the feature space, which is used within so-called random forest classifiers. Some works like [27] consider random forest as a generalization of bagging for decision trees, where the ensemble is formed by random resampling either the data set, the feature space, or the classifier parameter space. Hence, random forests are ensembles of decision trees. Each component of the ensemble is however trained on a subset of features with a lower number of components, i.e., lower dimensionality, as the original one. The particular feature components of each decision tree are randomly selected. The work in [2] is the only example of feature set random resampling in a BNCI system. This is done for the sake of comparing different methodologies, among which random forests are taken into account.

3.5.2.2 Spatial Partitioning

The selection of electrodes [31,45,46,53] seems to be a particular case of feature space partitioning in its application to EEG data. The spatial neighborhood among channels can be taken into account in the selection [46]. Furthermore [31] compares three different selection strategies: a priori selection of sensorimotor areas, heuristic selection, and bank selected by cross-fold validation for each subject. Lastly channel selection through a procedure denoted as Sequential Floating Forward Search is employed in [45]. All three approaches can be seen as a particular case of feature space subsampling applied on the EEG channels.

On the other hand the selection is done randomly in [53]. They select a fixed number of electrodes (four in the paper) per ensemble. Selection is done with a uniform distribution. They treat the problem of diversity in the classifiers, which fixes up the upper bound of the generalization error as mentioned in Sect. 3.2.3. The diversity of the ensemble lays on the random selection probability. One advantage of this method is its robustness w.r.t. artifacts because electrical signal values in the electrode are not taken into account in the selection process as claimed in [53]. If electrodes with faulty signal are detected, they can be easily deleted prior to the random selection.

3.5.3 *Signal Partitioning*

EEG data is defined in signal form. This signal nature allows data partitioning for ensemble generation both in the time [9, 37] and frequency [13, 14] domains. While time windowing is the preferred procedure for time partitioning, the frequency one is based on splitting the analysis in different bands through the application of filter banks. In both cases a component of the ensemble is lastly trained for each window/band.

3.6 Fusion Operators

Fusion operators are used in ensemble methodologies for realizing the mapping between the multi-dimensional space of the ensemble results into one dimension when integration is implemented through fusion (see Sect. 3.2.2). Decision profiles $DP(\mathbf{x}_i)$ [27, 29] constitute a good analytical tool for expressing the problem:

$$DP(\mathbf{x}_i) = \begin{bmatrix} d_{1,1}(\mathbf{x}_i) & \ldots & d_{1,j}(\mathbf{x}_i) & \ldots & d_{1,C}(\mathbf{x}_i) \\ & & \vdots & & \\ d_{i,1}(\mathbf{x}_i) & \ldots & d_{i,j}(\mathbf{x}_i) & \ldots & d_{i,C}(\mathbf{x}_i) \\ & & \vdots & & \\ d_{L,1}(\mathbf{x}_i) & \ldots & d_{L,j}(\mathbf{x}_i) & \ldots & d_{L,C}(\mathbf{x}_i) \end{bmatrix}, \tag{3.3}$$

where $\mathbf{x}_i$, denotes any point in the data set, L, the number of classifiers in the ensemble, and C, the number of classes. Thus the rows of this matrix represent the outputs of classifier i in the ensemble for each class, and the columns, the support of each classifier ensemble for class j. This allows us to define the fusion operation as the one aggregating the elements of the matrix column-wise, which delivers a vector with the combined support for each class j.

The fusion operation has received different names in the context of BNCI research. Hence it is denoted as final gating function as well [14]. Meta-classifier, which is the term used in [19], is somehow confusing because this term can be understood as the resampling strategy. Lastly other used terms are merging [34], combiner or classifier composer [44], although in this last case generalizes the concept of fusion operator by including algorithms for combining the results after classification concatenation (see Sect. 3.2.2).

The importance of this stage has been underscored in [34] within hybrid BCIs. Although this work recommends the usage of the weighted sum, other works like [52] reflect the importance of using more advanced fusion operators (see [4, 44, 50] for good catalogues). This is aligned with some works in the data fusion research field [5, 51].

The usage of alternative fusion operators is sometimes difficult to achieve given that some resampling strategies seem to have an associated fusion operator, e.g., bagging and random sampling use the majority voting operator, boosting, the weighted sum operator. However there is no fact making this association compulsory. One could think in changing the fusion operator and evaluating the obtained performance as done in [48]. In this context it is worth reminding that weighted operators, like weighted sum, fuzzy integrals, require a procedure for determining the weights to be applied. This is a difficult process but of "paramount" importance as commented in [14].

We can distinguish between fusion operators applied on sample-by-sample basis and those applied in the time domain. This second group reduces the information transfer rate of the result by smoothing the output streams.

3.6.1 Sample Based Fusion

Different operators have been used for fusing ensemble outputs sample by sample. We comment first on the usage of simple fusion operators: sum, product, minimum, maximum. They have been studied for pattern recognition from a very early stage of research [25]. In BNCI they have been used in [45], which employs the sum, [48], where the product is compared with majority vote, average, median, fuzzy integral, and decision templates, and [23], where the performance of the maximum operator is compared to this of the weighted sum and the average.

In spite of the recommendation in [40] to use the average operator in ensemble systems, its performance is not the best in the two comparisons where this operator was taken into account [23, 48]. In a further comparison [18] the average operator outperforms in just one out of three data sets. Other works using it are [13,14,19,53]. Moreover [41] claims to be using a double averaging procedure. The first one at the data level, which is the usual averaging for p300 analysis, and the other one, at the classification level, which is the ensemble fusion operator. However their implementation of this stage makes use of voting logic instead.[1] Voting logic is as well used in [18, 31, 38, 48]. In both fusion comparison works [18, 48] voting logic is the fusion operator outperforming the average and the product.

Weighted sum is suggested as a good alternative for the implementation of hybrid BCIs [34]. This operator is selected as well in [2, 23, 37, 55]. This is the operator usually applied for boosting [2]. Furthermore, [37] and [23] propose to relate the weights in the operator with the accuracy. In [23] this procedure lightly outperformed the other ones, although this is a subject-dependent fact.

Some more rare operators in BNCI are the Choquet fuzzy integral w.r.t. λ-fuzzy measures [17], decision templates [26], both evaluated in [48], and Kalman filtering [55]. In this context it is worth pointing out the relationship between Kalman filtering and the weighted sum operator. The fuzzy integral and the decision templates do not show a significant improvement of performance with respect to simpler ones in the ensemble taken into account in [48]. Lastly [55] uses an additional Kalman filtering in the time domain.

3.6.2 *Time Domain Fusion Operators*

In some BNCI applications a continuous output is needed. It can be smoothed by applying a fusion operator on consecutive samples of the output streams. This is applied at the cost of decreasing the information transfer rate, but practicable if the application allows for such a decrease.

The lowest error in prediction of movement trajectory is obtained in [55] by the Kalman filtering applied in the time domain. Kalman filtering applies a smoothing algorithm governed by some parameters following Bayesian statistics. While Kalman applies this smoothing for samples, neural networks, which are used as basis for comparison in that work, does not. This additional smoothing in the time domain might be the reason for better results.

Some other operators can be applied in the time domain fusion. Bayes classifiers are applied as well on several samples for improving the performance as done in [9]. Lastly a more simple operator like the average one can be used to fulfill this functionality [8].

3.7 Summary of Ensembles Obtained Results

We give in the following table a summary of the performance achieved by different ensembles in the analyzed literature (see Table 3.1). It is worth mentioning that we have only included works with clearly interpretable results and using similar

---

[1] Available at http://asi.insa-rouen.fr/~arakotom/code/bciindex.html

performance measures for the sake of comparison. As it can be observed in the table most of the achieved accuracies are in the range 80% to 95 %. This is a good performance. However most of the presented approaches are designed for distinguishing between two classes. This follows from the common practice in motor imagery systems, where the system is trained with more classes, usually four, but the two best discriminated are selected in the final performance description. When the number of classes increases, the performance seems to decrease. So this might be one of possible future topics deserving further research. The number of subjects in the data sets employed in the performance evaluation are usually small.

We have follow the criterium that methodologies for p300 attain the discrimination between attended and unattended stimuli, although the number of final classes, e.g., characters in the spelling matrix, is usually larger. It is worth pointing out in this particular BNCI modality the good results achieved by [41] and [45] on the same data set. Both approaches are rather different. The good performance might be related to the decrease in the dimensionality achieved by the ensemble application, and not so much to the particular type of applied ensemble.

Ensembles are applied as well on other BNCI modalities, namely slow cortical potentials, mental imagery, spike signal classification, mental state detection, and emotion recognition. Among them we see a good potential in the field of affective computing.

3.8 Final Remarks

Given the high-dimensionality of physiological signals, which are used for BNCI, and their large variability, the employment of classifier ensembles seems to perfectly fit in the application field. In this context it is worth connecting this variability with the partition strategy for generating each component of the ensemble, what has been done in different works in the literature.

Some works have focused the comparative performance evaluation of ensembles on the partitioning strategy [2, 52]. However the partitioning procedure of the ensemble is not always responsible for the bad performance of the ensemble, but more the classifier or classifiers embedded in the ensemble or the relationship between these two factors. For instance in [52] random sampling is stated to present a bad performance when using Support Vector Machines (SVM). This low performance is due in my opinion to the usage of a linear SVM and not to the usage of random sampling. A linear SVM is only justified if the dimensionality of the feature space is so large that it makes not necessary the kernel trick, whereby the feature space is projected into a space of larger dimensionality for allowing a linear separation. Random sampling reduces the dimensionality of the feature space and therefore makes questionable the usage of linear SVM within this resampling strategy. As a further example this same work [52] states that boosting loses in front of bagging and random sampling for the KNN classifier. This is comprehensible

Table 3.1 Summary of performances in % (*perfor.*) in the analyzed literature with respect to different data sets of a number of subjects (*subj*) and classes (*cls*). Best performing methodology in the reference given in column *ref* is reported. Employed data sets are either proprietary or come from different BCI competitions. In this last case the name is coded with latin numbers for the BCI competition name, and arabic numbers for the used data set of that particular competition, e.g., III3a stands for data set 3a of BCI competition III. BCI types are motor imagery (*motor*), actual movement data (*move*), mental imagery (*mental*), p300, self-regulation of slow cortical potentials (*SCP*), spike signals (*spikes*), emotion recognition (*affective*), and mental state classification (*state*). Reported parameters are: integration/fusion level (*integ./fus.level*), ensemble type, employed combiner, and employed partition strategy (*partitioning*). Following performance measures are reported: accuracy (*Acc*), Area Under the Curve (*AUC*), Kappa index (*Kap*), loss median over subjects (*loss*), Equal Error Rate (*EER*), and root mean square error (E_{rms})

Data set									
Name	BCI type	Subj.	Cls.	Perfor.[%]	Integ.|fus.level	Ensemble type	Combiner	Partitioning	Reference
III3a+IV2a	motor	14	2	Acc 82	feat. concat.	–	LDA	time	[9]
III3a	motor	3	4	Kap 52	classif. fusion	stacked	multiple	bagging	[8]
III3a	motor	3	4	Acc 74	classif. fusion	classifier	wei. sum	boosting	[2]
II3	motor	1	2	Acc 92	decis. fusion	classifier	maj. vote	feature	[48]
Propietary	motor	3	2	Acc 80	feat. concat.	–	NN	–	[39]
Propietary	motor	3	2	Acc 88	classif. fusion	classifier	average	parameter	[20]
Propietary	motor	80	2	AUC 90	feat. concat.	multi-channel	reg. LDA	spatial	[46]
Propietary	motor	3	2	Acc 83	decis. fusion	multi-channel	maj. vote	spatial	[31]
Proprietary	motor	83	2	loss 29	classif. concat.	classifier	SVM	frequency	[14]
Proprietary	motor	3	2	Acc 83	decis. fusion	classifier	maj. vote	spatial	[31]
II4	move	1	2	Acc 88	classif. fusion	classifier	maj. vote	feature	[18]
Proprietary	move	2	2	Acc 80	classif. fusion	classifier	average	feature	[19]
III2	p300	2	2	Acc 96	classif. fusion	classifier	average	class	[41]
III2	p300	2	2	Acc 95	classif. fusion	multi-channel	sum	spatial	[45]
Proprietary	p300	7	2	Acc 93	classif. fusion	classifier	wei. sum	random	[23]
Proprietary	p300	1	2	EER 10	classif. fusion	classifier	wei. sum	time	[37]
II1a	SCP	1	2	Acc 93	classif. fusion	classifier	maj. vote	feature	[18]
III5	mental	3	3	Acc 57	classif. fusion	multi-channel	average	spatial	[53]
Proprietary	state	1	5	Acc 89	classif. fusion	classifier	rule	class	[3]
Proprietary	spikes	syn	reg	E_{rms} 0.08	decis. fusion	stacked	Kalman/MLP	random	[55]
Proprietary	affective	1	2	Acc 74	classif. fusion	multi-modal	wei. sum	boosting	[30]

from the point of view that boosting might reinforce bad classifications due to mislabeling as reflected in [22, 32].

One further issue in the application of ensembles is related with the number of components to be generated [44]. It is important to analyze the performance variation w.r.t. this parameter. Hence [45] evaluates this issue within a random sampling without replacement. In this case the optimal number is four and performance decreases when augmenting it. The degradation in performance is caused by the fact that with each new classifier the number of samples to train is less because of the chosen resampling strategy (see Sect. 3.5.1.3). On the other hand [53] proposes a random selection of channels for generating the subsets. In this case the resampling strategy is based on spatial partitioning (see Sect. 3.5.2.2), which reduces the dimensionality of the feature space for each ensemble component. Therefore, the ensemble converges in performance from a number of components of 15–20 on. Hence stating that the more classifiers in the ensembles, the better, like [18] seems not to be well justified. Except for these three examples, no other experimental works in the BNCI literature take this factor into account.

The paper herein has given an extensive overview on the usage of ensembles in BNCI. Some conclusions in comparative studies might be misleading if the aforementioned final remarks are not taken into account. Moreover, the paper attained the unification of nomenclature by giving an overview of the different ensemble approaches presented in the BNCI literature heretofore. The work with different types of fusion operators and of further adapted partitioning strategies constitute open research issues. The analysis of the ensemble cardinality in relationship with the type of resample strategy used is also worth extending.

Acknowledgements The research works described herein have been partially funded by the EU under the FP7 as part of the AsTeRICS project (Grant Agreement 247730).

I would like to thank co-workers Ivan Cester, Anton Albajes-Eizagirre, and David Ibáñez for their collaboration in different research works around the usage of classifier ensembles for BNCI, and Steve Dunne for their valuable comments on the manuscript.

Brain and laptop graphics in Fig. 3.1 were downloaded from:
http://all-free-download.com/free-vector/vector-clip-art/brain_03_117577.html and
http://all-free-download.com/free-vector/vector-clip-art/ordinateur_portable_laptop_55955.html.

References

1. Abidi, M.A., Gonzalez, R.C.: Data fusion in robotics and machine intelligence. Academic Press, San Diego, CA, USA (1992). http://portal.acm.org/citation.cfm?id=149941
2. Alzoubi, O., Koprinska, I., Calvo, R.A.: Classification of Brain–Computer Interface Data. In: Proc. 7th Australasian Data Mining Conference, AusDM, pp. 123–132 (2008). http://citeseerx.ist.psu.edu/viewdoc/summary?doi=10.1.1.169.4737
3. Barbosa, A., Diaz, D., Vellasco, M., Meggiolaro, M., Tanscheit, R.: Mental Tasks Classification for a Noninvasive BCI Application. In: Alippi, C., Polycarpou M., Panayiotou, C., Ellinas G (eds) Artificial Neural Networks - ICANN 2009, Lecture Notes in Computer Science, vol. 5769, chap. 50, pp. 495–504 Springer, Berlin/Heidelberg, Berlin, Heidelberg (2009). DOI 10.1007/978-3-642-04277-5_50, http://dx.doi.org/10.1007/978-3-642-04277-5_50
4. Beliakov, G., Pradera, A., Calvo, T.: Aggregation Functions: A Guide for Practitioners (Studies in Fuzziness and Soft Computing), 1st edn. Springer, Berlin/Heidelberg, (2008). http://www.worldcat.org/isbn/3540737200
5. Bogdanov, A.V.: Neuroinspired architecture for robust classifier fusion of multisensor imagery. IEEE Trans. Geosci. Remote Sens. **46**(5), 1467–1487 (2008). DOI 10.1109/TGRS.2008.916214, http://dx.doi.org/10.1109/TGRS.2008.916214
6. Breiman, L.: Arcing classifiers. Ann. Stat. **26**(3), 801–849 (1998). http://links.jstor.org/sici?sici=0090-5364%28199806%2926%3A3%3C833%3ADAC%3E2.0.CO%3B2-M
7. Breiman, L.: Random forests. Mach. Learn. **45**(1), 5–32 (2001). DOI 10.1023/A:1010933404324, http://dx.doi.org/10.1023/A:1010933404324
8. Cester, I., Soria-Frisch, A.: Comparison of Feature Stages in a multi-classifier BCI. In: To be published in Proc. 5th International Brain–Computer Interface Conference, Graz (2011)

9. Coyle, D.: Neural network based auto association and time-series prediction for biosignal processing in brain–computer interfaces. IEEE Comput. Intell. Mag. **4**(4), 47–59 (2009). DOI 10.1109/MCI.2009.934560, http://dx.doi.org/10.1109/MCI.2009.934560
10. Dasarathy, B., Sheela, B.: A composite classifier system design: Concepts and methodology. Proc. IEEE **67**(5), 708–713 (1979). DOI 10.1109/PROC.1979.11321
11. Dornhege, G., del, J., Hinterberger, T., McFarland, D.J., Müller, K.R. (eds.): Toward Brain–Computer Interfacing (Neural Information Processing), 1st edn. MIT Press (2007). http://www.worldcat.org/isbn/0262042444
12. Duda, R.O., Hart, P.E., Stork, D.G.: Pattern Classification (2nd Edition), 2nd edn. New York: Wiley (2001) http://www.worldcat.org/isbn/0471056693
13. Fazli, S., Grozea, C., Danóczy, M., Blankertz, B., Müller, K.R., Popescu, F.: Ensembles of temporal filters enhance classification performance for ERD-based BCI systems. In: Proc. 4rd International Brain–Computer Interface Workshop and Training Course, pp 247–253 (2008). http://citeseerx.ist.psu.edu/viewdoc/summary?doi=10.1.1.139.6555
14. Fazli, S., Grozea, C., Danoczy, M., Blankertz, B., Popescu, F., Muller, K.R.: Subject independent EEG-based BCI decoding. In: Bengio, Y., Schuurmans, D., Lafferty, J., Williams, C.K.I., Culotta. A. (eds.) Advances in Neural Information Processing Systems, vol. 22, pp. 513–521 (2009a)
15. Fazli, S., Popescu, F., Danóczy, M., Blankertz, B., Müller, K.R., Grozea, C.: Subject-independent mental state classification in single trials. Neural Netw. **22**(9), 1305–1312 (2009b). DOI 10.1016/j.neunet.2009.06.003, http://dx.doi.org/10.1016/j.neunet.2009.06.003
16. Geurts, P.: Contributions to decision tree induction: bias/variance tradeoff and time series classification. PhD thesis, University of Liège (2002)
17. Grabisch, M., Nguyen, H.T., Walker, E.A.: Fundamentals of Uncertainty Calculi with Applications to Fuzzy Inference (Theory and Decision Library B), 1st edn. Kluwer Academic Publishers, Dordrecht (1994). http://www.worldcat.org/isbn/0792331753
18. Hammon, P.S., de Sa, V.R.: Preprocessing and meta-classification for brain–computer interfaces. IEEE Trans. Biomed. Eng. **54**(3), 518–525 (2007). DOI 10.1109/TBME.2006.888833, http://dx.doi.org/10.1109/TBME.2006.888833
19. Hammon, P.S., Makeig, S., Poizner, H., Todorov, E., de Sa, V.R.: Predicting reaching targets from human EEG. IEEE Signal Process. Mag. **25**(1), 69–77 (2008). DOI 10.1109/MSP.2008.4408443, http://dx.doi.org/10.1109/MSP.2008.4408443
20. Haselsteiner, E., Pfurtscheller, G.: Using time-dependent neural networks for EEG classification. IEEE Trans. Rehabil. Eng. **8**, 457–463 (2000). DOI 10.1109/86.895948
21. Jain, A., Chandrasekaran, B.: 39 Dimensionality and sample size considerations in pattern recognition practice. Handbook Stat. **2**, 835–855 (1982). DOI 10.1016/S0169-7161(82)02042-2, http://dx.doi.org/10.1016/S0169-7161(82)02042-2
22. Jain, A.K., Duin, R.P.W., Mao, J.: Statistical pattern recognition: a review. IEEE Trans. Pattern Anal. Mach. Intell. **22**(1), 4–37 (2000). DOI 10.1109/34.824819, http://dx.doi.org/10.1109/34.824819
23. Johnson, G.D., Krusienski, D.J.: Ensemble SWLDA Classifiers for the P300 Speller. In: Proc 13th International Conference on Human–Computer Interaction. Part II: Novel Interaction Methods and Techniques, pp. 551–557. Springer, Berlin, Heidelberg, (2009)
24. Kachenoura, A., Albera, L., Senhadji, L., Comon, P.: Ica: a potential tool for BCI systems. IEEE Signal Process. Mag. **25**(1), 57–68 (2008). DOI 10.1109/MSP.2008.4408442, http://dx.doi.org/10.1109/MSP.2008.4408442
25. Kittler, J., Hatef, M., Duin, R.P.W., Matas, J.: On combining classifiers. IEEE Trans. Pattern Anal. Mach. Intell. **20**(3), 226–239 (1998). DOI 10.1109/34.667881, http://dx.doi.org/10.1109/34.667881
26. Kuncheva, L.I.: "Fuzzy" versus "nonfuzzy" in combining classifiers designed by boosting. IEEE Trans. Fuzzy Syst. **11**(6), 729–741 (2003). DOI 10.1109/TFUZZ.2003.819842, http://dx.doi.org/10.1109/TFUZZ.2003.819842
27. Kuncheva, L.I.: Combining Pattern Classifiers: Methods and Algorithms. Hoboken, New Jersey: Wiley (2004)

28. Kuncheva, L.I.: Classifier Ensembles: Facts, Fiction, Faults and Future. In: Proc. 19th International Conference Pattern Recognition, ICPR'2008 (Plenary lecture) (2008) http://pages.bangor.ac.uk/~mas00a/papers/icpr2008plenary.ppt
29. Kuncheva, L., Bezdek, J.C., Duin, R.P.W.: Decision templates for multiple classifier fusion: an experimental comparison. Pattern Recogn. **34**, 299–314 (2001)
30. Kuncheva, L.I., Christy, T., Pierce, I., Mansoor, S.P.: Multi-modal Biometric Emotion Recognition using Classifier Ensembles. In: Proc 24th International Conference on Industrial, Engineering & Other Applications of Applied Intelligent Systems (2011)
31. Lei, X., Yang, P., Xu, P., Liu, T.J., Yao, D.Z.: Common spatial pattern ensemble classifier and its application in brain–computer interface. J. Electron. Sci. Tech. China **7**(1), 17–21 (2009)
32. Lotte, F., Congedo, M., Lécuyer, A., Lamarche, F., Arnaldi, B.: A review of classification algorithms for EEG-based brain–computer interfaces. J. Neural Eng. **4**(2), R1-R13, (2007). DOI 10.1088/1741-2560/4/2/R01, http://dx.doi.org/10.1088/1741-2560/4/2/R01
33. Luo, R.C., Kay, M.G. (eds.): Multisensor integration and fusion for intelligent machines and systems. Ablex Publishing Corp., Norwood, NJ, USA (1995). http://portal.acm.org/citation.cfm?id=212333
34. Millán, J.D., Rupp, R., Müller-Putz, G.R., Murray-Smith, R., Giugliemma, C., Tangermann, M., Vidaurre, C., Cincotti, F., Kübler, A., Leeb, R., Neuper, C., Müller, K.R.R., Mattia, D.: Combining Brain–Computer Interfaces and Assistive Technologies: State-of-the-Art and Challenges. Front. Neurosci. 4:161, R1-R33, (2010). DOI 10.3389/fnins.2010.00161, http://dx.doi.org/10.3389/fnins.2010.00161
35. Müller, K.R., Tangermann, M., Dornhege, G., Krauledat, M., Curio, G., Blankertz, B.: Machine learning for real-time single-trial EEG-analysis: From brain–computer interfacing to mental state monitoring. J. Neurosci. Methods **167**(1), 82–90 (2008). DOI 10.1016/j.jneumeth.2007.09.022, http://dx.doi.org/10.1016/j.jneumeth.2007.09.022
36. Oza, N., Tumer, K.: Classifier ensembles: Select real-world applications. Inf. Fusion **9**(1), 4–20 (2008). DOI 10.1016/j.inffus.2007.07.002, http://dx.doi.org/10.1016/j.inffus.2007.07.002
37. Parra, L., Christoforou, C., Gerson, A., Dyrholm, M., Luo, A., Wagner, M., Philiastides, M., Sajda, P.: Spatiotemporal Linear Decoding of Brain State. IEEE Signal. Process. Mag. **25**(1), 107–115 (2008). DOI 10.1109/MSP.2008.4408447, http://dx.doi.org/10.1109/MSP.2008.4408447
38. Pfurtscheller, G., Flotzinger, D., Kalcher, J.: Brain–Computer Interface-a new communication device for handicapped persons. J. Microcomputer Appl. **16**(3), 293–299 (1993). DOI 10.1006/jmca.1993.1030, http://dx.doi.org/10.1006/jmca.1993.1030
39. Pfurtscheller, G., Neuper, C., Flotzinger, D., Pregenzer, M.: x EEG-based discrimination between imagination of right and left hand movement. Electroencephalogr. Clin. Neurophysiol. **103**(6), 642–651 (1997). DOI 10.1016/S0013-4694(97)00080-1, http://dx.doi.org/10.1016/S0013-4694(97)00080-1
40. Polikar, R.: Ensemble based systems in decision making. IEEE Circuits Syst. Mag. **6**(3), 21–45 (2006)
41. Rakotomamonjy, A., Guigue, V.: BCI competition III: dataset II- ensemble of SVMs for BCI P300 speller. IEEE Trans. Biomed. Eng. **55**(3), 1147–1154 (2008). DOI 10.1109/TBME.2008.915728, http://dx.doi.org/10.1109/TBME.2008.915728
42. Rastrigin, L.A., Erenstein, R.H.: Metod kollectivnogo raspoznavaniya (Method of Collective Recognition, in Russian). Energiozdat, Moscow (1981)
43. Raudys, S., Jain, A.: Small sample size effects in statistical pattern recognition: recommendations for practitioners. IEEE Trans. Pattern Anal. Mach. Intell. **13**(3), 252–264 (1991). DOI 10.1109/34.75512
44. Rokach, L.: Ensemble-based classifiers. Artif. Intell. Rev. **33**(1), 1–39 (2010). DOI 10.1007/s10462-009-9124-7, http://dx.doi.org/10.1007/s10462-009-9124-7
45. Salvaris, M., Sepulveda, F.: Wavelets and ensemble of FLDs for P300 classification. In: Proc. 4th International IEEE/EMBS Conference on Neural Engineering, 2009. NER '09, pp. 339–342 (2009). DOI 10.1109/NER.2009.5109302, http://dx.doi.org/10.1109/NER.2009.5109302

46. Sannelli, C., Vidaurre, C., Müller, K.R., Blankertz, B.: CSP patches: an ensemble of optimized spatial filters. An evaluation study. J. Neural Eng. **8**(2), 025,012+ (2011). DOI 10.1088/1741-2560/8/2/025012, http://dx.doi.org/10.1088/1741-2560/8/2/025012
47. Sharkey, A.J. (ed.): Combining Artificial Neural Nets: Ensemble and Modular Multi-Net Systems, 1st edn. Springer, New York, Inc., Secaucus, NJ, USA (1999)
48. Shoaie Shirehjini, Z., Bagheri Shouraki, S., Esmailee, M.: Variant Combination of Multiple Classifiers Methods for Classifying the EEG Signals in Brain–Computer Interface. In: Sarbazi-Azad, H., Parhami, B., Miremadi, S.G., Hessabi, S. (eds.) Advances in Computer Science and Engineering, Communications in Computer and Information Science, vol. 6, chap. 59, pp. 477–484. Springer, Berlin, Heidelberg, (2009)
49. Skurichina, M., Duin, R.P.W.: Bagging, boosting and the random subspace method for linear classifiers. Pattern Anal. Appl. **5**(2), 121–135 (2002). DOI 10.1007/s100440200011, http://dx.doi.org/10.1007/s100440200011
50. Soria-Frisch, A.: Soft Data Fusion for Computer Vision. PhD thesis, TU Berlin (2004)
51. Soria-Frisch, A., Riera, A., Dunne, S.: Fusion operators for multi-modal biometric authentication based on physiological signals. In: Proc. 2010 IEEE International Conference on Fuzzy Systems (FUZZ), IEEE, pp. 1–7 (2010). DOI 10.1109/FUZZY.2010.5584121, http://dx.doi.org/10.1109/FUZZY.2010.5584121
52. Sun, S., Zhang, C., Zhang, D.: An experimental evaluation of ensemble methods for EEG signal classification. Pattern Recog. Lett. **28**(15), 2157–2163 (2007). DOI 10.1016/j.patrec.2007.06.018, http://dx.doi.org/10.1016/j.patrec.2007.06.018
53. Sun, S., Zhang, C., Lu, Y.: The random electrode selection ensemble for EEG signal classification. Pattern Recogn. **41**, 1680–1692 (2008) http://portal.acm.org/citation.cfm?id=1340830
54. Vallabhaneni, A., Wang, T., He, B.: Brain–computer interface. In: He, B., He, B. (eds.) Neural Engineering, Bioelectric Engineering, pp. 85–121. Springer, US (2005)
55. White, J.R., Levy, T., Bishop, W., Beaty, J.D.: Real-time decision fusion for multimodal neural prosthetic devices. PLoS ONE **5**(3), e9493+ (2010). DOI 10.1371/journal.pone.0009493, http://dx.doi.org/10.1371/journal.pone.0009493
56. Wolpaw, J.R., Birbaumer, N., McFarland, D.J., Pfurtscheller, G., Vaughan, T.M.: Brain–computer interfaces for communication and control. Clin. Neurophysiol. **113**(6), 767–791 (2002) http://view.ncbi.nlm.nih.gov/pubmed/12048038
57. Wolpert, D.H.: Stacked generalization. Neural Netw. **5**, 241–259 (1992). http://citeseerx.ist.psu.edu/viewdoc/summary?doi=10.1.1.56.15%33
58. Wu, W., Gao, X., Gao, S.: One-Versus-the-Rest(OVR) Algorithm: An Extension of Common Spatial Patterns(CSP) Algorithm to Multi-class Case. In: Proc. 27th Annual International Conference of the Engineering in Medicine and Biology Society, IEEE, pp. 2387–2390 (2005). DOI 10.1109/IEMBS.2005.1616947, http://dx.doi.org/10.1109/IEMBS.2005.1616947

Chapter 4
Improving Brain–Computer Interfaces Using Independent Component Analysis

Yijun Wang and Tzyy-Ping Jung

4.1 Introduction

In the past two decades, electroencephalogram (EEG)-based brain–computer interfaces (BCIs) have attracted much attention in the fields of neuroscience and neural engineering [3,23,46]. Researchers have made significant progress in designing and demonstrating usable BCI systems for the purpose of communication and control. Currently, the BCI community puts great effort into translating this technology from laboratory demonstrations to real-life products to help physically disabled people achieve improved quality of life [5,41]. Although many studies have been carried out to implement and evaluate demonstration systems in laboratory settings, developing practical BCI systems within a real-world environment still poses severe technical challenges.

In real-world applications, a BCI system must meet the requirements of convenient system use as well as robust system performance [40]. Recently, researchers have proposed different methods for improving the practicality of a BCI system in terms of hardware and software design. When working with a BCI product, researchers need to pay attention to two major issues: (1) ease-of-use, and (2) robustness of system performance. Current BCI research places increasing demand on advanced signal-processing techniques to improve system performance and ease-of-use. Among the different signal-processing techniques employed in current BCI systems, independent component analysis (ICA) is one of the most successful

Y. Wang (✉)
Institute for Neural Computation, University of California San Diego, San Diego, USA
e-mail: yijun@sccn.ucsd.edu

T.-P. Jung
Institute for Neural Computation, University of California San Diego, San Diego, USA

Institute of Engineering in Medicine, University of California San Diego, San Diego, USA
e-mail: jung@sscn.ecsd.edu

B. Z. Allison et al. (eds.), *Towards Practical Brain-Computer Interfaces*, Biological and Medical Physics, Biomedical Engineering, DOI 10.1007/978-3-642-29746-5_4,
© Springer-Verlag Berlin Heidelberg 2012

methods [28]. Due to its capability in decomposing scalp EEG signals into functionally independent brain activities and other non-neural activities, ICA has been widely applied to improve the signal-to-noise ratio (SNR) of task-related EEG signals in BCI systems. This study focuses on the use of ICA in current BCI systems. The goal of this study is twofold: (1) to investigate the feasibility of using ICA to improve BCI performance through reviewing the state-of-the-art BCI studies, and (2) to introduce our recent work on developing an ICA-based zero-training method for deriving EEG spatial filters in a motor imagery-based BCI. This study applied the extended infomax ICA algorithm [25] from the open-source EEGLAB toolbox [7] to multichannel EEG data.

4.2 ICA in EEG Signal Processing

Independent component analysis is a statistical method that aims to find linear projections of the observed data that maximize their mutual independence [16]. When applied to blind source separation (BSS), ICA aims to recover independent sources using multi-channel observations of mixtures of those sources. In the past two decades, ICA has been successfully used in processing biomedical signals including EEG, electrocardiogram (ECG), magnetoencephalogram (MEG), and functional magnetic resonance imaging (fMRI) signals [17]. In EEG signal processing, ICA has shown a good capability in separating the scalp EEG signals into functionally independent sources, such as neural components originating from different brain areas and artifactual components attributed to eye movements, blinks, muscle, heart, and line noise (Fig. 4.1). Due to its superiority in EEG source separation, ICA has been successfully applied to EEG research to reduce EEG artifact, enhance the SNR of task-related EEG signals, and facilitate EEG source localization [19, 20, 30, 38].

Given a linear mixing model, n-channel scalp EEG signals, $\mathbf{x} = [x_1, x_2 \ldots x_n]$ are generated by m independent sources $s = [s_1, s_2 \ldots s_m]$:

$$\mathbf{x} = \mathbf{A}\mathbf{s} \tag{4.1}$$

where $\mathbf{A}$ is the $n \times m$ mixing matrix in the model. After ICA, recovered source signals, $\mathbf{u}$, can be estimated by applying an unmixing matrix $\mathbf{W}(m \times n)$ to the observed EEG data $\mathbf{x}$:

$$\mathbf{u} = \mathbf{W}\mathbf{x} \qquad \mathbf{x} = \mathbf{W}^{-1}\mathbf{u} \tag{4.2}$$

where each row of $\mathbf{W}$ is a spatial filter for estimating an independent component (IC) and each column of $\mathbf{W}^{-1}$ consists of electrode weights (i.e., a spatial projection) of an independent component.

Figure 4.1 shows an example of ICA applied to 128-channel scalp EEG data recorded during a visually guided reaching task, which involved various kinds of movement artifacts [42]. It is apparently difficult to read the underlying neural activities from the scalp channel data, which include overlapped EEG signals and artifacts. For example, electrodes at the frontal area have very strong eye-movement

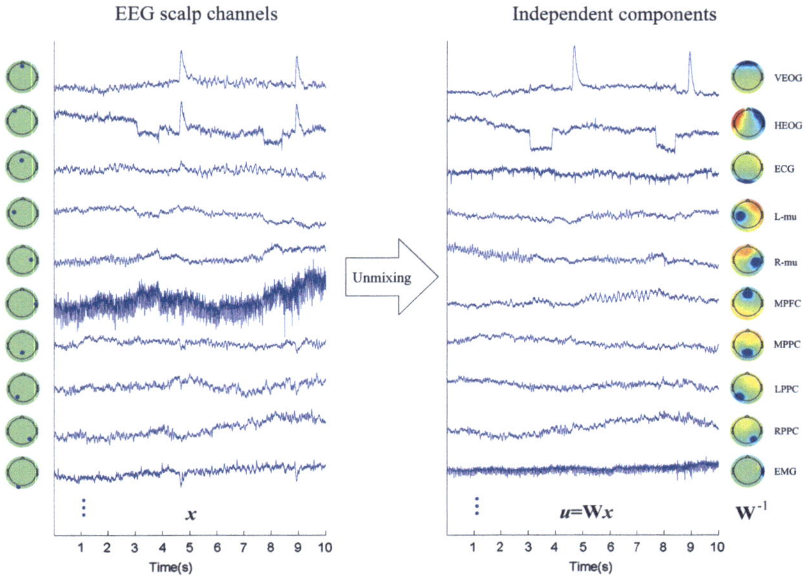

Fig. 4.1 Schematic overview of ICA decomposition of scalp EEG data (x). Activities of independent components (u), were obtained by applying an unmixing matrix $\mathbf{W}$ to x : $\mathbf{u} = Wx$. Each column of $\mathbf{W}^{-1}$, which consists of electrode weights, was shown as a scalp map and referred to as the spatial pattern of an IC. The spatial patterns ($\mathbf{W}^{-1}$) clearly showed scalp distributions of source activities of the ICs

artifacts, which seriously contaminated the midline theta activities over the prefrontal cortex area. In this example, ICA successfully separates scalp EEG signals into neural and non-neural independent source activities, which can be easily understood according to their spatio-temporal characteristics. As shown in Fig. 4.1, recovered independent brain activities include the left/right mu components over the sensorimotor areas (L-mu, R-mu), the midline prefrontal component (MPFC), and the posterior parietal components (MPPC, LPPC, RPPC). In addition, ICA also recovered the non-neural source activities including the vertical/horizontal Electrooculogram (VEOG/HEOG), ECG, and Electromyogram (EMG) components. This capability of decomposing scalp EEG signals into functionally independent sources makes ICA a potential tool for many applications in EEG-based BCIs.

4.3 ICA in BCI Systems

To better understand the state-of-the-art of ICA in BCI studies, this study first presents a survey of the literature. The articles selected for this survey were chosen from journal and conference research papers found in Google scholar using the

Table 4.1 Classification of ICA's applications in BCI studies

Application purpose	Study	BCI type
Removing EEG artifacts	Wang et al. [42]	Movement planning
	Halder et al. [12]	Motor imagery
	Ghanbari et al. [1]	Motor imagery
	Papadelis et al. [33]	Sleepiness monitoring
Enhancing SNR of task-related EEG signals	Xu et al. [47]	P300
	Serby et al. [36]	P300
	Li et al. [26]	P300
	Naeem et al. [32]	Motor imagery
	Delorme at al. [6]	Motor imagery
	Peterson [34]	Motor imagery
	Hung et al. [15]	Motor imagery
	Qin et al. [35]	Motor imagery
	Wang et al. [39]	Motor imagery
	Lee et al. [24]	VEP
	Hill et al. [14]	AEP
	Lin et al. [27]	Drowsiness monitoring
	Lan et al. [37]	Mental tasks
	Erfanian et al. [9]	Mental tasks
	Wang et al. [43]	Movement planning
	Hammon et al. [13]	Movement planning
Selecting optimal electrodes	Wang et al. [44]	VEP
	Lou et al. [29]	Motor imagery

following keywords: "ICA", "BCI", and "EEG". According to application purposes, 22 selected studies were categorized into three classes: (1) artifact removal [1,12,33, 42], (2) enhancement of SNR of task-related EEG signals [6,9,13–15,24,26,27,32, 34–37,39,43,47], and (3) selection of optimal electrodes [29,44]. In the applications of artifact removal and SNR enhancement, ICA was used to design spatial filters to remove task-irrelevant activities such as blinks and movement artifacts. The application of electrode selection aimed to reduce the number of electrodes needed in a BCI system based on the spatio-spectral characteristics of independent brain components. Table 4.1 lists details of these studies, including their application purpose, reference information, and types of BCI design. These studies cover most types of BCI designs including visual evoked potential (VEP), auditory evoked potential (AEP), P300 event-related potential (ERP), motor imagery, movement planning, mental tasks, and sleepiness/drowsiness monitoring. It is quite clear that most studies fall into the category of enhancing the SNR of task-related EEG signals. The method corresponding to each of the three categories is described in detail with example data in subsections below.

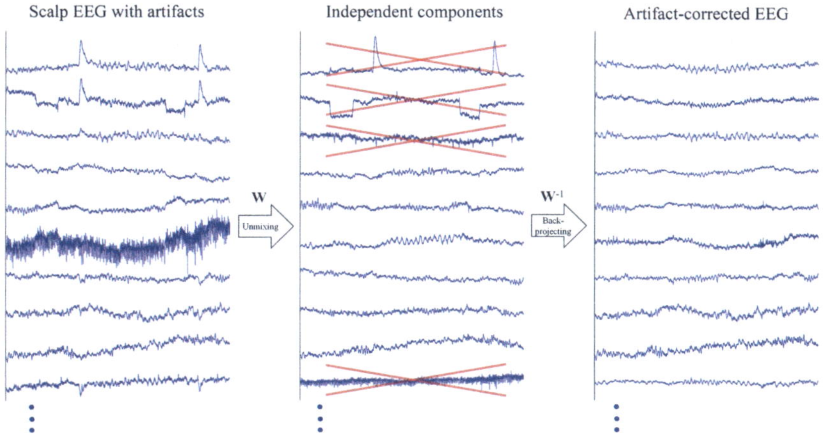

Fig. 4.2 Illustration of the ICA-based approach for artifact removal. Activities of ICs were obtained by applying ICA to scalp EEG data. Artifact (e.g., EOG, ECG, and EMG) ICs were identified and removed from the obtained ICs. Artifact-corrected EEG signals were obtained by only projecting the brain components back to the scalp electrodes

4.3.1 Artifact Removal

EEG signals are often contaminated by pervasive artifacts such as blinks and motions. These artifacts might seriously deteriorate the system performance of BCIs [11]. To make a BCI system more robust, movement and other artifacts need to be eliminated before the task-related EEG features can be extracted for classification. The superiority of ICA in EEG artifact removal has been well demonstrated by many studies [18]. In this application category, ICA aims to separate and eliminate the artifact-related non-neural activities from the EEG signals.

Wang et al. used ICA to correct EEG signals recorded in a movement-planning task, which involved a lot of eye and muscle movements [42]. The EEG signals encoding movement directions can be applied to predict the direction of an intended movement (e.g., reach and saccade) after removing artifact components arising from eye and muscle activities. In motor imagery-based BCIs, system performance (e.g., classification accuracy or the R-square values of features) was improved after removing EOG/EMG artifacts [1, 12]. In a drowsiness monitoring study [33], the ICA-based artifact removal was used as a routine approach to correct the EEG signals recorded in a driving task, which involved many head/body movements.

Figure 4.2 illustrates the procedures of ICA-based artifact removal. In this example, the scalp EEG data recorded during reach/saccade planning and execution were contaminated by artifacts [42]. The artifact-removal method consists of three procedures: (1) apply ICA to scalp EEG data, (2) identify and remove the artifact-related ICs, and (3) project EEG-related ICs back to scalp electrodes to reconstruct artifact-corrected EEG data. In general, identification of artifact ICs can be performed using prior knowledge of spatio-temporal characteristics in EEG artifacts. For example, the IC corresponding to horizontal eye movement has a two-dipole distribution with opposite polarities over the bilateral prefrontal areas (see Fig. 4.1). As shown in Fig. 4.2, the SNR of the EEG signals has been considerably improved after removing the artifact ICs including EOG, ECG, and EMG. In practice, online implementation of this approach can effectively improve the robustness of an online BCI system.

4.3.2 SNR Enhancement of Task-Related EEG Signals

Spatial filtering is one of the most important signal processing techniques employed in BCIs using multichannel EEG [31]. The basic principle of spatial filtering is to eliminate task-irrelevant signals through linearly weighting different channels, and thus, enhances the SNR of task-related EEG signals. Many multidimensional data-processing methods have been adopted in recent BCI studies. For example, the common spatial pattern (CSP) method [4], the canonical correlation analysis (CCA) [2], and ICA, have been successfully applied to the motor imagery, the SSVEP, and the P300-based BCIs respectively. In general, the ICA-based spatial filtering method has two advantages: (1) it is an unsupervised learning method and therefore no labeled data are required, and (2) it allows exploring the relationship between human behavior and the spatio-spectral pattern of an IC, facilitating the understanding of the specific neural mechanism. As listed in Table 4.1, the ICA-based spatial filtering has been widely applied to most types of BCIs including P300 [26, 36, 47], motor imagery [6, 15, 32, 34, 35, 39], VEP [24], AEP [14], drowsiness monitoring [27], mental tasks [9, 37], and movement planning [13, 43]. Generally, these studies aimed to enhance the SNR of task-related EEG signals by ICA so that the system performance (e.g., classification accuracy) can be improved. In practice, only a small number of task-related EEG ICs will be selected for obtaining spatial filters according to their capabilities for discriminating different tasks.

The major steps of the ICA-based spatial filtering approach include: (1) application of ICA to the training data, (2) identification of task-related ICs (i.e., the mu ICs in this case), and (3) the application of the corresponding spatial filters to EEG data before the training and testing steps of the classification process. Figure 4.3 shows an example of applying ICA-based spatial filters to enhance motor activities in a motor imagery-based BCI. IC activities with higher SNR of motor activities than the scalp channel data can be obtained by multiplying motor-related spatial filters to the scalp EEG data. As shown in Fig. 4.3, the spatial filters have the largest positive weights over the sensorimotor areas on both hemispheres with some negative weights around this area functioning as linear combinations to eliminate common background activities. The spatial patterns corresponding to the mu ICs show very typical dipolar distributions over the sensorimotor areas, indicating the source locations of mu activity modulated by motor imagery. In this example, visually cued motor imagery of right hand movement induced a contralateral

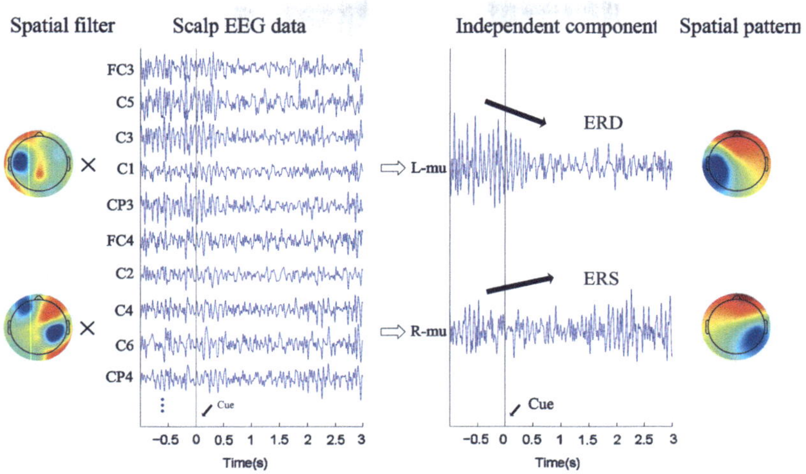

Fig. 4.3 ICA-based spatial filtering for extracting task-related brain activities during motor imagery. In this trial, the subject was instructed to imagine right-hand movement after a visual cue appeared at time 0. After ICA, two ICs with characteristic spatio-spectral patterns were selected as the motor components (L-mu and R-mu ICs), which were dominant by the mu rhythm in the frequency domain. The corresponding weighting vectors in the unmixing matrix (**W**) were selected to be used as spatial filters. *Arrows* indicate a contralateral ERD and an ipsilateral ERS, which can be more clearly observed in the IC activities than the scalp EEG channel data

event-related desynchronization (ERD) and an ipsilateral event-related synchronization (ERS) of the mu rhythm (indicated by arrows in Fig. 4.3), which were more clearly shown in IC activities than the unprocessed scalp EEG channel data.

4.3.3 Electrode Selection

An optimal selection of a small number of electrodes plays an important role in the design of a practical BCI system for real-life applications [40]. For example, in an SSVEP BCI, the goal of electrode selection is to achieve SSVEPs with a high SNR using a bipolar EEG channel consisting of a signal electrode and a reference electrode [44]. In practice, the electrode giving the strongest SSVEP, which is generally located in the occipital region, is selected as the signal electrode. The reference electrode is searched under the following criteria: its SSVEP should be weak, and its position should be close to the signal electrode so that its noise activity is similar to that of the signal electrode. In this way, a high SNR can be obtained with the bipolar channel because most of the spontaneous background activities are eliminated after the subtraction, while the SSVEP component is mostly retained.

Due to its superiority in decomposing independent brain sources, ICA can facilitate the electrode selection in BCIs. Wang et al. [44] developed an ICA-based

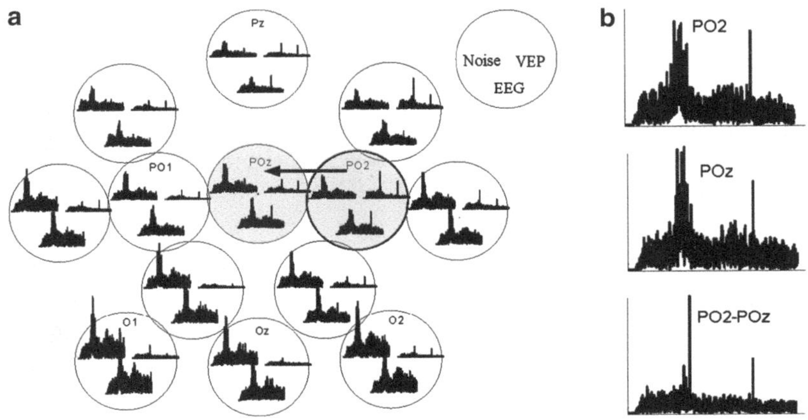

Fig. 4.4 (**a**) Illustration of the ICA-based approach for electrode selection in an SSVEP BCI. Thirteen electrodes around the occipital region were used in ICA to decompose EEG into SSVEP and background noise activities. For each channel, PSDs of scalp EEG, SSVEP and noise activities are put together for comparison. The *arrow* and *shaded areas* indicate the selected signal (PO2) and reference (POz) electrodes. (**b**) PSDs for monopolar channels PO2, POz, and the POz-PO2 bipolar channel (adapted from [44] with permission from IEEE)

approach for electrode selection in an SSVEP BCI. The detailed procedures are described as follows:

1. **ICA decomposition.** Thirteen-channel EEG signals $\boldsymbol{x}$ (with 13 Hz SSVEPs) between Pz and Oz (Fig. 4.4a) were selected as the input for ICA decomposition. Through ICA, 13 independent components were obtained as estimates of brain sources $\boldsymbol{s}$ including SSVEP components (signal) and other background EEG components (noise).
2. **Reconstruction of signal and noise.** The ICs with high SNR of SSVEP (i.e., the ratio of EEG power at 13 Hz to the EEG power in the rest of spectrum) were taken to be the true SSVEP-laden components and the remaining ICs were considered as background noise components. Through projecting the SSVEP sources and the noise sources back to the scalp electrodes, the SSVEP and noise activities at each electrode over the scalp can be separated.
3. **Selection of the signal electrode.** Power spectrum density (PSD) analysis was performed for calculating the SNR of the SSVEP. Figure 4.4a shows PSDs of original channel data and decomposed SSVEP and noise activities on all 13 scalp electrodes. The electrode giving the strongest SSVEP activity (i.e., PO2) was selected as the signal channel.
4. **Selection of the reference electrode.** The correlation of the SSVEP activity and the noise activity between electrodes was calculated. The ratio of the SSVEP correlation to the noise correlation between other electrodes and the signal electrode is the criterion for selecting the reference electrode. Electrodes with high noise correlation and low SSVEP correlation are good candidates.

Figure 4.4 shows an example of the proposed approach on one subject. As shown in Fig. 4.4a, the SSVEP of this subject is highly contaminated by spontaneous EEG signals. It is difficult to choose a good bipolar channel from the original EEG channel data. Through ICA decomposition, the distribution of SSVEP activities shows that PO2 has the most significant SSVEP. As indicated by an arrow in Fig. 4.4a, POz was selected as the reference channel due to its low SSVEP correlation and high noise correlation to PO2. Figure 4.4b proves that the PO2-POz bipolar channel can significantly enhance the SNR of SSVEP due to the elimination of the common noise activities.

Not limited to the SSVEP-based BCI, this approach could be easily adapted to other BCI systems. For example, Lou et al. [29] developed a similar approach for optimizing bipolar electrodes in a motor imagery-based BCI. In their study, ICA was used for separating background alpha rhythms from the sensorimotor mu rhythms. Typical bipolar leads between the sensorimotor areas and the prefrontal areas (e.g., C3-FCz and C4-FCz) were demonstrated most efficient for extracting the motor imagery induced power change of the mu rhythms.

4.4 ICA-Based Zero-Training-Training BCI

As mentioned above, EEG-based BCIs often use spatial filters to improve the SNR of task-related EEG activities [31]. To obtain robust spatial filters, large amounts of labeled data, which are often expensive and labor-intensive to obtain, need to be collected in a training procedure before online BCI control. Recently, several studies have developed zero-training methods using a session-to-session scenario in order to alleviate this problem [22]. To our knowledge, a state-to-state translation, which applies spatial filters derived from one state to another, has never been reported. This study proposes a state-to-state, zero-training method to construct spatial filters for extracting EEG changes induced by motor imagery. The unsupervised nature makes ICA a potential tool to obtain task-related spatial filters even from task-irrelevant data. In this study, ICA was separately applied to the multichannel EEG signals in the resting and the motor imagery states to obtain spatial filters specific for extracting the mu components. The resultant spatial filters were then applied to single-trial EEG to differentiate left- and right-hand imagery movements.

4.4.1 Experiment and Data Recording

Nine healthy right-handed volunteers (six males and three females, aged between 22 and 25) participated in the BCI experiments [45]. Figure 4.5 shows the paradigm for online motor imagery-based BCI control with visual feedback. The left- and right-hand movement imaginations were designated to control vertical cursor movement on the screen. The subject sat comfortably in an armchair, facing a computer screen

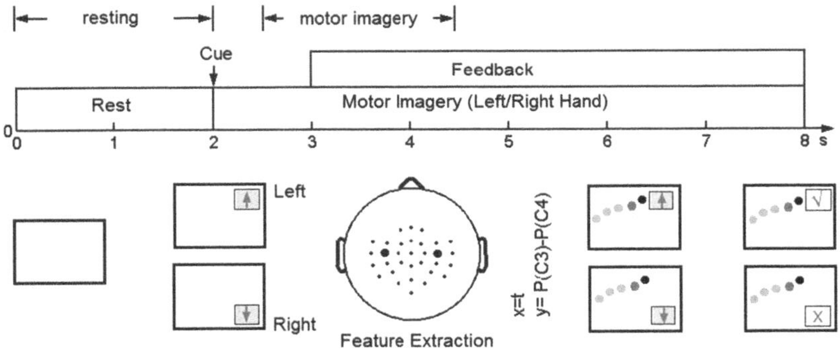

Fig. 4.5 Experiment paradigm for the motor imagery-based brain–computer interface

displaying visual feedback. The duration of each trial was 8 s. During the first 2 s, while the screen was blank, the subject was in the resting state. Immediately after these brief periods, a visual cue (arrow) was presented on the screen, indicating the imagery task to be performed. The arrows pointing upwards and downwards indicated the imagination of the left hand and the right hand movement, respectively. After 3 s, a cursor started to move at a constant speed from the left side to the right side of the screen. The vertical position of the cursor was determined by the power difference of mu rhythm between the left and right hemispheres (C3 and C4 electrodes). After 8 s, a true or false mark appeared on the screen to indicate the final result of the trial and the subject was asked to relax and wait for the next task.

Thirty-two-channel EEG signals referenced to the CMS-DRL ground were recorded using a BioSemi ActiveTwo system with the electrodes placed according to the 10–20 international system. The signals were digitized at 256 Hz and band-pass filtered (2–30 Hz) for further analysis. For each subject, the experiment consisted of four blocks, each including 60 trials (30 trials per class). There were 3–5 min of breaks between two consecutive blocks. A total of 240 trials (120 trials per class) were recorded for each subject.

4.4.2 *Method*

4.4.2.1 ICA Decomposition

As indicated in Fig. 4.5, the 0–2s and 2.5–4.5s segments in a trial were selected to represent the resting state and the motor imagery state, respectively. For each subject, ICA was performed on data under the two states separately. For each state, data of all trials were concatenated to a 480-s (240 trials $\times$ 2 s) long data segment. Because the size of data was very limited (480 s), to improve the robustness of ICA, 32-channel data were first projected to a 15-dimensional subspace using principal component analysis (PCA). Then, for each subject, ICA resulted in two sets of 15×32 spatial filters ($\mathbf{W}_{\text{rest}}$ and $\mathbf{W}_{\text{mi}}$) and 32×15 spatial projections ($\mathbf{W}_{\text{rest}}^{-1}$ and $\mathrm{W}_{\text{mi}}^{-1}$) corresponding to the resting and motor imagery.

4.4.2.2 ICA-Based Spatial Filters

In previous studies, ICA has shown its robustness in finding motor components, which have characteristic features in spatial and frequency domains [29]. This study used two criteria to identify the motor components: (1) the spatial pattern, which suggests the source location of the component, should be consistent with the scalp projection of the sensorimotor cortex on each hemisphere, and (2) the PSD of the component should match the typical spectral profile of the mu/beta rhythms. In practice, a motor component should fit both criteria. After identifying the two motor ICs, the corresponding weighting vectors in the unmixing matrix (W) were used as spatial filters for enhancing the sensorimotor mu/beta rhythms.

4.4.2.3 Resting-to-Work Translation

Suppose the two motor components in the resting state and the motor imagery state have strong similarities, it might be feasible to use the spatial filters obtained from the data in the resting state as estimates of the spatial filters for the motor imagery state. The proposed method can be described as follows:

$$\hat{\mathbf{W}}_{motor\_mi}^{-1} = \hat{\mathbf{W}}_{motor\_rest}^{-1} \qquad \mathbf{W}_{motor\_mi} = \mathbf{W}_{motor\_rest} \tag{4.3}$$

where $\mathbf{W}_{motor\_rest}$ and $\mathbf{W}_{motor\_mi}$ are motor-related spatial filters for the resting state and the motor imagery state respectively. Figure 4.6 illustrates the principle of the proposed method. In this paradigm, data in the resting state, which do not require the subject's attention or action, and the motor imagery state were totally non-overlapped. The spatial filters derived from the resting data were estimates of the spatial filters for the motor imagery data. In practice, the resting EEG data can be easily collected before a BCI session.

4.4.2.4 Feature Extraction and Classification

This study compares the classification performance of motor-imagery BCIs based on band-pass (8–30 Hz) power of the mu and beta rhythms extracted using four methods: (1) monopolar C3 and C4 electrodes, (2) spatial filtering based on ICA using the resting data, (3) spatial filtering based on ICA using the motor imagery data, and (4) CSP-based spatial filtering. After feature extraction, Fisher discriminant analysis (FDA) [8] was used to discriminate left and right hand motor imagery. The two-dimensional feature vector, which represented EEG power over

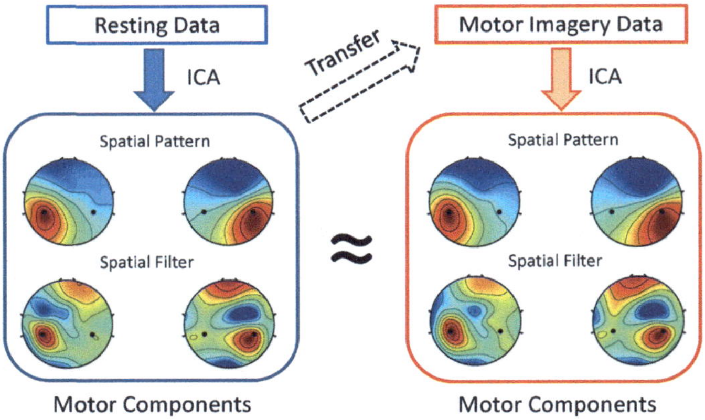

Fig. 4.6 Diagram of translating spatial filters from the resting state to the motor imagery state

the motor areas of two hemispheres, was fed into the FDA classifier for identifying the imagined hand. A 10 × 10-fold cross-validation was used to estimate the classification accuracy for each subject.

4.4.3 Results

Figure 4.7 shows spatial patterns of the motor components in the resting state and the motor imagery state for all subjects. All the components show a typical dipolar-like topography, which is widespread over the sensorimotor cortex on left or right hemisphere of the brain, and shows the highest amplitudes at C3 and C4 electrodes. To quantitatively evaluate the topographical similarity, this study calculated the correlations of spatial patterns of the motor components between the two states for each subject. The correlations were obtained by computing correlation coefficients of the 1 × 32 vectors. Spatial patterns (i.e., projections of the components to the scalp) between the resting and the motor imagery states were very comparable (mean correlation coefficients of 0.95 ± 0.05 and 0.94 ± 0.06 for left and right ICs) for all subjects.

The FDA classifier used the four different types of EEG features as inputs to classify single-trial motor-imagery movements. Table 4.2 summarizes the results of 10 × 10-fold cross-validation. A paired t-test across subjects was used to test the statistical significance of the differences between different feature extraction methods. As expected, compared to the monopolar method, all spatial-filtering methods achieved significantly higher classification accuracies (87.0 %, 85.9 %, and 86.4 % vs. 80.4 %, $p < 0.01$). The results of ICA trained with the motor imagery data were slightly better than those trained with the resting data (87.0 % vs. 85.9 %),

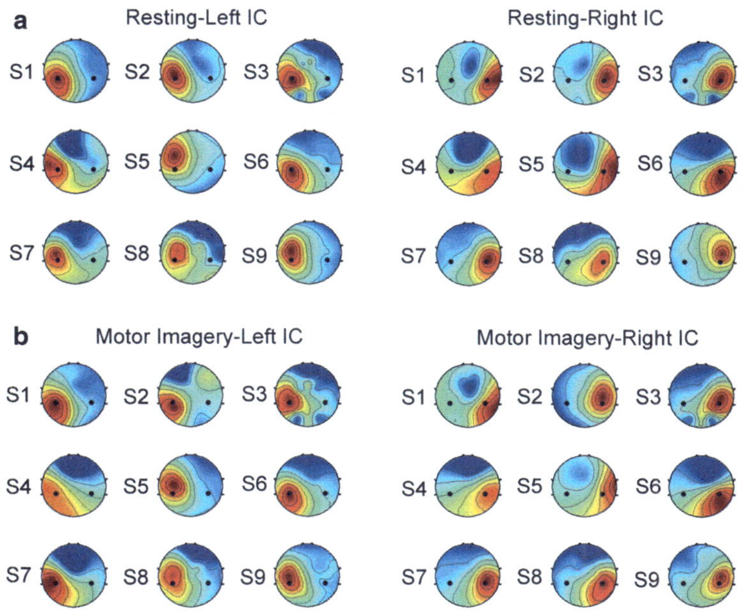

Fig. 4.7 Spatial patterns of the left and right motor components for all nine subjects (S1–S9): (**a**) spatial patterns of the resting state (*left panels*: left motor IC on the left hemisphere, *right panels*: right motor IC on the right hemisphere), (**b**) spatial patterns of the motor imagery state

Table 4.2 Classification accuracy (%) for all subjects using different feature extraction methods

	Method			
Subjects	Monopolar	ICA-mi	ICA-rest	CSP
S1	86	84	84	88
S2	66	70	70	72
S3	84	92	92	90
S4	86	94	88	93
S5	84	90	88	88
S6	93	96	96	92
S7	87	92	93	92
S8	85	97	95	95
S9	53	67	68	69
Mean	80.4 ± 12	87.0 ± 9	85.9 ± 11	86.4 ± 9

but the difference was not statistically significant ($p > 0.1$). The results of using CSP-filtered (based on motor-imagery data) were comparable with those using ICA trained with motor imagery data (86.4 % vs. 87.0 %, $p > 0.1$) and resting data (86.4 % vs. 85.9 %, $p > 0.1$). These findings demonstrated the effectiveness of translating resting spatial filters to classifying motor imagery EEG data using ICA.

4.5 Discussion and Conclusion

This chapter presents an up-to-date literature review on ICA in BCI applications by categorizing related studies into three classes (i.e., artifact removal, SNR enhancement of task-related EEG signals, and electrode selection) according to the roles of ICA. The basic principles and methodologies behind these applications have been fully illustrated through examples with real EEG data. This chapter also describes an extended application of the ICA-based spatial filter in the development of a zero-training method for a motor imagery-based BCI. In summary, this chapter shows that ICA can make a substantial contribution to the practical design of BCI systems.

Although the advantages of using ICA in EEG-based BCIs have been clearly shown in this chapter, most applications were developed and demonstrated only with offline data analysis. Among all examples presented in Table 4.1, only three studies performed online system implementation [12, 24, 36]. The study in [12] implemented an online-automated artifact removal technique for BCI using ICA. The P300-based BCI system developed in [36] adapted ICA-based filters obtained in previous offline sessions to current online sessions to enhance the P300 potentials. The VEP-based BCI system developed in [24] used predefined spatial templates to select VEP-related ICs after preforming ICA in near real time. Although these studies showed some functionality of online implementation of ICA in BCI systems, possibilities and practicalities of this approach still need further investigation.

Currently, researchers still face some technical challenges to truly implement ICA in online BCI systems. First, hardware and software must meet the computational requirements of ICA. In some situations, due to EEG nonstationarity [21], the ICA processing might need to be performed in near real time. Under the circumstances, adaptive algorithms can be used to reduce the computational complexity of ICA. Furthermore, the recent demand of mobile and wearable BCI systems poses more stringent limitations on their computational performance. A system-on-chip design [10] might be a practical solution to this problem. Second, automatic methods for identifying task relevant ICs need to be developed. In most studies, the IC identification was performed manually based on researchers' personal experiences. In real-time applications, this procedure will be labor-intensive and time-consuming, and therefore, decrease the system's practicality. Pattern recognition methods might be employed to realize automatic IC identification by comprehensively considering ICs' properties in time, frequency, and spatial domains. Third, stability and robustness of ICA based spatial filters always depend on the amount of training data. In an online BCI application, more training data require a longer user training time, thereby reducing the practicality of the BCI system. To alleviate this problem, a session-to-session translation, as well as the state-to-state translation method proposed in this chapter, might be a practical solution. Taken together, by solving these technical issues using advanced platform, signal processing, and machine learning techniques, ICA could make a substantial contribution to the development of practical online BCI systems.

Acknowledgements This work was supported by a gift fund from Abraxis Bioscience Inc. Research was also sponsored in part by Office of Naval Research, Army Research Office (under contract number W911NF-09-1-0510) and Army Research Laboratory (under Cooperative Agreement Number W911NF-10-2-0022). The views and the conclusions contained in this document are those of the authors and should not be interpreted as representing the official policies, either expressed or implied, of the Army Research Office, Army Research Laboratory, Office of Naval Research, or the US Government. The US Government is authorized to reproduce and distribute reprints for Government purposes notwithstanding any copyright notation herein. Melody Jung is appreciated for editorial assistance.

References

1. Asadi Ghanbari, A., Nazari Kousarrizi, M.R., Teshnehlab, M., Aliyari, M.: An evolutionary artifact rejection method for brain computer interface using ICA. Int. J. Elec. Comput. Sci. **9**, 461–466 (2009)
2. Bin, G., Gao, X., Yan, Z., Hong, B., Gao, S.: An online multi-channel SSVEP-based brain–computer interface using a canonical correlation analysis method. J. Neural Eng. **6**, 046002 (2009)
3. Birbaumer, N.: Brain–computer-interface research: Coming of age. Clin. Neurophysiol. **117**, 479–483 (2006)
4. Blankertz, B., Tomioka, R., Lemm, S., Kawanabe, M., Müller, K.R.: Optimizing spatial filters for robust EEG single-trial analysis. IEEE Signal Process. Mag. **25**, 41–56 (2008)
5. Brunner, P., Bianci, L., Guger, C., Cincotti, F., Schalk, G.: Current trends in hardware and software for brain–computer interfaces (BCIs). J. Neural Eng. **8**, 025001 (2011)
6. Delorme, A., Makeig, S.: EEG changes accompanying learned regulation of 12-Hz EEG activity. IEEE Trans. Neural Syst. Rehabil. Eng. **11**, 133–137 (2003)
7. Delorme, A., Makeig, S.: EEGLAB: an open source toolbox for analysis of single-trial EEG dynamics including independent component analysis. J. Neurosci. Meth. **134**, 9–21 (2004)
8. Duda, R.O., Hart, P.E., Stork, D.G.: Pattern Classification, 2nd edn. Wiley, New York (2000)
9. Erfanian, A., Erfani, A.: ICA-based classification scheme for EEG-based brain–computer interface: the role of mental practice and concentration skills. Proc 26th Int IEEE EMBS Conf, San Francisco, USA, 235–238 (2004)
10. Fang, W.C., Chen, C.K., Chua, E., Fu, C.C., Tseng, S.Y., Kang, S.: A low power biomedical signal processing system-on-chip design for portable brain-heart monitoring systems. Proc 2010 Int Conf Green Circuits and Systems (ICGCS), Shanghai, China, pp. 18–23 (2010)
11. Fatourechi, M., Bashashati, A., Ward, R., Birch, G.: EMG and EOG artifacts in brain computer interface systems: A survey. Clin. Neurophysiol. **118**, 480–494 (2007)
12. Halder, S., Bensch, M., Mellinger, J., Bogdan, M., Kubler, A., Birbaumer, N., Rosenstiel, W.: Online artifact removal for brain–computer interfaces using support vector machines and blind source separation. Comput. Intell. Neurosci. **2007**, 82069 (2007)
13. Hammon, P.S., Makeig, S., Poizner, H., Todorov, E., de Sa, V.R.: Predicting reaching targets from human EEG. IEEE Signal Process. Mag. **25**, 69–77 (2008)
14. Hill, N.J., Lal, T.N., Bierig, K., Birbaumer, N., Scholkopf, B.: Attentional modulation of auditory event-related potentials in a brain–computer interface. Proc IEEE International Workshop on Biomedical Circuits and Systems, Singapore, pp. 17–19 (2004)
15. Hung, C.I., Lee, P.L., Wu, Y.T., Chen, L.F., Yeh, T.C., Hsieh, J.C.: Recognition of motor imagery electroencephalography using independent component analysis and machine classifiers. Ann. Biomed. Eng. **33**, 1053–1070 (2005)
16. Hyvärinen, A., Oja, E.: Independent component analysis: algorithms and application. Neural Netw. **13**, 411–430 (2000)

17. James, C.J., Hesse, C.W.: Independent component analysis for biomedical signals. Physiol. Meas. **26**, R15–R39 (2005)
18. Jung, T.P., Makeig, S., Humphries, C., Lee, T.W., McKeown, M.J., Iragui, V., Sejnowski, T.J.: Removing electroencephalographic artifacts by blind source separation. Psychophysiology **37**, 163–78 (2000)
19. Jung, T.P., Makeig, S., McKeown, M.J., Bell, A.J., Lee, T.W., Sejnowski, T.J.: Imaging brain dynamics using independent component analysis. Proc. IEEE **89**, 1107–1122 (2001)
20. Kachenoura, A., Albera, L., Senhadji, L., Comon, P.: ICA: a potential tool for BCI systems. IEEE Signal Process. Mag. **25**, 57–68 (2008)
21. Krauledat, M.: Analysis of nonstationarities in EEG signals for improving brain–computer interface performance. PhD thesis, Technische Universität Berlin, Fakultät IVElektrotechnik und Informatik (2008)
22. Krauledat, M., Tangermann, M., Blankertz, B., Müller, K.R.: Towards zero training for brain–computer interfacing. PLoS ONE **3**, e2967 (2008)
23. Lebedev, M.A., Nicolelis, M.A.L.: Brain-machine interfaces: past, present and future. Trends Neurosci. **29**, 536–546 (2006)
24. Lee, P.L., Hsieh, J.C., Wu, C.H., Shyu, K.K., Chen, S.S., Yeh, T.C., Wu, Y.T.: The brain computer interface using flash visual evoked potential and independent component analysis. Ann. Biomed. Eng. **34**, 1641–1654 (2006)
25. Lee, T.W., Girolami, M., Sejnowski, T.J.: Independent component analysis using an extended infomax algorithm for mixed subgaussian and supergaussian sources. Neural Comput. **11**, 417–441 (1999)
26. Li, K., Sankar, R., Arbel, Y., Donchin, E.: Single trial independent component analysis for P300 BCI system. Proc 31th Int IEEE EMBS Conf, Minneapolis, USA, pp. 4035–4038, 2009
27. Lin, C.T., Wu, R.C., Liang, S.F., Chao, W.H., Chen, Y.J., Jung, T.P.: EEG-based drowsiness estimation for safety driving using independent component analysis. IEEE Trans. Circuits Syst. I **52**, 2726–2738 (2005)
28. Lotte, F., Congedo, M., Lecuyer, A., Lamarche, F., Arnaldi, B.: A review of classification algorithms for EEG-based brain–computer interfaces. J. Neural Eng. **4**, R1–R13 (2007)
29. Lou, B., Hong, B., Gao, X., Gao, S.: Bipolar electrode selection for a motor imagery based brain–computer interface. J. Neural Eng. **5**, 342–349 (2008)
30. Makeig, S., Westerfield, M., Jung, T.P., Townsend, J., Courchesne, E., Sejnowski, T.J.: Dynamic brain sources of visual evoked responses. Science **295**, 690–694 (2002)
31. McFarland, D.J., McCane, L.M., David, S.V., Wolpaw, J.R.: Spatial filter selection for EEG-based communication. Electroenceph. Clin. Neurophysiol. **103**, 386–394 (1997)
32. Naeem, M., Brunner, C., Leeb, R., Graimann, B., Pfurtscheller, G.: Seperability of four-class motor imagery data using independent components analysis. J. Neural Eng. **3**, 208–216 (2006)
33. Papadelis, C., Chen, X., Kourtidou-Papadeli, C., Bamidis, P.D., Chouvarda, I., Bekiaris, E., Maglaveras, N.: Monitoring sleepiness with on-board electrophysiological recordings for preventing sleep-deprived traffic accidents. Clin. Neurophysiol. **118**, 1906–1922 (2007)
34. Peterson, D.A.: Feature selection and blind source separation in an EEG-based Brain–Computer Interface. EURASIP J. Appl. Signal Process. **19**, 3128–3140 (2005)
35. Qin, L., Ding, L., He, B.: Motor imagery classification by means of source analysis for brain–computer interface applications. J. Neural Eng. **1**, 135–141 (2004)
36. Serby, H., Yom-Tov, E., Inbar, G.F.: An improved P300-based brain–computer interface. IEEE Trans. Neural Syst. Rehabil. Eng. **13**, 89–98 (2005)
37. Tian, L., Erdogmus, D., Adami, A., Pavel, M.: Feature selection by independent component analysis and mutual information maximization in EEG signal classification. Proc 2005 IEEE International Joint Conference on Neural Networks, Montreal, Canada, pp. 3011–3016, 2005
38. Vigario, R., Sarela, J., Jousmiki, V., Hamalainen, M., Oja, E.: Independent component approach to the analysis of EEG and MEG recordings. IEEE Trans. Biomed. Eng. **47**, 589–593 (2000)
39. Wang, S., James, C.J.: Extracting rhythmic brain activity for brain–computer interfacing through constrained independent component analysis. Comput. Intell. Neurosci. **2007**, 41468 (2007)

40. Wang, Y., Gao, X., Hong, B., Gao, S.: Practical designs of brain–computer interfaces based on the modulation of EEG rhythms. In: Graimann, B., Pfurtscheller, G. (eds.) Invasive and Non-Invasive Brain–Computer Interfaces. Springer, Berlin (2010)
41. Wang, Y., Gao, X., Hong, B., Jia, C., Gao, S.: Brain–computer interfaces based on visual evoked potentials: feasibility of practical system designs. IEEE EMB Mag. **27**, 64–71 (2008)
42. Wang, Y., Jung, T.P.: A collaborative brain–computer interface for improving human performance. PLoS ONE **6**, e20422 (2011)
43. Wang, Y., Makeig, S.: Predicting intended movement direction using EEG from human posterior parietal cortex. In: Schmorrow, D.D., Estabrooke, I.V., Grootjen, M. (eds.) Foundations of augmented cognition: Neuroergonomics and operational neuroscience (HCII 2009) pp. 437–446. Springer, Berlin (2009)
44. Wang, Y., Wang, R., Gao, X., Hong, B., Gao, S.: A practical VEP-based brain–computer interface. IEEE Trans. Neural Syst. Rehabil. Eng. **14**, 234–239 (2006)
45. Wei, Q., Wang, Y., Gao, X., Gao, S.: Amplitude and phase coupling measures for feature extraction in an EEG-based brain–computer interface. J. Neural Eng. **4**, 120–129 (2007)
46. Wolpaw, J.R., Birbaumer, N., McFarland, D.J., Pfurtscheller, G., Vaughan, T.M.: Brain–computer interfaces for communication and control. Clin. Neurophysiol. **113**, 767–791 (2002)
47. Xu, N., Gao, X., Hong, B., Miao, X., Gao, S., Yang, F.: BCI competition 2003-Data set IIb: enhancing P300 wave detection using ICA-based subspace projections for BCI applications. IEEE Trans. Biomed. Eng. **51**, 1067–1072 (2004)

Chapter 5
Towards Electrocorticographic Electrodes for Chronic Use in BCI Applications

Christian Henle, Martin Schuettler, Jörn Rickert, and Thomas Stieglitz

5.1 Introduction: From Presurgical Diagnostics to Movement Decoding

There is a long tradition of doctors trying to restore or replace lost body functions, caused by diseases or accidents, with the help of technical devices. A famous example was the knight Gottfried "Götz" von Berlichingen, who lost his right hand in a fight in 1504 which was afterwards replaced by a mechanical prosthetic, the so called "iron hand." In modern medicine brain–computer interfaces—BCI have been the object of intensive research in the last years, which now have to be proven in clinical studies to be available as certified medical devices for all patients. The idea behind this interface is to get a direct connection between the brain and a technical system obtaining a bidirectional data exchange both for recording

C. Henle (✉) · M. Schuettler
Laboratory for Biomedical Microtechnology, Department of Microsystems Engineering - IMTEK, University of Freiburg, Georges-Koehler-Allee 102, 79110 Freiburg, Germany

Cortec GmbH, Freiburg, Germany
e-mail: christian.henle@cortec-neuro.com;schuettler@ieee.org

J. Rickert
Bernstein Center Freiburg, University Freiburg, Freiburg, Germany

Cortec GmbH, Freiburg, Germany
e-mail: joern.rickert@cortec-neuro.com

T. Stieglitz
Laboratory for Biomedical Microtechnology, Department of Microsystems Engineering - IMTEK, University of Freiburg, Georges-Koehler-Allee 102, 79110 Freiburg, Germany

Bernstein Center Freiburg, University Freiburg, Freiburg, Germany

Cortec GmbH, Freiburg, Germany
e-mail: stieglitz@imtek.de

B. Z. Allison et al. (eds.), *Towards Practical Brain-Computer Interfaces*, Biological and Medical Physics, Biomedical Engineering, DOI 10.1007/978-3-642-29746-5_5,
© Springer-Verlag Berlin Heidelberg 2012

biomedical signals and feeding back information to the brain. During therapy brain functions could be influenced and during functional rehabilitation lost functions could be replaced. Independent of the technical realization every brain–computer interface translates electrical, magnetic or metabolic brain activity into control commands providing an application in real time. Next comes signal processing: analog amplification and filtering, digitalization and digital filtering. With the help of various algorithms, statistical characteristics are extracted out of the frequency domain of the processed signals to detect further certain, classified events. These events are used to control or drive assistive devices for communication or to perform movement tasks.

A BCI-system based on magnetic brain activity can be realized via magnetoencephalography (MEG), an expensive, non-invasive method using superconducting quantum interference devices (SQUIDS) [25]. Further non-invasive BCI-techniques based on the metabolic changes of the brain are real-time functional magnetic resonance imaging (rtfMRI) and functional near-infrared spectroscopy (fNIRS) [7, 42]. Recent developments, especially regarding the portability make fNIRS a promising BCI-System for the future [29]. For improving performance of learning with and without BCI's transcranial direct current stimulation (tDCS) has been investigated in several studies [10].

From the technical and physiological point of view there are following, three common ways to interface electrically the brain for a BCI:

- Electroencephalography (EEG), a non-invasive method using skin electrodes for recording mass activity. This technique does not allow direct electrical stimulation of nerve tissue.
- Electrocorticography (ECoG), an invasive method using epi- or subdural grids and/or strips for recording neural activity from the surface of the brain, in particular from the cortex. This method is also used for electrocortical stimulation mapping (ESM).
- Intracortical recording and/or stimulating, an invasive method using needle electrodes or needle-electrode arrays located inside the brain for recording activity from inside the brain, or for stimulating certain, local areas or individual neurons.

Figure 5.1 gives an overview about all BCI-Systems regarding the invasiveness, spatial resolution or integration density and the frequency of the measured signals.

This book chapter focuses on the second one of these neuro-technological interfaces based on the electrical signals of the brain. Here we present different manufacturing technologies, materials and designs for ECoG electrodes and review their applicability for BCIs. Signal processing will not be further explained in this chapter.

Regarding the history of electrocorticography, electrical recordings on rabbits' and monkeys' cortex were already carried out by Richard Caton in 1874 using two unpolarizable electrodes and a sensitive galvanometer [4]. About 50 years later, the first ECoG data of the human brain were obtained by Hans Berger on a patient with an already trephined skull [12]. He also developed the main field of application of

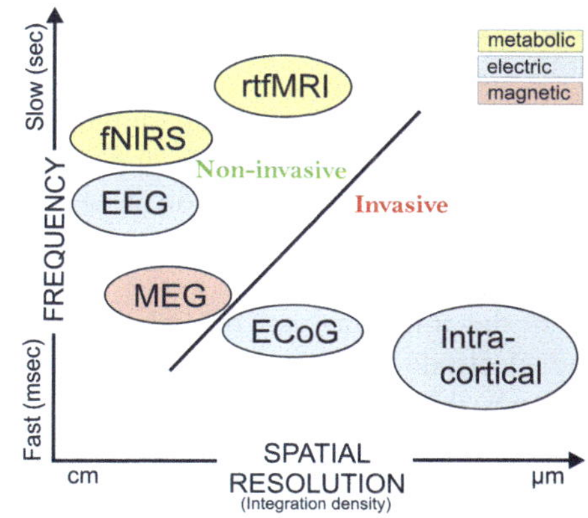

Fig. 5.1 Comparison of different BCI approaches, regarding the invasiveness, spatial resolution, frequency of the measured signals and functionality

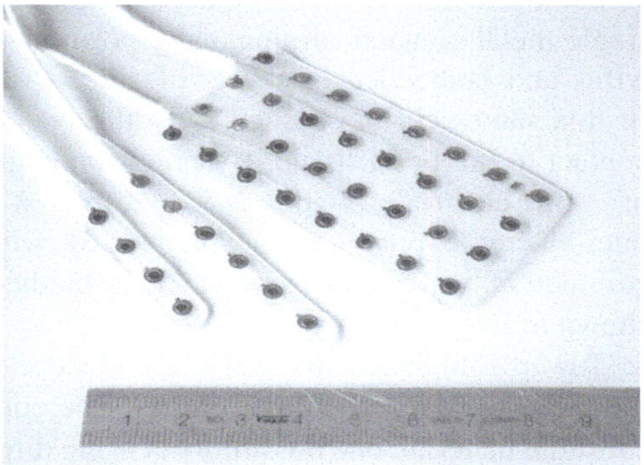

Fig. 5.2 Commercially available ECoG electrodes for subchronic implantation for presurgical epilepsy diagnosis

ECoG recordings, the localization of epileptogenic brain tissue. First carried out via intraoperative interictal ECoG for guiding cortical excision, for more than 30 years now presurgical epilepsy monitoring with subchronically implantable electrodes is a clinical standard. A typical commercially available ECoG electrode is shown in Fig. 5.2. Compared with noninvasive methods like EEG/MEG or MRI, presurgical epilepsy monitoring using ECoG electrodes turned out to be a reliable technique for localizing the epileptogenic brain tissue in many patients. Additionally ECoG electrodes are used for electrocortical stimulation mapping in these patients, which has proved to be a successful method for mapping primary brain functions like speech and movement to guide the surgical procedure.

Progress in computational and clinical neuroscience in the last two decades led to promising results in brain computer interface technology [2, 3, 20, 27]. Most of these studies focused on EEG or intracranial recording though. The proof of concept for ECoG-based BCIs followed a few years behind when Eric Leuthhardt and colleagues demonstrated online control of a cursor by a patient temporarily implanted with ECoG electrodes for localization of epileptic seizure foci [24]. In Sect. 5.2, we review the current status of ECoG-BCI research.

5.2 Approaches and Technologies for ECoG-Electrodes

Regarding ECoG electrodes, requirements for BCI technology are higher resolution, increased number of channels, the usage of chronically implantable materials and wireless data transfer. Based on these demands commercial companies and research groups all over the world have developed several ECoG electrodes with different properties (Table 5.1).

In the following, we describe selected examples of such electrodes in detail: the electrode manufacturing technology, designs, scaling limitations, materials and possible fields of application. Commercially available ECoG electrodes are mostly manufactured manually by precision engineering. Materials in clinically usable commercial electrode arrays are silicone rubber as insulation and substrate material and platinum or stainless steel as conductive material for the electrode contacts and the conductor paths. Regarding the manufacturing technology the single metal contacts are manually positioned and spot-welded to insulated micro wires and embedded in sheets of silicon rubber formed by injection moulding. In this sandwich-like system only the opened electrode contacts have electrical contact to the brain tissue. By precision engineering, maximum electrode density of 3 mm electrode pitches is achievable. A large ECoG electrode with a 5 mm inter-electrode pitch and a small array with maximum electrode density both commercially available from AD-Tech Medical Instrument Corporation, Racine are shown in Fig. 5.3.

Towards high-resolution ECoG electrodes (inter-electrode distance <2 mm) other fabrication technologies have to be considered. Based on silicone rubber as substrate material, one possibility is using directly micro wires as contact electrodes (Fig. 5.4a). The micro wires have a diameter of 70 μm and an insulation layer of 20 μm thickness. Between two sheets of silicone rubber the micro wires serve as conductor path as well as contact electrodes. At the contact sites they are bent to 90° and exposed removing the silicone rubber above. The 60 micro wire contacts are arranged within an area of $7{\times}7\ mm^2$ between two usual electrode contacts with a diameter of 4 mm. Still manually manufactured the contact spacing varies between 0.6 and 1.3 mm. The usage of more metal here leads to higher mechanical stiffness, because the metal dominates the mechanical properties compared to the substrate material (silicone rubber). Increased mechanical stiffness makes the electrode grid less adaptive to the curved surface of the cortex though and might increase the possibility of damage to the cortical surface often followed by bleeding.

This type of electrode with several different electrode contact arrangements was already used in clinical studies [44]. Although the mechanical stiffness of high-density micro wire arrays increases, this initial study with 24 patients showed an acceptable complication rate and indicated no increased risk associated with the use of high-density electrodes compared with standard subdural electrodes.

New manufacturing methods based on microsystem technologies meet all production orientated demands of high-density electrodes. Technologies like photolithography processes, physical deposition of metal layers in the nanometer range,

Table 5.1 Comparison of epicortical electrode array properties for BCI (modified after [43])

Manufacturer/ Reference	Electrode diameter in μm	Electrode pitch in μm	Conductor path pitch in μm	Number of electrodes	Dimensions of the array in mm
Ad-Tech [49]	2,000	3,000	Wire diameter: 70, isolation layer: 20	32	9×21
Craggs [50]	500	2,000	Polyimide insulated platinum wire; diameter: 76.2	60	200 mm² (diameter: 16 mm)
Tsytsarev et al. [51]	50	100	100	64	0.8×0.8
Malkin and Pendley [52]	50	100	100	400	10×10
Takahashi et al. [53]	80 square	225	50	69	2×2
Molina-Luna et al. [54]	100	640/750	35	72	6.1×4.6
Kitzmiller et al. [55]	200 square	400	Bonding wires	16	1.4×1.4
Hollenberg et al. [56]	150	900	100	64	6.5×6.5
Rubehn et al. [57]	1,000	2,000/3,000	30	252	35×60
Schuettler et al. [38]	600	1,200	100	29	8.3×7.0

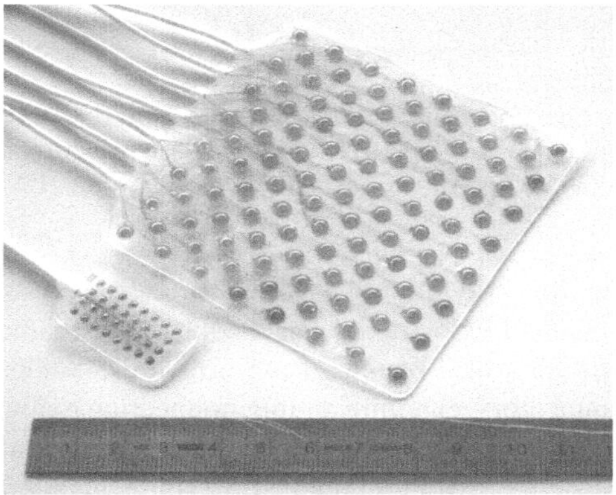

Fig. 5.3 Pushing the limits of commercially available ECoG Arrays: Large array with improved electrode density and small array with maximum electrode density

structuring via reactive ion etching in a particle-controlled cleanroom allow the manufacturing of electrode arrays with a thickness of 10 μm including integrated conductor paths and contact sites.

Based on polyimide as substrate- and isolation material and platinum as electrical conductor-and electrode material, an ECoG-array with 252 contact sites is shown in Fig. 5.4b. Not the contact site diameter of 1 mm required microsystem technologies but rather the permitted conductor path width of maximum 15 μm. The high flexibility realized by a thickness of approximately 10 μm and the finger-like structure allowed a good adjustment of the technical system to the brain. In preclinical studies, stable ECoG-signals were recorded over a period of 4 months [34].

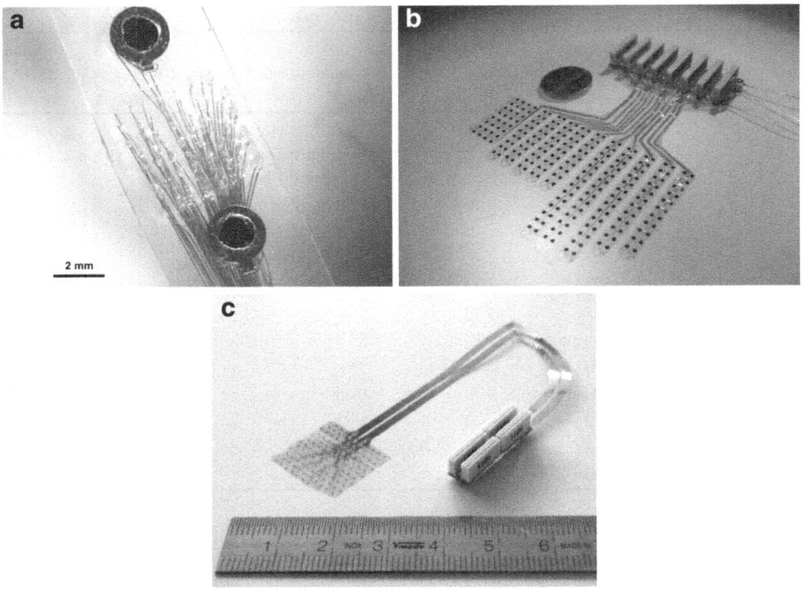

Fig. 5.4 Current approaches for high-resolution electrode arrays. (**a**) high resolution wires as additional option in clinical grid arrays (AD-Tech Medical Instrument Corporation), (**b**) polyimide based ECoG grid array [34], (**c**) polyimide based microelectrode array (IMTEK-BMT)

Another example for polyimide based micro-electrode arrays is shown in Fig. 5.4c. This 128-channel ECoG-array has a contact site diameter of 300 μm and an inter-electrode distance of 1.65 mm. Both arrays are soldered and fixed with epoxy to standard connectors, enabling an interconnection to amplifier systems.

Comparing the two approaches (silicone rubber, precision mechanics versus polyimide, microsystems technologies) referring manufacturing demands, microsystem technologies have several advantages. On the one hand electrode-contact-density can be increased and thinner electrode arrays are achievable; on the other hand it is possible to produce ECoG-electrodes in batch processes on wafers. However regarding application demands for BCI-systems, there is no assurance about chronic long-term stability and biocompatibility and no clinical experience. There is the big advantage in silicone rubber as substrate material: clinical experience and materials with approval for chronic implantation (USP class VI). Pace-makers with silicone rubber as encapsulation material were implanted since decades. Silicone rubber based neuro-stimulators in cochlear implants were implanted since 15 years, and there is a 25 year experience in pre-surgical epilepsy monitoring using silicone rubber based ECoG-electrodes.

5.3 ECoG Recordings in BCI Studies

Patients undergoing temporal implantation with ECoG-electrodes for localization of epileptic seizure foci provide a unique opportunity to study a BCI-system in humans. These implantations are carried out rarely but the patients stay implanted in the hospital for up to 3 weeks, during which they typically remain in bed. If the patients agree, this leaves time to connect them to a BCI-system and to study, for example the control of a computer cursor through their electrocorticogram.

This opportunity was realized rather recently after encouraging BCI-studies performed with different technologies, namely single-unit recordings and the electroencephalogram [26, 47]. First online BCI-studies with the ECoG demonstrated the control of a simple computer-cursor through changes in cortical rhythms induced by imagined or real movements like opening a hand or speaking a word in 2004 [24].

In comparison to EEG, ECoG enables to record from higher-frequencies (40–200 Hz or more, i.e., [31] with a higher spatial resolution on the order of at least mm [8] and is also much less susceptible to artifacts [1]. In comparison to single-units recorded intracortically with wire electrodes on the other hand, ECoG has potentially lower spatial resolution but for BCI-purposes the ECoG might have advantages in signal stability. ECoG does not penetrate the brain tissue and records signals from large neuronal populations that do not seem to get lost after tissue reactions and that do not need to be recalibrated [5] as it is common practice in chronic single-unit recordings.

Until the beginning of ECoG-BCI research it was unclear though, whether neuronal population signals recorded with ECoG electrodes could yield detailed information about voluntary movements as was known for quite some time for single-unit activity [18].

Today, numerous groups from all over the world have started to work on ECoG-based BCIs, and with the help of epilepsy patients they have found out that the ECoG encodes a wide spectrum of movement modalities. These comprise movement direction and movement of different fingers and different types of grasps [32]. Furthermore, additional neural signals suitable for BCIs based on cognitive signals or speech were successfully decoded using ECoG [23].

There are, however, a number of shortcomings inherent to BCI-studies performed with epilepsy patients: First of all, the patients are implanted only for a short time during which they typically have time for experiments only on a couple of days for a few hours each at most. Compared to the training time available in EEG-BCI studies or single-unit-BCI studies in monkeys this is much less. During this time the patients are often exhausted from the craniotomy required for the ECoG-electrode implantation. The electrode implantation localization is based on epilepsy diagnostics and not BCI-purposes. And finally the electrodes themselves are likewise optimized for the epilepsy diagnostics.

Some of these shortcomings have been addressed recently. For example the lack of long-term recordings was recently tackled in monkeys [5] and high-resolution grids have been employed in pilot trials [21]. In order to fully explore the potential of ECoG-BCI systems in the future we require the long-term implantation of high-resolution electrode grids whose recordings are then processed by online BCI-systems (for example, BCI2000, www.bci2000.org) for intensive training of prosthesis control. With regard to information content, the ideal implantation localization in the motor cortex, eventually guided by fMRI in the future [19], would certainly include locations within the Central Sulcus though. This is a very interesting, but much more difficult implantation localization for an ECoG-BCI, which was studied very little so far [48]. The ultimate test of such systems has to be done in paralyzed patients. Although there are pioneering studies in paralyzed patients [28], the availability of electrodes certified for permanent implantation in combination with a wireless recording system would greatly help to make these promising studies feasible.

5.4 High Channel ECoG Arrays for BCI

Current medical applications of most BCIs—no matter if EEG, ECoG or intra-cortical devices are used—do not need many independent information channels. They use either a "yes/no" signal like a switch that may be derived from different neuroscientific paradigms. The "brain switch" can be implemented by event related synchronisation/desynchronization, motor imagery for example to select letters or icons from a screen for communication or to switch between different states of a program control if artificial or paralyzed limbs shall be moved. Decoding of motor movement with so-called trajectory control needs more channels, at least eight for a sufficient description of two-dimensional movements [11] when intracortical needles are used. If three-dimensional trajectories have to be described, more channels are mandatory. Recording paradigms in EEG based BCI approaches showed that a larger number of electrodes in combination with machine-learning approaches [45] can reduce learning time to about 60 min to control a BCI.

Even though epicortical electrode arrays have been developed as a serious option for BCI a lot of research work had to be done to investigate optimal electrode spacing and size as to extract the maximum information out of the signals or to obtain helpful redundancy depending on the research paradigm. Having electrode arrays with different spacings and diameters together with either a uniform or a non-homogeneous distribution over the brain region of interest, opens questions in brain research concerning epicortical mass signals and local field potentials that could be addressed. A large number of electrodes over different areas of the brain are necessary to understand the interaction of the different brain areas and the nature of the signals that are the most robust ones for chronic implants. If one wants to detect movements of fingers from the primary motor cortex, the spatial resolution must be higher than for lower limb movement if the non-homogeneity of the homunculus is taken into account. If the dependence of influence of attention on the signal shall be included, multichannel electrode arrangements become even more complex since several brain areas have to be interfaced [9].

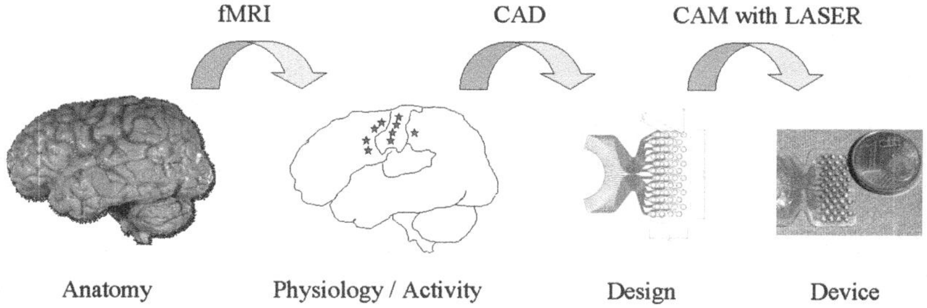

Fig. 5.5 Imaging data (fMRI) can be used to support computer aided design (CAD) and computer aided manufacturing (CAM) by laser fabrication of high channel ECoG. (Source of the brain picture: http://lbc.nimh.nih.gov/images/brain.jpg)

One important technical issue is that the number and spacing of electrodes shall not influence the mechanical properties of the implantable devices significantly. This is currently a drawback of established precision mechanics manufacturing technologies of commercially available electrode arrays for epicortical applications. Electrode size and especially the spacing between the electrodes influence the mechanical stiffness of the devices, not the mechanical properties of the substrate material. Manufacturing accuracy depends on the manual skills of the manufacturing person and limits the complexity of the devices.

Our approach tries to circumvent the challenges and limitations of traditional precision mechanics approaches. In the long term perspective (Fig. 5.5) we want to transfer the activity of a patient's brain that has been derived from non-invasive imaging directly into a device design in which certain design rules have been included. Using laser structuring, the data can be transferred directly to a computer aided manufacturing setup in an electrode array.

The manufacturing technology (for details see Sect. 5.4.1) uses thin metal sheets that do not significantly influence the device stiffness. The manufacturing is modular, allowing large devices with small feature sizes and several layers of metals as well as insulation layers. Flexibility and stretchability can be tailored by material selection and layer thickness. Final devices range from samples comparable to commercially available ones to highly complex arrays with more than 100 electrode sites and combinations of large medium size and small electrodes on a single device.

5.4.1 Manufacturing of Laser Structured Electrodes

The technology for micro fabricating electrode arrays by laser-processing (see above) has been developed [41], exclusively using medical grade materials such as silicone rubber, high-purity platinum foil or MP35N (Nickel–Cobalt base alloy) foil, with proven long-term stability and biocompatibility in the body and possibly facilitating the medical approval process. This manufacturing method allows quick

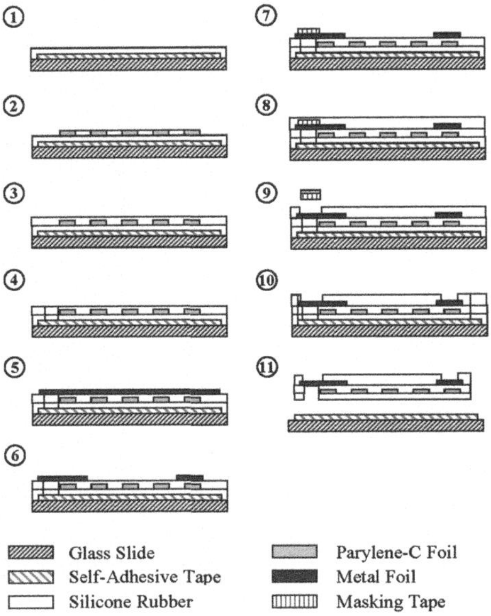

Fig. 5.6 Laser-structured electrode with large and small electrode sites (IMTEK-BMT)

and flexible fabrication of electrode arrays in many different shapes and sizes. Designs can be easily transferred from a CAD file into prototypes without the need of typical and expensive cleanroom technologies like photolithography. Silicon rubber layers down to a thickness of 25 μm can be fabricated by spin-coating n-heptane diluted silicone rubber. The electrodes and interconnects are structured by a laser out of a metal sheet that is sandwiched between layers of silicone rubber. Sufficient flexibility is warranted by creating meander-shaped electrode tracks [40]. The mechanical stability is increased by embedded, medical grade and laser-structured polymer foils [14]. The openings at the contact sites, which enable the electrical contact the brain tissue, were realized by removing the silicone by laser-structuring [37]. The complete manufacturing process of the electrode array is shown and described in Fig. 5.6. The currently used nanosecond Nd:YAG laser system with 1,064 nm wavelength allow medium scale integration of devices with minimum feature sizes and track pitches of 80 μm [17]. Studies with new picosecond Nd:YVO4 laser systems (355 nm wavelength) permits at least three times smaller feature sizes and allows 100 times faster material processing [22].

This foil system in combination with laser-structuring allows electrode arrays with mechanical properties which are almost independent of size, number and separation of the electrode sites. A design example for an epicortical BCI with two different resolution electrode arrays has been developed (Fig. 5.7): 64 electrode sites with a diameter of 2.4 mm and an electrode to electrode distance of 10 mm

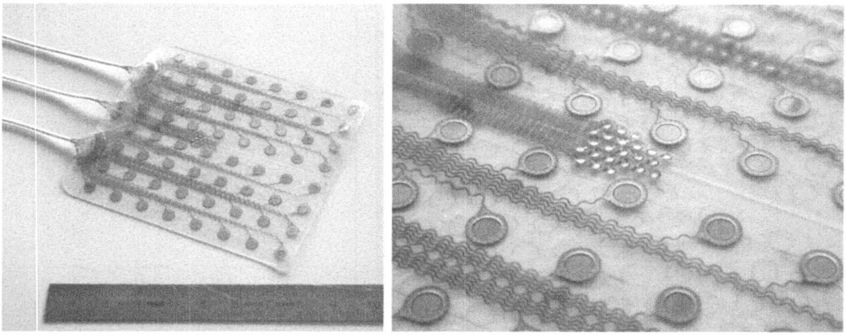

Fig. 5.7 Laser-structured electrode with large and small electrode sites (IMTEK-BMT)

and 23 integrated electrode sites with a diameter of 870 μm and an electrode to electrode distance of 1.68 mm have been manufactured. The interconnection to percutaneous cables has been realized by micro spot welding. The spot-welded parts were afterwards encapsulated in silicone rubber for isolation and mechanical stability. Several joining techniques are currently under investigation with respect to reliability, strength, reproducibility, production velocity and biocompatibility [36]. An alternative joining method with similar mechanical stability would be soldering. However this is controversial due to the cytotoxicity of the solder paste.

Information about the electrochemical properties of the electrode sites, impedance magnitude and phase shift of the large and small contacts are shown in Fig. 5.8. It is remarkable that the small platinum contacts show lower impedance values than the bigger MP35N contacts. On the one hand this is due to the different materials, on the other hand the laser ablation of the silicone rubber above is carried out in two different ways: For the big MP35N contacts only the boundary is laser-structured; the rest of the silicone is removed using tweezers. The silicone on the small contacts is ablated completely by laser, whereby a roughening of the platinum surface takes place that increases the electrochemically effective surface area [35].

5.4.2 Biological Evaluation/Results from First Studies

Before first human clinical trials can be scheduled, legal requirements that prepare the FDA approval or the CE mark have to be fulfilled. This work includes risk assessment (ISO 14971-Risk Assessment of Medical Devices) of the devices for the intended use and quality management during device manufacturing (ISO 13485) to reduce the hazard for the patient in novel applications of implantable medical devices. One milestone on the road approaching the approval is the proof of non-toxicity of the implant materials. The evaluation of the toxicity of implant materials

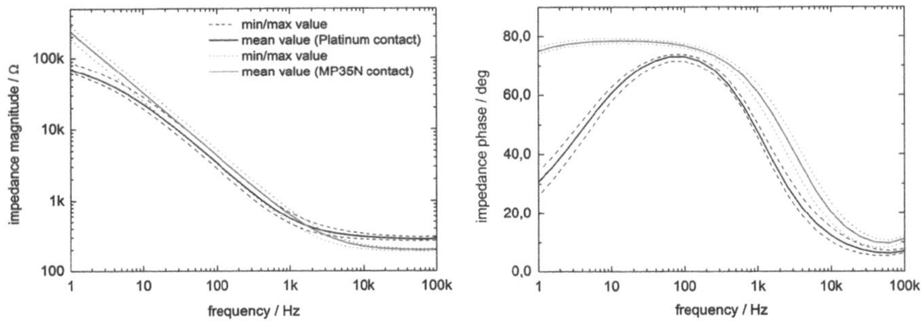

Fig. 5.8 Impedance magnitude and phase shift of the laser-structured electrodes with large and small electrode sites (IMTEK-BMT)

is described in the ISO 10993 standard series. Before human clinical trials, in vitro cytotoxicity followed by pre-clinical studies with appropriate animal models have to be carried out.

Although in the described laser-fabrication process of ECoG-electrodes only established materials like silicone rubber or platinum foil are used, biocompatibility can be changed by laser material processing. In fact, the laser micromachining of these materials cause morphological and chemical changes at the material surface [13]. Chemical investigations with X-ray photoelectron spectroscopy (XPS) on laser-structured silicone rubber and platinum foil showed the existence of by-products, mostly oxides of the component materials. Regarding silicone rubber predominantly silicon dioxide (SiO2) was measured and the platinum surface disclosed the presence of platinum oxides (PtOx).

Cytotoxicity studies on cell growth inhibition with L929 mouse fibroblasts showed no significant impact on cell growth caused by these by-products. Direct contact proliferation assay using L929 fibroblasts showed optimal cell spreading on positive control, tissue cultured plastic (TCP). On components, pure silicone rubber and pure platinum foil as well as on the electrode array, cells were more rounded indicating poorer interaction between the cells and substrate materials. None of the single electrode materials performed as well as the positive control, but silicone rubber and platinum foil, medically approved implant materials for decades, showed proliferation rates not significantly different to the laser-structured electrode samples constructed from these materials [13].

Similar results were obtained by Henle et al. regarding the laser technology and different used medical grade metal foils like platinum, stainless steel or MP35N foil, extraction tests and tests by direct contact carried out in accredited laboratories at mouse fibroblasts L929 resulted in no biological–toxicological reaction [15].

First pre-clinical in vivo trials with laser fabricated electrodes were carried out in rats. Regarding the functionality of the electrode over a period of 18 weeks, signal amplitudes and electrode-tissue impedance were investigated. The signal amplitudes were relatively stable with an amplitude of about 90–100 μV [6]. The electrode-tissue impedance between to electrodes with a distance of 3 mm increased

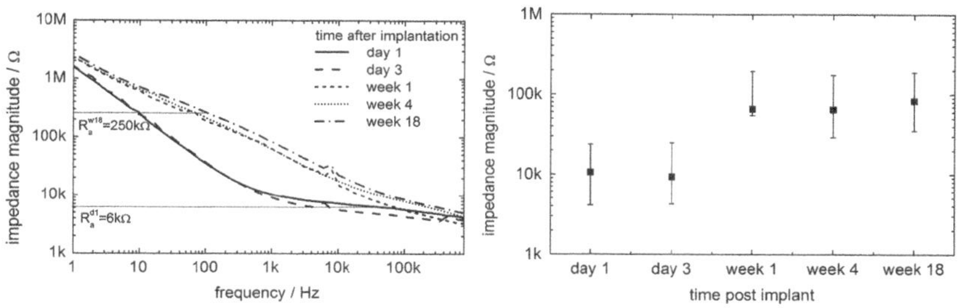

Fig. 5.9 Impedance spectra of in vivo measurements. Each *curve* is the average over 5 rats with 2 measurements per rat. Additionally, the minima and maxima of these measurements are shown at 1 kHz (*top right*) [16]

mainly within the first week and stayed relatively stable in the range of 70–90 kΩ at 1 kHz (Fig. 5.9) [16]. These impedance measurements showed typical behavior for chronically implanted micro electrodes. A change of impedance in the first week post-implantation followed by stabilization, has been already observed for intracortical electrodes [46]. The measurements also suggest that the encapsulation process is finished not later than one week after implantation.

Twenty-five weeks after implantation, histologically investigations of the brain tissue after explantation indicate that the microelectrodes were not cytotoxic as justified by no florid inflammations and no necrotic findings.

5.5 Towards Chronic Wireless Systems

The transfer from subchronical experiments to BCIs used in daily living requires that we overcome the concept of percutaneous cables. The direct mechanical coupling poses a risk on the patient who accidently applies pull forces to the cables, leading to a dislocation of the electrode array or even to damage of tissue and electrode array. There is a potential risk of discharging electrostatically collected charge through the electrode connector leading to unwanted brain stimulation. Besides these and other aspects like practicality, the most striking drawback of percutaneous wires is that they provide a channel for microorganisms entering the body, causing local inflammation or even worse body reactions. Although most aspects are addressed by the use of a bone-anchored percutaneous connector as commonly used in animal experiments [33] and successfully applied to a small number of patients [20], however the risk of infection is minimized sufficiently only when the skin is not chronically penetrated. This requirement is met by a fully implanted electronic system that transmits data recoded from the brain wirelessly through the skin to a receiver unit located outside the body.

The receiver unit is typically linked to a data processor (e.g., a personal computer) that runs the algorithm for extracting features from the recorded ECoG. It is usually powered by batteries, giving the user some freedom of motion, e.g., using an electrically powered wheel chair. The implanted recorder circuit that senses, amplifies and transmits the ECoG can either be powered by a battery which needs periodic replacement, by an inductive link using an external coil transmitting an alternating magnetic field aligned to an implanted receiver coil, or by a battery that is recharged, e.g., once per week, using an inductive link. Future applications might also utilize bio fuel cells, however, this power source is not available yet.

For data communication, digital or analogue modulation of a magnetic field can be used, which is a method obvious to use in combination with an inductively powered implant. However, the bandwidth (and hence: the data rate) of a modulated magnetic near field is practically limited by the carrier frequency (usually some MHz) and might not be sufficient for transmitting data from ECoG arrays with a large number of electrode contacts. Alternatively, data could be transmitted using the electromagnetic far field in the range of some 100 MHz. However, transmitting at this frequency usually is accompanied with high power consumption. A third method is the use of light transmitted through the skin and received outside the head. This method has moderate power consumption and permits high data transmission rates. A difficulty that remains is the integration of the implanted electronics into a suitable packaging concept.

Exposing complex electronics such as multichannel amplifiers and transmitters required for a wireless BCI to the harsh environment of the body can quickly cause catastrophic failure in electronic circuitry. As a consequence the electronics have to be packaged in a water-tight package prohibiting any water vapor from the body reaching the semiconductors. Such a hermetic package is traditionally built either from metal, preferably titanium (sometimes: stainless steel), or from ceramic (e.g., alumina, zirconia). A major challenge in applying traditional hermetic implant packaging technologies to wireless BCIs is the demand for a large number of electrical feedthroughs. Feedthroughs connect the electronics inside the implant package to the electrode contacts outside the package, providing a path for the electrical current to flow while being impermeable for moisture. A conventional hermetic feedthrough is made of a metal pin protruding through an electrically insulating glass of ceramic bead, which is framed by small metal bulkhead. The bulkhead is brazed or welded to the package wall. Most commercially successful implants use this sort of electrical feedthrough. Depending on the application, one package can have up to 16 or sometimes even more electrical feedthroughs. However, future BCIs might require signals from 100 or more electrode contacts. Currently, there is no commercially available implant package that provides this amount of feedthroughs, potential technical solutions to this problem are still under investigation, e.g., 360 feedthroughs in a miniature implant package based on screen printed and laser patterned alumina [39] or 232 feedthroughs in a micro package using high temperature co-fired ceramics [30]. These technological developments are fundamental to the fabrication and clinical evaluation of chronically implanted, wireless brain computer interfaces based on ECoG electrode within the upcoming 12 years. These BCIs will be linked to computers (e.g., attached to wheelchair), allow the patients the use of computers for communication via email, SMS or

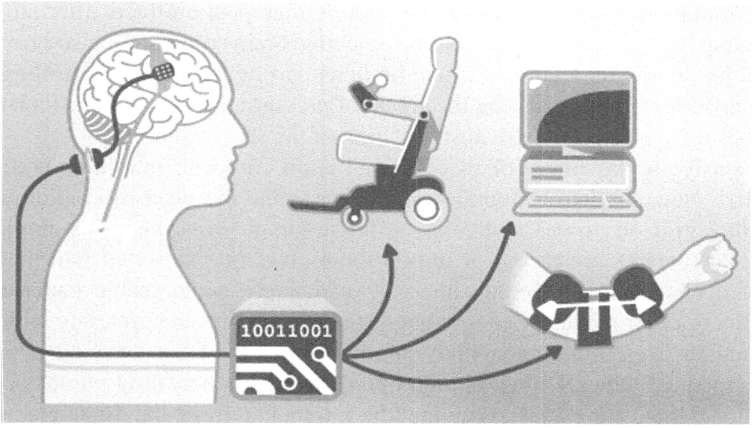

Fig. 5.10 Illustration for a wireless BCI-system linked to various applications (http://cortec-neuro.com)

artificial speech generation (Fig. 5.10). Furthermore, aids for controlling the home environment (stereo, light, television set) can be controlled by thought. Whether the acuity and speed is sufficient for safe and useful control of an artificial, robotic limb can be achieved within two years remains to be investigated.

Future devices to be developed within 2–8 years might have some of the following desirable features: The electrode arrays, if still large in area are mechanically compliant to the curvature of the brain, either by the properties of the materials used or by designing the arrays as finger-like structures [34], see Fig. 5.4b. The electrode contacts can be individually addressed as recording or stimulation sites, allowing adaptation to changes, either introduced by neural plasticity or by alteration of the electrode grid. The latter might be due to electrode breakage, contact insulation by local tissue growth or by electrode grid migration. In order to obtain a good adaptability as well as reliability of the neural interface, the electrode grid comprises a very large number of electrode contacts (some 100) of which only a few are used while the actual selection of the contacts undergoes continuous revision. Some applications might require the combination of ECoG electrode grids with penetrating electrodes that allow recording from neural ensembles in a depth of up to a few millimeters. If such BCIs are powered by a battery that is inductively recharged, e.g., overnight, high bandwidth data communication can be realized using electrical far fields, bridging distances of some 10 m, allowing the direct control of some appliances in the household, like light switches, radiators, television sets, computer games, etc. Numerous issues on data security have to be resolved before this application can be realized. Despite controlling the environment, medical doctors might use the recorded ECoG for diagnostic purposes, potentially coupled with information harvested from other sensors integrated into the implanted device, like glucose sensors, neural transmitter sensors, temperature, acceleration sensors, etc. Recording from the brain will be accompanied by brain stimulation. Chronic brain stimulation has the potential of life-long therapeutic effects after stroke or progressing neurodegenerative diseases. Selective brain stimulation also provides a channel for supplying feedback to the brain as part of a BCI for controlling, e.g., robotic orthoses, e.g., translating the signal of pressure sensors of an artificial hand prosthesis into selective electrical stimulation of the sensory cortex.

Obviously, the far future of ECoG based brain computer interfaces is difficult to predict. New surgical techniques might overcome the need of craniotomy for implanting grid electrodes and might offer minimal invasiveness by implanting arrays that (partly) stretch out or unfold themselves once inserted into the body. Implanted electronics powered either by inductively rechargeable batteries, bio fuel cells, or other means of energy harvesting communicate wirelessly with other implanted devices, building a body-internal network of sensors and actuators. Other implanted network devices might be actuators such as drug pumps or nerve stimulators, that, e.g., after receiving the command from the BCI, electrically activate peripheral nerves for moving a paralyzed limb. In return, peripheral nerve interfaces record signal from natural sensors (e.g., skin receptors) and communicate that information to the BCI which delivers a particular electrical stimulus pattern to the sensory cortex, allowing the perception of natural sensation.

The development of smart phones during the past few years proved that the combination of new technology—hardware (wireless data transmission in combination with multi-sensor technology) as well as software (cloud computing and pattern recognition, voice, faces, locations)—allows the realization of new applications we were not able to imaging just a few years ago, causing ethics-related discussions worldwide. When introducing highly selective brain computer interfaces that are able to communicate wirelessly with other devices, technological developments become thinkable, confronting individuals as well as their society with ethical questions of monumental consequence. Although, this development will take place much slower since the number of users and hence the pressure on companies developing more advanced devices is much smaller, one always has to reconsider the (current) purpose of BCIs: To alleviate physical disability in patients.

Acknowledgements Part of the work that is presented here has been funded by the German Federal Ministry for Education and Research (BMBF) in the grants Go Bio (313891) and the Bernstein Focus Neurotechnology Freiburg-Tuebingen "The hybrid brain" (01GQ0830).

References

1. Ball, T., Kern, M., Mutschler, I., Aertsen, A., Schulze-Bonhage, A.: Signal quality of simultaneously recorded invasive and non-invasive EEG. Neuroimage **46**(3), 708–716 (2009)
2. Birbaumer, N., Ghanayim, N., Hinterberger, T., Iversen, I., Kotchoubey, B., Kübler, A., Perelmouter, J., Taub, E., Flor, H.: A spelling device for the paralysed. Nature **398**(6725), 297–298 (1999)
3. Carmena, J.M., Lebedev, M.A., Crist, R.E., O'Dohety, J.E., Santucci, D.M., Dimitrov, D.F., Patil, P.G., Henriquez, C.S., Nicolelis, M.A.L.: Learning to control a brain-machine interface for reaching and grasping by primates. PLoS Biol. **1**(2), 193–208 (2003)

4. Caton: Brit. med. J. **2**(278) (1875), Ref. Zbl Physiol **4**(25) (1890). Bechterew.: (1902) Die Energie des Lebenden Organismus, pp 102. Wiesbaden, Germany
5. Chao, Z.C., Nagasaka, Y., Fujii, N.: Long-term asynchronous decoding of arm motion using electrocorticographic signals in monkey. Front Neuroeng. **3**(3), 1–10 (2010)
6. Cordeiro, J., Henle, C., Raab, M., Meier, W., Sieglitz, T., Schulze-Bonhage, A., Rickert, J.: Micromanufactured electrodes for cortical field potentials recording: in vivo study. In: IFMBE Proceedings, vol. 22, pp. 2375–2378 (2008)
7. Coyle, S.M., Ward, T.E., Markham, C.M.: Brain-computer interface using a simplified functional near-infrared spectroscopy system. J. Neural. Eng. **4**(3), 219–226 (2007)
8. Freeman, W.J., Holmes, M.D., Burke, B.C., Vanhatalo, S.: Spatial spectra of scalp EEG and EMG from awake humans. Clin. Neurophysiol. **114**, 1053–1068 (2003)
9. Fries, P., Womelsdorf, T., Oostenveld, R., Desimone, R.: The effects of visual stimulation and selective visual attention on rhythmic neuronal synchronization in macaque area V4. J. Neurosci. **28**(18), 4823–4835 (2008)
10. Fritsch, B., Reis, J., Martinowich, K., Schambra, H.M., Yuanyuan, J., Cohen, L.G., Lu, B.: Direct current stimulation promotes bdnf-dependent synaptic plasticity potential implications for motor learning. Neuron **66**(2), 198–204 (2010)
11. Georgopoulos, A.P., Kalaska, J.F., Massey, J.T.: Spatial trajectories and reaction times of aimed movements: effects of practice, uncertainty, and change in target location. J. Neurophysiol. **46**(4), 725–743 (1981)
12. Gloor, P.: Hans Berger and the discovery of the electroencephalogram. Electroencephalogr. Clin. Neurophysiol. **28**, 1–36 (1969)
13. Green, R.A., Ordonez, J.S., Schuettler, M., Poole-Warren, L.A., Lovell, N.H., Suaning, G.J.: Cytotoxicity of implantable microelectrode arrays produced by laser micromachining. Biomaterials **31**(5), 886–893 (2010)
14. Henle, C., Hassler, C., Kohler, F., Schuettler, M.: Mechanical characterization of neural electrodes based on PDMS-parylene C-PDMS sandwiched system. In: Proceedings of the 33rd Annual International Conference of the IEEE Engineering in Medicine and Biology Society, pp. 640–643 (2011)
15. Henle, C., Madjarow, A., Schuettler, M., Stieglitz, T.: Evaluation on cytotoxicity on laser-fabricated neural implants: different manufacturing processes and materials. Biomed. Tech. **55**(Suppl. 1), 229–231 (2010)
16. Henle, C., Raab, M., Cordeiro, J.G., Doostkam, S., Schulze-Bonhage, A., Stieglitz, T., Rickert, J.: First long-term in vivo study on subdurally implanted micro-ECoG electrodes, manufactured with a novel laser technology. Biomed. Microdevices **13**(1), 59–68 (2011)
17. Henle, C., Schuettler, M., Ordonez, J.S., Stieglitz, T.: Scaling limitations of laser-fabricatednerve electrode arrays. In: Proceedings of the International Conference of the IEEE Engineering in Medicine and Biology Society, pp. 4208–4211 (2008)
18. Hepp-Reymond, M.C.: Neurosciences, functional organization of motor cortex and its participation in voluntary movements. Comp. Prim. Biol. **4**, 501–624 (1988)
19. Hermes, D., Vansteensel, M.J., Albers, A.M., Bleichner, M.G., Benedictus, M.R., Mendez Orellana, C., Aarnoutse, E.J., Ramsey, N.F.: Functional MRI-based identification of brain areas involved in motor imagery for implantable brain-computer interfaces. J. Neural. Eng. **8**(2), 025007 (2011)
20. Hochberg, L.R., Serruya, M.D., Friehs, G.M., Mukand, J.A., Saleh, M., Caplan, A.H., Branner, A., Chen, D., Penn, R.D., Donoghue, J.P.: Neuronal ensemble control of prosthetic devices by a human with Tetraplegia. Nature **442**(7099), 164–171 (2006)
21. Kellis, S.S., House, P.A., Thomson, K.E., Brown, R., Greger, B.: Human neocortical electrical activity recorded on nonpenetrating microwire arrays: applicability for neuroprostheses.Neurosurg. Focus **27**(1), E9 (2009)
22. Kohler, F., Schuettler, M., Ordonez, J.S., Stieglitz, T.: Laser microfabrication of neural electrode arrays: comparison of nanosecond and picosecond laser technology. In: Proceedings of the IFESS (2011)

23. Leuthardt, E.C., Gaona, C., Sharma, M., Szrama, N., Roland, J., Freudenberg, Z., Solis, J., Breshears, J., Schalk, G.: Using the electrocorticographic speech network to control a brain-computer interface in humans. J. Neural. Eng. **8**(3), 036004 (2011)
24. Leuthardt, E.C., Schalk, G., Wolpaw, J.R., Ojemann, J.G., Moran, D.W.: A brain–computer interface using electrocorticographic signals in humans. J. Neural Eng. **1**, 63–71 (2004)
25. Mellinger, J., Schalk, G., Braun, C., Preissl, H., Rosenstiel, W., Birbaumer, N., Kübler, A.: An MEG-based brain-computer interface (BCI). Neuroimage **36**(3), 581–593 (2007)
26. Moran, D.: Evolution of brain–computer interface: action potentials, local field potentials and electrocorticograms. Curr. Opin. Neurobiol. **20**, 741–745 (2010)
27. Moritz, C.T., Perlmutter, S.I., Fetz, E.E.: Direct control of paralysed muscles by cortical neurons. Nature **456**(7222), 639–642 (2008)
28. Murguialday, A.R., Hill, J., Bensch, M., Martens, S., Halder, S., Nijboer, F., Schoelkopf, B., Birbaumer, N., Gharabaghi, A.: Transition from the locked in to the completely locked-in state: a physiological analysis. Clin. Neurophysiol. **122**(5), 925–933 (2011)
29. Nagaoka, T., Sakatani, K., Awano, T., Yokose, N., Hoshino, T., Murata, Y., Katayama, Y., Ishikawa, A., Eda, H.: Development of a new rehabilitation system based on a brain–computer interface using near-infrared spectroscopy. Adv. Exp. Med. Biol. **662**, 497–503 (2010)
30. Ordonez, J., Keller, M., Schuettler, M., Stieglitz, T., Wilder, J., Guenther, T., Suaning, G.J.: High-Density Hermetic Feedthroughs for Multi-Channel Implantable Electronics. Proc. Conf. Tech. Assis. Rehab. (2011). IGEPrint Internet, ISSN 2192-161X
31. Pfurtscheller, G., Graimann, B., Huggins, J.E., Levine, S.P., Schuh, L.A.: Spatiotemporal patterns of beta desynchronization and gamma synchronization in corticographic data during self-paced movement. Clin. Neurophysiol. **114**, 1226–1236 (2003)
32. Pistohl, T., Schulze-Bonhage, A., Aertsen, A., Mehring, C., Ball, T.: Decoding natural grasp types from human ECoG. Neuroimage **59**(1), 248–260 (2012)
33. Rousche, J., Petersen, R.S., Battiston, S., Gianotta, S., Diamond, M.E.: Examination of the Spatial and Temporal Distribution of Sensory Cortical Activity Using a 100-electrode Array. J. Neurosci. Meth. **90**, 57–66 (1999)
34. Rubehn, B., Bosman, C., Oostenveld, R., Fries, P., Stieglitz, T.: A MEMS-based flexible multichannel ECoG-electrode array. J. Neural. Eng. **6**(3), 036003 (2009)
35. Schuettler, M.: Electrochemical properties of platinum electrodes in vitro: comparison of six different surfaces qualities. In: Proceedings of the International Conference of the IEEE Engineering in Medicine and Biology Society, pp. 186–189 (2007)
36. Schuettler, M., Henle, C., Ordonez, J.S., Meier, W., Guenter, T., Stieglitz, T.: Interconnection technologies for laser-patterned electrode arrays. In: Proceedings of the International Conference of the IEEE Engineering in Medicine and Biology Society, pp. 3212–3215 (2008)
37. Schuettler, M., Henle, C., Ordonez, J.S., Suaning, G.J., Lovell, N.H., Stieglitz, T.: Patterning of silicone rubber for micro-electrode array fabrication. In: Proceedings of the IEEE Conference on Neural Engineering, pp. 53–56 (2007)
38. Schuettler, M., Ordonez, J.S., Henle, C., Oh, D., Gilad, O., Holder, D.S.: A flexible 29 channel epicortical electrode array. In: Proceedings of the 13th Annual Conference of the International Functional Electrical Stimulation Society, pp. 232–234 (2008)
39. Schuettler, M., Ordonez, J.S., Santisteban, T.S., Schatz, A., Wilde, J., Stieglitz, T.: Fabrication and test of a hermetic miniature implant package with 360 electrical feedthroughs. In: Proceedings of the International Conference of the IEEE Engineering in Medicine and Biology Society, vol. 578, pp. 1585–1588 (2010)
40. Schuettler, M., Pfau, D., Ordonez, J.S., Henle, C., Woias, P., Stieglitz, T.: Stretchable tracks for laser-machined neural electrode arrays. In: Proceedings of the International Conference of the IEEE Engineering in Medicine and Biology Society, pp. 1612–1615 (2009)
41. Schuettler, M., Stiess, S., King, B.V., Suaning, G.J.: Fabrication of implantable microelectrode arrays by laser cutting of silicone rubber and platinum foil. J. Neural. Eng. **2**(1), 121–128 (2005)
42. Sitaram, R., Caria, A., Birbaumer, N.: Hemodynamic brain–computer interfaces for communication and rehabilitation. Neural Netw. (2009) **22**, 1320–1328

43. Stieglitz, T., Rubehn, B., Henle, C., Kisban, S., Herwik, S., Ruther, P., Schuettler, M.: Brain–computer interfaces: An overview of the hardware to record neural signals from the cortex. In: Verhaagen, J., Hol, E.M., Huitinga, I., Wijnhold, J., Bergen, A.B., Boer, G.J., Swaab, D.F. (eds.) NeurotherapyProgress in Restorative Neuroscience and Neurology. Prog. Brain Res. **175**, 297–315 (2009)
44. Van Gompel, J.J., Stead, S.M., Giannini, C., Meyer, F.B., Marsh, W.R., Fountain, T., So, E., Cohen-Gadol, A., Lee, K.H., Worrell, G.A.: Phase I trial: safety and feasibility of intracranial electroencephalography using hybrid subdural electrodes containing macro- and microelectrode arrays. Neurosurg. Focus. **25**(3), E23 (2008)
45. Vidaurre, C., Sannelli, C., Müller, K.R., Blankertz, B.: Machine-Learning-Based Coadaptive Calibration for Brain–Computer Interfaces. Neural Comput. (2010) [Epub ahead of print]
46. Williams, J.C., Hippensteel, J.A., Dilgen, J., Shain, J.W., Kipke, D.R.: Complex Impedance Spectroscopy for Monitoring Tissue Responses to Inserted Neural Implants. J. Neural Eng. **4**(4), 410–423 (2007)
47. Wolpaw, J.R., Birbaumer, N., McFarland, D.J., Pfurtscheller, G., Vaughan, T.M.: Brain–computer interfaces for communication and control. Clin. Neurophysiol. **113**, 767–791 (2002)
48. Yanagisawa, T., Hirata, M., Saitoh, Y., Kato, A., Shibuya, D., Kamitani, Y., Yoshimine, T.: Neural decoding using gyral and intrasulcal electrocorticograms. Neuroimage **45**(4), 1099–2106 (2009)
49. Product Catalogue 2012. http://www.adtechmedical.com
50. Craggs, M.D.: The cortical control of limb prostheses. Ph.D., University of London (1974)
51. Tsytsarev, V., Taketani, M., Schottler, F., Tanaka, S., Hara, M.: A new planar multielectrode array: Recording from a rat auditory cortex. J. Neural. Eng. **3**(4), 293–298 (2006)
52. Malkin, R.A., Pendley, B.D.: Construction of a very high-density extracellular electrode array. Am. J. Physiol. Heart C, **279**, 437–442 (2000)
53. Takahashi, H., Ejiri, T., Nakao, M., Nakamura, N., Kaga, K., Herve, T.: Microelectrode array on folding polyimide ribbon for epidural mapping of functional evoked potentials. IEEE Trans. Biomed. Eng. **50**(4), 510–516 (2003)
54. Molina-Luna, K., Buitrago, M.M., Hertler, B., Schubring, M., Haiss, F., Nisch, W., Schulz, J.B., Luft, A.R.: Cortical stimulation mapping using epidurally implanted thin-film microelectrode arrays. J. Neurosci. Meth. **161**, 118–125 (2007)
55. Kitzmiller, J., Beversdorf, D., Hansford, D.: Fabrication and testing of microelectrodes for small-field cortical surface recordings. Biomed. Microdevices **8**, 81–85 (2006)
56. Hollenberg, B.A., Richards, C.D., Richards, R., Bahr, D.F., Rector, D.M.: A MEMS fabricated flexible electrode array for recording surface field potentials. J. Neurosci. Meth. **153**, 147–153 (2006)
57. Rubehn, B., Bosman, C., Oostenveld, R., Fries, P., Stieglitz, T.: A MEMS-based flexible multichannel ECoG-electrode array. J. Neural. Eng. **6**(3), 036003 (2009)

Part II
Devices, Applications and Users

Chapter 6
Introduction to Devices, Applications and Users: Towards Practical BCIs Based on Shared Control Techniques

Robert Leeb and José d.R. Millán

6.1 Introduction

In this chapter we will first provide a short introduction into the topic of devices, applications, and users. Practical brain–computer interfaces (BCI) should allow users not only to control a cursor on the screen, but provide opportunities to interact through real world applications. The research and development in the direction of new applications are especially important since BCIs are no longer only used by healthy subjects under controlled conditions in laboratory environments, but by patients controlling applications at their homes.

There are five major BCI application areas in which disabled individuals could greatly benefit from advancements in BCI technology, namely, "Communication and Control," "Motor Substitution," "Entertainment," "Motor Recovery," and "Mental State Monitoring." The performance of these applications can be improved by novel hybrid BCIs architectures, which are a synergetic combination of a BCI with other residual input channels. These architectures explore the BCI as part of a multimodal multi-channel system and offer a more intuitive, robust and natural way of interaction. Moreover, it has been recently shown that not only the BCI research and applications can benefit from human–computer interaction (HCI) techniques but also the reverse. More precisely, the BCIs can extract cognitive-relevant information from the user (e.g. recognition of error processing) that could be used to improve standard interactions. Such passive monitoring offers potential benefits for both patients and healthy subjects. Furthermore another area of research, interesting for healthy subjects, are BCI controlled or support games; by augmentation of the operation capabilities or by allowing multi-task operations.

R. Leeb · J.d.R. Millán (✉)
Chair in Non-Invasive Brain-Machine Interface, École Polytechnique Fédérale de Lausanne, Lausanne, Station 11, CH-1015, Switzerland
e-mail: robert.leeb@epfl.ch; jose.millan@epfl.ch

B. Z. Allison et al. (eds.), *Towards Practical Brain-Computer Interfaces*, Biological and Medical Physics, Biomedical Engineering, DOI 10.1007/978-3-642-29746-5_6,
© Springer-Verlag Berlin Heidelberg 2012

Most of the applications presented in the literature operate either software oriented, like mentally writing text via a virtual keyboard on a screen, or could be more hardware oriented, like controlling a small mobile tele-presence robot or wheelchair. These typical applications require a very good and precise control channel to achieve performances comparable to healthy users without a BCI. However, current day BCIs offer low throughput information and are insufficient for the full dexterous control of these complex applications. Techniques like shared control can enhance the interaction to a similar level, despite the fact that BCI is not such a perfect control channel. In such a control scheme, the responsibilities are then shared between the user in giving high-level commands and the system in executing fast and precise low-level interactions. For example, let us consider driving a wheelchair in a home environment (scattered with obstacles like chairs, tables, doors ...) that requires precise control to navigate through rooms. In the shared control framework, the user issues via the BCI the high level commands such as left, right and forward, which are then interpreted by the wheelchair controller based on the contextual information from its sensors. Based on these interpretations, the wheelchair can perform intelligent maneuvers (e.g. obstacle avoidance, guided turnings). Shared control is helping on a direct interaction with the environment but is conveying a different principle than autonomous control. In autonomous control high-level commands which are more abstract (e.g. drive to the kitchen or the living room) are issued and then executed autonomously by the robotic device without interaction of the user, till the selected target is reached.

Different types of BCIs exist and various methods can be used to acquire brain activity, but since the electroencephalogram (EEG) is the most practical modality [59]—if we want to bring BCI technology to a large population—this chapter will focus on EEG based BCIs only. Nevertheless, brain activity can be measured through non-electrical means as well, such as through magnetic and metabolic changes, which can be also measured non-invasively. Magnetic fields can be recorded with magnetoencephalography (MEG), while brain metabolic activity (reflected in changes in blood flow) can be observed with positron emission tomography (PET), functional magnetic resonance imaging (fMRI), and optical imaging (NIRS). Unfortunately, such alternative techniques require sophisticated devices that can be operated only in special facilities (except for NIRS). Moreover, techniques for measuring blood flow have long latencies compared to EEG systems and thus are less appropriate for interaction, although they may provide good spatial resolution. Besides EEG, electrical activity can also be measured through invasive means such as ElectroCorticogram (ECoG) or intracranial recordings. Both methods require surgery to implant electrodes. The relative advantages and disadvantages of currently available noninvasive and implanted (i.e., invasive) methodologies are discussed in [117]. Since these surgical procedures are only possible for some patient groups (such as persons with intractable epilepsy or severe motor disorders), this chapter will not discuss invasive BCIs here.

6.2 Current and Emerging User Groups

The classic user group in BCI research is severely disabled patients: persons who are unable to communicate through other means [10]. However, recent progress in the field of BCI technology shows that BCIs could also be helpful to less disabled users. New user groups are emerging as new devices and applications develop and improve. Rehabilitation of disorders has gained a lot of attention recently, especially for users with other disabilities such as stroke, addiction, autism, ADHD and emotional disorders [2,9,54,83,89]. Furthermore, BCIs could also help healthy users in specific situations, such as when conventional interfaces are unavailable, cumbersome, or do not provide the needed information [1].

Such passive monitoring offers potential benefits for both patients and healthy subjects. Furthermore another area of research, interesting for healthy subjects, are BCI controlled or support games; by augmentation of the operation capabilities or by allowing multi-task operations [60]. Millán et al. [92] recently validated a BCI for space applications. Another recent extension of BCI for healthy users is in the field of biometrics. Since the brainwave pattern of every person is unique, a person authentication based on BCI technology could use EEG measures to help authenticate a user's identity, either by mental tasks [58] or reactive frequency components [83].

Many new BCI devices and applications have recently been validated mostly with healthy users, such as control of smart home or other virtual environment [37, 51, 100], games [47, 64, 68, 76], orthosis or prosthesis [15, 72, 85], virtual or real wheelchairs [20,30,50], and other robotic devices [6,34]. We can even turn the BCI shortcomings into challenges [55, 77], by e.g. explicitly requiring a gamer to issue BCI commands to solve a task. Thereby far from perfect control "solutions" are more interesting and challenging. These and other emerging applications adumbrate dramatic changes in user groups. Instead of being devices that only help severely disabled users and the occasional curious technophile, BCIs could benefit a wide variety of disabled and even healthy users. Several chapters in this book present applications for disabled and healthy users, along with discussion of the different application interface and environment challenges (Chaps. 7, 8, 10 and 11).

6.3 BCI Devices and Application Scenarios

The main focus of this part is on applications for disabled people, but some applications can benefit both disabled and healthy users. Five big application areas have been identified [65] and are addressed below where disabled individuals could greatly benefit from advancements in BCI technology, namely, "Communication & Control," "Motor Substitution," "Entertainment and Gaming," "Motor Rehabilitation and Recovery," and "Mental State Monitoring." Furthermore, recent trends help overcome BCI limitations with hybrid BCIs [65, 86] and shared control techniques.

Another challenge is in improving the performance and reliability of current BCI systems. EEG-based BCIs can be characterized by noisy input signals (low signal-to-noise ratio) and low-bit-rate outputs. Modern human–computer interaction (HCI) principles have shown promise. They can explicitly take into account the noisy and lagged nature of the BCI control signals to adjust the dynamics of the interaction as a function of the reliability of user's control capabilities. Based on these principles the first outstanding and very intuitive interface was designed for a virtual keyboard and is called "Hex-O-Spell" [69, 115]. The issue of novel and smart application interfaces is addressed in the parallel Chap. 9 in more detail.

Finally, new EEG hardware also aims at making BCI more practical for daily home use. Smaller amplifiers, standardized systems and dry electrodes that require minimal preparation are necessary. Novel devices like dry or water-based electrodes are gaining attention [35, 96, 118]. Several companies have introduced dry or water-based systems, but objective studies that compare different systems are only beginning to emerge. The issue is addressed in more detail in Chaps. 15 and 16.

6.3.1 Communication and Control

BCI could enable severely disabled individuals to communicate with other people and to control their environment. The first communication with a locked-in patient was established by Birbaumer [10]. Several studies aimed to show the feasibility and to compare the performances with healthy subjects using either slow cortical potentials [45] or cognitive evoked potentials like P300 [87] or motor imagery (MI) [46]. Further research has shown that persons, even those suffering from severe disabilities, may interact with computers by only using their brain—in the extreme case using the brain channel as a single switch, just like a hand mouse. Research on establishing communication functions were mostly focused on writing (spelling) applications and surfing (browsing) the internet.

Several spelling devices based on the voluntarily modulation of brain rhythms have been demonstrated. These systems can operate synchronously [10, 79] or asynchronously [61, 62, 69, 81, 98, 115]. Mostly binary choices of the BCI were used to select letters, e.g., in a procedure where the alphabet was iteratively split into halves. The big disadvantage of all these systems is that the writing speed is very slow. Particularly relevant is the spelling system called Hex-O-Spell [115], which illustrates how a normal BCI can be significantly improved by state-of-the-art human–computer interaction principles, although the text entry system is still controlled only by one or two input signals (based on motor imagery). The principle of structuring the character locations based on an underlying language model speeds up the writing process.

Other kinds of BCI spelling devices, especially those mostly used by disabled people, are based on the detection of potentials that are evoked by external stimuli. The most prominent is the approach that elicits a P300 component [22]. All characters are presented in a matrix. The symbol on which the user focuses her/his attention can be predicted from the brain potentials that are evoked by random flashing of rows and columns. Similar P300-based spelling devices have extensively been investigated and developed since then (e.g., [74, 87, 102, 103]). Also steady-state visual evoked potentials (SSVEP) can be used for virtual keyboards. Either each character of the alphabet or each number on a numpad is stimulated with its own frequency and can be selected directly [32], or additional stimulation boxes (like arrows) are placed aside the keyboard and are used for navigating left/right/up/down and selecting the letter [112].

The first application to access the Internet via the BCI was a very simple solution, by displaying web pages for a fixed amount of time ("Descartes" [41]), but later browsers allowed a more flexible selection of links ("Nessi" [7]). The challenge of selecting a large amount of links with only a limited amount of BCI commands (mostly two) can be overcome by applying scanning techniques, which allow a sequential switching or auto-switching between them. Even functions like zoom in/out, scroll up/down, go back/forward can be added in the user interface and selected by the BCI via a hierarchical approach [81]. Nevertheless, users reported that the correct selection can be quite demanding. More recently, browsers based on the P300 have been developed by different groups. In the first one, all possible links are tagged with characters and a normal character P300 matrix (6×6 matrix) was used on a separate screen for selecting [67]. In a more recent approach, an active overlay was placed over the web site that elicited the P300 by directly highlighting the links. Hence, switching between the stimulation device and the browsing screen was not necessary [94]. The overlay has to be automatically generated for each website since the links appear on different places for each site.

6.3.2 *Motor Substitution: Grasp Restoration*

The restoration of grasp functions in spinal cord injured (SCI) patients or patients suffering from paralysis of upper extremities typically rely on Functional Electrical Stimulation (FES). In this context, the term neuroprosthesis is used for FES systems that seek to restore a weak or lost grasp function when controlled by physiological signals.

Some of these neuroprostheses are based on surface electrodes for external stimulation of muscles of the hand and forearm [38, 57, 108]. Others, like the Freehand system (NeuroControl, Cleveland, US), uses implantable neuroprostheses to overcome the limitations of surface stimulation electrodes concerning selectivity and reproducibility [42], but this system is no longer available on the market.

Pioneering work by the groups in Heidelberg and Graz showed that a BCI could be combined with an FES-system with surface electrodes [84]. In this study, the restoration of a lateral grasp was achieved in a spinal cord injured subject. The subject suffered from a complete motor paralysis with missing hand and finger function. The patient could trigger sequential grasp phases by imagining foot movements. After many years of using the BCI, the patient can still control the

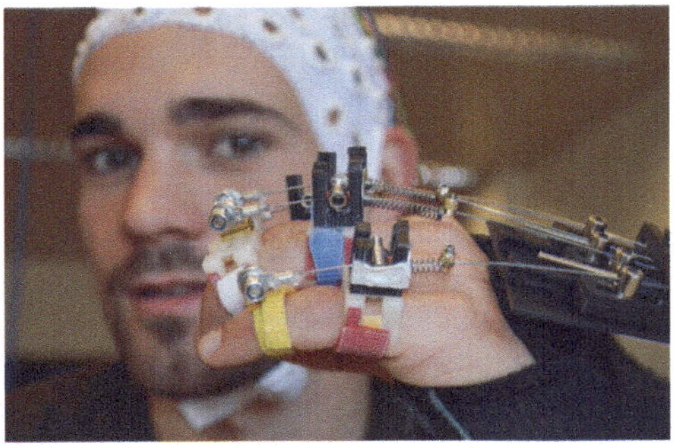

Fig. 6.1 Picture of BCI subject with an adaptable passive hand orthosis. The orthosis is capable of producing natural and smooth movements when coupled with FES. It evenly synchronizes (by bendable strips on the back) the grasping movements and applied forces on all fingers, allowing for naturalistic gestures and functional grasps of everyday objects

system, even during conversation with other persons. The same procedure could be repeated with another tetraplegic patient who was provided with a Freehand system [70]. All currently available FES systems for grasp restoration can only be used by patients with preserved voluntary shoulder and elbow function, which is the case in patients with an injury of the spinal cord below C5. So neuroprostheses for the restoration of forearm function (like hand, finger and elbow) require the use of residual movements not directly related to the grasping process. To overcome this restriction, a new method of controlling grasp and elbow function with a BCI was introduced recently [73]. Thereby a low number of pulse-width coded brain patterns are used to control sequentially more degrees of freedom [71].

BCIs have been used to control not only grasping but also other complex tasks like writing. Millan's group used the motor imagery of hand movements to stimulate the same hand for a grasping and writing task [106]. Thereby the subjects had to split his/her attention to multitask between BCI control, reaching, and the primary handwriting task itself. In contrast with the current state of the art, an approach in which the subject was imagining a movement of the same hand he controls through FES was applied. Moreover, the same group developed an adaptable passive hand orthosis (see Fig. 6.1), which evenly synchronizes the grasping movements and applied forces on all fingers [52]. This is necessary due to the very complex hand anatomy and current limitations in FES-technology with surface electrodes, because of which these grasp patterns cannot be smoothly executed. The orthosis support and synchronize the movement of the fingers stimulated by FES for patients with upper extremity palsy to improve everyday grasping and to make grasping more ergonomic and natural compared to the existing solutions. Furthermore, this orthosis also avoids fatigue in long-term stimulation situations, by locking the position of the fingers and switching the stimulation off [52].

6.3.3 Entertainment and Gaming

The area of entertainment has typically had a lower priority in BCI work, compared to more functional activities such as basic communication or control tasks. Several studies explored BCI's for controlling games [27, 44, 47, 61, 75, 76, 90, 105] and virtual reality (VR) environments [5, 48–51, 56, 95, 100]. For able-bodied users the usage of BCI for gaming has become more widely used. They could use the BCI as an (extra) interaction method with the game. Nevertheless, patients have mentioned entertainment as one of their needs, although it is indeed a need with a lower priority [120]. Moreover, BCI's may be used to assess the user's cognitive or emotional state in real-time and use that information to opportunely adapt human–computer interaction [75, 119]. A recent overview of HCI, BCI and Games can be found in [91]. More information about Games is given in this book Chaps. 11 and 13 and about Virtual Reality in Chap. 10.

6.3.4 Motor Rehabilitation and Motor Recovery

The use of BCI protocols to promote recovery of motor function by encouraging and guiding plasticity phenomena occurring after stroke (or more generally after brain injury) has been proposed recently [40, 78]. Discussion is currently underway over several factors including: the extent to which patients have detectable brain signals that can support training strategies; which brain signal features are best suited for use in restoring motor functions and how these features can be used most effectively; and what are the most effective BCI approaches for BCIs aimed at improving motor functions (for instance, what guidance should be provided to the user to maximize training that produces beneficial changes in brain signals). Preliminary findings suggested that event-related EEG activity time-frequency maps of event-related EEG activity and their classification are proper tools to monitor MI related brain activity in stroke patients and to contribute to quantify the effectiveness of MI [4, 8, 88, 104]. Preliminary studies on stroke patients using BCI found that the best signals were depicted over the ipsilateral (unaffected hemisphere) [15]. Finally, the idea that BCI technology can induce neuroplasticity has received remarkable support from the community based on invasive detection of brain electrical signals [65].

Furthermore, the continuous monitoring of mental tasks execution based on BCI techniques could support the positive effects of standard therapies. We could show that a combination of time-frequency analysis and topographic analysis of the EEG identifies and tracks task-related patterns of brain activity emerging during a single BCI session [8]. Six stroke patients executed Motor Imagery of the affected and unaffected hands: EEG sites were ranked depending on their discriminant power (DP) at different time instants and the resulting discriminant periods were used as a prior to extract EEG Microstates. Results show that the combination of these two techniques can provide insights about specific motor-related processes happening at a fine grain temporal resolution. Such events, represented by EEG microstates, can be tracked and used both to quantify changes of underlying neural structures and to provide feedback to patients and therapists.

6.3.5 Mental State Monitoring

Another area of recent research is in the recognition of the user's mental states (mental workload, stress level, tiredness, attention level) and cognitive processes (awareness to errors made by the BCI), which will facilitate interaction and reduce the user's cognitive effort by making the BCI assistive device react to the user. For instance, in case of high mental workload or stress level, the dynamics and complexity of the interaction will be simplified, or the system will trigger the switch to stop brain interaction and move on to muscle-based interaction (see also Sect. 6.3.6). As another example, in case of detection of excessive fatigue, the telepresence mobile robot or wheelchair will take over complete control and move autonomously to its base station close to the user's bed. Pioneering work in this area deals with the recognition of mental states (such as mental workload [43], attention levels [36] and fatigue [111]) and cognitive processes such as anticipation [31] and error-related potentials [11, 24–26] from EEG. In the latter case, Ferrez and Millán [25, 26] have shown that errors made by the BCI can be reliably recognized and corrected, thus yielding significant improvements in performance. Recently the areas of cognitive monitoring and implicit human–computer interaction are also phrased as passive BCI's in literature [33, 119].

6.3.6 Hybrid BCI

Despite the progress in BCI research, the level of control is still very limited compared to natural communication or existing assistive technology products (AP). Practical Brain–Computer Interfaces for disabled people should allow them to use all their remaining functionalities as control possibilities. Sometimes these people have residual activity of their muscles, most likely in the morning when they are not exhausted. In such a hybrid approach, where conventional APs (operated using some residual muscular functionality) are enhanced by BCI technology, leads to what is called a hybrid BCI (hBCI).

As a general definition, a hBCI is a combination of different input signals including at least one BCI channel [65, 86]. Thus, it could be a combination of two BCI channels but, more importantly, also a combination of a BCI and other biosignals (such as EMG, etc.) or special AT input devices (e.g., joysticks, switches, etc.). There exist a few examples of hybrid BCIs. Some are based on multiple brain signals. One of such hBCIs is the combination of motor imagery (MI)-based BCI with error potential (ErrP) detection and correction of false mental commands [26]. A second example is the combination of MI with steady state visual evoked potentials (SSVEP) explored in some offline studies [3, 14]. Other hBCIs combine brain and other biosignals. For instance, Scherer et al. [99] combined a standard SSVEP BCI with an on/off switch controlled by heart rate variation. Here the focus is to give users the ability to use the BCI only when they want or need to use it. Alternatively, and following the idea of enhancing people's residual capabilities with a BCI, Leeb et al. [53] fused electromyographic (EMG) with EEG activity, so that the subjects could achieve a good control of their hBCI independently of their level of muscular fatigue. Finally, EEG signals could be combined with eye gaze [21]. Pfurtscheller et al. [86] recently reviewed preliminary attempts, and feasibility studies, to develop hBCIs combining multiple brain signals alone or with other biosignals. Millán et al. [65] reviewed the state of the art and challenges in combining BCI and assistive technologies. For a more detailed review see Chap. 18.

6.4 Practical BCIs Based on Shared Control Techniques: Towards Control of Mobility

Another area where BCI technology can support motor substitution (see Sect. 6.3.2) is in assisting user's mobility. Users could move directly through brain-controlled wheelchairs or by mentally driving a tele-presence mobile robot—equipped with a camera and a screen—to join relatives and friends located elsewhere and participate in their activities.

Driving a wheelchair or a robot in a natural environment demands a fine and quick responding control signal. Unfortunately BCIs are limited by a low information transfer rate, because of the inherent properties of the EEG. Therefore the requirements and the skills don't match at all. Nonetheless, researchers have demonstrated the feasibility of mentally controlling complex robotic devices from EEG. A key factor to do so is the use of smart interaction designs, which in the field of robotics corresponds to shared control [16, 28, 114]. In the case of neuroprosthetics, Millán's group has pioneered the use of shared control that takes the continuous estimation of the operator's mental intent and provides assistance to achieve tasks [30, 63, 66, 109].

Generally in a shared autonomy framework, the BCI's outputs are combined with information about the environment (obstacles perceived by the robot sensors) and the robot itself (position and velocities) to better estimate the user's intent. Some broader issues in human–machine interaction are discussed in [28], where the H-Metaphor is introduced, suggesting that interaction should be more like riding a horse, with notions of "loosening the reins," allowing the system more autonomy. Shared autonomy (or shared control) is a key component of future hybrid BCI systems, as it will shape the closed-loop dynamics between the user and the brain-actuated device so tasks can be performed as easily as possible and effectively.

As mentioned above, the idea is to integrate the user's mental commands with the contextual information gathered by the intelligent brain-actuated device, so as to help the user to reach the target or override the mental commands in critical situations. In other words, the actual commands sent to the device and the feedback to the user will adapt to the context and inferred goals. In such a way, shared control can make target-oriented control easier, can inhibit pointless mental commands (e.g. driving zig-zag), and can help determine meaningful motion sequences (e.g., for a neuroprostheses). A critical aspect of shared control for BCI is coherent feedback—the behavior of the robot should be intuitive to the user and the robot should unambiguously understand the user's mental commands. Otherwise, people find it difficult to form mental models of the neuroprosthetic device.

Furthermore, thanks to the principle of mutual learning, where the user and the BCI are coupled together and adapt to each other, humans learn to operate the brain-actuated device very rapidly, in a few hours normally split between a few days [64]. Examples of shared control applications are neuroprostheses such as robots and wheelchairs [30, 63, 66, 109, 113], as well as smart virtual keyboards [69, 115, 116] and other AT software with predictive capabilities. Underlying all assistive mobility scenarios, there is the issue of shared autonomy. The crucial design question for a shared control system is: who—man, machine or both—gets control over the system, when, and to what extent?

6.4.1 Tele-Presence Robot Controlled by Motor-Disabled People

Applying the above mentioned principle of shared control allows BCI subjects to drive a mobile tele-presence platform remotely in a natural office environment. Normally this would be a complex and frustrating task, especially since the timing and speed of interaction is limited by the BCI. Furthermore, the user has to pay attention to the BCI and the tele-presence screen and also remember the place where he is and where he wants to go. Many difficulties are starting from the variability of an unknown remote environment to the reduced vision field through the control camera. In this scenario, shared control facilitates navigation in two ways. On one hand, shared control takes care of the low-level details (such as obstacle detection and avoidance for safety reasons). On the other hand, it can interpret the user's intentions to reach possible targets (such as persons or objects the user wants to approach).

Although the whole field of neuroprosthetics target disabled people with motor impairments as end-users, all successful demonstrations of brain-controlled robots or neuroprosthetics, except [70], have been actually carried out with either healthy human subjects or monkeys. In the work [109], we report the results with two patients (suffering from myopathy and spinal cord injury) who mentally drove a tele-presence robot from their clinic more than 100 km away and compare their

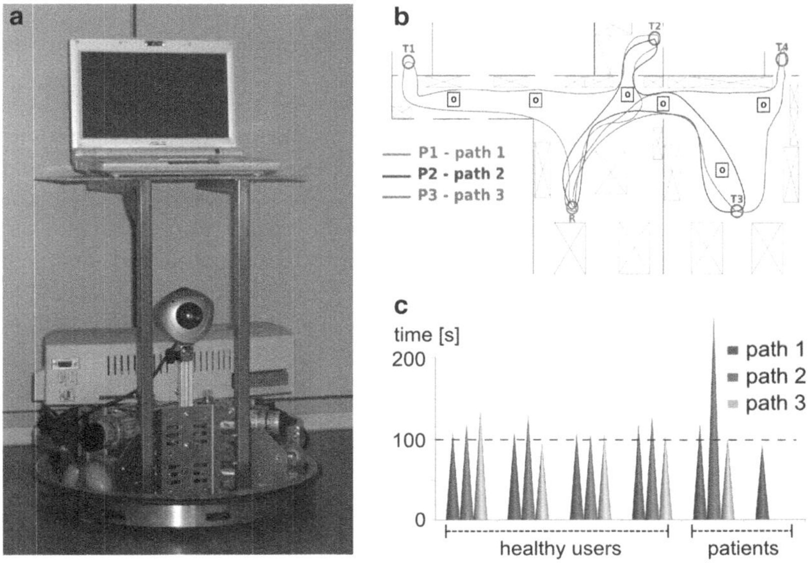

Fig. 6.2 (**a**) Tele-presence robot. (**b**) Layout of the experimental environment with the four target positions (T1, T2, T3, T4), start position (R). *Lines* (P1, P2, P3) indicate possible paths. (**c**) Time in seconds required to complete the task for each subject (four healthy and two patients) and for each of the three paths

performances to a set of healthy users carrying out the same tasks. Remarkably, the system functioned effectively although the patients had never visited the location where the tele-presence robot was operating.

The robot in this project is based on Robotino™ by FESTO a small circular mobile platform. The robot is equipped with nine infrared sensors that can detect obstacles up to ∼15 cm and a webcam that can also be used for obstacle detection. Furthermore, a notebook with an integrated camera was added on top of the robot for tele-presence purposes (see Fig. 6.2a).

The subject's task was to bring the robot to four predefined target positions within the natural working space. The space contains natural obstacles (i.e. desks, chairs, furniture, people) and six additional objects in the middle of the "normal" pathways (see Fig. 6.2b). The same paths were followed with BCI control and with manual control (i.e. button presses). Furthermore, shared control was either applied or not. The used implementation of shared control is based on the dynamical system concept coming from the fields of robotics and control theory [101]. Two dynamical systems have been created which control two independent motion parameters: the angular and translation velocities of the robot. The systems can be perturbed by adding attractors or repellors in order to generate the desired behaviors. The dynamical system implements the following navigation modality. The default device behavior is to move forward at a constant speed. If repellors or attractors are added to the system, the motion of the device changes in order to avoid the obstacles or reach the targets. At the same time, the velocity is determined according to the proximity of the repellors surrounding the robot.

The time and number of commands needed were previously reported for healthy users [109] and recently for patients [110]. Remarkably, the patients performed similar to the healthy users who were familiar with the environment. Shared control also helped all subjects (including novel BCI subjects or users with disabilities) to complete a rather complex task in similar time and with similar number of commands to those required by manual commands without shared control (see Fig. 6.2c). More details are given in [109, 110]. Thus, we argue that shared control reduces subjects' cognitive workload as it: (a) assists them in coping with low-level navigation issues (such as obstacle avoidance and allows the subject to focus the attention on his final destination) and thereby (b) helps BCI users to maintain attention for longer periods of time (since the amount of BCI commands can be reduced and their precise timing is not so critical).

6.4.2 BCI Controlled Wheelchair

In the case of brain-controlled robots and wheelchairs, Millán's group has pioneered the development of a shared autonomy approach within the European MAIA project. This research effort estimated the user's mental intent asynchronously and provided appropriate assistance for wheelchair navigation, which greatly improved BCI driving performance [30, 66, 109, 113]. Although asynchronous spontaneous BCIs seem to be the most natural and suitable alternative, there are a few examples of synchronous evoked BCIs for wheelchair control [39, 93]. The systems are based on the P300, so the system flashes the possible predefined target destinations several times in a random order. The stimulus that elicits the largest P300 is chosen as the target. Then, the intelligent wheelchair reaches the selected target autonomously. Once there, it stops and the subject can select another destination—a process that takes around 10 s.

Here, we describe our recent work, during which a subject controlled the movement of its wheelchair by thought. Like the aforementioned study (see Sect. 6.4.1 and [109]), we applied shared control techniques. The user asynchronously sent high-level commands (with the help of a motor-imagery based BCI) to achieve the desired goals, while short-term low-level interaction for obstacle avoidance was done by the shared control (see Fig. 6.3).

In our shared control paradigm, the wheelchair pro-actively slows down and turns to avoid obstacles as it approaches them. For that reason the wheelchair was equipped with proximity sensors and two webcams for obstacle detection. Using the computer vision algorithm described in [17], we constructed a local 10 cm resolution occupancy grid [13], which was then used by the shared control module for local planning. Generally the vision zone was divided into three zones. Obstacles detected in the left or right zone triggered rotation of the wheelchair, whereas

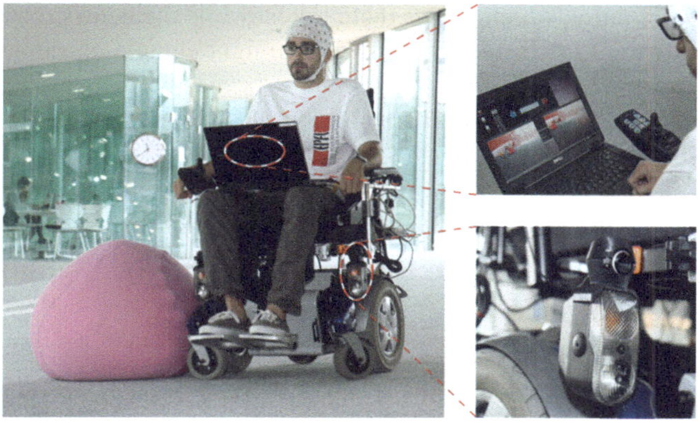

Fig. 6.3 Picture of a subject sitting in the BCI controlled wheelchair. On the *right side* two close-ups show (*below*) the webcams for obstacle detection and (*above*) the identified obstacles highlighted in *red* which will be avoided by the shared control

obstacle in center (in front) slowed it down. We also implemented a docking mode, additionally to the obstacle avoidance. Therefore we considered any obstacle to be a potential target, provided it was located directly in front of the wheelchair. Consequently, the user was able to *dock* to any "obstacle," be it a person, table, or even a wall (Note: the choice of using cheap webcams and not using an expensive laser rangefinder was taken to facilitate the development of affordable and useful assistive devices. If we want to bring the wheelchair to patients, the additional equipment should not cost more than the wheelchair itself).

Four healthy subjects (aged 23–28) participated successfully in this study. The task was to enter an open–plan environment, through a doorway, dock to two different desks, whilst navigating around natural obstacles and finally reach the corridor through a second doorway. Controlling a wheelchair with discrete commands alone, while going through a doorway or docking (attaching) to a table, is very difficult and demanding.

We want to highlight that, in this study we increased not only the complexity of the task, but also the potential stressfulness of the situation, since the user was co-located with the robotic device that he or she was controlling and was subject to many external factors. This means the user had to put trust in the shared control system and expected that negative consequences (e.g. a crash) could result in the system failing (although an experimenter was always in control of a fail-safe emergency stop button).

We also observed that, to drive a brain-controlled wheelchair or robot, subjects not only need effective BCI control, but also need to quickly deliver the appropriate mental command at the correct time. Otherwise, they will miss key maneuvers and fail to complete the task efficiently. In our experience, fast decision making is critical and depends on the proficiency of the subject as well as on his/her attention level. Along the same line, another critical ability that BCI subjects must exhibit is intentional non-control, which allows them to rest while the neuroprosthesis is in a state they don't want to change (e.g., moving straight along a corridor). However, once again, we have shown that subjects have been able to overcome all these difficulties and, with the help of a shared control system, were able to navigate safely and effectively in a realistic, open–plan environment.

6.5 Adaptation of Gesture Recognition Systems Using EEG Error Potentials

Improving performance in both humans and artificial systems relies on recognizing erroneous behavior or decisions. A wealth of studies have focus on neural activity correlated to erroneous actions or feedback [23, 107] and the possible use of the so-called error-related potentials in non-invasive brain–computer interfaces [80,97]. Remarkably, besides showing the existence of these potentials during BCI operation it has also been shown that they can be used for correcting BCI decisions [25,26], or to adapt artificial systems [18,82].

These studies have been typically performed during control of simulated devices where the subject is asked to limit his movements in order to avoid artifact contamination of the EEG signals. It is therefore not yet clear whether this type of signal can be detected or exploited in less restrictive conditions. In order to address this, we propose a hybrid system where the brain activity conveys information about the subject's cognitive and perceptual state, while control commands are delivered using faster, more efficient channels (e.g. residual muscular activity). In this work, we study the possibility of decoding EEG error-related signals during gesture-based human computer interaction and using them to improve the performance of the HCI system [19]. An illustration of the proposed approach is presented in Fig. 6.4a.

Seven male subjects took part in the experiment, during which they played a "memory game" consisting of finding pairs of images in 4×4 matrix. Subjects used five hand gestures to control a cursor to select and flip the images. These gestures are recognized using a light-barrier frame and the cursor movement (500 ms after the end of the gesture) provides feedback about whether the command was correctly recognized by the interface. Gesture recognition errors were artificially added to induce ErrPs (error rates varied from 5% to 33%). Each recording was composed of 7 sessions, with two memory games each. The overall recording for one subject contains around 2,700 gestures, and its total duration was about 2 h.

We assessed the theoretical improvement of the gesture recognition system when the detection of the error-related potentials is integrated into the system. In particular, we assume that a subject independent gesture classifier has been previously obtained and the EEG decoding signals are used to adapt this classifier to a specific new user. Accordingly, trials that are not classified as errors would signal that the last gesture was correctly recognized, and can be used as examples to further train the current classifier in a supervised manner [29]. EEG signals were classified using a Bayesian filtering technique [12], and hand acceleration was used for gesture

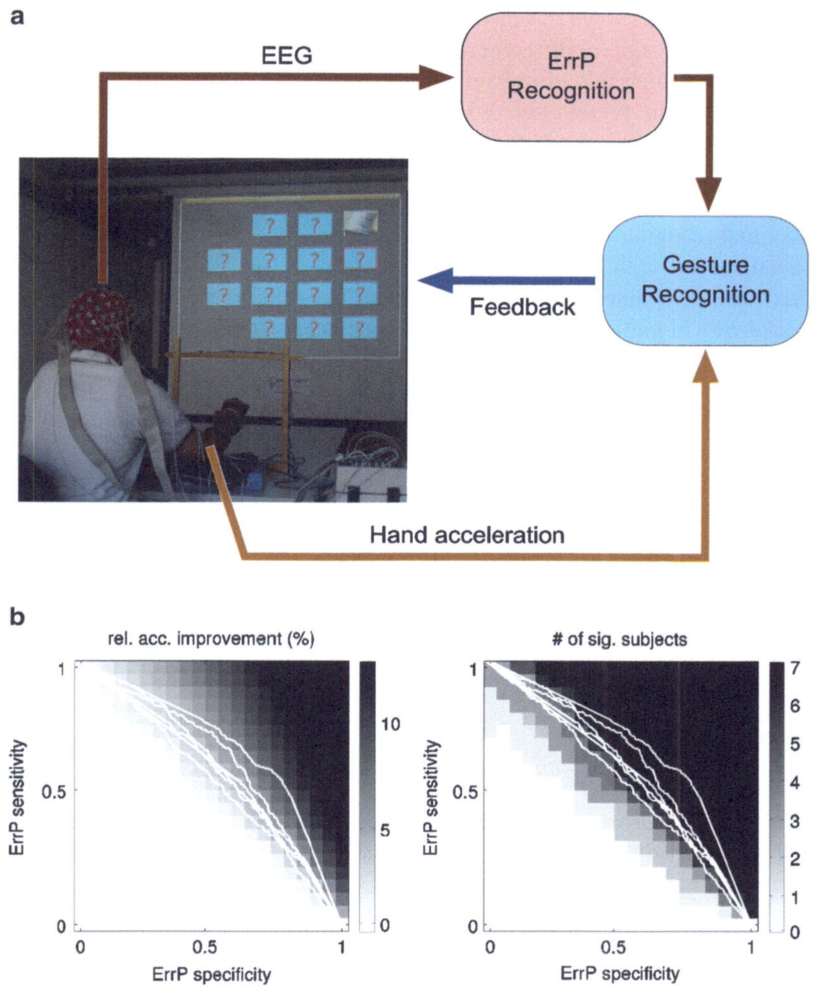

Fig. 6.4 EEG based adaptation of gesture recognition system. (**a**) Experimental setup. (**b**) Potential improvement in gesture recognition for different EEG classification performance in the ROC space, as well as the number of subjects for which this improvement is statistically significant. *White lines* correspond to the actual ErrP classification accuracy obtained using the Bayesian filter

recognition using a kNN classifier. We simulate the potential improvement in the gesture recognition for different levels of accuracy in the ErrP classification. This simulation gives us an estimation of the capabilities for adaptation of a general recognition system, and can be used to estimate the possible improvement given the EEG classification accuracy. Reported results correspond to the average over 20 repetitions across all subjects.

These results are shown in Fig. 6.4b where the gray level map shows the relative accuracy improvement for different regions of the ROC space, as well as the number of subjects for which this improvement is significantly different than the subject-independent gesture recognition system ($p < 0.05$). EEG classification accuracy for all subjects is superimposed (white lines) showing that, despite the high level of noise of the signals, they convey enough information to significantly increase the performance of the gesture recognition system for most subjects. In the present case, the maximal recognition improvement is around 6.4%, while the maximal possible improvement (i.e. with perfect EEG recognition) is of 16.8%. More details see [19].

This work constitutes an example of a hybrid system where user movements are combined with cognition-related information decoded from EEG. Given the current protocol, subjects are expected to move during the experiment, thus releasing some of the constraints that are commonly imposed in BCI setups. Overall the work provides preliminary evidence that brain-generated signals can complement other communication channels to improve performance during realistic interaction. Future work will be devoted to further explore other denoising techniques that may increase the SNR of EEG in this type of scenarios, and the deployment of hybrid BCI systems outside research laboratory conditions. In addition, further studies will also be performed to assess the influence of error rate in the evoked EEG activity and its classification performance, as well as the online application of the presented framework.

6.6 Conclusion

In this chapter we gave a broad overview of current brain–computer interface trends, current and emerging user groups, and applications and devices. We emphasized communication and control (especially for spelling and internet browsing), motor substitution by functional electrical stimulation, entertainment and games, motor recovery especially after stroke, mental state monitoring and recent developments in hybrid BCIs.

We then presented examples of how shared control can help in overcoming some of the BCI limitations and help in developing more practical BCIs especially towards the control of mobility (either a tele-presence robot or a wheelchair). We showed results from healthy users and users with disabilities, which were able to perform a rather complex tele-presence navigation task. Remarkably, although the patients had never visited the location where the tele-presence robot was operating, their performances were similar to a group of healthy users who were familiar with the environment. Furthermore, the help of shared control allowed all subjects to complete task in similar time and with similar number of commands to those required by manual commands without shared control. Thus, we argue that shared control reduces subjects' cognitive workload as it: (a) assists them in coping with low-level navigation issues (such as obstacle avoidance) and (b) helps BCI users to keep attention for longer periods of time.

We expect even faster progress in the next years, since the BCI field is still gaining attention from funding agencies and companies. More practical and powerful tools for disabled people will develop. Furthermore, BCIs can benefit from other signals and human–computer interaction techniques, and vice-versa. BCIs can be used to extract cognitive-relevant information to improve standard interactions, which is becoming increasingly interesting for healthy users.

Acknowledgements The research leading to these results has received funding from the European Union Seventh Framework Programme ([FP7/2007-2013] under grant agreement TOBI: Tools for Brain-Computer Interaction (FP7-224631) and Opportunity: Activity and Context Recognition with Opportunistic Sensor Configuration (ICT-225938). The dissemination was supported by the European ICT coordination and support action Future BNCI (FP7-248320). This paper only reflects the authors' views and funding agencies are not liable for any use that may be made of the information contained herein.

References

1. Allison, B., Graimann, B., Gräser, A.: Why use a BCI if you are healthy? In: Proceedings of BRAINPLAY 2007 "Playing with your brain", pp. 7–11 (2007a)
2. Allison, B.Z., Wolpaw, E.W., Wolpaw, J.R.: Brain–computer interface systems: Progress and prospects. Expert Rev. Med. Devices **4**(4), 463–474 (2007b)
3. Allison, B., Brunner, C., Kaiser, V., Müller-Putz, G., Neuper, C., Pfurtscheller, G.: Toward a hybrid brain–computer interface based on imagined movement and visual attention. J. Neural Eng. **7**(2), 026007 (2010)
4. Ang, K., Guan, C., Chua, K., Ang, B., Kuah, C., Wang, C., Phua, K., Chin, Z., Zhang, H.: A large clinical study on the ability of stroke patients to use an EEG-based motor imagery brain–computer interface. Clin. EEG Neurosci. **42**(4), 253–258 (2011)
5. Bayliss, J.D.: Use of the evoked potential P3 component for control in a virtual apartment. IEEE Trans. Neural Syst. Rehabil. Eng. **11**(2), 113–116 (2003)
6. Bell, C.J., Shenoy, P., Chalodhorn, R., Rao, R.P.N.: Control of a humanoid robot by a noninvasive brain–computer interface in humans. J. Neural Eng. **5**, 214–220 (2008)
7. Bensch, M., Karim, A.A., Mellinger, J., Hinterberger, T., Tangermann, M., Bogda, M., Rosenstiel, W., Birbaumer, N.: Nessi: an EEG-controlled web browser for severely paralyzed patients. Comput. Intell. Neurosci. **2007**, 71,863 (2007)
8. Biasiucci, A., Chavarriaga, R., Hamner, B., Leeb, R., Pichiorri, F., De Vico Fallani, F., Mattia, D., Millán JdR.: Combining discriminant and topographic information in BCI: Preliminary results on stroke patients. In: 5th International IEEE EMBS Conference on Neural Engineering. (2011)
9. Birbaumer, N., Cohen, L.G.: Brain-computer interfaces: communication and restoration of movement in paralysis. J. Physiol. **579**, 621–636 (2007)
10. Birbaumer, N., Ghanayim, N., Hinterberger, T., Iversen, I., Kotchoubey, B., Kübler, A., Perelmouter, J., Taub, E., Flor, H.: A spelling device for the paralysed. Nature **398**(6725), 297–298 (1999)
11. Blankertz, B., Dornhege, G., Schäfer, C., Krepki, R., Kohlmorgen, J., Müller, K., Kunzmann, V., Losch, F., Curio, G.: Boosting bit rates and error detection for the classification of fast-paced motor commands based on single-trial EEG analysis. IEEE Trans. Neural Syst. Rehabil. Eng. **11**(2), 127–131 (2003)
12. Bollon, J.M., Chavarriaga, R., Millán JdR., Bessière, P.: EEG error-related potentials detection with a Bayesian filter. In: Proc. 4th International IEEE/EMBS Conference on Neural Engineering NER '09, pp. 702–705. (2009)

13. Borenstein, J., Koren, Y.: The vector field histogram – fast obstacle avoidance for mobile robots. IEEE Trans. Rob. Autom. **7**(3), 278–288 (1991)
14. Brunner, C., BZAllison, Krusienski, D., Kaiser, V., Müller-Putz, G., Pfurtscheller, G., Neuper, C.: Improved signal processing approaches in an offline simulation of a hybrid brain–computer interface. J. Neurosci. Methods **188**(1), 165–173 (2010)
15. Buch, E., Weber, C., Cohen, L.G., Braun, C., Dimyan, M.A., Ard, T., Mellinger, J., Caria, A., Soekadar, S., Fourkas, A., Birbaumer, N.: Think to move: a neuromagnetic brain–computer interface (BCI) system for chronic stroke. Stroke **39**, 910–917 (2008)
16. Carlson, T., Demiris, Y.: Human-wheelchair collaboration through prediction of intention and adaptive assistance. In: Proceedings of the IEEE International Conference on Robotics and Automation (ICRA), Pasadena, CA, pp. 3926–3931 (2008)
17. Carlson, T., Monnard, G., Millán, J.: Vision-based shared control for a BCI wheelchair. In: Tools for Brain Computer Interaction Workshop (TOBI Workshop 2), Rome, Italy (2010)
18. Chavarriaga, R., Millán, J.: Learning from EEG error-related potentials in noninvasive brain–computer interfaces. IEEE Trans. Neural Syst. Rehabil. Eng. **18**(4), 381–388 (2010)
19. Chavarriaga, R., Biasiucci, A., Förster, K., Roggen, D., Tröster, G., Millán, JdR.: Adaptation of Hybrid Human-Computer Interaction Systems using EEG Error-Related Potentials. In: 32nd Annual International Conference of the IEEE Engineering in Medicine and Biology Society (EMBC'10) (2010)
20. Cincotti, F., Mattia, D., Aloise, F., Bufalari, S., Schalk, G., Oriolo, G., Cherubini, A., Marciani, M.G., Babiloni, F.: Non-invasive brain–computer interface system: towards its application as assistive technology. Brain Res. Bull. **75**, 796–803 (2008)
21. Danoczy, M., Fazli, S., Grozea, C., Müller, K., Popescu, F.: Brain2robot: A grasping robot arm controlled by gaze and asynchronous EEG BCI. In: Proc. 4th Int. BCI Workshop & Train. Course (2008)
22. Donchin, E., Spencer, K.M., Wijesinghe, R.: The mental prosthesis: assessing the speed of a P300-based brain–computer interface. IEEE Trans. Neural Syst. Rehabil. Eng. **8**, 174–179, (2000)
23. Falkenstein, M., Hoormann, J., Christ, S., Hohnsbein, J.: ERP components on reaction errors and their functional significance: A tutorial. Biol. Psychol. **51**(2-3), 87–107 (2000)
24. Ferrez, P.W., Millán, J.: You are wrong!—Automatic detection of interaction errors from brain waves. In: Proceedings of the 19th International Joint Conference on Artificial Intelligence, Edinburgh, UK (2005)
25. Ferrez, P.W., Millán, J.: Error-related EEG potentials generated during simulated brain–computer interaction. IEEE Trans. Biomed. Eng. **55**, 923–929 (2008a)
26. Ferrez, P.W., Millán, J.: Simultaneous real-time detection of motor imagery and error-related potentials for improved BCI accuracy. In: Proc 4th Intl. Brain–Computer Interface Workshop and Training Course, Graz, Austria (2008b)
27. Finke, A., Lenhardt, A., Ritter, H.: The MindGame: A P300-based brain–computer interface game. Neural Netw. **22**, 1329–1333 (2009)
28. Flemisch, O., Adams, A., Conway, S., Goodrich, K., Palmer, M., Schutte, P.: The H-Metaphor as a guideline for vehicle automation and interaction. Tech. Rep. NASA/TM–2003-212672, NASA (2003)
29. Förster, K., Biasiucci, A., Chavarriaga, R., Millán JdR., Roggen, D., Tröster, G.: On the use of brain decoded signals for online user adaptive gesture recognition systems. In: Pervasive 2010 The Eighth International Conference on Pervasive Computing, (2010)
30. Galán, F., Nuttin, M., Lew, E., Ferrez, P.W., Vanacker, G., Philips, J., Millán, JdR.: A brain-actuated wheelchair: Asynchronous and non-invasive brain–computer interfaces for continuous control of robots. Clin. Neurophysiol. **119**(9), 2159–2169 (2008)
31. Gangadhar, G., Chavarriaga, R., Millán, JdR.: Fast recognition of anticipation related potentials. IEEE Tran. Biomed. Eng. **56**(4), 1257–1260 (2009)
32. Gao, X., Xu, D., Cheng, M., Gao, S.: A BCI-based environmental controller for the motion-disabled. IEEE Trans. Neural Syst. Rehabil. Eng. **11**, 137–140 (2003)

33. George, L., Lécuyer, A.: An overview of research on "passive" brain–computer interfaces for implicit human–computer interaction. In: International Conference on Applied Bionics and Biomechanics (ICABB), Venice, Italy (2010)
34. Graimann, B., Allison, B., Mandel, C., Lüth, T., Valbuena, D., Gräser, A.: Non-invasive brain–computer interfaces for semi-autonomous assistive devices. In: Schuster, A. (ed.) Robust intelligent systems, pp. 113–137. Springer, London (2008)
35. Guger, C., Krausz, G., Edlinger, G.: Brain–computer interface control with dry EEG electrodes. In: Proceedings of the 5th Int. Brain–Computer Interface Conference, Verlag der Technischen Universität Graz, Graz, Austria, pp. 316–320 (2011)
36. Hamadicharef, B., Zhang, H., Guan, C., Wang, C., Phua, K., Tee, K., Ang, K.: Learning EEG-based spectral-spatial patterns for attention level measurement. In: Proc. 2009 IEEE Int. Symp. Circuits Syst. (2009)
37. Holzner, C., Guger, C., Edlinger, G., Gronegress, C., Slater, M.: Virtual smart home controlled by thoughts. In: Enabling Technologies: Infrastructures for Collaborative Enterprises, 2009. W ETICE '09. 18th IEEE International Workshops on, pp. 236–239 (2009)
38. Ijzermann, M., Stoffers, T, Klatte, M., Snoeck, G., Vorsteveld, J., Nathan, R.: The NESS handmaster orthosis: Restoration of hand function in C5 and stroke patients by means of electrical stimulation. J. Rehab. Science **9**, 86–89 (1996)
39. Iturrate, I., Antelis, J., Kubler, A., Minguez, J.: A noninvasive brain-actuated wheelchair based on a p300 neurophysiological protocol and automated navigation. IEEE Trans. Robot. **25**(3), 614–627 (2009)
40. Jeannerod, M.: Neural simulation of action: a unifying mechanism for motor cognition. NeuroImage **14**, S103–S109 (2001)
41. Karim, A.A., Hinterberger. T., Richter, J., Mellinger, J., Neumann, N., Flor, H., Kübler, A., Birbaumer, N.: Neural internet: Web surfing with brain potentials for the completely paralyzed. Neurorehabil. Neural Repair **20**(4), 508–515 (2006)
42. Keith, M.W., Hoyen, H.: Indications and future directions for upper limb neuroprostheses in tetraplegic patients: a review. Hand Clin. **18**(3), 519–528, viii (2002)
43. Kohlmorgen, J., Dornhege, G., Braun, M.L., Blankertz, B., Müller, K., Curio, G., Hagemann, K., Bruns, A., Schrauf, M., Kincses, W.: Improving human performance in a real operating environment through real-time mental workload detection. In: Dornhege, G., Millán, J., Hinterberger, T., McFarland, D., Müller, K. (eds.) Toward Brain–Computer Interfacing, pp. 409–422. MIT press, Cambridge, MA (2007)
44. Krepki, R., Blankertz, B., Curio, G., Müller, K.R.: The Berlin Brain-Computer Interface (BBCI)—towards a new communication channel for online control in gaming applications. Multimed. Tools Appl. **33**, 73–90 (2007)
45. Kübler, A., Neumann, N., Wilhelm, B., Hinterberger, T., Birbaumer, N.: Predictability of brain–computer communication. J. Psychophysiol. **18**(2-3), 121–129 (2004)
46. Kuebler, A., Nijboer, F., Mellinger, J., Vaughan, T.M., Pawelzik, H., Schalk, G., McFarland, D.J., Birbaumer, N., Wolpaw, J.R.: Patients with als can use sensorimotor rhythms to operate a brain–computer interface. Neurology **64**(10), 1775–1777 (2005)
47. Lalor, E., Kelly, S., Finucane, C., Burke, R., Smith, R., Reilly, R.B., McDarby, G.: Steady-state vep-based brain computer interface control in an immersive 3-d gaming environment. EURASIP J Appl. Signal Process. **19**, 3156–3164 (2005)
48. Lécuyer, A., Lotte, F., Reilly, R., Leeb, R., Hirose, M., Slater, M.: Brain–computer interfaces, virtual reality, and videogames. Computer **41**(10), 66–72 (2008)
49. Leeb, R., Keinrath, C., Friedman, D., Guger, C., Scherer, R., Neuper, C., Garau, M., Antley, A., Steed, A., Slater, M., Pfurtscheller, G.: Walking by thinking: the brainwaves are crucial, not the muscles! Presence (Camb.) **15**, 500–514 (2006)
50. Leeb, R., Friedman, D., Müller-Putz, G.R., Scherer, R., Slater, M., Pfurtscheller, G.: Self-paced (asynchronous) BCI control of a wheelchair in virtual environments: a case study with a tetraplegics. Comput. Intell. Neurosci. **2007**, 79,642 (2007a)
51. Leeb, R., Lee, F., Keinrath, C., Scherer, R., Bischof, H., Pfurtscheller, G.: Brain-computer communication: motivation, aim and impact of exploring a virtual apartment. IEEE Trans. Neural Syst. Rehabil. Eng. **15**, 473–482 (2007b)

52. Leeb, R., Gubler, M., Tavella, M., Miller, H., Millán, J.: On the road to a neuroprosthetic hand: A novel hand grasp orthosis based on functional electrical stimulation. In: Proc. Annual International Conference of the IEEE Engineering in Medicine and Biology Society EMBC 2010, pp. 146–149 (2010)
53. Leeb, R., Sagha, H., Chavarriaga, R., Millán, J.: A hybrid brain–computer interface based on the fusion of electroencephalographic and electromyographic activities. J. Neural Eng. **8**(2), 025,011 (2011)
54. Lim, C.G., Lee, T.S., Guan, C., Sheng Fung, D.S., Cheung, Y.B., Teng, S.S.W., Zhang, H., Krishnan, K.R.: Effectiveness of a brain–computer interface based programme for the treatment of adhd: A pilot study. Psychopharmacol. Bull. **43**(1), 73–82 (2010)
55. Lotte, F.: Brain–computer interfaces for 3D games: Hype or hope? In: Foundations of Digital Games, ACM New York, USA, 325–327 (2011)
56. Lotte, F., Langhenhove, A.V., Lamarche, F., Ernest, T., Renard, Y., Arnaldi, B., Lécuyer, A.: Exploring large virtual environments by thoughts using a brain–computer interface based on motor imagery and high-level commands. Presence (Camb.) **19**(1), 54–70 (2010)
57. Mangold, S., Keller, T., Curt, A., Dietz, V.: Transcutaneous functional electrical stimulation for grasping in subjects with cervical spinal cord injury. Spinal Cord **43**(1), 1–13 (2005)
58. Marcel, S., Millán, J.: Person authentication using brainwaves (EEG) and maximum a posteriori model adaptation. IEEE Trans. Pattern Anal. Mach. Intell. **29**, 743–748 (2007)
59. Mason, S.G., Bashashati, A., Fatourechi, M., Navarro, K.F., Birch, G.E.: A comprehensive survey of brain interface technology designs. Ann. Biomed. Eng. **35**(2), 137–169 (2007)
60. Menon, C., de Negueruela, C., Millán, J., Tonet, O., Carpi, F., Broschart, M., Ferrez, P., Buttfield, A., Tecchio, F., Sepulveda, F., Citi, L., Laschi, C., Tombini, M., Dario, P., Rossini, P.M., de Rossi, D.: Prospects of brain-machine interfaces for space system control. Acta Astronaut. **64**, 448–456 (2009)
61. Millán, J.: Adaptive brain interfaces. Commun. ACM **46**(3), 74–80 (2003)
62. Millán, J., Renkens, F., Mouriño, J., Gerstner, W.: Brain-actuated interaction. Artif. Intell. **159**, 241–259 (2004a)
63. Millán, J., Renkens, F., Mouriño, J., Gerstner, W.: Noninvasive brain-actuated control of a mobile robot by human EEG. IEEE Trans. Biomed. Eng. **51**(6), 1026–1033 (2004b)
64. Millán, J., Ferrez, P.W., Galán, F., Lew, E., Chavarriaga, R.: Non-invasive brain-machine interaction. Intern. J. Pattern Recognit. Artif. Intell. **22**(5), 959–972 (2008)
65. Millán, J., Rupp, R., Müller-Putz, G., Murray-Smith, R., Giugliemma, C., Tangermann, M., Vidaurre, C., Cincotti, F., Kübler, A., Leeb, R., Neuper, C., Müller, K., Mattia, D.: Combining brain–computer interfaces and assistive technologies: State-of-the-art and challenges. Front. Neurosc. **4**, 161 (2010)
66. Millán JdR., Galán, F., Vanhooydonck, D., Lew, E., Philips, J., Nuttin, M.: Asynchronous non-invasive brain-actuated control of an intelligent wheelchair. In: Proc. 31st Annual Int. Conf. IEEE Eng. Med. Biol. Soc., pp 3361–3364 (2009)
67. Mugler, M.E., Bensch, Halder, S., Rosenstiel, W., Bogdan, M., Birbaumer, N., Kübler, A.: Control of an internet browser using the P300 event-related potential. Int. J. Bioelectromagn. **10**, 56–63 (2008)
68. Müller, K., Tangermann, M., Dornhege, G., Krauledat, M., Curio, G., Blankertz, B.: Machine learning for real-time single-trial EEG-analysis: From brain–computer interfacing to mental state monitoring. J. Neurosci. Methods **167**, 82–90 (2008)
69. Müller, K.R., Blankertz, B.: Toward noninvasive brain–computer interfaces. IEEE Signal Proc. Mag. **23**, 125–128 (2006)
70. Müller-Putz, G., Scherer, R., Pfurtscheller, G., Rupp, R.: EEG-based neuroprosthesis control: A step towards clinical practice. Neurosci. Lett. **382**, 169–174 (2005)
71. Müller-Putz, G., Scherer, R., Pfurtscheller, G., Neuper, C.: Temporal coding of brain patterns for direct limb control in humans. Front. Neuroprosthetics **4**, 00034 (2010)
72. Müller-Putz, G.R., Pfurtscheller, G.: Control of an electrical prosthesis with an SSVEP-based BCI. IEEE Trans. Biomed. Eng. **55**, 361–364 (2008)

73. Müller-Putz, G.R., Scherer, R., Pfurtscheller, G.: Control of a two-axis artificial limb by means of a pulse width modulated brain switch. In: European Conference for the Advancement of Assistive Technology (2007)
74. Nijboer, F., Sellers, E.W., Mellinger, J., Jordan, M.A., Matuz, T., Furdea, A., Halder, S., Mochty, U., Krusienski, D.J., Vaughan, T.M., Wolpaw, J.R., Birbaumer, N., Kübler, A.: A P300-based brain–computer interface for people with amyotrophic lateral sclerosis. Clin. Neurophysiol. **119**(8), 1909–1916 (2008)
75. Nijholt, A.: BCI for games: A 'state of the art' survey. In: Proc. 7th Int. Conf. Entertain. Comp. (ICEC '08), pp. 225–228 (2009)
76. Nijholt, A., Tan, D., Allison, B., Millán, J., Graimann, B., Jackson, M.: Brain–computer interfaces for HCI and games. In: Proceedings ACM CHI 2008 (2008)
77. Nijholt, A., Plass-Oude Bos, D., Reuderink, B.: Turning shortcomings into challenges: Brain–computer interfaces for games. Entertain. Comput. **1**(2), 85–94 (2009)
78. Nilsen, D., Gillen, G., Gordon, A.: Use of mental practice to improve upper-limb recovery after stroke: A systematic review. Am. J. Occup. Ther. **64**(5), 695–708 (2010)
79. Obermaier, B., Müller, G.R., Pfurtscheller, G.: "Virtual keyboard" controlled by spontaneous EEG activity. IEEE Trans. Neural Syst. Rehabil. Eng. **11**, 422–426 (2003)
80. Parra, L.C., Spence, C.D., Gerson, A.D., Sajda, P.: Response error correction—A demonstration of improved human-machine performance using real-time EEG monitoring. IEEE Trans. Neural Syst. Rehabil. Eng. **11**(2), 173–177 (2003)
81. Perdikis, S., Leeb, R., Liboni, N., Coincenot, L., Giugliemma, C., Millán, J.: Bci for augmenting communication capabilities of disabled people. In: Proceedings of the TOBI Workshop **2010**, Integrating Brain–Computer Interfaces with Conventional Assistive Technology, p 17 (2010)
82. Perrin, X., Chavarriaga, R., Colas, F., Siegwart, R., Millán, J.: Brain-coupled interaction for semi-autonomous navigation of an assistive robot. Rob. Auton. Syst. **58**(12), 1246–1255 (2010)
83. Pfurtscheller, G., Neuper, C.: Future prospects of ERD/ERS in the context of brain–computer interface (BCI) developments. In: Neuper, C., Klimesch W (eds) Event-Related Dynamics of Brain Oscillations, Progress in Brain Research, vol. 159, pp. 433–437. Elsevier, London (2006)
84. Pfurtscheller, G., Müller, G.R., Pfurtscheller, J., Gerner, H.J., Rupp, R.: "Thought"-control of functional electrical stimulation to restore handgrasp in a patient with tetraplegia. Neurosci. Lett. **351**, 33–36 (2003)
85. Pfurtscheller, G., Müller-Putz, G., Scherer, R., Neuper, C.: Rehabilitation with brain–computer interface systems. IEEE Computer Mag. **41**, 58–65 (2008)
86. Pfurtscheller, G., Allison, B., Bauernfeind, G., Brunner, C., Solis Escalante, T., Scherer, R., Zander, T., Müller-Putz, G., Neuper, C., Birbaumer, N.: The hybrid BCI. Front. Neurosci. **4**, 42 (2010)
87. Piccione, F., Giorgi, F., Tonin, P., Priftis, K., Giove, S., Silvoni, S., Palmas, G., Beverina, F.: P300-based brain computer interface: Reliability and performance in healthy and paralysed participants. Clin. Neurophysiol. **117**(3), 531–537 (2006)
88. Pichiorri, F., Cincotti, F., Fallani, F., Pisotta, I., Morone, G., Molinari, M., Mattia, D.: Towards a brain computer interface-based rehabilitation: from bench to bedside. In: Proceedings of the 5th Int. Brain–Computer Interface Conference, Verlag der Technischen Universität Graz, Graz, Austria, pp. 268–271 (2011)
89. Pineda, J., Brang, D., Hecht, E., Edwards, L., Carey, S., Bacon, M., Futagaki, C., Suk, D., Tom, J., Birnbaum, C., Rork, A.: Positive behavioral and electrophysiological changes following neurofeedback training in children with autism. Res. Autism Spectr. Disord. **2**(3), 557–581 (2008)
90. Pineda, J.A., Silverman, D.S., Vankov, A., Hestenes, J.: Learning to control brain rhythms: making a brain–computer interface possible. IEEE Trans. Neural Syst. Rehabil. Eng. **11**, 181–184 (2003)

91. Plass-Oude Bos, D., Reuderink, B., van de Laar, B., Gürkök, H., Mühl, C., Poel, M., Nijholt, A., Heylen, D.: Brain–computer interfacing and games. In: Tan, D., Nijholt, A. (eds.) Brain–Computer Interfaces. Applying our Minds to Human–Computer Interaction, chap. 10, pp. 149–178. Springer, London (2010)
92. d R Millán, J., Ferrez, P., Seidl, T.: Validation of brain-machine interfaces during parabolic flight. Int. Rev. Neurobiol. **86**, 189–197 (2009)
93. Rebsamen, B., Guan, C., Zhang, H., Wang, C., Teo, C., Ang, M., Burdet, E.: A brain controlled wheelchair to navigate in familiar environments. IEEE Trans. Neural Syst. Rehabil. Eng. **18**(6), 590–598 (2010)
94. Riccio, A., Leotta, F., Bianchi, L., Aloise, F., Zickler, C., Hoogerwerf, E.J., Kübler, A., Mattia, D., Cincotti, F.: Workload measurement in a communication application operated through a p300-based brain–computer interface. J. Neural Eng. **8**(2), 025,028 (2011)
95. Ron-Angevin, R., Diaz-Estrella, A., Velasco-Alvarez, F.: A two-class brain computer interface to freely navigate through virtual worlds. Biomed. Tech. (Berl.) **54**(3), 126–133 (2009)
96. Saab, J., Battes, B., Grosse-Wentrup, M.: Simultaneous EEG recordings with dry and wet electrodes in motor-imagery. In: Proceedings of the 5th Int. Brain–Computer Interface Conference, Verlag der Technischen Universität Graz, Graz, Austria, pp. 312–315 (2011)
97. Schalk, G., Wolpaw, J.R., McFarland, D.J., Pfurtscheller, G.: EEG-based communication: Presence of an error potential. Clin. Neurophysiol. **111**(12), 2138–2144 (2000)
98. Scherer, R., Müller, G.R., Neuper, C., Graimann, B., Pfurtscheller, G.: An asynchronously controlled EEG-based virtual keyboard: Improvement of the spelling rate. IEEE Trans. Biomed. Eng. **51**(6), 979–984 (2004)
99. Scherer, R., Müller-Putz, G.R., Pfurtscheller, G.: Self-initiation of EEG-based brain–computer communication using the heart rate response. J. Neural Eng. **4**, L23–L29 (2007)
100. Scherer, R., Lee, F., Schlögl, A., Leeb, R., Bischof, H., Pfurtscheller, G.: Toward self-paced brain–computer communication: navigation through virtual worlds. IEEE Trans. Biomed. Eng. **55**, 675–682 (2008)
101. Schöner, G., Dose, M., Engels, C.: Dynamics of behavior: Theory and applications for autonomous robot architectures. Robot. Auton. Syst. **16**, 213–245 (1995)
102. Sellers, E.W., Krusienski, D.J., McFarland, D.J., Vaughan, T.M., Wolpaw, J.R.: A p300 event-related potential brain–computer interface (BCI): The effects of matrix size and inter stimulus interval on performance. Biol. Psychol. **73**(3), 242–252 (2006)
103. Silvoni, S., Volpato, C., Cavinato, M., Marchetti, M., Priftis, K., Merico, A., Tonin, P., Koutsikos, K., Beverina, F., Piccione, F.: P300-based brain–computer interface communication: Evaluation and follow-up in amyotrophic lateral sclerosis. Front. Neurosci. **3**, 60 (2009)
104. Silvoni, S., Ramos-Murguialday, A., Cavinato, M., Volpato, C., Cisotto, G., Turolla, A., Piccione, F., Birbaumer, N.: Brain–computer interface in stroke: A review of progress. Clin. EEG Neurosci. **42**(4) pp. 245–252 (2011)
105. Tangermann, M., Krauledat, M., Grzeska, K., Sagebaum, M., Vidaurre, C., Blankertz, B., Müller, K.: Playing pinball with non-invasive BCI. In: Proceedings of NIPS (2008)
106. Tavella, M., Leeb, R., Rupp, R., Millán, J.: Towards natural non-invasive hand neuroprostheses for daily living. In: Proc. 32rd Annual International Conference of the IEEE Engineering in Medicine and Biology Society EMBC 2010, pp. 126–129 (2010)
107. Taylor, S.F., Stern, E.R., Gehring, W.J.: Neural systems for error monitoring: Recent findings and theoretical perspectives. Neuroscientist **13**(2), 160–172 (2007)
108. Thorsen, R., Spadone, R., Ferrarin, M.: A pilot study of myoelectrically controlled FES of upper extremity. IEEE Trans. Neural Syst. Rehabil. Eng. **9**, 161–168 (2001)
109. Tonin, L., Leeb, R., Tavella, M., Perdikis, S., del R Millán, J.: The role of shared-control in BCI-based telepresence. In: Proc. of 2010 IEEE International Conference on Systems, Man and Cybernetics (2010)
110. Tonin, L., Carslon, T., Leeb, R., Millán, J.: Brain-controlled telepresence robot by motor-disabled people. In: Proc. Annual International Conference of the IEEE Engineering in Medicine and Biology Society EMBC 2011 (2011)

111. Trejo, L., Kochavi, R., Kubitz, K., Montgomery, L., Rosipal, R., Matthews, B.: EEG-based estimation of cognitive fatigue. In: Proc. SPIE, vol. 5797 (2005)
112. Valbuena, D., Sugiarto, I., Gräser, A.: Spelling with the Bremen brain–computer interface and the integrated SSVEP. In: Proc 4th Intl. Brain–Computer Interface Workshop and Training Course, Graz, Austria (2008)
113. Vanacker, G., Millán, J., Lew, E., Ferrez, P., Galán Moles, F., Philips, J., Van Brussel, H., Nuttin, M.: Context-based filtering for assisted brain-actuated wheelchair driving. Comput. Intell. Neurosci. **2007**, ID 25,130 (2007)
114. Vanhooydonck, D., Demeester, E., Nuttin, M., Van Brussel, H.: Shared control for intelligent wheelchairs: An implicit estimation of the user intention. In: Proc. 1st Int. Workshop Advances in Service Robot., pp. 176–182 (2003)
115. Williamson, J., Murray-Smith, R., Blankertz, B., Krauledat, M., Müller, K.: Designing for uncertain, asymmetric control: Interaction design for brain–computer interfaces. Int. J. Hum. Comput. Stud. **67**(10), 827–841 (2009)
116. Wills, S., MacKay, D.: DASHER—an efficient writing system for brain–computer interfaces? IEEE Trans. Neural. Syst. Rehabil. Eng. **14**, 244–246 (2006)
117. Wolpaw, J.R., Loeb, G.E., Allison, B.Z., Donchin, E., do Nascimento, O.F., Heetderks, W.J., Nijboer, F., Shain, W.G., Turner, J.N.: BCI meeting 2005–workshop on signals and recording methods. IEEE Trans. Neural. Syst. Rehabil. Eng. **14**, 138–141 (2006)
118. Zander, T., Lehne, M., Ihme, K., Jatzev, S., Correia, J., Kothe, C., Picht, B., Nijboer, F.: A dry EEG-system for scientific research and brain–computer interfaces. Front. Neurosci. **5**, 53 (2011)
119. Zander, T.O., Kothe, C.: Towards passive brain–computer interfaces: applying brain–computer interface technology to human-machine systems in general. J. Neural Eng. **8**(2), 025,005 (2011)
120. Zickler, C., Donna, V.D., Kaiser, V., Al-Khodairy, A., Kleih, S., Kuebler, A., Malavasi, M., Mattia, D., Mongardi, S., Neuper, C., Rohm, M., Rupp, R.: Brain computer interaction applications for people with disabilities: Defining user needs and user requirements. In: AAATE, p. 5 (2009)

Chapter 7
Brain Computer Interface for Hand Motor Function Restoration and Rehabilitation

Donatella Mattia, Floriana Pichiorri, Marco Molinari, and Rüdiger Rupp

7.1 Introduction

Brain–computer interfaces (BCIs) are technical systems that provide a direct connection between the human brain and a computer [99]. Such systems are able to detect thought-modulated changes in electrophysiological brain activity and transform such changes into control signals. Most of the BCI systems rely on brain signals that are recorded non-invasively by placing electrodes on the scalp (electroencephalogram, EEG). At present, these EEG-based BCI systems can function in most environments with relatively inexpensive equipment and thus, offer the possibility for practical BCIs to gain relevance in the rehabilitation field. One type of EEG-based BCIs exploits the modulation of sensorimotor rhythms (SMRs). These rhythms are oscillations in the EEG occurring in the alpha (8–12 Hz) and beta (18–26 Hz) bands and can be recorded over the sensorimotor areas. Their amplitude typically decreases during actual movement and similarly during mental rehearsal of movements (motor imagery; MI) [61,69]. Several studies have shown that people can learn to modulate SMR amplitude by practicing MI of simple movements, e.g. hand/foot movements, to control output devices [14]. This process occurs in

---

D. Mattia (✉) · F. Pichiorri
Clinical Neurophysiology, Neuroelectrical Imaging and BCI Lab, Fondazione Santa Lucia, IRCCS, Via Ardeatina, 306, 00179, Rome, Italy
e-mail: d.matt@hsantalucia.itia; f.pichiorri@hsantalucia.it

M. Molinari
Spinal Cord Injury Unit, Fondazione Santa Lucia, IRCCS, Via Ardeatina, 306, 00179, Rome, Italy
e-mail: m.molinari@hsantalucia.it

R. Rupp (✉)
Spinal Cord Injury Center, Heidelberg University Hospital, Schlierbacher Landstr. 200a, 69118 Heidelberg, Germany
e-mail: Ruediger.Rupp@med.uni-heidelberg.de

B. Z. Allison et al. (eds.), *Towards Practical Brain-Computer Interfaces*, Biological and Medical Physics, Biomedical Engineering, DOI 10.1007/978-3-642-29746-5_7,
© Springer-Verlag Berlin Heidelberg 2012

a close-loop where the system recognizes the SMR amplitude changes evoked by MI and these changes are instantaneously fed back to the users. This neurofeedback procedure based on operant conditioning enables BCI users to control their SMR activity and thus the system.

The BCI related research in the field of rehabilitation of hand motor function is mainly concentrating on two applications. In the first application, the BCI provides for a new channel to operate neuroprostheses for restoring permanent lost hand functions after a spinal cord injury (SCI). In the second more recent application, the BCI has been explored as a training tool to encourage the recovery of hand motor functions after stroke. In case of restoration of a lost or restricted hand function after SCI, a neuroprosthesis based on Functional Electrical Stimulation (FES) may be used to execute the intended movements of the hand and arm. The FES plays a special role as a BCI controlled actuator since it cannot only be used as an energy efficient way of physiologically activating muscles, but also as an effective therapeutic tool for prevention of muscle atrophy, maintenance of joint mobility and generation of enriched proprioceptive feedback into the central nervous system. In case of motor training of the hand function in stroke survivors, several actuating systems have been proposed to aid motor task practice and training such as virtual reality feedback systems [50], robotic assistive devices and FES systems for supporting the desired movements [1, 76]. The opportunity to introduce the BCI to operate the devices highlighted above would provide the neural substrate for promoting the adaptive neuroplasticity and thus facilitate the functional recovery after stroke. This chapter will be devoted to the provision and discussion of the state of the art of the combination of BCI and FES technology to restore hand motor function in SCI and to promote hand motor recovery after stroke.

7.2 Restoration of Hand Motor Functions in SCI: Brain-Controlled Neuroprostheses

The bilateral loss of the grasp function associated to a complete or nearly complete lesion of the cervical spinal cord severely limits the affected individuals' ability to live independently and retain gainful employment post injury. Thus, it represents a tremendous reduction in the patients' quality of life.

The incidence of spinal cord injuries (SCI) in industrial countries is around 40 new cases every year per million population with an increasing percentage with non-traumatic origin [93]. In Europe an estimated number of 330.000 people are suffering from a spinal cord injury with 11.000 new injuries per year [65], of which 40 % are tetraplegic with paralyses not only of the lower extremities and hence restrictions in standing and walking but also of the upper extremities resulting in limitations of the grasp function.

Any improvement of a lost or limited functions is highly desirable not only from the patients point of view [3, 86] but also for economical reasons [63]. Together with the fact that tetraplegic patients are often young persons due to sport and diving accidents, modern rehabilitation medicine aims at the compensation of the individual functional deficits and restoration of the function particularly of the grasp function.

Regeneration of the adult spinal cord following injury is extremely limited and so far cannot be enhanced by any pharmaceutical therapy [88]. During the last years remarkable progress has been made to unveil the mechanisms responsible for failed regeneration of spinal nerve fibers. This gain in knowledge led to the design of therapeutic strategies aimed to limit the tissue scar, to enhance the proregeneration versus the inhibitory environment, and to replace tissue loss, including the use of stem cells [44]. They have been successfully tested in several animal models. However, a lot of work remains to be done to ascertain whether any of these therapies can safely improve outcome after human SCI [45, 92].

7.2.1 Functional Electrical Stimulation of the Upper Extremity

Today, the only possibility of restoring permanently restricted or lost functions to a certain extend in case of missing surgical options [33] is the application of the Functional Electrical Stimulation (FES). Over the last 20 years several FES systems with different levels of complexity have been developed and introduced into the clinical environment [77]. These FES systems deliver short current impulses (unipolar pulse width < 1 ms) eliciting action potentials on the efferent nerves, which generate contractions of the innervated, yet paralyzed muscles of the hand and the forearm [94]. On this basis FES artificially compensates for the loss of voluntary muscle control. In individuals with a chronic SCI a profound disuse atrophy of the paralyzed muscles occurs, which leads to a severely decreased fatigue resistance and capability for force generation. This atrophy can be reversed by an low frequency-FES training even many years after the SCI. The time needed for achieving a meaningful fatigue resistance and force is depending on the individual status of the muscles and ranges from weeks to months [29].

Additionally, the stimulation impulses generate a profound afferent input to the spinal cord by direct activation of sensory nerve fibers and by indirect activation of proprioceptive fibers triggered by muscle contractions. In individuals with either an incomplete SCI or a complete SCI with a relevant zone of partial preservation below the level of lesion the intensive sensory feedback to the motor cortex forms the basis for guiding neuroplastic changes for functional improvement [34]. When using the FES in a restorative setup the easiest way of improving a weak or lost grasp function is the application of multiple surface electrodes (Fig. 7.1). Examples for stimulation systems based on these electrodes are the commercially available H200 (formerly called Handmaster, Bioness Ltd., Ridderkerk, Netherlands, Europe) [1], the ActiGrip®-system [76] and the Bionic Glove [2, 75]. Generally, the major advantage of these non invasive systems is that they can be offered to patients for temporary application at a very early stage of primary rehabilitation. This offers the

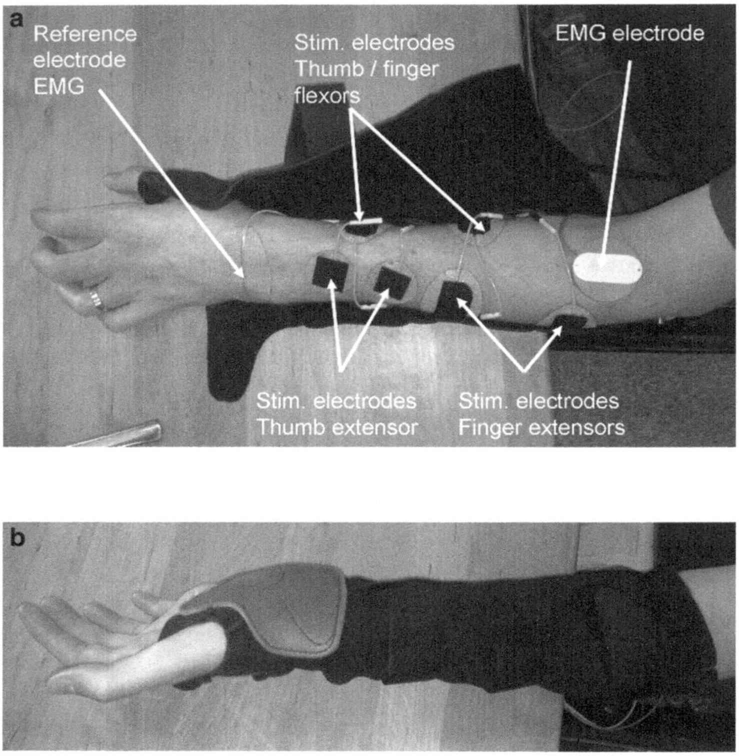

Fig. 7.1 Surface electrode configuration for restoration of a key grip (**a**) and electrode fixation soft orthosis with integrated leather protection of the palm needed for manual wheelchair propulsion (**b**). The EMG electrode in (**a**) is used for recording the residual myographic activity of the M. brachioradialis, which is used as a proportional control signal for the grasp stimulation

possibility to successfully apply FES as an adjunct therapy to occupational therapy and guiding neuroplastic changes in SCI patients with some preserved functions in the upper extremities [6, 35, 78].

In individuals with a chronic SCI and a permanent loss or restriction of the hand function the surface electrode systems have the disadvantage of insufficient selectivity in terms of stimulating individual muscles, difficulties with daily reproduction of movements, limited excitability of deeper muscle groups and pain sensations. Additionally, patients describe the placement of the electrodes as complicated [41]. Since surface electrodes tend to drop off over time an adjunct fixation mechanism in form of a sleeve (Fig. 7.1) or an orthosis is needed, which users often rate uncomfortable or not cosmetically acceptable.

Since these are major limitations when using the systems in everyday life, implantable neuroprostheses for permanent restoration of motor functions have been developed. Implantable devices range from the BION [49], a small single channel microstimulator that is injectable through a cannula, to a stimulus router system [26], an implantable electrode that picks up the current from surface electrodes, to a multichannel implantable stimulator [85], to a modular networked and wirelessly controlled system for stimulation and sensing [98]. Implantable systems inherently bare the risk of infections and risks associated to the surgery. Complex revision surgeries are necessary in case of a failure of any implanted component. Though it has been shown that these events occur rather rare [40], it has to be communicated to patients, who decide to receive an implant.

One of the implantable grasp neuroprosthesis—the Freehand system—achieved commercialization in 1997, and has been successfully used by over 300 C5/C6 individuals with SCI throughout the world and is therefore the most widespread implantable neuroprosthesis for restoration of the grasp function [36]. While the first systems have now been operating for over 15 years, the commercialization of the system stopped in 2001 not for clinical, but for financial reasons. Freehand users control hand grasp through operation of an external joystick, controlled by the movement of the opposing non-paralyzed shoulder, which through a radio frequency powered and controlled implanted stimulator, delivers electrical stimulation [82]. Importantly, a multi-center trial of the Freehand system based on 51 C5/C6 patients quantitatively demonstrated its functional efficacy [67] and economic benefits [16]. Building from this success, the implantable FES technology is undergoing significant design improvements, e.g., the implementation of implanted rechargeable power and wireless telemetry to allowing the setup of systems without any external power supply. Nevertheless, it has to be clearly stated that the degree of functional restoration by the currently available neuroprostheses either based on surface or implantable electrodes is rather limited. Even with the most sophisticated systems the restoration of only one or two grasp patterns is possible, which does not include the independent activation of single fingers or joints [98]. Additionally, the movements and forces generated by FES are less graduated when compared to the physiological condition. This is in particular the case when low forces for fine control need to be produced with surface electrodes.

Most of the current neuroprosthesis for the upper extremity have only been used for grasp restoration in individuals with SCI and preserved voluntary shoulder and elbow function. Only a few experimental studies showed the feasibility of supporting the elbow function in high lesioned subjects with SCI [15]. These systems have not been tested in real world conditions during activities of daily living, since due to the weight of the upper limb and the non-physiologic synchronous activation of the paralyzed muscles through external electrical pulses a rapid muscle fatigue occurs. A major problem in FES-based restoration of the grasp function is the occurrence of a combined lesion of the spinal fiber tracts and motoneurons in subjects with cervical spinal cord lesions [19, 57]. The denervated and flaccid muscles may be stimulated directly by the application of high charge stimulation pulses. However, the contractions produced by the FES are not effective enough in terms of force development and fatigue resistance to be used for a meaningful time [37, 38]. To overcome these limitations a combination of FES and an orthosis with actively driven or at least de-/lockable joints called "FES-hybrid orthosis" is proposed. In general, an orthosis is a mechanical device that fits to a limb and corrects a pathological joint function. An actively driven orthosis supports the joints movements with active drives, e.g., an electrical motor or a pneumatic actuator. The disadvantages of these exoskeletons are their mechanical complexity, limited possibility for use in activities of all day living and their need for a sufficient power supply. Therefore, these systems are mainly intended to be applied in users, in which sufficient movements cannot be generated by FES. If sufficient joint movements can be generated by FES a more efficient solution is the application of an orthosis with a lockable and delockable joint. In its released state this joint allows for free movements and keeps a fixed joint position in the locked state. The latter helps to avoid fatigue of the stimulated muscles needed to maintain a stable joint position.

Both types of FES-hybrid orthoses may lead to an expansion of the potential users of an upper extremity neuroprosthesis in the future [84].

7.2.2 Combining BCI and FES Technology

Through the last decade it has become obvious that the user interface of all current FES devices is not optimal in the sense of natural control, relying on either the movement or the underlying muscle activation from a non-paralyzed body part to control the coordinated electrical stimulation of muscles in the paralyzed limb [39, 51]. In the case of individuals with a high, complete SCI and the associated severe disabilities not enough residual functions are preserved that can be used for control. This has been a major limitation in the development of a reaching neuroprostheses for individuals with a loss not only of hand and finger but also of elbow and shoulder function.

Several BCI approaches mainly based on steady-state visual-evoked potentials (SSVEPs) have been introduced as a substitute for traditional control interfaces for control of an abdominal FES system [27] or for control of a wrist and hand orthosis [64]. Another exciting application is the use of a BCI to detect voluntary movement intentions in the presence of arm tremor for control of a compensatory FES [81].

Beyond these applications, BCIs have enormous implications providing natural control of a grasping and reaching neuroprosthesis control in particular in individuals with a high SCI by relying on volitional signals recorded from the brain directly involved in upper extremity movements.

The ultimate goal would be to establish a technical bypass around the lesion of the spinal cord (Fig. 7.2) and to provide neuroprosthetic users with a natural control, enabling them to accomplish movements in a fluid and transparent way. The first steps into this direction have already been undertaken involving persons with SCI [54].

In a pioneering work the BCI group in Graz and the FES group in Heidelberg addressed the general problem of the influence of a FES induced hand movement on the EEG-signals of the motor cortex in healthy subjects [56]. Event-related beta EEG changes were studied during wrist movements induced by FES, whereas active

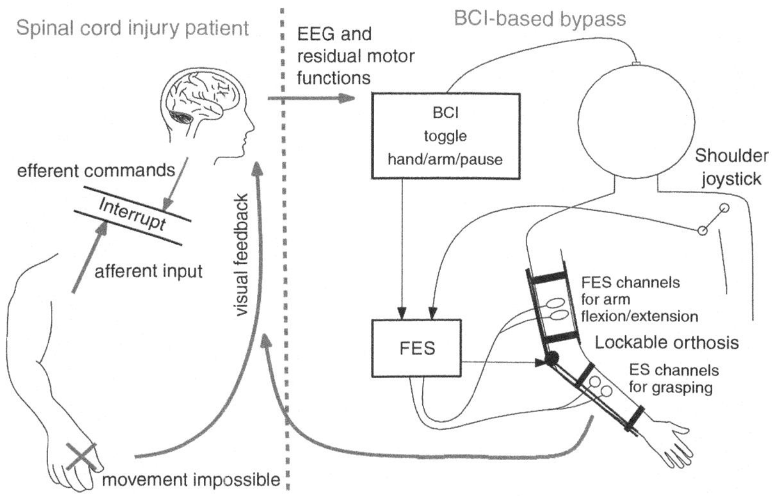

Fig. 7.2 Schematic overview of the realization of a technical bypass of a cervical spinal cord injury by using a hybrid BCI-controlled FES-orthosis consisting of a mechanically lockable elbow joint and an FES system

and passive hand movements were investigated as control conditions. Immediately after the beginning of the FES movement, a prominent EEG desynchronization (ERD, [69]) was found, followed by a beta ERS (event-related synchronization) similar to that observed after active or passive wrist movements. The main difference between active and stimulation-induced movements was that in the latter case no ERD of the beta rhythms (sensorimotor rhythms) over the primary motor cortex was detectable prior to movement-onset. These findings suggest that the sensorimotor processing during FES involves some of the processes which are also involved in voluntary hand movements.

Based on these results they performed an experiment in which they were able to prove for the first time that a BCI control of an FES-system based on surface electrodes is feasible [70]. In this single case study the restoration of a lateral grasp was achieved in a tetraplegic spinal cord injured subject, who suffers from a complete motor paralysis with missing hand and finger function. The patient has been able to move through a predefined sequence of grasp phases by the imagination of foot movements controlled the system with a 100 % accuracy. He has reached this performance level already prior to the experiment by some years of training with a motor imagery (MI) based BCI [71] and have maintained it for almost a decade by continuation of the training in regular intervals [23].

A second feasibility experiment has been performed by the same group, in which a short-term BCI-training has been applied in another tetraplegic individual. This subject has been using a Freehand system for several years. After three days of training the patient was able to control the grasp sequence of the implanted neuroprosthesis sufficiently [53]. A major result of this experiment was that

Fig. 7.3 A tetraplegic user holding a cup and drinking with the use of a BCI controlled FES hand and elbow orthosis

the artifacts in the EEG signals caused by the implanted device were much higher than those occurring in the experiment with the surface stimulation system. A possible explanation for this effect may be the presence of higher electromagnetic interferences caused by the larger inter-electrode distance of the implantable system, in which the stimulation electrodes are placed in the forearm and the common reference electrode is placed under the chest. This issue needs further systematic investigation to successfully establish BCI-controlled neuroprosthesis in the future.

With the introduction of FES-hybrid orthoses (Fig. 7.3) in the European Integrated project TOBI (Tools for Brain Computer Interaction) by the Heidelberg group it becomes more and more important to increase the number of degrees of freedom that can be controlled by the BCI. Therefore a new method has been setup by the Graz group to control the grasp as well as elbow function with a single BCI-system. It is based on pulse-width coded brain patterns to sequentially control more degrees of freedom while utilizing a minimum number of EEG electrodes [52]. A second approach to increase the number of degrees of control is to distinguish between different types of imagined movements of the same limb. The feasibility of distinguishing of different wrist motor imageries using EEG signals has recently been shown [30]. Another prerequisite for a natural BCI control of a neuroprosthesis is the independence of an imagined and performed movement of the same limb. A first study with healthy subjects performed by the Lausanne BCI group has successfully been able to show that a MI of hand movements can be used to trigger a FES of the same hand for a grasping and writing task [89].

Despite the tremendous progress that has been made in the last few years there are still a lot of open issues that have to be targeted for a successful application of BCI controlled neuroprosthesis in human tetraplegic individuals. One of the major limitations of the human work is that the results were obtained either in healthy subjects or in selected users with SCI. This raises the question, to which extend the published results can be generalized to a larger user population. In a recent study it has been demonstrated that in the majority of paraplegic users motor imagery induced EEG patterns can be discriminated in the first training session. However, this is not the case in tetraplegic subjects, in which extensive training sessions are necessary to achieve a sufficient BCI performance at least in some patients [68]. Also subjects with SCI show a diffuse and broad distributed ERD/ERS pattern during attempted foot movements in contrast to the focal beta ERD/ERS pattern during attempted foot movement of healthy subjects. Furthermore, no significant ERD/ERS patterns during passive foot movement have been found in the group of the paraplegic individuals [55].

The performance of a non invasive BCI as a neuroprosthesis control interface is rather poor compared to traditional control interfaces based on either the movement or the underlying muscle activation from non-paralyzed body parts [32, 83]. This applies not only to the limited number of possible commands per minute, but also their nature which is mainly digital ("brain-switch"). Furthermore, the latency and low number of degrees of freedom are major drawbacks for real-time, complex neuroprosthesis control [47]. The latter may be overcome with implantable systems, which have not yet reached a maturity beyond the experimental level [66]. As a final step towards routine use of BCI controlled neuroprostheses it has to be clearly proven that the experimental results, which have been obtained in an artificial lab environment, can be replicated under real world conditions without a significant degradation of performance and stability.

7.3 Rehabilitation of Hand Motor Functions After Stroke: BCI-Based Add-On Intervention

Among the possible applications of BCI technology, the neurorehabilitation of stroke survivors is gaining constantly growing interest among researchers and gathering a considerable amount of resources in the field. Over the last 3 years, about 50 scientific papers have been published in which the application of BCI in stroke rehabilitation has been investigated. These studies have been appeared either as preliminary studies on healthy subjects [28,59] or as case report [7, 17] and small clinical trials in patients [8, 58, 80].

The reasons for this growing interest are manifold. To begin with, stroke is one of the most prevalent neurological conditions worldwide and one of the leading causes of motor impairment in the population [97]. Moreover, a recent review conducted worldwide shows that the burden of stroke is high and is likely to increase in future decades [24]. Such a large prevalence broadens the catchment area of BCI technology to an otherwise unthinkable number of potential users.

Secondly, the application of BCI technology in stroke rehabilitation (either alone or combined with neuroprosthetic devices) offers an interesting means by which to encourage the activity-dependent neuroplasticity in the attempt to guide the spontaneous plastic changes occurring in the brain after stroke and hence, to lead to a better motor recovery outcome. In this regards, studies using functional neuroimaging and neurophysiological techniques have revealed robust examples of regional cortical plasticity associated with sensorimotor gains and training protocols (for review, see [18, 20, 21, 96]). For instance, multiple evidence indicate that non-primary motor areas (such as supplementary motor area, dorsal premotor cortex) might contribute significantly to movement of paretic limb after stroke, while the degree of activation of the contralesional primary sensorimotor areas would be associated with an impairment of behavioral performances after stroke.

Lastly, most of the well-established rehabilitation strategies targeting the upper limb in stroke patients (constraint induced movement therapy, bilateral arm training) rely on a certain degree of residual motor ability which unfortunately, excludes from clinical trials a considerable number of stroke survivors with no residual motor activity especially in upper limb leaving them with few rehabilitative options [46].

Hence, the use of BCI systems relying on motor related brain activity, recorded either by EEG, magnetoencephalography (MEG) or near infrared spectroscopy (NIRS), could offer a valuable tool to support training and practice in the neurorehabilitation of stroke patients even in the absence of residual motor activity [18,20,96].

7.3.1 BCI in Stroke Rehabilitation: A State-of-the-Art

The currently available few EEG-based BCI studies involving stroke survivors provide for encouraging results that however still preclude the formulation of a unambiguous conclusion.

In the study by Buch and colleagues [8], no improvement in functional clinical scores was observed despite of the successful motor imagery (MI)-based BCI training performed by chronic stroke patients. Conversely, in a study involving a larger number of patients undergoing a MI-based BCI training combined with robotic therapy, a significant improvement in motor function was reported after training as compared to baseline [4]. However, no significant differences were observed when comparing this condition with the neurorobotic training alone [4]. Interestingly, a recent multicentric study performed in stroke patients ($n = 127$ patients) with long-term upper limb deficits showed that robot-assisted therapy did not significantly improve motor function as compared with usual care or intensive care after 12 weeks [48]. Few other descriptive cases also reported improvements in rehabilitation outcome measures after BCI training alone [7, 17]. Until now no study designed on stroke patient group was able to point out significant improvements in any of the motor outcome measure as compared to a control condition, despite of the positive trend reported [80].

Neurophysiological indexes of brain reorganization associated with training have been described in one stroke patient [9]. Functional connectivity among motor related brain areas as measured by functional magnetic resonance imaging (fMRI) and diffusor tensor imaging revealed an enhanced activity on ipsilesional dorsal premotor region and supplementary motor area (SMA) after training. The correlation of BCI performances with fMRI activation in SMA was also showed in healthy subjects [31]. Moreover, evidence of a long-lasting change in the motor cortical responsiveness characterized by an increase in the motor cortical excitability measured by means of transcranial magnetic stimulation (TMS) technique was shown to occur after one month of MI-based BCI training in healthy subjects [73].

Two different strategies for the application of BCI in stroke rehabilitation are currently under investigations, and both are targeting the activity-dependent brain plasticity and its modulation [18]. The first strategy foresees the use of BCIs to train patients to produce more "normal" brain activity that is, to re-establish those brain responses physiologically associated to motor overt/covert performances. The hypothesis behind this top-down approach is that more physiological brain activity reflects more "normal" brain function, possibly resulting in an improvement of motor control. The plausibility of this strategy is supported by extensive evidence from animals and human beings that appropriate conditioning regimens can change brain signal features. The second strategy is to use BCIs to operate devices which are capable to assist movements. This bottom-up strategy is supported by the evidence that practicing or observing movements that are as close to normal as possible might help to improve motor function and help to guide the flow of sensory input generated by the assisted movements to the appropriate brain regions. The BCI-driven practice of assisted movements will foster plastic changes in the central nervous system leading to better motor function [18].

This latter approach was explored in a study that applied a BCI paradigm for post-stroke rehabilitative purposes [8]. In this study, eight chronic stroke patients with no residual finger function in their affected hand underwent a MEG-based BCI training during which they were asked to modulate their sensorimotor rhythms (SMR) by performing the MI of their affected hand, in order to operate a mechanical orthosis that passively flexed or extended their fingers. The MEG features that best discriminated the imagery tasks from the rest condition were chosen as control features regardless of their location on the scalp (either depicted over the lesioned or the intact hemisphere). This approach in the selection of control features draws from previous BCI applications (e.g. communication and control), in which the best discriminative features are eventually chosen in order to achieve the highest control accuracy [99].

Other authors have adopted a more selective strategy in identifying the control features, based on the assumption that the BCI training should reinforce those brain signals which are as close as possible to "normal" motor cortical activity (according to the top-down strategy mentioned above). For instance, in Daly et al., features were selected by comparing the EEG activity generated from MI of the affected hand to that generated from MI of the unaffected limb [17]. In other studies the control signal was collected from the ipsilesional hemisphere only (contralateral to the imagined movement of the affected hand) [4, 7, 9]. There is evidence that adaptive changes occurring after stroke may result in an increased recruitment of contralesional motor areas overcoming the more normal responsiveness of brain areas contralateral to the limb involved in the motor task [20]. Therefore, collecting and reinforcing the signal from the stroke hemisphere has the objective of contrasting this "taking over" of the contralateral unaffected hemisphere.

7.3.2 FES in Stroke Rehabilitation of Upper Limb

Several evidences from the neurorehabilitation literature have suggested that electrical stimulation might promote recovery of movement and functional ability after stroke. FES has been long applied in stroke patients for gait rehabilitation [90], however much less information is currently available on its use in the upper limb for hand motor recovery. The rationale behind the application of electrical stimulation in stroke is that afferent stimuli might have a beneficial impact on brain reorganization occurring after injury [20]. The electrical stimulation of muscles can be triggered by voluntary electromyographic activity [11] or in a push-button modality [91]. The first approach foresees the coupling of voluntary activity and afferent stimulation via FES and it seems very promising if one considers that the association of peripheral stimuli and the conjoint brain activity has been proved to enhance neuroplasticity [87]. A 2006 Cochrane review including 24 trials concluded that there was still insufficient data to inform clinical use of electrical stimulation for neuromuscular retraining, and that more research was needed in order to identify the most effective type of stimulation, the appropriate timing and dosage of the treatment [74]. Electrical stimulation has been compared with no treatment [10–12, 76, 79], with conventional therapy [25], or with a placebo intervention [13, 42]. At least one aspect of functional motor ability was found to improve in all comparisons. Moreover, good acceptability rated among stroke patients as regards pain, discomfort or adverse effects was reported. In most studies stimulation of the upper limb muscles regarded wrist and finger extensors [10–12], with the aim of contrasting the spontaneous flexion spasticity of the upper limb that is almost constantly observed in stroke patients. Nevertheless in some studies the flexor muscles in the forearm were stimulated [43]. In one study, electrical stimulation was applied to forearm muscles both in the flexor and extensor sides to exercise finalized movements of the affected hand in order to hold and release objects [76]. The authors reported a significant decrease of spasticity as measured by the Ashworth scale [5] only in the higher functioning group (least affected) [76]. A more recent study reported a significant improvement in the clinical scores of acute stroke patients who received FES treatment for rehabilitation of reaching and grasping as an "add-on" to their conventional therapy with respect to the group of patients who received conventional therapy alone. No significant changes where observed in chronic stroke patients receiving a similar intervention [91]. Clinical improvements derived from FES therapy were also analyzed in parallel with fMRI and TMS assessments. Interesting results have been provided, supporting the role of neuroplasticity in the rehabilitation protocols based on FES. An increased activation in motor related brain areas (fMRI) accompanied by an enhanced intracortical facilitation measured by TMS positively correlated with functional improvement following a 8-week FES training in nine stroke patients [95].

7.3.3 Combining BCI and FES Technology in Rehabilitation Clinical Setting: An Integrated Approach

In the attempt to design an effective format for BCI technology aiming to operate into a real rehabilitative setting, the European Integrated project TOBI (Tools for Brain Computer Interaction; www.tobi-project.org) partners have considered as crucial to define the developing principles in close collaboration with professionals of stroke rehabilitation. Many studies have shown that mental practice can reduce motor impairment and improve functional recovery of the upper limb in stroke patients, and motor imagery is often applied in conventional therapy as a strategy to access the motor system after damage resulting from stroke [62]. One of the greatest restrictions in the systematic application of this strategy is the impossibility to objectively monitor the patients' adherence to the therapist's instruction and the lack of information on the correctness of the performed mental task. From this perspective, BCI technology might provide the therapist with an instrument capable to objectively monitor MI. In this "user-center" approach, the end-users are the rehabilitation experts, and the patient who is enabled to practice MI in a setting which is comparable to a traditional rehabilitation session. According to this, Pichiorri and colleagues [72] have proposed a BCI system in which the EEG activity of the patient is fed back to the therapist who is provided with instant information on the responsiveness of the patient's brain activity to the mental task, and then he is able to guide the patient during the exercise (Fig. 7.4). The feedback to the patient is provided by the continuous interaction with the therapist and by a discrete reward at the end of each successful trial. The incorporation of the therapist is an absolute novelty in the BCI training framework normally contemplating the user and the BCI system and it represents a solution which provides both the therapist and the patient with an instrument for the first to monitor and for the second to practice the MI training. Moreover, such setting by resembling a rehabilitation session increases the acceptability by both end-users and thus, it fosters the transfer of BCI technology usage from the research laboratory to the clinical environment. This transition will ultimately support the goal of proving that the benefits derived from the BCI-supported MI training observed in a small pool of stroke volunteers, could be consolidated in a large clinical trial.

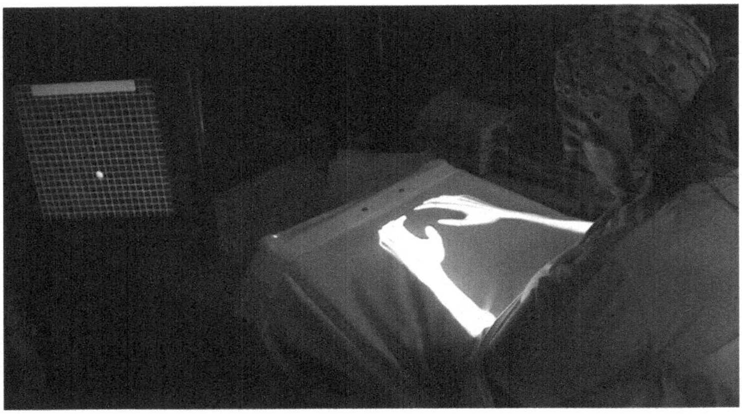

Fig. 7.4 Training session with the proposed BCI prototype currently under validation as an "add-on" rehabilitative intervention at the Fondazione Santa Lucia (Rome, Italy) within the TOBI project. The patient is trained to gain control of his visual hand representation by imaging hand movements (either closing or opening) and he receives as a feedback the congruent movements of the visual hand (successful trial). The therapist (*bottom, left corner*) is fed back with the real-time movement of a cursor on a screen that is actually controlled by the patient EEG relevant features

Pursuing this ultimate objective, it is decisive that to implement a BCI training in such a way that the MI task required to operate the system would be performed in a congruent, ecological setting. With respect to this, the feedback provided to the patient during the exercise is of utmost relevance in order to keep him focused on the required task. Prasad and colleagues [80] reported that most of the stroke patients confronted with the MI-based BCI paradigm expressed the need for more interesting, challenging and immersive scenarios. In their paradigm, the BCI training was conducted with a computer game-like feedback, in which patients were asked to move a ball on the screen by means of their right/left hand MI and place it into a target basket. It is common experience that in a MI-based BCI training, the MI task initially adopted to modulate EEG rhythms might become less important as users may achieve control of the system in an "automatic" fashion [60, 99]. Moreover, it has been shown in healthy subjects that plastic changes induced by the BCI training depend on the MI strategy adopted by the subject. Pichiorri and colleagues demonstrated that only those subjects who adopted a goal-oriented hand-grasping imagination strategy showed significant training-induced changes in the TMS functional map of the hand muscles and the brain network organization derived from EEG signals [73]. In this rehabilitative application the BCI training is not just a means to acquire good control of the system to efficiently send a specific command to the outer world. Here, the training itself and its effects on brain motor circuits are the final objective of the BCI application which is supposed to encourage the innate tendency of the brain to adapt to a lesion and thus, improve motor function of the stroke patients in their daily activities performed without the BCI system. For this reasons it is believed that during the BCI training the patients have to be immerged

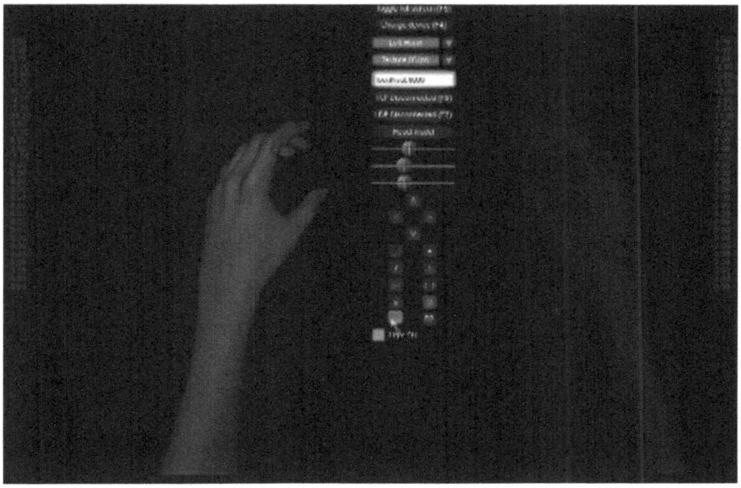

Fig. 7.5 Patient's feedback is realized by means of a dedicated software which allows the therapist to reconstruct the image of the patient's own hand, by adjusting size, appearance orientation, and other characteristics of the hand texture

in a setting which helps them to keep their attention focused on the required task and reminds them constantly of the final objective of the training they are undergoing. Accordingly, the feedback they receive has to be congruent with the imaginative task they are asked to perform. This issue has been approached by introducing a simple and straightforward feedback to the patient, in which a visual representation of the own hands is projected on a blanket resting on the own real hands. The reward at the end of each successful trial is the projected hand actuating the movement that the patient is asked to imagine (Figs. 7.4 and 7.5).

The congruence of the feedback with the imagined movement is even more important when the feedback is not only a visual perception of movement, but is accompanied by a real movement of the patient's limb, actuated either by a means of a robot device or by FES of the hand muscles. The integration of BCI technology and FES has been experimented in neurological conditions other than stroke, such as SCI (see previous section of this chapter) and more recently to control involuntary movements like tremor [81]. Also, it has been argued that enriched sensory feedback may facilitate the decoding of movement intention and thus improve the system performance [28]. A case report of a chronic stroke patient undergoing a BCI-controlled FES training proved feasibility and showed recovery of volitional isolated finger movements after a 3 weeks of FES-BCI combined training [17]. This nascent approach of rehabilitative training still requires significant efforts to bring it into a structured clinical trial to prove its efficacy in post-stroke motor recovery.

Following the rationale that enriched feedback can facilitate stroke patients to practice mental rehearsal of motor actions by means of a BCI system operated in an "enlarged loop" (i.e. with the therapist presence), the integration of a FES device can open the avenue for a more "comprehensive" BCI-driven rehabilitative device which is designed to reinforce overall individual patient's sensorimotor experience by having voluntary (covert and/or overt) access to the affected hand. Such device is currently under testing within the EU project TOBI, according to a stage III pilot study design [22]. Up to now, ten unilateral, first ever, stroke patients consecutively enrolled from a rehabilitation clinic have undergone a one-month BCI training with the BCI system described above which is installed in the rehabilitation hospital ward. Prior training, an extensive neurophysiological screening based on a multimodal approach which includes high-density EEG and TMS techniques is performed to evaluate the motor cortical responsiveness of individual patients during the imagery and/or the attempt/execution of simple hand movements of the affected and healthy hand. This screening provides for the EEG patterns generated from the lesioned hemisphere that best correlates with covert and overt motor performances of the affected hand. TMS was applied to verify the compliance of patients in performing the type of MI inducing a modulation of the motor evoked potential (MEP) recorded from the affected hand muscles. In case of absence of reproducible and stable MEP (namely, when the corticospinal tract is completely interrupted) the compliance to the task is defined during MI of the contralateral healthy hand. Stroke impairment is assessed by means of several standardized clinical and functional scales. All neurophysiological and clinical measurements are repeated after the MI-based BCI training and contrasted against those obtained from a to a control group of stroke patients who undergo a MI training without the support of the BCI system. So far, all patients were able to perform the practice of MI of affected hand by reinforcing only the modulation of the EEG desynchronization of the SMRs depicted over the ipsilesional scalp electrodes. These EEG patterns are used to control the movement (either opening or closing) of the visual representation of their own affected hand through the BCI system and with the therapist's verbal feedback (see Fig. 7.4). Preliminary findings indicate that one month of such training led to a persistent change (increase) of the EEG patterns generated from the motor areas of the ipsilesional hemisphere (Fig. 7.6). This change modulation is accompanied by an improvement of the functional motor scales (Fugl–Meyer score related to the affected upper limb). Finally, the BCI system and its related training has received a high level of acceptability by the patients and the professionals as evaluated by means of a set of questionnaires specifically applied within the TOBI project.

7.4 Conclusion and Expectations for the Future

The impairment of the upper extremity function in neurological conditions like spinal cord injury or stroke leads to a severe reduction of the patients' quality of life. Thus, the recovering and restoration even of partial arm and hand function have a significant impact on their independence and is highly desired by caregivers and patients. So far, the remarkable progress of non-invasive BCIs based on the modulation of EEG sensorimotor rhythms by either the execution or imagination of

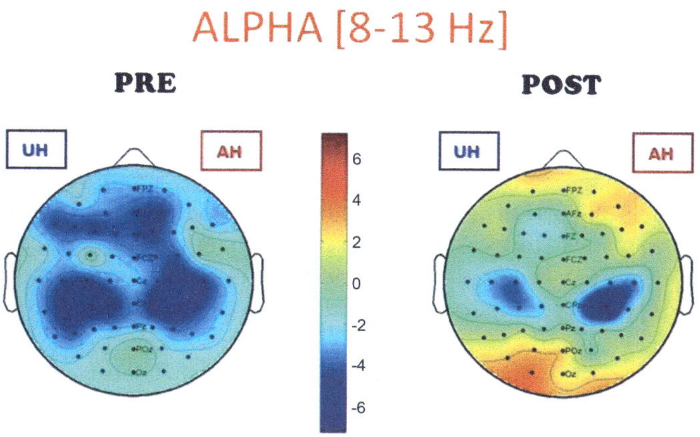

Fig. 7.6 Topographical representation of the statistical (t-test) contrast between power spectral density (PSD) values in Alpha (8–13 Hz) band, estimated at rest and during motor imagery of the affected hand. PSD values were calculated on 30 rest trials and 30 motor imagery trials. The color bar codes for t values, being negative for a decrease in PSD during the task. On the *left* and *right side* respectively, maps recorded before and after a one-month BCI training of a stroke patient with a lesion on the right hemisphere (*AH*, affected hemisphere; *UH*, unaffected hemisphere)

movements has provided a solid step towards the feasibility of their application not only under laboratory conditions but also in real world regimen. There is growing evidence that BCIs may serve as a valuable tool in rehabilitation of motor impairments. This applies either to substitution of a permanently lost upper extremity function of spinal cord injured individuals by BCI-controlled neuroprosthesis or to promoting the intrinsic recovery of stroke survivors by BCI triggered visual or proprioceptive (like FES) generated feedback. However, as with any other emerging field in neuroscience more knowledge has to be created about the usefulness of the BCI controlled rehabilitation approaches in a large user population.

At its current stage BCIs for control of neuroprotheses have a rather limited performance compared to other user interfaces based on residual motor functions. Nevertheless, BCIs can provide an additional control or feedback channel, if embedded into a user-centered, individually adapted user interface. Therefore novel ideas like the hybrid BCI integrating command signals from a BCI and several other sources are a key prerequisite for a successful introduction of BCIs neuroprosthetic applications.

As for the post-stroke rehabilitation intervention, BCIs appears a potential technology to augment stroke-related physiological processes and thus to facilitate neuroplasticity phenomena. Nevertheless, the translation of an idea from basic system neuroscience research into clinical practice has just begun and several gaps in our understanding need to be filled. One of the first unsolved issues is to define to what extent the stroke patients have detectable and suitable brain signals that can support the appropriate use of the BCI aiming at improving motor functions.

Future studies should also anticipate some issues related to clinical trial design such as optimization of components and interventions, definition of appropriated outcome measures and identification of end users who may benefit. With similar relevance, the participation and satisfaction of patients should be taken into high consideration and quantified.

The eventual value of BCI technology for improving motor function in individuals with neurological disorders depends on an advancement in technological development that must foresee a multidisciplinary interaction and collaboration between engineers, physicians, therapists and patients moving on together for a solid technology supporting restoration and functional recovery of hand motor function.

Acknowledgements This work is supported by the European ICT Programme Project FP7-224631 (TOBI—Tools for Brain Computer Interaction). This chapter only reflects the authors' views and funding agencies are not liable for any use that may be made of the information contained herein.

References

1. Alon, G., Levitt, A.F., McCarthy, P.A.: Functional electrical stimulation enhancement of upper extremity functional recovery during stroke rehabilitation: a pilot study. Neurorehabil. Neural Repair **21**(3), 207–215 (2007). DOI 10.1177/1545968306297871
2. Alon, G., McBride, K.: Persons with C5 or C6 tetraplegia achieve selected functional gains using a neuroprosthesis. Arch. Phys. Med. Rehabil. **84**(1), 119–124 (2003). DOI 10.1053/apmr.2003.50073
3. Anderson, K.D.: Targeting recovery: priorities of the spinal cord-injured population. J. Neurotrauma **21**(10), 1371–1383 (2004)
4. Ang, K.K., Guan, C., Chua, K.S., Ang, B.T., Kuah, C., Wang, C., Phua, K.S., Chin, Z.Y., Zhang, H.: Clinical study of neurorehabilitation in stroke using EEG-based motor imagery brain–computer interface with robotic feedback. Conf. Proc. IEEE Eng. Med. Biol. Soc. **2010**, 5549–5552 (2010). DOI 10.1109/IEMB.S.2010.5626782
5. Ashworth, B.: Preliminary trial of carisoprodol in multiple sclerosis. Practitioner **192**, 540–542 (1964)
6. Backus, D.: Exploring the potential for neural recovery after incomplete tetraplegia through nonsurgical interventions. PM R **2**(12 Suppl 2), S279–285 (2010). DOI S1934-1482(10)01196-2
7. Broetz, D., Braun, C., Weber, C., Soekadar, S.R., Caria, A., Birbaumer, N.: Combination of brain–computer interface training and goal-directed physical therapy in chronic stroke: a case report. Neurorehabil. Neural Repair **24**(7), 674–679 (2010). DOI 1545968310368683
8. Buch, E., Weber, C., Cohen, L.G., Braun, C., Dimyan, M.A., Ard, T., Mellinger, J., Caria, A., Soekadar, S., Fourkas, A., Birbaumer, N.: Think to move: a neuromagnetic brain–computer interface (BCI) system for chronic stroke. Stroke **39**(3), 910–917 (2008). DOI STROKEAHA107.505313
9. Caria, A., Weber, C., Brotz, D., Ramos, A., Ticini, L.F., Gharabaghi, A., Braun, C., Birbaumer, N.: Chronic stroke recovery after combined BCI training and physiotherapy: a case report. Psychophysiology **48**(4), 578–582 (2011). DOI 10.1111/j.1469-8986.2010.01117.x
10. Cauraugh, J.H., Kim, S.: Two coupled motor recovery protocols are better than one: electromyogram-triggered neuromuscular stimulation and bilateral movements. Stroke **33**(6), 1589–1594 (2002)

11. Cauraugh, J.H., Kim, S.B.: Chronic stroke motor recovery: duration of active neuromuscular stimulation. J. Neurol Sci. **215**(1–2), 13–19 (2003). DOI S0022510X03001692
12. Cauraugh, J., Light, K., Kim, S., Thigpen, M., Behrman, A.: Chronic motor dysfunction after stroke: recovering wrist and finger extension by electromyography-triggered neuromuscular stimulation. Stroke **31**(6), 1360–1364 (2000)
13. Chae, J., Bethoux, F., Bohine, T., Dobos, L., Davis, T., Friedl, A.: Neuromuscular stimulation for upper extremity motor and functional recovery in acute hemiplegia. Stroke **29**(5), 975–979 (1998)
14. Cincotti, F., Mattia, D., Aloise, F., Bufalari, S., Schalk, G., Oriolo, G., Cherubini, A., Marciani, M.G., Babiloni, F.: Non-invasive brain–computer interface system: towards its application as assistive technology. Brain Res. Bull. **75**(6), 796–803 (2008). DOI 10.1016/j.brainresbull.2008.01.007
15. Crago, P.E., Memberg, W.D., Usey, M.K., Keith, M.W., Kirsch, R.F., Chapman, G.J., Katorgi, M.A., Perreault, E.J.: An elbow extension neuroprosthesis for individuals with tetraplegia. IEEE Trans. Rehabil. Eng. **6**(1), 1–6 (1998)
16. Creasey, G.H., Kilgore, K.L., Brown-Triolo, D.L., Dahlberg, J.E., Peckham, P.H., Keith, M.W.: Reduction of costs of disability using neuroprostheses. Assist. Technol. **12**(1), 67–75 (2000). DOI 10.1080/10400435.2000.10132010
17. Daly, J.J., Cheng, R., Rogers, J., Litinas, K., Hrovat, K., Dohring, M.: Feasibility of a new application of noninvasive Brain Computer Interface (BCI): a case study of training for recovery of volitional motor control after stroke. J. Neurol Phys. Ther. **33**(4), 203–211 (2009). DOI 10.1097/NPT.0b013e3181c1fc0b
18. Daly, J.J., Wolpaw, J.R.: Brain–computer interfaces in neurological rehabilitation. Lancet Neurol **7**(11), 1032–1043 (2008). DOI S1474-4422(08)70223-0
19. Dietz, V., Curt, A.: Neurological aspects of spinal-cord repair: promises and challenges. Lancet Neurol. **5**(8), 688–694 (2006). DOI S1474-4422(06)70522-1
20. Dimyan, M.A., Cohen, L.G.: Neuroplasticity in the context of motor rehabilitation after stroke. Nat. Rev. Neurol. **7**(2), 76–85 (2011). DOI nrneurol.2010.200
21. Dobkin, B.H.: Training and exercise to drive poststroke recovery. Nat. Clin. Pract. Neurol. **4**(2), 76–85 (2008). DOI 10.1038/ncpneuro0709
22. Dobkin, B.H.: Progressive staging of pilot studies to improve phase III trials for motor interventions. Neurorehabil. Neural Repair **23**(3), 197–206 (2009). DOI 10.1177/1545968309331863
23. Enzinger, C., Ropele, S., Fazekas, F., Loitfelder, M., Gorani, F., Seifert, T., Reiter, G., Neuper, C., Pfurtscheller, G., Müller-Putz, G.: Brain motor system function in a patient with complete spinal cord injury following extensive brain–computer interface training. Exp. Brain Res. **190**(2), 215–223 (2008). DOI 10.1007/s00221-008-1465-y
24. Feigin, V.L., Lawes, C.M., Bennett, D.A., Barker-Collo, S.L., Parag, V.: Worldwide stroke incidence and early case fatality reported in 56 population-based studies: a systematic review. Lancet Neurol. **8**(4), 355–369 (2009). DOI 10.1016/S1474-4422(09)70025-0
25. Francisco, G., Chae, J., Chawla, H., Kirshblum, S., Zorowitz, R., Lewis, G., Pang, S.: Electromyogram-triggered neuromuscular stimulation for improving the arm function of acute stroke survivors: a randomized pilot study. Arch. Phys. Med. Rehabil. **79**(5), 570–575 (1998). DOI S0003-9993(98)90074-0
26. Gan, L.S., Prochazka, A.: Properties of the stimulus router system, a novel neural prosthesis. IEEE Trans. Biomed. Eng. **57**(2), 450–459 (2010). DOI 10.1109/TBME.2009.2031427
27. Gollee, H., Volosyak, I., McLachlan, A.J., Hunt, K.J., Graser, A.: An SSVEP-based brain–computer interface for the control of functional electrical stimulation. IEEE Trans. Biomed. Eng. **57**(8), 1847–1855 (2010). DOI 10.1109/TBME.2010.2043432
28. Gomez-Rodriguez, M., Peters, J., Hill, J., Scholkopf, B., Gharabaghi, A., Grosse-Wentrup, M.: Closing the sensorimotor loop: haptic feedback facilitates decoding of motor imagery. J. Neural Eng. **8**(3), 036005 v. DOI S1741-2560(11)65262-2
29. Gordon, T., Mao, J.: Muscle atrophy and procedures for training after spinal cord injury. Phys. Ther. **74**(1), 50–60 (1994)

30. Gu, Y., Dremstrup, K., Farina, D.: Single-trial discrimination of type and speed of wrist movements from EEG recordings. Clin. Neurophysiol. **120**(8), 1596–1600 (2009b). DOI S1388-2457(09)00343-5
31. Halder, S., Agorastos, D., Veit, R., Hammer, E.M., Lee, S., Varkuti, B., Bogdan, M., Rosenstiel, W., Birbaumer, N., Kubler, A.: Neural mechanisms of brain–computer interface control. Neuroimage **55**(4), 1779–1790 (2011). DOI 10.1016/j.neuroimage.2011.01.021
32. Hart, R.L., Kilgore, K.L., Peckham, P.H.: A comparison between control methods for implanted FES hand-grasp systems. IEEE Trans. Rehabil. Eng. **6**(2), 208–218 (1998)
33. Hentz, V.R., Leclercq, C.: (eds.) Surgical Rehabilitation of the Upper Limb in Tetraplegia. W.B. Saunders, London, Edingburgh, New York (2002)
34. Hoffman, L.R., Field-Fote, E.C.: Functional and corticomotor changes in individuals with tetraplegia following unimanual or bimanual massed practice training with somatosensory stimulation: a pilot study. J. Neurol. Phys. Ther. **34**(4), 193–201 (2010). DOI 10.1097/NPT.0b013e3181fbe692
35. Kapadia, N.M., Zivanovic, V., Furlan, J.C., Craven, B.C., McGillivray, C., Popovic, M.R.: Functional electrical stimulation therapy for grasping in traumatic incomplete spinal cord injury: randomized control trial. Artif. Organs **35**(3), 212–216 (2011). DOI 10.1111/j.1525-1594.2011.01216.x
36. Keith, M.W., Hoyen, H.: Indications and future directions for upper limb neuroprostheses in tetraplegic patients: a review. Hand Clin. **18**(3), 519–528, viii (2002)
37. Kern, H., Carraro, U., Adami, N., Hofer, C., Loefler, S., Vogelauer, M., Mayr, W., Rupp, R., Zampieri, S.: One year of home-based daily FES in complete lower motor neuron paraplegia: recovery of tetanic contractility drives the structural improvements of denervated muscle. Neurol Res. **32**(1), 5–12 (2010a). DOI 10.1179/174313209X385644
38. Kern, H., Carraro, U., Adami, N., Hofer, C., Loefler, S., Vogelauer, M., Mayr, W., Rupp, R., Zampieri, S.: One year of home-based daily FES in complete lower motor neuron paraplegia: recovery of tetanic contractility drives the structural improvements of denervated muscle. Neurol. Res. **32**(1), 5–12 (2010b). DOI 10.1179/174313209X385644
39. Kilgore, K.L., Hoyen, H.A., Bryden, A.M., Hart, R.L., Keith, M.W., Peckham, P.H.: An implanted upper-extremity neuroprosthesis using myoelectric control. J. Hand Surg. Am. **33**(4), 539–550 (2008). DOI S0363-5023(08)00011-7
40. Kilgore, K.L., Peckham, P.H., Keith, M.W., Montague, F.W., Hart, R.L., Gazdik, M.M., Bryden, A.M., Snyder, S.A., Stage, T.G.: Durability of implanted electrodes and leads in an upper-limb neuroprosthesis. J. Rehabil. Res. Dev. **40**(6), 457–468 (2003)
41. Kilgore, K.L., Scherer, M., Bobblitt, R., Dettloff, J., Dombrowski, D.M., Godbold, N., Jatich, J.W., Morris, R., Penko, J.S., Schremp, E.S., Cash, L.A.: Neuroprosthesis consumers' forum: consumer priorities for research directions. J. Rehabil. Res. Dev. **38**(6), 655–660 (2001)
42. Kimberley, T.J., Lewis, S.M., Auerbach, E.J., Dorsey, L.L., Lojovich, J.M., Carey, J.R.: Electrical stimulation driving functional improvements and cortical changes in subjects with stroke. Exp. Brain Res. **154**(4), 450–460 (2004). DOI 10.1007/s00221-003-1695-y
43. King, T.I., II: The effect of neuromuscular electrical stimulation in reducing tone. Am. J. Occup. Ther. **50**(1), 62–64 (1996)
44. Kwon, B.K., Okon, E.B., Plunet, W., Baptiste, D., Fouad, K., Hillyer, J., Weaver, L.C., Fehlings, M.G., Tetzlaff, W.: A systematic review of directly applied biologic therapies for acute spinal cord injury. J. Neurotrauma **28**(8), 1589–1610 (2011). DOI 10.1089/neu.2009.1150
45. Kwon, B.K., Sekhon, L.H., Fehlings, M.G.: Emerging repair, regeneration, and translational research advances for spinal cord injury. Spine **35**(21 Suppl), S263–S270 (2010). DOI 10.1097/BRS.0b013e3181f3286d
46. Langhorne, P., Coupar, F., Pollock, A.: Motor recovery after stroke: a systematic review. Lancet Neurol. **8**(8), 741–754 (2009). DOI 10.1016/S1474-4422(09)70150-4
47. Lauer, R.T., Peckham, P.H., Kilgore, K.L., Heetderks, W.J.: Applications of cortical signals to neuroprosthetic control: a critical review. IEEE Trans. Rehabil. Eng. **8**(2), 205–208 (2000)

48. Lo, A.C., Guarino, P.D., Richards, L.G., Haselkorn, J.K., Wittenberg, G.F., Federman, D.G., Ringer, R.J., Wagner, T.H., Krebs, H.I., Volpe, B.T., Bever, C.T., Jr., Bravata, D.M., Duncan, P.W., Corn, B.H., Maffucci, A.D., Nadeau, S.E., Conroy, S.S., Powell, J.M., Huang, G.D., Peduzzi, P.: Robot-assisted therapy for long-term upper-limb impairment after stroke. New England J. Med. **362**(19), 1772–1783 (2010). DOI 10.1056/NEJMoa0911341
49. Loeb, G.E., Davoodi, R.: The functional reanimation of paralyzed limbs. IEEE Eng. Med. Biol. Mag. **24**(5), 45–51 (2005)
50. Merians, A.S., Poizner, H., Boian, R., Burdea, G., Adamovich, S.: Sensorimotor training in a virtual reality environment: does it improve functional recovery poststroke? Neurorehabil. Neural Repair **20**(2), 252–267 (2006). DOI 10.1177/1545968306286914
51. Moss, C.W., Kilgore, K.L., Peckham, P.H.: A novel command signal for motor neuroprosthetic control. Neurorehabil. Neural Repair.25(9), 847–54 (2011). DOI 1545968311410067
52. Müller-Putz, G.R., Scherer, R., Pfurtscheller, G., Neuper, C.: Temporal coding of brain patterns for direct limb control in humans. Front. Neurosci. 4, 34 (2010). DOI 10.3389/fnins.2010.00034
53. Müller-Putz, G.R., Scherer, R., Pfurtscheller, G., Rupp, R.: EEG-based neuroprosthesis control: a step towards clinical practice. Neurosci. Lett. **382**(1-2), 169–174 (2005). DOI S0304-3940(05)00300-9
54. Müller-Putz, G.R., Scherer, R., Pfurtscheller, G., Rupp, R.: Brain–computer interfaces for control of neuroprostheses: from synchronous to asynchronous mode of operation. Biomed. Tech. (Berl.) **51**(2), 57–63 (2006). DOI 10.1515/BMT.2006.011
55. Müller-Putz, G.R., Zimmermann, D., Graimann, B., Nestinger, K., Korisek, G., Pfurtscheller, G.: Event-related beta EEG-changes during passive and attempted foot movements in paraplegic patients. Brain Res. **1137**(1), 84–91 (2007). DOI S0006-8993(06)03605-5
56. Müller, G.R., Neuper, C., Rupp, R., Keinrath, C., Gerner, H.J., Pfurtscheller, G.: Event-related beta EEG changes during wrist movements induced by functional electrical stimulation of forearm muscles in man. Neurosci. Lett. **340**(2), 143–147 (2003). DOI S0304394003000193
57. Mulcahey, M.J., Smith, B.T., Betz, R.R.: Evaluation of the lower motor neuron integrity of upper extremity muscles in high level spinal cord injury. Spinal Cord **37**(8), 585–591 (1999)
58. Muralidharan, A., Chae, J., Taylor, D.M.: Extracting Attempted Hand Movements from EEGs in People with Complete Hand Paralysis Following Stroke. Front. Neurosci. **5**, 39 (2011). DOI 10.3389/fnins.2011.00039
59. Nagaoka, T., Sakatani, K., Awano, T., Yokose, N., Hoshino, T., Murata, Y., Katayama, Y., Ishikawa, A., Eda, H.: Development of a new rehabilitation system based on a brain–computer interface using near-infrared spectroscopy. Adv. Exp. Med. Biol. **662**, 497–503 (2010). DOI 10.1007/978-1-4419-1241-1_72
60. Neumann, N., Hinterberger, T., Kaiser, J., Leins, U., Birbaumer, N., Kubler, A.: Automatic processing of self-regulation of slow cortical potentials: evidence from brain–computer communication in paralysed patients. Clin. Neurophysiol. **115**(3), 628–635 (2004). DOI 10.1016/j.clinph.2003.10.030
61. Neuper, C., Scherer, R., Reiner, M., Pfurtscheller, G.: Imagery of motor actions: differential effects of kinesthetic and visual-motor mode of imagery in single-trial EEG. Brain Res. Cogn. Brain Res. **25**(3), 668–677 (2005). DOI 10.1016/j.cogbrainres.2005.08.014
62. Nilsen, D.M., Gillen, G., Gordon, A.M.: Use of mental practice to improve upper-limb recovery after stroke: a systematic review. Am. J. Occup. Ther. **64**(5), 695–708 (2010)
63. NSCISC.: The 2006 annual statistical report for the model spinal cord injury care systems. National SCI Statistical Center (2006)
64. Ortner, R., Allison, B.Z., Korisek, G., Gaggl, H., Pfurtscheller, G.: An SSVEP BCI to control a hand orthosis for persons with tetraplegia. IEEE Trans. Neural Syst. Rehabil. Eng. **19**(1), 1–5 (2011). DOI 10.1109/TNSRE.2010.2076364
65. Ouzký M.: Towards concerted efforts for treating and curing spinal cord injury (2002), report of the Social, Health and Family Affairs Committee of the Council of Europe, Doc. 9401, available under http://assembly.coe.int/documents/workingdocs/doc02/edoc9401.htm (last access 3rd of July 2012)

66. Patil, P.G., Turner, D.A.: The development of brain-machine interface neuroprosthetic devices. Neurotherapeutics **5**(1), 137–146 (2008). DOI 10.1016/j.nurt.2007.11.002
67. Peckham, P.H., Keith, M.W., Kilgore, K.L., Grill, J.H., Wuolle, K.S., Thrope, G.B., Gorman, P., Hobby, J., Mulcahey, M.J., Carroll, S., Hentz, V.R., Wiegner, A.: Efficacy of an implanted neuroprosthesis for restoring hand grasp in tetraplegia: a multicenter study. Arch. Phys. Med. Rehabil. **82**(10), 1380–1388 (2001). DOI S0003-9993(01)45286-5
68. Pfurtscheller, G., Linortner, P., Winkler, R., Korisek, G., Müller-Putz, G.: Discrimination of motor imagery-induced EEG patterns in patients with complete spinal cord injury. Comput. Intell. Neurosci., Article ID 104180 (2009). DOI 10.1155/2009/104180
69. Pfurtscheller, G., Lopes da Silva, F.H.: Event-related EEG/MEG synchronization and desynchronization: basic principles. Clin. Neurophysiol. **110**(11), 1842–1857 (1999)
70. Pfurtscheller, G., Müller, G.R., Pfurtscheller, J., Gerner, H.J., Rupp, R.: 'Thought'–control of functional electrical stimulation to restore hand grasp in a patient with tetraplegia. Neurosci. Lett. **351**(1), 33–36 (2003a). DOI S0304394003009479
71. Pfurtscheller, G., Neuper, C., Muller, G.R., Obermaier, B., Krausz, G., Schlogl, A., Scherer, R., Graimann, B., Keinrath, C., Skliris, D., Wortz, M., Supp, G., Schrank, C.: Graz-BCI: state of the art and clinical applications. IEEE Trans. Neural Syst. Rehabil. Eng. **11**(2), 177–180 (2003b). DOI 10.1109/TNSRE.2003.814454
72. Pichiorri, F., Cincotti, F., de Vico Fallani, F., Pisotta, I., Morone, G., Molinari, M., Mattia, D.: Towards a brain computer Interface-based rehabilitation: from bench to bedside. . 5th International BCI Conference Proceedings, Graz, Austria (2011a)
73. Pichiorri, F., De Vico Fallani, F., Cincotti, F., Babiloni, F., Molinari, M., Kleih, S.C., Neuper, C., Kubler, A., Mattia, D.: Sensorimotor rhythm-based brain–computer interface training: the impact on motor cortical responsiveness. J. Neural Eng. **8**(2), 025020 (2011b). DOI S1741-2560(11)66162-9
74. Pomeroy, V.M., King, L., Pollock, A., Baily-Hallam, A., Langhorne, P.: Electrostimulation for promoting recovery of movement or functional ability after stroke. Systematic Review and Meta-Analysis *Stroke. 2006; 37: 2441-2442.* Cochrane Database Syst. Rev. (2), CD003241 (2006). DOI 10.1002/14651858.CD003241.pub2
75. Popovic, D., Stojanovic, A., Pjanovic, A., Radosavljevic, S., Popovic, M., Jovic, S., Vulovic, D.: Clinical evaluation of the bionic glove. Arch. Phys. Med. Rehabil. **80**(3), 299–304 (1999). DOI S0003-9993(99)90141-7
76. Popovic, M.B., Popovic, D.B., Sinkjaer, T., Stefanovic, A., Schwirtlich, L.: Clinical evaluation of Functional Electrical Therapy in acute hemiplegic subjects. J. Rehabil. Res. Dev. **40**(5), 443–453 (2003)
77. Popovic, M.R., Popovic, D.B., Keller, T.: Neuroprostheses for grasping. Neurol. Res. **24**(5), 443–452 (2002a)
78. Popovic, M.R., Thrasher, T.A., Adams, M.E., Takes, V., Zivanovic, V., Tonack, M.I.: Functional electrical therapy: retraining grasping in spinal cord injury. Spinal Cord **44**(3), 143–151 (2006). DOI 3101822
79. Powell, J., Pandyan, A.D., Granat, M., Cameron, M., Stott, D.J.: Electrical stimulation of wrist extensors in poststroke hemiplegia. Stroke **30**(7), 1384–1389 (1999)
80. Prasad, G., Herman, P., Coyle, D., McDonough, S., Crosbie, J.: Applying a brain–computer interface to support motor imagery practice in people with stroke for upper limb recovery: a feasibility study. J. Neuroeng. Rehabil. **7**, 60 (2010). DOI 1743-0003-7-60
81. Rocon, E., Gallego, J.A., Barrios, L., Victoria, A.R., Ibanez, J., Farina, D., Negro, F., Dideriksen, J.L., Conforto, S., D'Alessio, T., Severini, G., Belda-Lois, J.M., Popovic, L.Z., Grimaldi, G., Manto, M., Pons, J.L.: Multimodal BCI-mediated FES suppression of pathological tremor. Conf. Proc. IEEE Eng. Med. Biol. Soc. **2010**, 3337–3340 (2010). DOI 10.1109/IEMBS.2010.5627914
82. Rupp, R., Gerner, H.J.: Neuroprosthetics of the upper extremity–clinical application in spinal cord injury and challenges for the future. Acta. Neurochir. Suppl. **97**(Pt 1), 419–426 (2007)
83. Rupp, R., Müller-Putz, G.R., Pfurtscheller, G., Gerner, H.J., Vossius, G.: Evaluation of control methods for grasp neuroprostheses based on residual movements, myoelectrical activity and

cortical signals. Biomed. Tech. (Berl.) **53**(Suppl. 1), 2 (2008)
84. Schill, O., Wiegand, R., Schmitz, B., Matthies, R., Eck, U., Pylatiuk, C., Reischl, M., Schulz, S., Rupp, R.: OrthoJacket: an active FES-hybrid orthosis for the paralysed upper extremity. Biomed. Tech. (Berl.) **56**(1), 35–44 (2011). DOI 10.1515/BMT.2010.056
85. Smith, B., Peckham, P.H., Keith, M.W., Roscoe, D.D.: An externally powered, multichannel, implantable stimulator for versatile control of paralyzed muscle. IEEE Trans. Biomed. Eng. **34**(7), 499–508 (1987)
86. Snoek, G.J., MJ, I.J., Hermens, H.J., Maxwell, D., Biering-Sorensen, F.: Survey of the needs of patients with spinal cord injury: impact and priority for improvement in hand function in tetraplegics. Spinal Cord **42**(9), 526–532 (2004). DOI 10.1038/sj.sc.3101638
87. Stefan, K., Kunesch, E., Cohen, L.G., Benecke, R., Classen, J.: Induction of plasticity in the human motor cortex by paired associative stimulation. Brain **123** (Pt 3), 572–584 (2000)
88. Tator, C.H.: Review of treatment trials in human spinal cord injury: issues, difficulties, and recommendations. Neurosurgery **59**(5), 957–982; discussion 982–957 (2006). DOI 10.1227/01.NEU.0000245591.16087.89
89. Tavella, M., Leeb, R., Rupp, R., Millan del, J.R.: Towards natural non-invasive hand neuroprostheses for daily living. Conf. Proc. IEEE Eng. Med. Biol. Soc. **2010**, 126–129 (2010). DOI 10.1109/IEMBS.2010.5627178
90. Thrasher, T.A., Popovic, M.R.: Functional electrical stimulation of walking: function, exercise and rehabilitation. Ann. Readapt. Med. Phys. **51**(6), 452–460 (2008). DOI S0168-6054(08)00092-5
91. Thrasher, T.A., Zivanovic, V., McIlroy, W., Popovic, M.R.: Rehabilitation of reaching and grasping function in severe hemiplegic patients using functional electrical stimulation therapy. Neurorehabil. Neural Repair **22**(6), 706–714 (2008). DOI 1545968308317436
92. Thuret, S., Moon, L.D., Gage, F.H.: Therapeutic interventions after spinal cord injury. Nat. Rev. Neurosci. **7**(8), 628–643 (2006). DOI 10.1038/nrn1955
93. van den Berg, M.E., Castellote, J.M., Mahillo-Fernandez, I., de Pedro-Cuesta, J.: Incidence of spinal cord injury worldwide: a systematic review. Neuroepidemiology **34**(3), 184–192; discussion 192 (2010). DOI 000279335
94. van den Honert, C., Mortimer, J.T.: The response of the myelinated nerve fiber to short duration biphasic stimulating currents. Ann. Biomed. Eng. **7**(2), 117–125 (1979)
95. von Lewinski, F., Hofer, S., Kaus, J., Merboldt, K.D., Rothkegel, H., Schweizer, R., Liebetanz, D., Frahm, J., Paulus, W.: Efficacy of EMG-triggered electrical arm stimulation in chronic hemiparetic stroke patients. Restor. Neurol. Neurosci. **27**(3), 189–197 (2009). DOI W57810637650U3X1
96. Wang, W., Collinger, J.L., Perez, M.A., Tyler-Kabara, E.C., Cohen, L.G., Birbaumer, N., Brose, S.W., Schwartz, A.B., Boninger, M.L., Weber, D.J.: Neural interface technology for rehabilitation: exploiting and promoting neuroplasticity. Phys. Med. Rehabil. Clin. N. Am. **21**(1), 157–178 (2010). DOI S1047-9651(09)00061-8
97. Warlow, C., van Gijn, J., Dennis, M., Wardlaw, J., Bamford, J., Sandercock, P., Rinkel, G., Langhorne, P., Sudlow, C., Rothwell, P.: Stroke: Practical management. 3rd edn. Blackwell, Oxford (2008)
98. Wheeler, C.A., Peckham, P.H.: Wireless wearable controller for upper-limb neuroprosthesis. J. Rehabil. Res. Dev. **46**(2), 243–256 (2009)
99. Wolpaw, J.R., Birbaumer, N., McFarland, D.J., Pfurtscheller, G., Vaughan, T.M.: Brain–computer interfaces for communication and control. Clin. Neurophysiol. **113**(6), 767–791 (2002). DOI S1388245702000573

Chapter 8
User Centred Design in BCI Development

Elisa Mira Holz, Tobias Kaufmann, Lorenzo Desideri, Massimiliano Malavasi, Evert-Jan Hoogerwerf, and Andrea Kübler

Valeria: A case scenario. Valeria was born 35 years ago and diagnosed with congenital spastic tetraparesis, a severe paralysis of all four limbs. Completely dependent on assistive technology (AT), she initiated contact to a local centre for Assistive Technology. Today Valeria is looking back on many years of progress in which she and experts at the AT centre developed assistive solutions that are individually tailored to her needs. The outcome of this progress is impressive: Valeria has autonomous control over numerous applications on her personal computer, such as email and browser applications. And even more, with joint effort her home environment was transferred into an autonomous apartment, which she recently moved into. Such a "smart" apartment enables her to live on her own, almost independently from family and care givers. By using a voice activated remote control she can control the blinds, lighting and entrances to her home as well as her motorized bed. Altogether, most of the devices she is controlling today were selected and implemented with her support as only she would know her evolving individual needs.

Valeria's case demonstrates a concept that is applied by most Assistive Technology (AT) Centres worldwide: the user at the centre. In other words, the importance of involving the user in any step of the AT selection process is emphasized and starts with a preliminary analysis of the needs and wishes followed by the selection of the most appropriate assistive devices and is finalized with system implementation.

E.M. Holz (✉) · T. Kaufmann · A. Kübler
University of Würzburg, Department of Psychology I, Marcusstr. 9-11,
97070 Würzburg, Germany
e-mail: elisa.holz@uni-wuerzburg.de; tobias.kaufmann@uni-wuerzburg.de; andrea.kuebler@uni-wuerzburg.de

L. Desideri · M. Malavasi · E.-J. Hoogerwerf
AIAS Bologna onlus, Ausilioteca AT Centre, Corte Roncati, Via S.Isaia 90, 40123 Bologna, Italy
e-mail: ldesideri@ausilioteca.org; mmalavasi@ausilioteca.org; hoogerwerf@ausilioteca.org

B. Z. Allison et al. (eds.), *Towards Practical Brain-Computer Interfaces*, Biological and Medical Physics, Biomedical Engineering, DOI 10.1007/978-3-642-29746-5_8,
© Springer-Verlag Berlin Heidelberg 2012

The perceived usefulness of AT depends on different factors, such as the characteristics of the needs, environmental conditions, presence of care-givers and on user's requirements. For example, a person working full time in an office will have completely different requirements in terms of computer control than a person with a different daily occupation.

In this chapter, written in joint collaboration between a BCI research team and a clinical team of AT experts, we will illustrate why user-centred design is essential in BCI research. We will introduce the user centred approach and illustrate how it can be realized in BCI research and development, from patient enrollment up to implementation of individually tailored solutions. We will give insight into established standards for user involvement and methods used by a large-scale integrating BCI project, funded by the European Union (TOBI; Tools for Brain–Computer Interaction, http://www.tobi-project.org/). Finally, we will give an overview of different approaches of BCI deployment in the AT field, from BCI only to solutions combining BCI with other existing technology.

Altogether, we will show why the involvement of potential end-users in all stages of the development cycle is of utmost importance to develop technologies that will fulfill users' needs and requirements.

8.1 Technology Based Assistive Solutions for People with Disabilities

8.1.1 Understanding and Defining Disability

In the past decade, the International Classification of Functioning, Disability and Health (ICF) has integrated into a unified model of disability the two existing models usually seen as antithetical to each other, i.e., the medical model, for which the impairment of the person is the only cause of his or her limited participation in daily activities, and the social model, for which the impairment of the person obstructs his or her participation in life situations due to environmental barriers.

"Disability" is thus the outcome of the interaction between persons with impairment and the environmental and attitudinal barriers they may face (World Health Organization, 2001) [53].

8.1.2 Assistive Technologies and BCI

The term *Assistive Technologies (AT)* identifies a field that designs and develops solutions for helping people with disabilities in manifold daily life situations. In general, people with disabilities, supported by clinicians, can face their functional limitations by adopting three different typologies of intervention [14]. *Remediation* techniques are exclusively focused on changing the person because they target the problem at the "impairment" level in an effort to correct it and to promote normal functioning. *Adaptation* techniques involve modifications to the environment to permit a person with functional impairment access. *Compensation* techniques involve the use of facilitators to circumvent the functional impairment.

As illustrated by the ICF approach, disability may emerge not only due to the presence of a physical impairment, but even due to the absence of alternative strategies to accomplish certain tasks. The objective of compensation techniques could be to provide technological solutions generally referred to as assistive technologies (AT) or enabling technologies.

Any new device that aims at increasing the potential of people with disabilities will have to operate in an existing technological and economic field. Advancements in any new AT must not only be measured by the number of new technologies available, but also by the amount of people that "effectively use AT for activities and participation" [2]. Therefore, in any approach to AT, the users have to be considered, as well as the processes of efficiently selecting, providing and implementing solutions.

The AT field is changing constantly due to individual, environmental, technological, social and political conditions. On the one hand, technology creates access and participation and can, thus, be a precious alley in reducing the gap between people with disabilities and mainstream society. On the other hand, the potential of new technology might remain unexploited, if not developed and designed according to the users' needs.

A further complicating factor is that in most cases AT users require not only a single piece of equipment, but a personalized solution, possibly including mainstream devices, devices specifically designed for people with disabilities, appropriate software and services, which can be summarized as "technology based assistive solutions" (ETNA project http://www.etna-project.eu). This requires a careful analysis of needs, available solutions and possible individual adaptation. The more complex the case, the more professional support is needed for rendering the user as independent as possible, underlining the need of AT centres where AT experts and practitioners work (see case scenario). Figure 8.1 gives an overview of the dimensions that have to be considered in the design of an appropriate AT solution for a person, either whether this is based on existing or entirely new technologies.

BCI is a product of the technical progress of the last approximately 20 years that offers new possibilities in the field of AT. Intensive BCI basic research and initial testing with patients [4] made it possible that we can now consider establishing BCI in the AT sector. BCIs can be seen as a powerful tool for the motor disabled user because it circumvents the impaired motor functions independent of the etiology of disability. Particularly for people with severe muscular diseases leading to complete loss of muscular control (e.g., amyotrophic lateral sclerosis, ALS), BCIs may be the only remaining possibility to access technology based assistive solutions, i.e., BCI may serve as input channel for assistive technology (see [32] for a review). But also for people with residual muscular control, BCIs can serve as an alternative control channel combined with other input channels into a hybrid solution. A hybrid solution allows the users to switch between at least two control channels, e.g.,

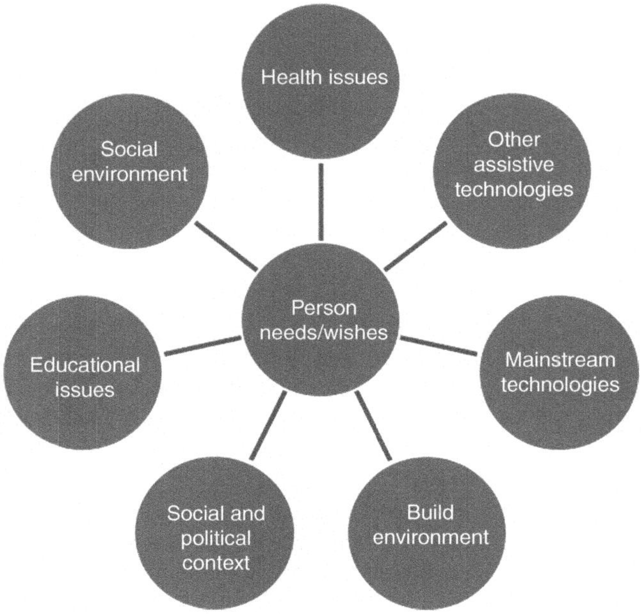

Fig. 8.1 Principal dimensions to be considered in designing technological solutions for and with people with disabilities. The correct assessment of these factors positively affects the acceptance of any AT solution by the end users

muscular and BCI control [32, 37]. Specifically, if residual muscular activity is depleted, BCI input could maintain AT based interaction. Although initial studies with people with disabilities show promising results (for a review, see [27,57]), BCIs still have to be integrated into daily life solutions. Due to continuously evolving environmental and individual requirements, the goal of BCI research should be to create a general controller, i.e., a BCI system that can be connected to various individually tailored applications or assistive devices, that requires only little set-up and preparation time, and is reliable in various settings and usable for users in various conditions.

8.2 User Centred BCI Development

8.2.1 User Centred Design Principles

User Centred Design (UCD) is an approach that supports the entire developmental process of new products or services with user-centred activities, in order to create applications which are easy to use and that are of added value to the targeted users. (International Standards Organization, 2010).

Table 8.1 Principles of Human-Centred-Design (HCD) defined by ISO (International Standards Organization, 2010)

1	Include a clear understanding of user's tasks and environmental requirements
2	Encourage an early and active involvement of users
3	Be driven and refined by user centred evaluation
4	Iterate developmental stages for identification of optimal design solutions
5	Incorporate the whole user experience
6	Encourage multi-disciplinary design

Table 8.2 Activities involved in system development lifecycle defined by ISO (International Standards Organization, 2010)

1	Understand and specify the context of use
2	Specify the user requirements
3	Produce design solutions to meet user requirements
4	Evaluate the designs against requirements

UCD is fully integrated with knowledge translation and technology transfer models, where researchers and developers collaborate with end-users, industry and other stakeholders to evolve from problem identification to solution validation [24]. A lack of user involvement or incomplete definition of user requirements are the main reasons for the failure of ICT (Information and Communication Technology) system development [45]. Success of new AT is largely determined by how well it meets the needs of the intended users.

The ISO standard on human centred design (HCD) processes for interactive systems defines the following principles of HCD (International Standards Organization, 2010; see Table 8.1). The same standard defines four linked activities which are conducted during a system development lifecycle (see Table 8.2).

A good overview of possible tools and strategies to involve users and to capture their feedback is provided by the web resource http://www.usabilitynet.org. User participation can be designed in accordance with the Living Lab model[1] (see Table 8.3). Living Labs are permanent communities of users who are involved interactively in product innovation at various stages of the design/development/validation and marketing process. Their feedback is collected by means of various socio-ethnographic research methods (focus groups, interviews, discussion groups, surveys, testing, polls, etc.).

The Living Labs concept is particularly appropriate for AT centres, which, if conducted according the above mentioned principles, can be considered permanent Living Labs for inclusion and participation. On the basis of established relationships between clients and AT centres, AT can be optimally and continuously adapted to

[1] Study on the potential of the Living Labs approach, including its relation to experimental facilities for future Internet related technologies. A study by ALTEC on behalf of the EC DG information, Society and Media. Directorate F—Emerging Technologies and Infrastructures. March 2009. http://ec.europa.eu/information_society/activities/livinglabs/docs/study/study_final_%20↩ report_d4_1_el.pdf; Retrieved, 11/09/2010

Table 8.3 Features of Living Labs

1	Different stakeholders working together for innovation
2	Open innovation concept: sharing and spreading
3	Real life testing environment: seamless and spontaneous interaction between people and technologies (and environments)
4	User centric approach to innovation: people's feedback is put at the core, especially at the beginning

Table 8.4 Conditions in AT development need to be fulfilled

1	Multidisciplinary team work
2	Participative design
3	Communication and a shared language between actors
4	A holistic approach to pull all required strings

the individual. Collaboration between users, AT experts and BCI researchers in a user centred design process can thus become a learning process for all [15].

To address these requirements and to realize the above mentioned activity stages, several framework conditions in AT development need to be fulfilled (see Table 8.4).

8.2.2 Working with End-Users in BCI Research

Development of new BCI technologies is often a balancing act between testing new BCI devices with people without particular functional limitations, usually referred to as healthy subjects, on the one hand, and testing with persons with disabilities, potential end-users on the other. Furthermore, the point in time has to be determined when a prototype is ready to be tested with end-users. Although development of BCI mainly pursues the goal of establishing devices for people with severe motor impairment, a large majority of studies in the field of BCI reports on data from healthy samples only [27]. Testing in healthy subjects is inevitable for initial evaluation of developments and modifications to the BCI device, but not sufficient, as new technologies which are designed have to be evaluated by the potential end-users themselves. The initial testing in healthy subjects is valuable though, since patients' effort can be kept at minimum by evaluating systems with healthy controls until achieving satisfactory results (e.g., testing of reliability of the BCI system). Yet thereafter, testing of BCI devices with motor impaired end-users is mandatory for conclusions on the usability of the BCI for the target user groups. For example, new flash patterns and stimuli in the P300-BCI could lead to high performance in healthy subjects, but be useless for patients with constraints in eye movement [8,18,50]. Yet, only few studies confirm their results obtained from healthy subjects with users and even fewer evaluate usability on a quantitative and qualitative level. Evaluation of BCI prototypes in a user-centred approach will facilitate bringing technology from basic research toward possible end users.

Below, methods and standards in user-centred BCI development used by the EU-project TOBI will be described.

8.2.2.1 The TOBI Project

The TOBI project aims at developing and evaluating BCI prototypes by integrating the end-user into the evaluation process from the very beginning. Therefore, the project not only comprises basic research groups from several European universities, but also Assistive Technology Centres, Rehabilitation Hospitals and Industrial Partners. Such joint effort and close collaboration provides the above described framework conditions.

Following the user-centred approach, a first step of TOBI was to identify the requirements and needs of AT users [55]. The results guided development and design of new prototypes. These prototypes were then tested with healthy subjects followed by evaluation with end-users in their home environment [57]. This user-centred approach will not only foster, but is mandatory for bringing the prototypes toward the market, i.e., to a broader range and number of end-users. To foster marketability, TOBI continuously organizes workshops to which industrial partners are invited for discussion. By that, researchers can be sensitized to the requirements of a sellable product and industrial partners can be updated about BCI development and thus, give feedback of the current potential for commercial use and what further steps have to be taken to go to scale.

8.2.2.2 Addressing the Four Stages of Developmental Activity

As described in Table 8.2, the user-centred developmental process can be subdivided into four stages.

Stage 1: Understand and specify the context of use
Available solutions from AT market were evaluated and compared how and to what extent BCI applications could be considered as useful supplements or alternatives in the AT field. The "hybrid BCI" in which a BCI is integrated into a device that uses different physiological signals or motor execution for interaction and control is a direct consequence of this approach. Further, in TOBI, BCI-systems were considered to have an impact on the life of severely disabled people in the four areas: (1) Communication and Control, (2) Motor substitution, (3) Entertainment and (4) Motor Recovery.

Stage 2: Specify the user requirements
At the beginning of the developmental process, needs and requirements of potential BCI end-users have to be assessed. In TOBI this was realized with questionnaires and semi-standardized interviews asking end-users for satisfaction with their actual AT devices regarding different aspects of independence. Evaluation of the questionnaires revealed that between 16% and 30% severely impaired end-users were not satisfied with their AT in the areas environment, communication, internet access and

manipulation. Considering the employment of a new AT solution the users indicated "functionality" (e.g., effectiveness) as the most important aspect, followed by "ease of use" and "possibility of independent use" as the most important aspects [55].

Stage 3: Produce design solutions to meet user requirements
In the next step, early-state prototypes were designed along the identified user requirements. In TOBI, these early-state prototypes were then tested with healthy subjects, resulting in first prototypes that could be brought to end-users in the field [57]. By iteratively improving prototype releases, the resulting final prototype is expected to be a generic device that can be individually tailored to the patient's requirements.

Stage 4: Evaluate the design against requirements
From the iterative process with first release prototypes, the final prototypes are developed and then again tested and evaluated by end-users. After this final feedback, the prototype can merge into the first AT product which may be considered for the market.

8.2.2.3 User Enrolment and Management of Their Expectations

To select potential end-users for the evaluation process, people with severe motor impairment were screened according to several inclusion/exclusion criteria. Main inclusion criteria within TOBI were: age above 18, severe motor disability, understanding of spoken language and context, ability to give informed consent and to communicate unambiguous feedback. Main exclusion criteria were: properties that prevent EEG acquisition, e.g., wounds on scalp or dermatitis, or influence EEG signals, e.g., medication. Those users that met inclusion criteria were invited to a meeting in which general information about BCI technologies and BCI applications was provided and questions were answered. If potential end-users were interested in taking part in the study, they were further contacted for a session in which information was provided about the aims of a specified study, the effort participation will require duration and details about the prototype to be tested. Such sessions were valued by the end-users, but were restricted to those who were able to reach the premises. Some patients may be so severely disabled that such a transport would have been too strenuous and were thus, informed individually at their home. In this educational process it was particularly important to clarify expectations of the end-user regarding the benefits and limits of the study [35]. The AT Centres paid the end users for their participation in the study.

It has to be unambiguously communicated that patients may not directly benefit from the studies and that the time of interaction between them and the research team will be limited [35]. This is of utmost importance as prototype testing involves a particularly vulnerable patient group that strongly depends on support by a social network and health care professionals. Finally, if end-users consent to participating, sessions for user training and prototype testing are scheduled.

8.2.2.4 Patient Training

The number and duration of training sessions differ between paradigms and prototypes. Generally, at the beginning of training and testing, so-called screening runs adapt the BCI to the brain activation pattern of each individual patient. Depending on the stability of such patterns, screening runs have to be repeated within one session or across sessions. Experience with the so-called P300-BCI is very encouraging as almost all patients reach high accuracy albeit sometimes at the cost of speed (for a review, see [18, 19]). Results with the sensorimotor rhythm based BCI (SMR-BCI) are not so clear cut as cortical or sub-cortical lesions such as after stroke or traumatic brain injury or widespread cortical degeneration, such as seen in patients with amyotrophic lateral sclerosis, may hamper voluntary regulation of cortical activity which is required to actively alter the activation patterns in sensorimotor areas [22]. If cue-related activation patterns can be readily detected, machine learning approaches can improve performance impressively [5, 6, 29–31, 51], however, in patients such patterns may not be detectable and operant conditioning approaches may be the method of choice [22]. In the TOBI project patient training is based on both procedures, machine learning and operant conditioning, or a mixture of both.

8.2.2.5 Evaluation Tools

The usability of the prototypes was evaluated according to the ISO 9241-210. Core variables of the evaluation process are effectiveness (accuracy), efficiency (information transfer rate and subjective workload) and satisfaction with the prototype [39,57]. In this evaluation process, the users' satisfaction with an assistive device was assessed with an extended and for BCI technologies adapted version of the QUEST 2.0 (Quebec User Evaluation of Satisfaction with assistive Technology, [9]). Satisfaction was assessed with regard to the properties dimensions, weight, adjustment, safety, comfort, ease of use, effectiveness and professional services. To adapt the QUEST 2.0 to the specificities of BCI use, items were added to evaluate reliability, speed, learnability, and aesthetic design. We refer to this extended QUEST 2.0 as TUEBS 1.0. In case participants indicated low scores in the TUEBS 1.0, they were specifically asked for an explanation. As we were interested in the effort end-users have to invest, we also measured the subjective workload with the NASA-TLX [13] on six subscales: temporal demand, physical demand, mental demand, performance, effort and frustration that the user experienced by completing the different tasks. Finally, we assessed device satisfaction with a visual analogue scale ranging from 1 "not at all satisfied" to 10 "absolutely satisfied." Patients were also invited to share their opinion about the BCI prototype in an open interview (see also user experience evaluation in Chap. 10).

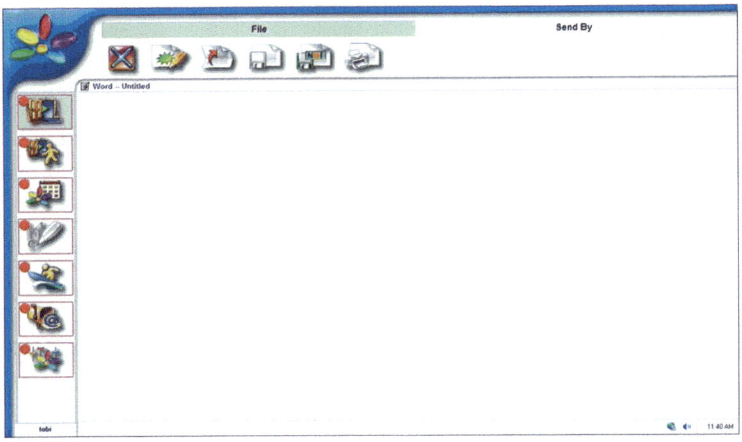

Fig. 8.2 Qualilife application with overlaid P300-stimulation (*red dots* flash close to the selectable items [57])

8.2.2.6 Examples of TOBI Prototypes Evaluated by Patients

In the following, we provide examples of TOBI communication and entertainment prototypes, which were the first BCI prototypes that were evaluated by end-users in terms of their effectiveness, efficiency and satisfaction. Specifically, we provide results of: (1) P300-Qualilife-communication prototype and (2) Brain-Painting application.

The P300-Qualilife-communication prototype is the first application that combines BCI with commercial accessibility AT software Qualiworld (QW, Software by Qualilife Lugano, Switzerland). The adapted visual P300-stimulation is superimposed as a graphical layer on the buttons for functions and commands provided by the Qualilife application (see Fig. 8.2). This AT-software offers communication and control functions such as word processing, emailing and internet browsing. The P300-Qualilife-communication prototype has been tested and evaluated by four potential end-users [57]. The evaluation results show that the end-users performed very well (accuracies between 70% and 100%) on the communication and internet tasks with an ITR of 4.03–8.57 bits/min. With respect to the subjective workload the users stated that mental demand followed by temporal demand contributed most to their workload. One of these end-users was a BCI novice and his mental workload was high in the first session, but decreased remarkably in the second, and the patient stated that he was surprised how quickly he became familiar with the application. Overall, the users were quite satisfied with the applications. However, there were specific issues which contributed to dissatisfaction. They were not very satisfied with adjustment ("it takes too long to adjust the EEG cap/electrodes"), speed (item selection is "too slow"), comfort (wearing of the cap is "uncomfortable"), ease of use ("it takes too long to set-up the system," EEG cap/electrodes and

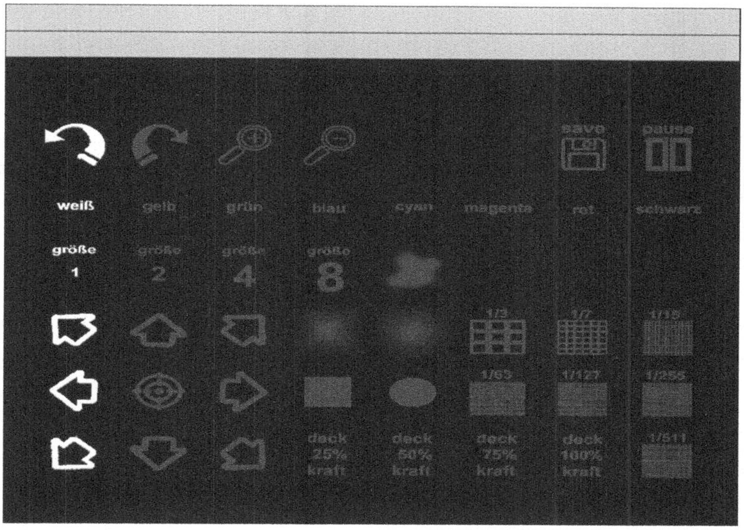

Fig. 8.3 Brain Painting Matrix with selectable symbols, e.g., for shape, color, size, zooming, and cursor position

calibration of signals), and effectiveness ("error correction is time-consuming"). Thus, for daily use BCI needs to be improved, specifically with regards to speed and comfort/adjustment.

A further TOBI prototype is the Brain Painting application [23, 34]. Brain painting is based on the P300-BCI matrix (often referred to as P300-Speller), in which letters of the matrix are replaced by symbols for, e.g., shape, color, size, zooming, cursor position etc. (see Fig. 8.3). It allows the user to create paintings and thus provides a new means of non-verbal communication. This application was tested and evaluated by four severely disabled end-users [56]. All end-users performed very well, with accuracy between 80% and 100%. One user, who earned her living with painting before she was diagnosed with ALS, stated "I am deeply moved to tears. I have not been able to paint for more than 5 years. Today I again had butterflies in my stomach, a feeling that I have missed so much, so much. I was so sad, I was plagued by fears of loss, I was in shock because I could not paint. For me the picture I have created is so very typical me, no other paints like I do (which is not to say that I am the best). No one else paints in my style, and despite 5 years of absence, I'm simply an artist again; I'm back to life!." End-users were satisfied with the Brain Painting application and evaluation results regarding usability are comparable to the Qualilife application. They indicated that they enjoyed the painting and that they liked to use it 1–2 times per week at home, but that they would be afraid of skin problems due to the electrode gel. Like in the Qualilife application, users indicated that for daily use the application would need to improve its comfort (EEG cap, electrode gel and cables), speed and ease of use (simpler set-up of hardware/software).

In summary, first evaluation results indicate that severely disabled end-users can operate the BCI prototypes effectively (high accuracy) and efficiently (high ITR and low subjective work load). Patients enjoyed using BCI as an entertainment tool (Brain Painting) and indicated that they would use both prototypes in daily life provided improvements. Major issues hampering daily life use of BCI are the process of setting up the system, the electrode cap and speed of communication. Encouraging progress is made with regards to all these aspects: semi-dry and dry electrodes are being developed [52, 54]. Communication speed is improved by evaluating and altering stimulus presentation (e.g., [12,17,28,40,41,46,48,49]) and improving the classification procedure (e.g., [5, 6, 20, 21] for a review, see [26]). Set-up is improved by introducing standard platforms (e.g., provided by TOBI), wireless data transfer, and facilitation and automation of set-up components such as calibration and adaptation to the individual users. This progress will pave the way for BCI home-use.

8.3 BCI for Supporting or Replacing Existing AT Solutions

Bruno: A case scenario.
Three years ago, Bruno was diagnosed with Amyotrophic lateral sclerosis, a neurodegenerative disease, leading to severe impairment up to entire loss of muscular control. In the first stage of his disease, Bruno was able to control a computer using residual hand movement. When control over his hand vanished, he learned to use a mouth mini joystick and again as this solution failed, he switched to an eye-tracker. Unfortunately any time soon he will also find it difficult and fatiguing to control his eye muscles. BCI may then be the remaining option due to its independence from muscular control.

This scenario demonstrates how individual needs and requirements can change along with the individual progress of a disease. In his case a non-visual BCI system, auditory [10, 16, 42, 43] or tactile [7], could be a solution.

However, using a BCI at the end of a cascade of AT devices is not the only option. We may consider introducing a BCI driven solution for control in an earlier phase of the disease. This would avoid that the user has to undergo a cognitively and attentionally demanding training when mental and physical capacities may already be heavily hampered. Naturally, an end-user would always stick to the best working, least effortful and most reliable device at hand. BCIs may not fulfill these criteria in all situations; however, as explained above, it is also possible to combine BCI technology with AT technology in a hybrid approach [32, 37]. Thus, in the early stage of disease, BCIs could simply assist and be used for training of later control. And when losing motor control, a BCI could possibly take over control functions adapted to the individual progress of a disease.

Such iterative processes of adapting toward BCI control might already now be a reasonable option due to promising results in BCI development.

8.3.1 *Benefit in Different Fields*

In general, severely disabled people may benefit from the use of assistive technologies for communication, access of ICT and environmental interaction. These areas, however, do not cover all aspects of human functioning which can be compensated by an AT intervention. However, we will limit our brief discussion to these three aspects that can be seen as crucial in human life and will show how BCI could support AT access by adding a new independent input channel, either with a hybrid approach, or by replacing other traditional AT devices.

8.3.1.1 Communication

Severe motor disability is often associated with communication difficulties. For this reason, applications of Alternative and Augmentative Communication (AAC) techniques are often considered as most important for this user group [44]. We can divide communication needs into three levels of rising complexity:

Level 1: Call for attention, i.e., the possibility to call for help or interaction.
For some users with locked-in syndrome or those with progressive muscular diseases, BCI could be an important possibility to preserve these basic communication functions, particularly in case difficulties are encountered with standard AT devices (i.e., fatigue or even if muscular control is completely lost). Regarding level 1, a P300-Speller may be considered that enables selection of different functions of signaling and calling for someone or for expressing basic needs (for a review, see [18]), e.g., by integrating buttons for "help" in the P300-communication matrix (visual, auditory or tactile). Another option would be the use of SMR-BCI for one-class-choices, e.g., for the single function "emergency call." Currently however, both options cannot yet be realized with a BCI due to the lack of practicability and reliability of state-of-the-art BCI.

Level 2: Written communication.
Different spelling applications have been designed for BCI based written communication. Severely disabled users were able to write with a P300-Speller [36, 57] or with a BCI based on slow cortical potentials [4, 35].

Level 3: Real time conversation.
Using BCI for rapid communication is inherently limited by several factors: (1) Communication is limited to selecting letters consecutively, (2) the speed of letter selection is limited to the latency of the classified brain potential, and (3) accuracy of letter selection is often dependent on the number of brain responses used for averaging. For example, in the P300-BCI usually several ERP responses are averaged, leading to a maximum of about ten selections per minute. However, using symbols as well as predefined messages and/or predictive text-entry function instead of typing the word letter by letter may increase BCI efficiency.

8.3.1.2 Access to ICT

Besides the communication needs, another important necessity for everyone, disabled and non-disabled persons alike, is to have full access to ICT. This is fundamental for being integrated in modern society: learning, working, leisure, hobbies, and generally participating in social activities are linked to ICT. The AT market offers a lot of specific hardware and software solutions, exclusively developed for specific needs. The simplest solutions are keyboards, mice and trackballs with a specific design (bigger or smaller keys, simplified layouts, fingers shields, specific color schemes etc.). If it is impossible for users to efficiently use their upper limbs, solutions are designed to be used with alternative movements like head trackers, mouth or chin mini-joysticks, eye trackers etc. In other cases, the best solution may have to rely on a residual movement for an on/off activation of a sensor: the ICT device in these cases has to be equipped with a specific scan-based access user interface. BCIs can enable such access to ICT for the user with severe motor impairment [3, 33, 57].

8.3.1.3 Interacting with the Environment

Interaction with our daily environment is a key aspect for reaching real independence in some areas of our life, and this is particularly important for people with severe motor disabilities. From very simple devices up to full featured smart homes the key is to find a simple, yet efficient and reliable human–environment interface.

BCIs may possibly constitute such a key instrument: For example many researchers investigated BCI based wheelchair control (e.g., [11, 25, 38]), environmental control (e.g., [1]) and for entertainment applications (e.g., [34]). Regarding entertainment, BCIs have been used for controlling gaming applications [29, 47], which can be important for severely disabled users to interact with other people by playing games together.

8.4 Conclusion

BCI technologies have the potential to provide new AT-solutions for severely disabled end users. In the cycle of BCI development, the user-centred approach plays an important role since it adopts early collaboration between BCI researchers, AT practitioners and potential end-users. User needs and requirements are assessed at the very beginning and the usability of the AT is evaluated by end-users. In the EU-project TOBI, BCI prototypes were developed on the basis of a user-centered approach. The testing and evaluation with end-users in the field provides encouraging results such that end-users considered BCIs as an option for AT control, provided that the set-up, including the EEG cap, and speed will be considerably improved. Only by developing BCI solutions on the basis of UCD principles,

BCI technologies can be designed that match the users' needs, requirements and expectations, paving the way for that breakthrough that many people are hoping for.

References

1. Aloise, F., Schettini, F., Aricò, P., Leotta, F., Salinari, S., Mattia, D., Babiloni, F., et al.: P300-based brain–computer interface for environmental control: an asynchronous approach. J. Neural Eng. **8**(2), 025025 (2011). DOI 10.1088/1741-2560/8/2/025025
2. Association for the Advancement of Assistive Technology in Europe. Strategic Development, 2009. Public document on www.aaate.net. Retreived 9/07/2012
3. Bensch, M., Karim, A.A., Mellinger, J., Hinterberger, T., Tangermann, M., Bogdan, M., Rosenstiel, W., Birbaumer, N.: Nessi: An EEG-controlled web browser for severely paralyzed patients. Comput. Intell. Neurosci. 71863 (2007).
4. Birbaumer, N., Ghanayim, N., Hinterberger, T., Iversen, I., Kotchoubey, B., Kübler, A., Perelmouter, J., et al.: A spelling device for the paralysed. Nature **398**(6725), 297–298 (1999). DOI 10.1038/18581
5. Blankertz, B., Sannelli, C., Halder, S., Hammer, E.M., Kübler, A., Müller, K-R., Curio, G., et al.: Neurophysiological predictor of SMR-based BCI performance. NeuroImage **51**(4), 1303–1309 (2010). DOI 10.1016/j.neuroimage.2010.03.022
6. Blankertz, B., Lemm, S., Treder, M., Haufe, S., Müller K-R.: Single-trial analysis and classification of ERP components – A tutorial. NeuroImage **56**(2), 814–825 (2011). DOI 10.1016/j.neuroimage.2010.06.048
7. Brouwer A-M., van Erp, J.B.F.: A tactile P300 brain–computer interface. Front. Neurosc. **4**, 19 (2010). DOI 10.3389/fnins.2010.00019
8. Brunner, P., Joshi, S., Briskin, S., Wolpaw, J.R., Bischof, H., Schalk, G.: Does the "P300" speller depend on eye gaze? J. Neural Eng. **7**(5), 056013 (2010). DOI 10.1088/1741-2560/7/5/056013
9. Demers, L., Weiss-Lambrou, R., Ska, Quebec, B.: User Evaluation of Satisfaction with Assistive Technology (QUEST version 2.0) an outcome measure for assistive technology devices (2002). Institut for Matching Person and Technology. Webster, NY
10. Furdea, A., Halder, S., Krusienski, D.J., Bross, D., Nijboer, F., Birbaumer, N., Kübler, A.: An auditory oddball (P300) spelling system for brain–computer interfaces. Psychophysiology **46**(3), 617–625 (2009). DOI 10.1111/j.1469-8986.2008.00783.x
11. Galán, F., Nuttin, M., Lew, E., Ferrez, P.W., Vanacker, G., Philips, J., Millán, JdR.: A brain-actuated wheelchair: asynchronous and non-invasive Brain–computer interfaces for continuous control of robots. Clin. Neurophysiol. **119**(9), 2159–2169 (2008). DOI 10.1016/j.clinph.2008.06.001
12. Guger, C., Daban, S., Sellers, E., Holzner, C., Krausz, G., Carabalona, R., Gram, F., et al.: How many people are able to control a P300-based brain–computer interface (BCI)? Neurosci. Lett. **462**(1), 94–98 (2009). DOI 10.1016/j.neulet.2009.06.045
13. Hart, S.G., Staveland, L.E.: Development of NASA-TLX (Task Load Index): Results of experimental and theoretical research. In: Hancock, P.A., Meshkati, N. (eds.), Human mental workload, pp. 139–183. North-Holland, Amsterdam (1988)
14. Henderson, S., Skelton, H., Rosenbaum, P.: Assistive devices for children with functional impairments: impact on child and caregiver function. Dev. Med. Child Neurol. **50**(2), 89–98 (2008).
15. Hoogerwerf, E.J., Mongardi, S., Staiger-Sälzer, P., Zickler, C.: BCI research and user involvement: the role of independent AT centres in the TOBI project. In: Assistive Technology – technology transfer: Proceedings of the AAATE Workshop 2010, Sheffield, UK, 4-5/10/2010 (2010)

16. Höhne, J., Schreuder, M., Blankertz, B., Tangermann, M.: A novel 9-class auditory ERP paradigm driving a predictive text entry system. Front. Neurosci. **5**, 99 (2011). DOI 10.3389/fnins.2011.00099
17. Jin, J., Horki, P., Brunner, C., Wang, X., Neuper, C., Pfurtscheller, G.: A new P300 stimulus presentation pattern for EEG-based spelling systems. Biomedizinische Technik. Biomed. Eng. **55**(4), 203–210 (2010). DOI 10.1515/BMT.2010.029
18. Kleih, S.C., Kaufmann, T., Zickler, C., Halder, S., Leotta, F., Cincotti, F., Aloise, F., Riccio, A., Herbert, C., Mattia, D., Kübler, A.: Out of the frying pan into the fire-the P300-based BCI faces real-world challenges. Prog. Brain Res. **194**, 27–46 (2011).
19. Krausz, G., Ortner, R., Opisso, E.: Accuracy of a brain computer interface (P300 Spelling Device) used by people with motor impairments. Stud. Health Technol. Inform. **167**, 182–186 (2011).
20. Krusienski, D.J., Sellers, E.W., McFarland, D.J., Vaughan, T.M., Wolpaw, J.R.: Toward enhanced P300 speller performance. J. Neurosci. Methods **167**(1), 15–21 (2008). DOI 10.1016/j.jneumeth.2007.07.017
21. Krusienski, D.J., Sellers, E.W., Cabestaing, F., Bayoudh, S., McFarland, D.J., Vaughan, T.M., Wolpaw, J.R.: A comparison of classification techniques for the P300 Speller. J. Neural Eng. **3**(4), 299–305 (2006)
22. Kübler, A., Nijboer, F., Mellinger, J., Vaughan, T.M., Pawelzik, H., Schalk, G., McFarland, D.J., Birbaumer, N., Wolpaw, J.R.: Patients with ALS can use sensorimotor rhythms to operate a brain–computer interface. Neurology **64**(10), 1775–1777 (2005). DOI 10.1212/01.WNL.0000158616.43002.6D
23. Kübler, A., Furdea, A., Halder, S., Hösle, A.: Brain Painting BCI meets art. In: Müller-Putz, G.R., Brunner, C., Leeb, R., furtscheller, G., Neuper C. (eds.) 4th International Brain–Computer Interface Workshop and Training Course, pp. 361–366. Graz University of Technology, Austria: Verlag der Technischen Universität Graz (2008)
24. Lane, J.P.: Delivering on the 'D' in R&D: Recommendations for Increasing Transfer Outcomes from Development Projects. SEAT Center and ATIA (2010)
25. Leeb, R., Friedman, D., Müller-Putz, G.R., Scherer, R., Slater, M., Pfurtscheller, G.: Self-paced (asynchronous) BCI control of a wheelchair in virtual environments: a case study with a tetraplegic. Comput. Intell. Neurosci. 79642 (2007). DOI 10.1155/2007/79642
26. Lotte, F., Congedo, M., Lécuyer, A., Lamarche, F., Arnaldi, B.: A review of classification algorithms for EEG-based brain–computer interfaces. J. Neural Eng. **4**(2), R1–R13 (2007). DOI 10.1088/1741-2560/4/2/R01
27. Mak, J.N., Arbel, Y., Minett, J.W., McCane, L.M., Yuksel, B., Ryan, D., Thompson, D., et al.: Optimizing the P300-based brain–computer interface: current status, limitations and future directions. J. Neural Eng. **8**(2), 025003 (2011). DOI 10.1088/1741-2560/8/2/025003
28. McFarland, D.J., Sarnacki, W.A., Townsend, G., Vaughan, T., Wolpaw, J.R.: The P300-based brain–computer interface (BCI): effects of stimulus rate. Clin. Neurophysiol. **122**(4), 731–737 (2011).
29. Millán, JdR.: Adaptive brain interfaces. Commun. ACM. **46**(3), 74–80 (2003)
30. Millán, JdR., Ferrez, P.W., Galán, F., Lew, E., Chavarriaga, R.: Non-invasive brain-machine interaction. Intern. J. Pattern Recognit. Artif. Intell. **22**(5), 959–972 (2008)
31. Millán, JdR., Mouriño, J.: Asynchronous BCI and local neural classifiers: An overview of the adaptive brain interface project. IEEE Trans. Neural Syst. Rehabil. Eng. **11**(2), 159–161 (2003)
32. Millán, J.D., Rupp, R., Müller-Putz, G.R., Murray-Smith, R., Giugliemma, C., Tangermann, M., Vidaurre, C., Cincotti, F., Kübler, A., Leeb, R., Neuper, C., Müller, K.R., Mattia, D.: Combining brain–computer interfaces and assistive technologies: State-of-the-art and challenges. Front. Neurosci. **7**, 4 (2010)
33. Mugler, E.M., Ruf, C.A., Halder, S., Bensch, M., Kübler, A.: Design and implementation of a P300-based brain–computer interface for controlling an internet browser. IEEE Trans. Neural Syst. Rehabil. Eng. 18(6), 599–609 (2010). DOI 10.1109/TNSRE.2010.2068059
34. Münßinger, J.I., Halder, S., Kleih, S., Furdea, A., Raco, V., Hösle, A., Kübler, A.: Brain painting: First evaluation of a new brain–computer interface application with ALS-patients and healthy volunteers. Front. Neurosci. **4**, 182 (2010). DOI 10.3389/fnins.2010.00182

35. Neumann, N., Kübler, A.: Training locked-in patients: a challenge for the use of brain–computer interfaces. IEEE Trans. Neural Syst. Rehabil. Eng. **11**(2), 169–172 (2003). DOI 10.1109/TNSRE.2003.814431
36. Nijboer, F., Sellers, E.W., Mellinger, J., Jordan, M.A., Matuz, T., Furdea, A., Halder, S., Mochty, U., Krusienski, D.J., Vaughan, T.M., Wolpaw, J.R., Birbaumer, N., Kübler, A.: A P300-based brain–computer interface for people with amyotrophic lateral sclerosis. Clin. Neurophysiol. **119**(8), 1909–1916 (2008)
37. Pfurtscheller, G., Allison, B.Z., Brunner, C., Bauernfeind, G., Solis-Escalante, T., Scherer, R., Zander, T.O., Müller-Putz, G., Neuper, C., Birbaumer, N.: The hybrid BCI. Front. Neurosci. **21**(4), 30 (2010)
38. Pires, G., Castelo-Branco, M., Nunes, U.: Visual P300-based BCI to steer a wheelchair: a Bayesian approach. Conference Proceedings: Annual International Conference of the IEEE Engineering in Medicine and Biology Society. IEEE Engineering in Medicine and Biology Society. Conference, 2008, 658–661 (2008). DOI 10.1109/IEMBS.2008.4649238
39. Riccio, A., Leotta, F., Bianchi, L., Aloise, F., Zickler, C., Hoogerwerf, E.J., et al.: Workload measurement in a communication application operated through a P300-based brain–computer interface. J. Neural Eng. **8**(2), 025028 (2011)
40. Salvaris, M., Sepulveda, F.: Visual modifications on the P300 speller BCI paradigm. J. Neural Eng. **6**(4), 046011 (2009). DOI 10.1088/1741-2560/6/4/046011
41. Sellers, E.W., Kübler, A., Donchin, E.: Brain–computer interface research at the University of South Florida Cognitive Psychophysiology Laboratory: the P300 Speller. IEEE Trans. Neural Syst. Rehabil. Eng. **14**(2), 221–224 (2006). DOI 10.1109/TNSRE.2006.875580
42. Schreuder, M., Blankertz, B., Tangermann, M.: A new auditory multi-class brain–computer interface paradigm: spatial hearing as an informative cue. PloS One 5(4), e9813 (2010). DOI 10.1371/journal.pone.0009813
43. Schreuder, M., Rost, T., Tangermann, M.: Listen, you are writing! Speeding up online spelling with a dynamic auditory BCI. Front. Neurosci. **5**, 112 (2011). DOI 10.3389/fnins.2011.00112
44. Smith, E., Delargy, M.: Clinical review: Locked-in syndrome. Br. Med. J. **330**(7488), 406–409 (2005)
45. Standish Group. CHAOS Report. Boston: The Standish Group International Inc. (2009)
46. Takano, K., Komatsu, T., Hata, N., Nakajima, Y., Kansaku, K.: Visual stimuli for the P300 brain–computer interface: A comparison of white/gray and green/blue flicker matrices. Clin. Neurophysiol. **120**(8), 1562–1566 (2009). DOI 10.1016/j.clinph.2009.06.002
47. Tangermann, M.W., Krauledat, M., Grzeska, K., Sagebaum, M., Vidaurre, C., Blankertz, B., Vidaurre, C., Müller, K.R.: Playing pinball with non-invasive BCI. In Proc. Advances in Neural Information Processing Systems 21 (NIPS); Vancouver, Canada (2008)
48. Tangermann, M., Schreuder, M., Dähne, S., Höhne, J., Regler, S., Ramsey, A., Queck, M., Williamson, J., Murray-Smith, R.: Optimized stimulation events for a visual ERP BCI. Int. J. Bioelectromagn **13**(3), 119–120 (2011).
49. Townsend, G., LaPallo, B.K., Boulay, C.B., Krusienski, D.J., Frye, G.E., Hauser, C.K., Schwartz, N.E., Vaughan, T.M., Wolpaw, J.R., Sellers, E.W.: A novel P300-based brain–computer interface stimulus presentation paradigm: moving beyond rows and columns. Clin. Neurophysiol. **121**(7), 1109–1120 (2010). DOI 10.1016/j.clinph.2010.01.030
50. Treder, M.S., Blankertz, B.: (C)overt attention and visual speller design in an ERP-based brain–computer interface. Behav. Brain Funct. **6**, 28 (2010). DOI 10.1186/1744-9081-6-28
51. Vidaurre, C., Sannelli, C., Müller, K.-R., Blankertz, B.: Machine-learning-based coadaptive calibration for brain–computer interfaces. Neural Comput. **23**(3), 791–816 (2010). DOI 10.1162/NECO_a_00089
52. Volosyak, I., Valbuena, D., Malechka, T., Peuscher, J., Gräser, A.: Brain–computer interface using water-based electrodes. J. Neural Eng. **7**(6), 066007 (2010). DOI 10.1088/1741-2560/7/6/066007
53. World Health Organization. International Classification of Functioning, Disability and Health: ICF. Geneva: World Health Organization (2001)

54. Zander, T.O., Lehne, M., Ihme, K., Jatzev, S., Correia, J., Kothe, C., Picht, B., Nijboer, F.: A dry EEG-system for scientific research and brain–computer interfaces. Front. Neurosci. **5**, 53 (2011). DOI 10.3389/fnins.2011.00053
55. Zickler, C., Di Donna, V., Kaiser, V., Al-Khodairy, A., Kleih, S., Kübler, A., Malavasi, M., Mattia, D., Mongardi, S., Neuper, C., Rohm, M., Rupp, R., Staiger-Sälzer, P., Hoogerwerf, E.J.: Brain Computer Interaction Applications for People with Disabilities: Defining User Needs and User Requirements. In: Emiliani, P.L., Burzagli, L., Como, A., Gabbanini, F., Salimen, A.L. (eds.) Assistive Technology from Adapted Equipment to Inclusive Environments. AAATE 2009, Assistive Technology Research Series, vol. 25, pp. 185–189. IOS, Amsterdam (2009)
56. Zickler, C., Hösle, A., Franz, D., Staiger-Sälzer, P., Busch, G., Kübler.: A Brain Painting: Evaluation of performance and satisfaction by end users with severe disabilities. Poster presented at the TOBI workshop II, Rome, Italy, pp. 2–3 (2010)
57. Zickler, C., Riccio, A., Leotta, F., Hillian-Tress, S., Halder, S., Holz, E.M., Staiger-Sälzer, P., Hoogerwerf, E.J., Desideri, L., Mattia, D., Kübler, A.: A brain–computer interface as input channel for a standard assistive technology software. Clin. EEG Neurosci. **42**(4), 236–244 (2011).

Chapter 9
Designing Future BCIs: Beyond the Bit Rate

Melissa Quek, Johannes Höhne, Roderick Murray-Smith, and Michael Tangermann

9.1 Introduction

The scope of this chapter is limited to applications where a Brain–Computer Interface (BCI) is used as an explicit interaction technique. In other words, we refer here to BCI as input which is voluntarily controlled by the user, rather than as an implicit interaction as in for mental or cognitive state monitoring. Designing applications using BCI as an explicit input technique for users with severe disability depends on understanding the control signals and how users can interact with systems using these controls. Although designing for able-bodied users has a different set of challenges, the BCI has to "add value" in both cases. Over the past 20 years of BCI research and design, the basic control functions have been realized by the collaboration of engineers, psychologists, machine learners and end users. These basic functions provide us with the freedom to design future BCI applications which are reliable in long-term use, easy to learn and set up, aesthetically pleasing, and have the potential to improve the lives of their users.

BCI can be thought of as an input technology which takes properties of other emerging input technologies to the extreme. The term "extreme" is used because the BCI interaction is much slower, noisier and more error-prone compared to other input devices, and lacks *proprioceptive feedback*. Because of these unusual characteristics, a theoretical framework which successfully analyses current BCI systems provides a springboard for developing and refining theories and practices within Human Computer Interaction (HCI). Although still important, research in

M. Quek (✉) · R. Murray-Smith
School of Computing Science, University of Glasgow, Scotland
e-mail: melissa@dcs.gla.ac.uk; rod@dcs.gla.ac.uk

J. Höhne · M. Tangermann
BBCI Lab, Berlin Institute of Technology, Germany
e-mail: j.hoehne@tu-berlin.de; michael.tangermann@tu-berlin.de

B. Z. Allison et al. (eds.), *Towards Practical Brain-Computer Interfaces*, Biological and Medical Physics, Biomedical Engineering, DOI 10.1007/978-3-642-29746-5_9,
© Springer-Verlag Berlin Heidelberg 2012

BCI is moving beyond improving the bit rate of selection tasks to building whole systems that are enjoyable to use. This involves improving the usability and user experience of BCI applications, and requires taking into account the whole system rather than single isolated intention detection events. The following sections provide an overview of some factors we consider important for such a broader view on the design of future BCI applications.

Section 9.2 introduces the specific characteristics and problems of BCI in comparison with other HCI application fields. Section 9.3 emphasizes the focus on neuroergonomic principles in addition to usability principles especially for paradigms using Event-Related Potentials (ERP). Section 9.4 looks at how control can be shared between the user and computer. Section 9.5 looks at the structure of the application. Section 9.6 looks at how to involve end users, given the specific requirements and constraints of BCI. Section 9.7 presents analytic tools for investigating interaction designs.

9.2 Control Characteristics of BCI

In a BCI, brain signals produced by the user are directly interpreted by the computer. In contrast to most other HCI control paradigms, one major aspect of all BCI control paradigms is that there is no proprioceptive feedback. The user does not perceive the aspects of his/her brain signals which are measured by the EEG, but only perceives the feedback from the BCI. Thus, the user does not know the exact input to which the computer is responding. This uncertainty may create a series of problems in interaction design which are specific to BCIs. This is especially pronounced in the case of BCIs with low accuracy, since the user cannot know the reason for the malfunctioning interface: the input, the computer's interpretation of the input, or a combination of both.

A second difficulty associated with BCI is that the error rate is high in comparison with other input technologies. Although a typical user improves over time, a selection accuracy of 70 % has been considered acceptable for BCI use, which is rather low compared to traditional input technologies. This is not unique to BCI—other input technologies based on machine learning, for example, gestural interaction [72], also suffer from this, although there are few people who depend on such systems for interaction—they tend to be "luxury" items which allow amusing or rapid access to a small number of features, as a complement to higher throughput, more reliable input mechanisms such as keyboards. In the same way, the HCI community in these areas have typically focussed on improving the performance of these input technologies, without considering whether and how systems can be built to take into account the property of high error rate. With input technologies that have been widely studied and established (e.g. mouse, keyboard, touch input), selection error is usually very low. There is thus a huge scope for research into how users can interact with error-prone systems.

Further difficulties in BCI are associated with the amount of *measurement noise* and *uncertainty* in the system. There are several sources of uncertainty: together with the user's internal state (attention etc.), the EEG signals change over time (non-stationarity), they are furthermore prone to muscle artifacts, and the amount of class discriminative information which is extracted to drive the interface varies within and between users (see also Sect. 9.5).

9.2.1 Issues Specific to BCI Paradigms

Since brain signals acquired with EEG are very weak, noisy and non-stationary, there is an entire research area in signal processing [40, 41, 68, 69] and classification [2, 8, 30, 44] aiming to derive a stable control signal from the complex brain activity that enables a reliable BCI control. Various types of EEG signals and paradigms have been successfully used to drive a BCI. The two most important of them will be shortly discussed in the following, as they exhibit rather different control characteristics:

9.2.1.1 Self-Driven Paradigms

It is possible to extract signals that correlate with mental states that are voluntarily produced by the user (self-driven paradigms). The most common of these is the imagination of motor execution, where the user imagines repeated movements of a body part, for example, their hands, feet or tongue. Other mental states include imaging a cube rotating, or performing complex calculations mentally. It is then possible to train a classifier that separates the resulting EEG features. All applications based on self-driven paradigms have to deal with a very limited number of control signals: although multi-class (e.g. [48]) and multi-dimensional (e.g. [19]) paradigms have also been demonstrated, most successful paradigms are based on two mental states (i.e. imagination of left hand vs right hand). Important issues include that the paradigm is a learned skill which may take some time to acquire, that there is a delay between the imagination of movement and its detection due to the classification time window and switching between mental states, and that there must be appropriate mapping from the mental states to application controls. For example, mapping of right hand imagery to a "turn right" command is more intuitive than mapping visual rotation of a cube to the same command. Current techniques also struggle to distinguish between control states, where the user wishes to issue a command, and an idle state where the user does not wish to issue any commands. This parallels the segmentation problem in gesture recognition [60] and is currently and active area of research (e.g. [18]).

9.2.1.2 Evoked Potentials

Control signals with higher dimensionality and higher communication speed can be obtained by paradigms that present different stimuli and evaluate the corresponding neural response (Evoked Potentials, EP). Thereby, one can make use of neural correlates of attention, that are mostly found in two types of signals: steady-state evoked potentials (SSEP) and event related potentials (ERPs). When applying BCI paradigms based on ERPs or SSEPs, the user is constantly perceiving numerous stimuli which are presented either visually [17,67], via the auditory channel [33,56] or on the somatosensory pathway [11,12]. While attending to one stimulus only and ignoring all others, the target stimulus elicits an EEG response that can be separated from the EEG response of non-target stimuli.

From the user's point of view, paradigms using EPs differ strongly from those that are self-driven. Paradigms based on EPs are generally faced with the danger of overloading the senses of the user and forcing a fast sequence of stimulations and control events upon him. The constant perception of (visual, auditory, or somatosensory) stimuli may cause the user to become overwhelmed or befuddled. Using such interfaces might be uncomfortable, and the stimuli might not be aesthetically pleasing. To address the problem of unpleasant stimuli on the one hand and to increase BCI performance on the other hand, the stimulation principles should follow neuroergonomic principles (see Sect. 9.3).

9.2.2 Approaches to Overcoming the Limitations of BCI

Despite the problems of error and noise in the system, BCI applications have successfully been developed and used by able-bodied [73] and disabled [37] users. Much work has been put into developing systems for text entry input, for example, which take advantage of language models and optimal search trees. Section 9.4 and Chap. 6 of this book discuss the role of shared control in overcoming these limitations.

Users can be "deceived" into thinking that they have more control over a system than they actually do, as people are optimistic in their perception of how much control they have over a system. The phenomenon is called the *Illusion of control* [38]. This fact has been used in entertainment applications which provide some novel control but are not very accurate. The first commercially available "mind reading" devices like Mindflex by Mattel are discussed in [74]. The very successful Nintendo Wii controller often has very limited control based on accelerometer inputs, but users, especially new users, may not realise this, as the interaction makes sense within a particular game context. Often, a richer set of responses is generated by the user than is necessary for input detection, but which gives the illusion of a richer immersion in the game. For example, in the Nintendo WiiSports tennis game new players tend to interact with the system using flamboyant gestures and large swinging hand motions as they mistakenly understand the system to be requiring the same movements as in the physical sports, whereas much smaller movements can have the same effect. Such features can bring improved initial engagement and immersion in a task, which might be of benefit in training, but there is a trade-off with the effort required to use a system in the longer-term.

Recent approaches also combine BCI with other biosignals (such as EMG), aiming for a HCI with increased reliability and stability. Those approaches are called hybrid BCI [42]. Other solutions attempt to identify mental states that can be used to improve reliability of the signal or allow for error correction (e.g. error potentials [7]).

9.3 BCI: From Usability Research to Neuroergonomic Optimization

For optimizing task performance, Nielsen [45] proposed to focus on the usability components of *learnability* (how quickly novice users can learn to use a system), *efficiency* (how quickly expert users can perform tasks), *memorability* (how well users can gain control of an interface after not having used it for a period of time), *errors* and *satisfaction*. Even though Nielsen's concept has been criticised in recent years for lack of enhancing the overall user experience, these five components are widely used within the HCI design (especially in web design) community [58].

For the special case of a BCI-controlled application that is based on event related potentials (ERP), the optimization of overall task performance via the components *efficiency*, *errors* and *satisfaction* lead to a rather domain-specific target: the stimuli. During the use of an application (e.g. a text entry system) these stimuli are presented continuously and in quick succession. The stimuli are utilized to evoke brain activity that is informative with respect to the user's intention. Stimulus characteristics are thus at the core of an ERP application and their influence on the measurable (via EEG) evoked neuronal activity becomes subject of a *neuroergonomic* stimulus design approach.

The search space for neuroergonomically optimized stimuli, however, is extremely large: stimulus parameters can vary along many dimensions, with duration, intensity, stimulus timing and sequence aspects being only the most general ones. By selecting a specific stimulus modality (e.g. visual, auditory, haptic), an enormous extra amount of modality-specific parameters are to be decided upon by the designer. To add to the misery, some effects are rather subject-specific.

9.3.1 Existing Literature on Determinants for ERP

Generalized models that describe the mechanism for specific stimulus parameters in detail (or even their interplay) are not known to the authors. When it comes to the influence of cognitive and biological determinants of specific ERP components, the situation seems to be better. The neuropsychology of stimulus modality, intensity, sequence effects, etc. but also of gender, handedness, age etc. has well been studied [49, 51, 52, 54]. These results, however, were derived under well-controlled lab-conditions, used larger stimulus onset asynchrony (SOA) values and two or three stimulus classes only. Apart from a very few recent examples that will be introduced later, they report basic research from neurophysiology but do not take specific requirements of BCI into consideration. They focus, for example, on single aspects of an ERP component (e.g. the latency of P300) instead of the class discriminative information between attended and unattended stimuli (e.g via signal-to-noise ratio (SNR), area under ROC curve, classification accuracy etc). The latter is not only important for BCI, but it is even spread over a number of ERP components. Under these circumstances, it is not clear how the reported influence of the studied determinants generalizes to the rapid multi-class setups that are prevalent in BCI. The studies can, however, provide a starting point for more BCI-focussed investigations, where the goal is to obtain high class-discrimination. This goal seems to be extremely important, as it directly affects the *efficiency* of BCI control and the rate and severity of *errors*. Indirectly, it influences the level of *satisfaction*. Neuroergonomic stimulus design can attempt to improve the quality of evoked brain responses with respect to the class-discriminability or signal-to-noise ration (SNR) in a number of obvious ways: brain responses should be different for target and non-target stimuli, overall strong in order to contrast against background EEG activity, low in within-class variance etc. Less obvious aspects are of similar importance for a robust long-term efficiency: sustained brain responses are required that show minimal habituation over time and which are robust with respect to changes in the unavoidable surrounding perceptual influences. In combination, stimuli should be used that robustly result in high classification accuracy per BCI decision or (as a trade-off) in quick class decisions that use only a small number of repetitions.

So far, only a limited number of studies HAVE investigated details of stimulus design in the context of BCI, but the level of improvement that could be gained by optimized stimuli is already promising. The first attempts for stimulus optimization were described by Hill et al. [29], comparing the standard visual flash stimulus (color intensification) with a flip stimulus, where a letter was stimulated by a rotation in its background. They found that the BCI performance for the flip stimulus (virtual rotation) was higher than for the flash stimulus. It is generally possible to modify the type of stimulus individually. As already described by Allison and Pineda [1], Hill et al. [29] and others, the choice of stimulus type strongly affects the BCI performance in visual paradigms. Comparable work for auditory ERP studies were done by Schreuder et al. [55], Halder et al. [27].

To investigate the influence of stimulus characteristics, an offline study was performed with 13 able-bodied users performing a visual ERP paradigm with row-column highlighting in a grid with $6 \times 6 = 36$ entries. The highlighting effect varied in six conditions: (1) brightness enhancement, (2) scaling, (3) rotation, (4) color inversion, (5) masking with a grid, and (6) a combination of effects (1,2,3,5), see Fig. 9.1a. Conditions were presented block-randomized and with a constant stimulus onset asynchrony (SOA) of 225 ms.

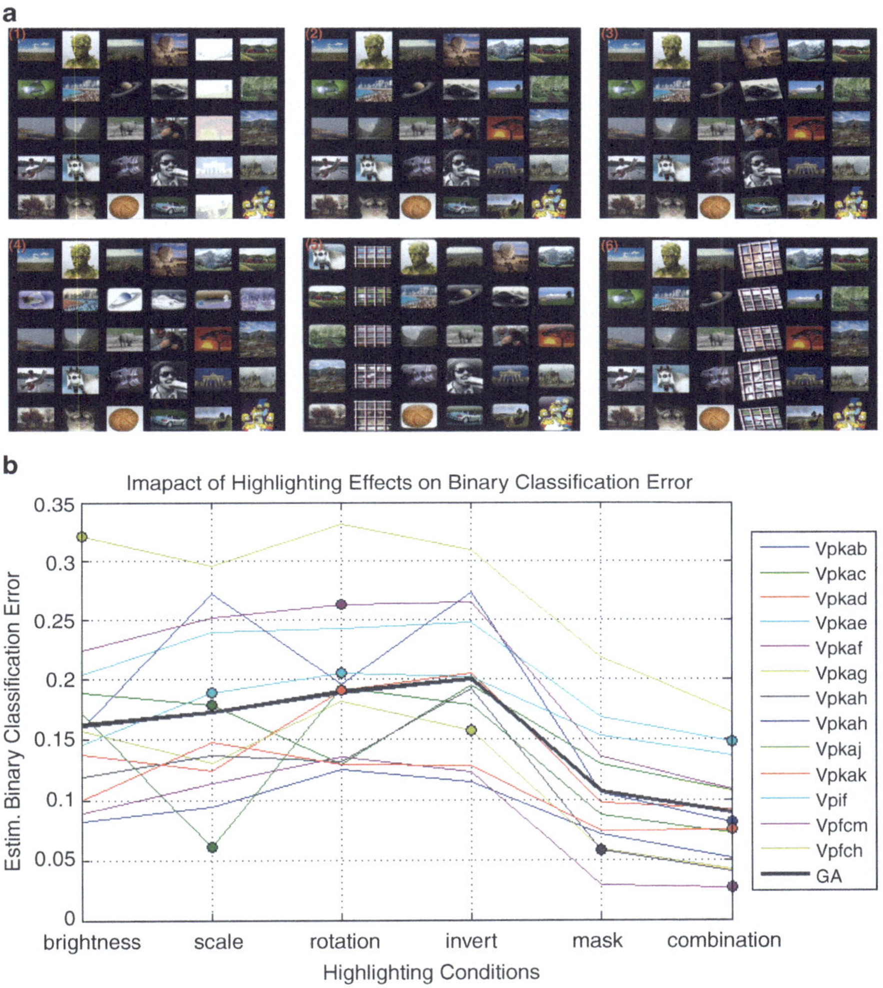

Fig. 9.1 (**a**) Visualization of the six highlighting conditions brightness, scale, rotation, invert, mask and combination (ordered from left to right). (**b**) Estimated binary classification error of 13 users performing a visual ERP paradigm in the six conditions.

Figure 9.1b visualizes the binary classification error (representing the inverse of BCI performance) of calibration data collected. The uniformity of the results is astonishing: the conditions mask and combination perform best for all but one user, indicating that those conditions should generally be used to obtain best performance. As the subjects did not get feedback about the classification error during the experiment (the analysis was done post-hoc only), it is interesting to investigate if they were aware that the different conditions had different effects on their brain response. For this reason, the users were asked which condition they considered best for long-term use (answers are marked with a circle in Fig. 9.1). It can be seen that the individually favored condition often performs poorly in terms of classification error. Thus, if users could choose the type of stimulus themselves, they would choose a highlighting effect with poor performance in ~50 % of the cases.

9.3.1.1 Interaction of Neuroergonomic Optimization with Other Usability Goals

Obviously a faster or more reliable control interface can improve the user's level of satisfaction with an application. While the optimization of stimuli is important for increased control performance, other usability goals must not be lost out of sight. For example, the various rather indirect influences of stimulus parameters on the user's comfort level are unclear and will need evaluation. Important research questions to tackle are:

- Which stimuli lead to optimal learning curves for the discrimination and attention task?
- How can we design good stimuli to be familiar, pleasant and constant?
- Which stimuli show a low obtrusiveness level (are they recognized by a present third party? Do they disturb?)?
- How much should stimuli be allowed to interfere (perceptually and in terms of cognitive processing) with other sources of information?

9.3.1.2 Example: Comparison of User Ratings (Satisfaction)

Coming back to the study introduced in the above example, some aspects of user satisfaction were probed in addition to the analysis of pure classification performance: To collect information about how a stimulus condition was perceived subjectively, the 13 participants were asked after the EEG recording to provide ratings (among others) for how motivating, how clearly perceivable and how non-exhausting a stimulus class was, using a visual-analog-scale (VAS).

As the conditions "mask" and "combination" were found to lead to a great improvement of the binary classification error, and "combination" being the best condition on average, it is worth taking a closer look at the VAS ratings for this condition. From Fig. 9.2, it can be observed that all three VAS ratings are negatively correlated with the binary classification error. Subjects, for example, that rated the "combination" effect more motivating than other subjects are typically able to use the BCI paradigm with lower error. VAS ratings can be obtained for a new BCI user quite easily and without an EEG recording. Although based on offline data analysis only (in contrast to the more relevant online performance), such ratings may serve as powerful predictors for a subject's estimated classification error on the calibration data. Based on the error prediction, the complexity of the BCI application interface could be adapted in advance for the individual user. For poorly performing

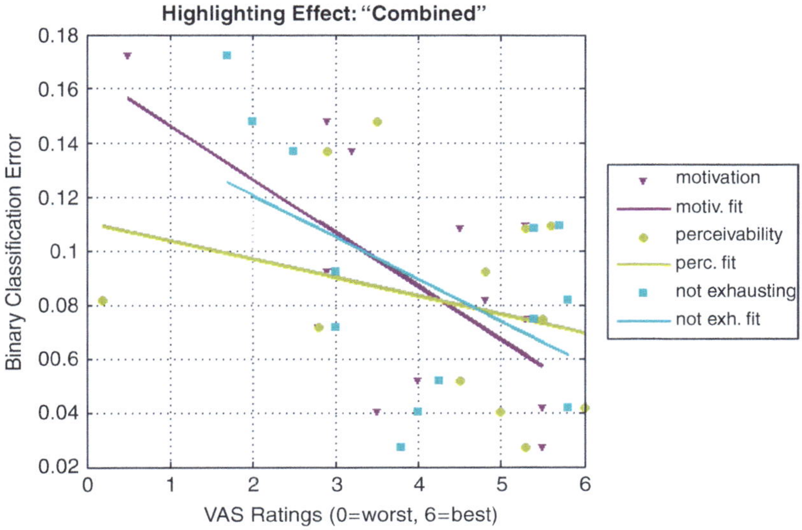

Fig. 9.2 VAS ratings vs. estimated binary classification error. For 13 subjects, the x-axis marks the subjective level of motivation, the level of perceivability of the stimuli and the rating how non-exhausting the subjects considered the "combined" highlighting effect. The y-axis marks the estimated binary classification error of single target vs. non-target EEG epochs

stimulation conditions (e.g. color inversion), however, a similar correlation between VAS and classification error could not be observed. Furthermore it is not clear, to what extent these or similar subjective ratings can be utilized to determine the best stimulation parameter from a set of alternatives for a new user.

It remains an ongoing research goal to investigate how BCIs need to be designed to co-optimize the three aspects: (1) motivation to use the applications, (2) high-level perception of the applications, and (3) neuro-ergonomics for desirable brain responses.

For long-term use, the level of cognitive workload that a user has to invest to control the BCI application should be kept low. Influencing factors might be the clarity or distinctiveness of stimuli, their intensity, whether they grab the user's attention or if they are annoying over time. Intense stimuli might be suitable to elicit strong ERP responses, but load an increased long-term burden of high workload on the user.

9.3.2 Aesthetics, Interaction Metaphors, Usability and Performance

The focus on usability in HCI came from the need to improve task performance for work-related applications. However, in recent years there has been a shift within the HCI community not just to improve usability but also the user experience [65]. This has paralleled the shift of technology usage from being used merely for work purposes to being consumer-focused. Tractinsky et al. [66] showed that a high level of perceived aesthetics of an interface is viewed as also being highly usable, regardless of the actual usability of the system. Norman [46] also proposed that products which elicit a positive emotion with regard to aesthetics are more likely to be used. Since the performance of a BCI system is often less than ideal, improving the aesthetic qualities of such a system is important in order to maximize the users' perception of and motivation to use the system.

However, aesthetics should not just be "eye candy" but should be part of creating a convincing interaction metaphor, with its rapid communication of state, indicating action affordances and helping learnability. In mainstream interfaces, aesthetic feedback is used to provide information about a system to the user. For example, a progress bar showing how much of a file is left to be downloaded can indicate that the system is still doing something. The sound of rubbish being thrown into a bin, accompanied by a "drag and drop" metaphor of a file into the waste basket provides the user with a sense of closure that the file has been deleted. Specific aspects of BCIs which can improve interaction performance (rather than just perceived usability) include presentation of stimuli which are easily interpretable and which motivate, delight and engage the user, and which are rich enough to provide feedback about the effects of the user's brain signals. They should make the system state clear to the user, and give them feedback about what they have selected or are about to select. The feedback could be across multiple channels beyond visual feedback, including audio, vibrotactile, and perhaps even smell.

Users' subjective experiences can differ from the performance of the system. For example, in investigating a single-switch scanning input system, Felzer et al. [20, 21] found that a user was faster with automatic scanning (where the scanner automatically moves on to the next selection) than self-paced scanning (where the user decides when to move on to the next selection), but made more errors. The user reported that the automatic scanning was more frustrating, and that he was surprised that he was faster with the auto scan mode as he felt more in control with self-paced scanning.

In our own experience, we find that some people prefer a paradigm like the rotate-extend (REx, a generalization of the hex-o-spell paradigm [71], Fig. 9.5), where the pace of interaction can be slower, to a discrete binary paradigm which might objectively have higher throughput. The paradigm appears to work particularly well where the 2-class classifier is biased to (tends towards) one class: the biased class is used to rotate an arrow round the centre of a circle, while the other, control, class works to extend the arrow at the correct point of time in order to select a segment. Several possible reasons for some users' preference include that the control method of switching between mental states feels easier or more natural, that there is a possibility of going round the circle again if a target is missed and hence although slower, be more accurate, and that the pace of the interaction feels more comfortable or relaxed. In the context of an application, this may allow users to feel less pressured by the system to make a quick binary decision. The example illustrates again the need for research into how enhancing other control features in addition to the performance of a BCI is important for improving the user experience, and how BCI-specific problems (e.g. the presence of a biased classifier) can be turned into features.

9.4 Shared Control

Shared control involves the co-operative control of some process between a system and a human, where autonomy is smoothly distributed between the system and the human, possibly in a time-varying manner. Shared control can be desired due to different sensing abilities in user and automated systems, speed and safety requirements, as part of a user's learning problems, or simply to reduce the effort required to control a system. In BCI, where the input channel is impoverished, the inputs of the user are too valuable to naively interpret in a mechanistic fashion. For instance, in a robot control system, a simple mapping of a brain-controlled cursor movement to robot movement is far too slow and error-prone to be practical. Similar arguments apply to text entry with a brain-controlled cursor that selects letters from a virtual keyboard. Instead, the user's actions are interpreted partly as direct control signals and partly as indications of the user's higher level intentions. The system attempts to "intelligently" infer what the user wants to do, based on knowledge about the task, and make changes to the response of the system. It uses prior information about likely or sensible behaviours (e.g., smoothness constraints, obstacle avoidance, predicted words, music file features) based on other in-built knowledge and contextual information which the system can draw upon.

The handover of autonomy between the user and the system is determined by how certain the system is about the user's intention. When the user can precisely express detailed intention, the system will follow. As communication breaks down and the system is unable to reliably and quickly infer the user's intention, the system will fall back to prior behaviours. These behaviours need not be static and can depend on knowledge of the environment, such as robot cameras to estimate likely obstacles or language models for text entry. Shared or hybrid controls can also be used to combat fatigue associated with a particular control channel or level of control, allowing the user to "dip in" to direct control as they feel appropriate during interaction [10, 63, and Chap. 6].

In Flemisch et al.'s [22] influential paper, the *H-metaphor* is introduced, which suggests the relationship between a rider and a horse as metaphor for shared control. The horse can navigate obstacles around it without rider attention; the rider can vary the level of control over the horse by tightening or loosening the reins. With tight reins, the horse obeys precisely and immediately, whereas when looser reins are used, the horse acts partially autonomously. The horse treats the tightness of the reins as an indication of the user's certainty about control actions. Although this was developed at NASA to deal with advances in cockpit automation [24], many of the core ideas can be brought over to other areas of interaction design where users can

Fig. 9.3 The combination of direct user input and system knowledge into a single controllable liquid blob. The system can display the user input state and the effect of environmental constraints as shadows to a blob whose form represents the mutual control state from the fusion of user commands and prior and external knowledge (*left*). An example of this approach used in browsing a map of music files (*right*)

regulate the current level of control, via "loosening the reins." This can be seen as an example of hierarchical control where the user can change the level of control they are currently are active at. The challenge for BCI is therefore to enable the user to "take up the reins" when in good control and he feels like it; and progressively (but non-intrusively) "take over" when the person has bad control or when the person wishes to relax.

An open research question is in which ways can we enable the system to "help" the user at a level that is invisible to the user and at which the user still feels they are in control? Note the comparison to the earlier discussion of the *illusion of control*—designers need to make an ethical and practical decision about when and how transparent to be about when the user controls the system and when not. If the system does what the user wants it to do, the user has the feeling that they are in control, or they are the one controlling the system. This has an obvious impact on the learnability of BCI systems (if a user thinks they are successfully controlling it on their own, but actually all the control came from autonomous systems, they cannot improve their performance). It may also have an impact on the user experience, as during the application of BCI systems it might be that users who depend on a BCI for communication especially want the feeling of control in a system.

The liquid metaphors explored as part of the Tools for Brain-Computer Interaction (TOBI) research project show how shared control systems can be visualised http://www.tobi-project.org/, with appropriate representations of uncertainty. Figure 9.3 shows how this could work. The user's input can be visualised as a moving "blob," whose area gives an impression of the associated uncertainty. This can be combined with system knowledge about likely intended actions (e.g. obstacle avoidance), and the result displayed as another blob, which deforms and flows according to both the user input and the system's changing beliefs about intended actions. Displaying the user's raw input as a kind of "shadow blob," makes it easy to see how actions are being sensed and then re-interpreted in a shared control setting. This system is being used in a music player to combine a user input with music knowledge from the automatic systems.

9.5 Creating an Effective Application Structure: A 3-Level Task

A human being can be thought of as a control system with an unreliable input signal, often making mistakes, slips or errors in interacting with machines [15]. As such, computer interfaces should be designed such that they:

- Prevent errors whenever possible.
- Deactivate invalid commands.
- Make errors easy to detect and show users what they have done.
- Allow undoes, reverse, correct errors easily [35].

This is especially true in BCI systems where the inherent error in the input signal is higher than for other input methods. An example for such an application structure in this context can be found in the Rotate-extend (REx) control illustrated in the Hex-o-Spell text entry system [71].

So far, BCI studies have largely focussed on the selection accuracy of single events, while the dynamics of serial selection tasks or within the whole application has rarely been taken into account (text entry BCI applications that use language models are a positive exception). While it seems obvious that application learnability is of paramount importance for usability, and that the learning user has to be taken into account, BCI systems show the necessity for one further, very basic level of learning. In the following, the potential of all three levels of learnability and adaptive behaviour are visited.

9.5.1 Low Level: BCI Control Signal

Learning on the level of the BCI system is often performed by machine learning methods, that are used to establish (initially, by learning regularities from calibration data) and maintain (during the use of applications, by adaptation strategies) a robust transduction of the user's brain signals into control signals for an application. On this low-level, a BCI-system is learning about the user, and how to detect the regular patterns as well as the non-stationarities of his brain activity. Advanced BCI systems may in addition fulfill somewhat more complex, but still relatively basic tasks. They can for example be used to monitor mental background states (e.g. levels of workload or fatigue) and gain information that is not used for control but that indirectly can support the detection of control commands or take influence on e.g. the complexity of the BCI application.

Systems can also keep track of the amount of evidence accumulated over time during the use of a BCI application in order to provide different control alternatives. For example, decisions can be taken earlier by following dynamic stopping criteria (e.g. for ERP paradigms [57]), or the speed-accuracy trade-off can be utilized beneficially in special situations. This latter case immediately leads to the next level of learning and adaptation: the application.

9.5.2 Mid Level: Application

On the application level, the system should adapt to the user's behaviour rather than to his brain signal characteristics (the latter should hopefully be stabilized already by actions taken at the low level). Obvious examples of such mid-level adaptation strategies are text entry systems, that update the statistics of a supporting language model by taking previously written text into account (cp. the discussion in [33]). As the number of control signals per time is extremely limited in BCI, an efficient menu structure, the availability of shortcuts, the avoidance of errors and the simple recovery of errors is important for every kind of BCI application. Although users expect consistency and predictability in a user interface, some research shows that acceptance of adaptive systems depends on the order of presentation [59], while Gajos et al. [23] found that providing accurate hints in an adaptive toolbar was more important than predictability of the toolbar in terms of performance and perceived usability. Applied to extreme conditions like in BCI, a trade-off between consistency and efficiency might be possible and desirable for the user: future applications could learn form past user behaviour and increase efficiency by, for example bootstrapping the menu structure or introducing new shortcuts. This is important as a match between the user requirements and system functions is paramount to user satisfaction [47].

9.5.3 High Level: User

While using a BCI system, the user is learning about the application. This might affect the input characteristics. In a BCI driven by motor imagery, for example, the user continuously learns how to perform best, by e.g. optimizing the timing for motor imagery and relaxation. For paradigms driven by evoked potentials, an example would be that the user is learning about the application structure. This leads to an individually more efficient way to interact with the application. The application can be designed such that it is able to cope with these dynamics.

In general HCI frameworks, user interfaces should aim to optimize the control system (menu hierarchy) as much as possible, when the cost of input is high. Current operating systems still have some way to go in optimizing system control for assistive technologies, as there is often an implicit assumption that input mechanisms are reliable with high throughput, and there is little consideration of the perceptual difficulties of mainstream HCI technologies such as visual menu structures. The problems are exacerbated for BCI users where the cost of input is high.

9.6 Engaging End Users and the Role of Expectation

Designing interaction requires participation or evaluation by target end users since designers and developers often have different ideas and assumptions about the target group with respect to the target users' requirements and mental model about an application or interface [15]. General design guidelines and principles can help in development and design, but even in applications using typical input technologies, the requirements and experience of users are sometimes not intuitive to designers – let alone for novel input technologies such as BCI. In this section, we describe the need to choose appropriate evaluation methods for different user groups, taking into account the impact of user expectation on the methods and tools used (see Chap. 8 of this book for a more in depth discussion on the topic).

BCI user groups can be distinguished according to their physical abilities:

1. Users with no physical disability may be interested in using BCI for gaming or other conditions where physical movement is restricted. An interesting area of research here is in using cognitive workload monitoring to evaluate usability of interfaces [28]. User feedback can be collected through interviews and questionnaires during tasks or after sessions. Evaluation techniques that aim to find out about aspects of user experience are in early stages of development, see also Chaps. 11 and 13 on game evaluation.
2. Users with severe physical disabilities may wish to use BCI as a secondary input, switching from muscle to BCI input on the onset of muscle fatigue. User feedback can sometimes be collected through interviews and questionnaires depending on how easily they can communicate through other means, and it is important to condense the amount of responses required since responses will take far longer to acquire and users will tire easily.
3. Users who are locked-in (having no residual muscle control) or almost locked-in (having very limited residual muscle control), may need to use BCI as a method for communication. User feedback here is restricted to questionnaires, while access to such people is limited. Since HCI evaluation techniques typically require multiple participants, and performing a large testbed of trials is not possible, case studies have been used to elicit feedback and requirements from this group of users [36].

Motivation to use a system is dependent on expectation: able-bodied users are likely to be impatient with the inferior control properties of BCI, while for disabled users, previous experience of mainstream and assistive technologies can have a huge influence on the acceptance of new technologies. For example, someone who was highly competent with mainstream input devices and operating systems before a disability occurred might expect an assistive technology to enable him/her to attain a level of performance or autonomy similar to what they had been used to. They can often become disappointed or disillusioned when they realise that the input device will take some time to learn and be slower and more difficult to control. For BCI, this effect is augmented as the current state-of-the-art is far worse than the usual assistive technologies such as single switch devices.

Increasingly we are finding that motivation is an important factor for users learning to use a BCI paradigm [14, 39], as well as for wanting to use the system. For example, Mönßinger et al. [43] found that disabled users have a higher level of motivation than able-bodied users to use a BCI painting application. As user requirements and expectations can differ between able-bodied and disabled users [62], more work is needed to find out how user expectations can be shaped or primed to increase motivation to use BCI applications. In addition to increasing user performance and accuracy, increasing positive affect could prove to help the user overlook or better tolerate a low bitrate of communication, improving the perceived usability and overall experience of the system.

9.7 Investigating Interaction: Prototyping and Simulation

In designing applications for users, the gap between designers' *understanding* of end users and end users' *actual* requirements, abilities and perceptions means that designers need to engage with users to establish how usable or desirable their design is to actual users. A prototype is an object or system that simulates or represents some limited aspect of a future system in order to obtain feedback from the intended users. Prototypes can be used to investigate the role a system will play in users' lives, how the functionality should work, and what it should look and feel like [34]. Different tools and techniques can be used depending on the what the designer wishes to find out [61]. Although BCI is currently expensive, and time consuming to set up and use for evaluation, methods for engaging users prior to real BCI development and testing has been under utilized. These can reduce the time-costs of engaging with end users, and allow researchers and designers with limited access to BCIs and end user groups to participate. In this section, we explain the value of using prototyping and simulation to explore BCI interaction even without an EEG cap, as the known properties of control can be simulated or animated.

9.7.1 Low Fidelity Prototyping to Expose User Requirements

Low fidelity prototypes are those which look "cheap and cheerful" and whose development do not take much time. They can be used to evaluate initial designs as they have the advantage of reducing the cost of development, and have been shown to provide virtually the same feedback as usability evaluations with high fidelity prototypes [70]. A challenge for developing prototypes for BCI is that demonstrating the control characteristics are an important part of the interaction,

Fig. 9.4 Screen shot of video of a paper-prototype of a scanning-based BCI music player

while the usual low fidelity prototypes that involve pointing and selecting do not intuitively allow for this. Engaging end users with these prototypes has not yet received any attention in the BCI literature.

9.7.1.1 Example: Paper Prototype of a Scanning Interface

In 2010, we developed a prototype scanning interface using paper and cardboard (Fig. 9.4). The prototype was intended to gain insight into how users would want to use a BCI-controlled music player given the error properties of BCI. Participants were six users with minor—severe disabilities and one participant described as locked-in. As users were likely to be familiar with a scanning system from using other AT systems, a scanning-based system was used to demonstrate some features of the music player. The aim was to show how a motor imagery-based music player would work, and to highlight the problems with error and time taken to achieve goals that might be encountered in this system.

We created video scenarios depicting possible behaviours of the system. To address the issue of there being no true asynchronous control (i.e. a selection to the system would always be made after some period of time), users were shown a video prototype of a music player that started or stopped playing, or skipped to the next or previous track even though the listener was not trying to control anything. They were first of all asked to comment on whether they would use this music player, followed by whether they preferred one of several options, with each option having its own video simulation:

1. Do a sequence of selections to activate the player (in this example—left, right, left, left).
2. Remove some functions like back/next that would abruptly change the music being played more often.
3. Create the playlist but someone else can decide when to start and stop the music.

In general, participants thought it was ok for the music player to make mistakes where the music would start and stop randomly. Participants showed a range of tolerance to and preference for the different options, with one accepting all the solutions as better than the initial presentation of the player, even if someone else could decide when to start and stop the music—as long as he could choose when this function was enabled, while another rejected all the options. In response to (2), two of the participants suggested making these functions possible but harder to do. One participant decided that the initial option where the music started and stopped randomly was still the best, while three thought the best option was (1). A couple of participants indicated that they would be content to use BCI to create the playlist, then remove the BCI cap while listening to music. The discussions highlighted that flexibility and individual preference are major factors in developing interfaces with error-prone control; thus the ability to customize applications is paramount.

9.7.2 High Fidelity Simulations for Design and Development

One potential way of developing prototypes for BCI that represent the control characteristics is to build a simulator. Simulation in the BCI literature usually refers to running offline analysis on some raw EEG data in order to improve or explore classification techniques (e.g. [18,25,64]). Here, we refer to simulation as modelling the control of a system in order to tell us something about the interaction between the human and the machine. In this sense, simulation in HCI and BCI tends to focus either on prediction of task performance via offline analysis, or on the feel of the input via online analysis. Offline analyses of interfaces using mainstream input technologies incorporate research in cognitive psychology [15]. In AT research, some work has been carried out on estimating task performance using perceptual, cognitive and motor models of an individual [5,6]. These usually involve models of motor performance which cannot readily be applied to BCI. Bensch et al. [4] used a model to predict how many transitions it would take to select certain menu items given an error rate, but did not compare this to actual use of the system.

"Online" simulation of disability includes simulating problems that may be faced by the elderly [31, 32], simulation of deficiencies in visual perception [3], and more recently, simulation of aphasia [26]. Such simulations, if used correctly [9], can enable designers and stakeholders to understand the constraints and opportunities for development. Cincotti et al. [12, 13] used a noisy mouse input representing the noisy input of BCI to explore tactile feedback, showing that tactile feedback could compensate for visual feedback under high visual workload conditions.

Plass-Oude Bos et al. [50] asked users to imagine different mental states to control an input, showing that users preferred different mental states depending on the accuracy of detection. Other than these examples, simulation seems to be an under-used, but potentially highly valuable tool for BCI research.

Low-level simulation is not often used to make predictions about the actual performance of a user interface. One exception is the EASE tool which simulates the interaction of users with motor disabilities [16]. Using this tool, it was found that adaptive word prediction is useful only for typing speeds less than five to eight words per minute. We propose that a similar approach is useful for BCI research, where low-level simulation of the control characteristics of BCI can be used to investigate aspects of application control that have previously been discussed such as how users respond to error, delay, and the speed-accuracy trade-off. This is especially useful for investigating interaction with a wide range of individual differences in control properties. Quek et al. [53] showed that the delay and error properties could be simulated for different users using a simple model of the interaction. The next step is to model sequential interactions, perhaps where errors create more errors. Using such a tool, we hope to be able to combine the predictive capabilities of simulation with online use of the system. Our goal is to lower the pre-requisite knowledge and tools for non-BCI-specialists wanting to develop and design applications for BCIs.

Importantly, our simulator replaces real BCI input for testing BCI applications without the need to wear the cap. In our experience, problems with the application interfaces are discovered as soon as it receives real BCI input, indicating a lack of understanding of how the interaction would flow once the BCI is connected. Here we present an example which shows that knowledge gained from testing user interfaces with a simulator can inform design and be used to debug applications before testing with real BCI.

9.7.2.1 Example: Application Design and Development Using Simulators

In developing a BCI-controlled music player, we are able to employ an iterative style of development through the use of a simulator. In this case, we decided to use the rotate-extend (REx) paradigm to control the functions of a music browser (Fig. 9.5). One error was that the first segment can easily be selected unintentionally. A possible solution is to make sure that enough time is given at the start (before the feedback starts moving) for the user to prepare the correct mental state. Other solutions include making sure that the segment most likely to produce false positive errors is one of low risk (can be easily undo-able, e.g. play or pause), or ensuring that the first part of the wheel is not selectable. We were also able to experiment with ways of enabling an "intentional non-control," or idle, state. When the user selects "play," the wheel selector locks so that music can be listened to without interruption. Depending on the control properties of an individual on a given day, the parameters of the application can be tuned to make it easier or more difficult to lock and unlock the music player.

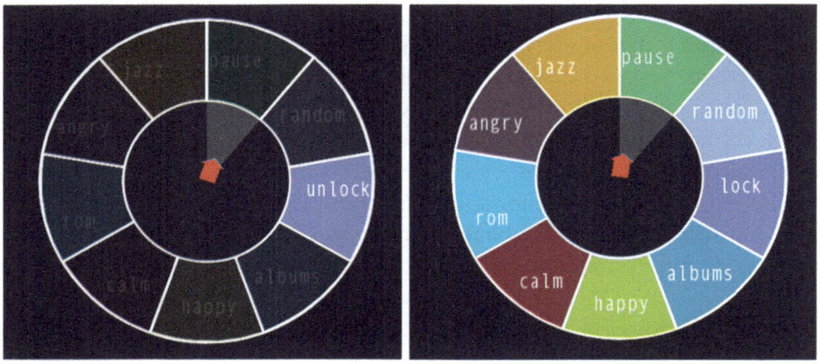

Fig. 9.5 Music player selection wheel interface based on the hex-o-spell paradigm [71]. One mental class is used to rotate the arrow in the centre of the circle, while the other class is used to extend the arrow in order to select a control segment. *Left*: while music is playing, the music player is in a locked state: the "unlock" segment must be selected in order to reactivate the player. *Right*: the selection wheel in an unlocked state where selection of any segment is possible

9.8 Conclusion

The future of BCIs is in integration. Already an interdisciplinary area, the field of BCI must be and is already starting to inform and be informed by various disciplines outside neuroscience and engineering, specifically HCI, control theory and design. BCI researchers should start to pin down the characteristics that are similar to and different from other assistive technologies, and from other emerging input technologies that deal with uncertain, noisy inputs which may provide either implicit (e.g. context awareness, bio sensing) or explicit (e.g. gesture) control. Where BCI input characteristics are similar to other more established methods, we should embrace what has been learned from these and identify areas where shared knowledge is currently lacking. Where there are differences in interaction design that are unique to direct communication with devices using brain activity, we need to further develop BCI-specific design principles and guidelines.

The extreme nature of current BCI input is well-suited to highlight the conceptual gaps in the foundations of human–computer interaction research, and will stimulate the creation of new frameworks.

Some effort is necessary to integrate what we already know about low-level brain-signal characteristics, neuroergonomics, user expectations and motivation, individual differences etc. into whole systems that are enjoyable to use. Researchers should thus focus not only on improving the communication rate of BCIs, but also on improving the user experience of systems which use BCI. This will be even more important in future applications of BCI to able-bodied users, where the user experience will need to be acceptable for users to engage with the technology at all.

References

1. Allison, B.Z., Pineda, J.A.: Effects of SOA and flash pattern manipulations on ERPs, performance, and preference: implications for a BCI system. Int. J. Psychophysiol. **59**(2), 127–140 (2006)
2. Babiloni, F., Cincotti, F., Lazzarini, L., Millan, J., Mourinõ, J., Varsta, M., Heikkonen, J., Bianchi, L., Marciani, M.G.: Linear classification of low-resolution EEG patterns produced by imagined hand movements. IEEE Trans. Rehabil. Eng. **8**(2), 186–188 (2000)
3. Ball, S., Rousell, J.: Virtual Disability: Simulations as an Aid to Lecturers' Understanding of Disability. Computers Helping People with Special Needs, pp. 624–624 (2004)
4. Bensch, M., Karim, A., Mellinger, J., Hinterberger, T., Tangermann, M., Bogdan, M., Rosenstiel, W., Birbaumer, N.: Nessi: an EEG-controlled web browser for severely paralyzed patients. Comp. Intell. Neurosci. **2007**, 71,863 (2007)
5. Biswas, P., Robinson, P.: Simulation to predict performance of assistive interfaces. In: Proceedings of the 9th International ACM SIGACCESS Conference on Computers and Accessibility, ACM, pp. 227–228 (2007)
6. Biswas, P., Robinson, P.: Automatic evaluation of assistive interfaces. In: Proceedings of the 13th International Conference on Intelligent User Interfaces, ACM, pp. 247–256 (2008)
7. Blankertz, B., Dornhege, G., Schäfer, C., Krepki, R., Kohlmorgen, J., Müller, K.R., Kunzmann, V., FLosch, Curio, G.: Boosting bit rates and error detection for the classification of fast-paced motor commands based on single-trial EEG analysis. IEEE Trans. Neural Syst. Rehabil. Eng. **11**(2), 127–131 (2003)
8. Blankertz, B., Lemm, S., Treder, M.S., Haufe, S., Müller, K.R.: Single-trial analysis and classification of ERP components – a tutorial. Neuroimage **56**, 814–825 (2011)
9. Burgstahler, S., Doe, T.: Disability-related simulations: If, when, and how to use them in professional development. Rev. Disabil. Stud. **1**(2), 4–17 (2004)
10. Carlson, T., Demiris, Y.: Human-wheelchair collaboration through prediction of intention and adaptive assistance. In: Proceedings of the IEEE International Conference on Robotics and Automation (ICRA), pp. 3926–3931 (2008)
11. Chatterjee, A., Aggarwal, V., Ramos, A., Acharya, S., Thakor, N.: A brain–computer interface with vibrotactile biofeedback for haptic information. J. Neuroeng. Rehabil. **4**, 40 (2007)
12. Cincotti, F., Kauhanen, L., Aloise, F., Palomaki, T., Caporusso, N., Jylanki, P., Mattia, D., Babiloni, F., Vanacker, G., Nuttin, M., Marciani, M.G., del R Millán, J.: Preliminary experimentation on vibrotactile feedback in the context of mu-rhythm based BCI. In: Annual International Conference of the IEEE Engineering in Medicine and Biology Society. IEEE Engineering in Medicine and Biology Society. Conference, pp. 4739–4742 (2007a)
13. Cincotti, F., Kauhanen, L., Aloise, F., Palomäki, T., Caporusso, N., Jylänki, P., Mattia, D., Babiloni, F., Vanacker, G., Nuttin, M., Marciani, M.G., del R Millán, J.: Vibrotactile feedback for brain–computer interface operation. Intell. Neurosci. **2007**, 7–7 (2007b)
14. Curran, A., Stokes, M.J.: Learning to control brain activity: A review of the production and control of EEG components for driving brain–computer interface (BCI) systems. Brain Cogn. **51**(3), 326–336 (2003)
15. Dix, A., Finlay, J., Abowd, G., Beale, R.: Human–computer interaction. Prentice Hall, Hemel Hempstead (2004)
16. Fait, H., Mankoff, J.: EASE: A Simulation Tool for Accessible Design. Computer Science Division, University of California (2003)
17. Farwell, L.A., Donchin, E.: Talking off the top of your head: toward a mental prosthesis utilizing event-related brain potentials. Electroencephalogr. Clin. Neurophysiol. **70**, 510–523 (1988)
18. Fazli, S., Danóczy, M., Popescu, F., Blankertz, B., Müller, K.R.: Using rest class and control paradigms for brain computer interfacing. In: Proc. of the 10th Int. Work-Conference on Artificial Neural Networks: Part I: Bio-Inspired Systems: Computational and Ambient Intelligence, Heidelberg, IWANN '09, pp. 651–665. Springer, Berlin (2009)

19. Felton, E.A., Radwin, R.G., Wilson, J.A., Williams, J.C.: Evaluation of a modified fitts law brain–computer interface target acquisition task in able and motor disabled individuals. J. Neural Eng. **6**(5), 056,002 (2009)
20. Felzer, T., Strah, B., Nordmann, R.: Automatic and self-paced scanning for alternative text entry. In: Proc. IASTED Int. Conf. on Telehealth/Assistive Technologies, Telehealth/AT '08, pp. 1–6. ACTA, Anaheim, CA, USA (2008)
21. Felzer, T., Nordmann, R., Rinderknecht, S.: Scanning-based human–computer interaction using intentional muscle contractions. Universal Access in Human–Computer Interaction Intelligent and Ubiquitous Interaction Environments, pp. 509–518 (2009)
22. Flemisch, O., Adams, A., Conway, S., Goodrich, K., Palmer, M., Schutte, P.: The H-Metaphor as a guideline for vehicle automation and interaction. Tech. Rep. NASA/TM–2003-212672, NASA (2003)
23. Gajos, K.Z., Everitt, K., Tan, D.S., Czerwinski, M., Weld, D.S.: Predictability and accuracy in adaptive user interfaces. In: Proc. of 26th annual SIGCHI conference on Human factors in computing systems, CHI '08, pp. 1271–1274. ACM, New York, NY, USA (2008)
24. Goodrich, K., Schutte, P., Flemisch, F., Williams, R.: Application of the H-mode, a design and interaction concept for highly automated vehicles, to aircraft. In: Proc. IEEE Digital Avionics Syst. Conf., pp. 1–13 (2006)
25. Guger, C., Schlogl, A., Neuper, C., Walterspacher, D., Strein, T., Pfurtscheller, G.: Rapid prototyping of an EEG-based brain–computer interface (BCI). IEEE Trans. Neural Syst. Rehabil. Eng. **9**(1), 49–58 (2001)
26. Hailpern, J., Danilevsky, M., Harris, A., Karahalios, K., Dell, G., Hengst, J.: Aces: promoting empathy towards aphasia through language distortion emulation software. In: Proc. of the 2011 annual conference on Human factors in computing systems, CHI '11, pp. 609–618. ACM, New York, NY, USA (2011)
27. Halder, S., Rea, M., Andreoni, R., Nijboer, F., Hammer, E.M., Kleih, S.C., Birbaumer, N., Kübler, A.: An auditory oddball brain–computer interface for binary choices. Clin. Neurophysiol. **121**(4), 516–523 (2010)
28. Heger, D., Putze, F., Schultz, T.: Online workload recognition from EEG data during cognitive tests and human-machine interaction. In: Proc. 33rd annual German conference on Advances in artificial intelligence, KI'10, pp. 410–417. Springer, Berlin, Heidelberg (2010)
29. Hill, J., Farquhar, J., Martens, S., Bießmann, F., Schölkopf, B.: Effects of Stimulus Type and of Error-Correcting Code Design on BCI Speller Performance. Adv. Neural Inf. Process. Syst. **21**, 665–672 (2009)
30. Hill, N., Lal, T.N., Tangermann, M., Hinterberger, T., Widman, G., Elger, C.E., Schölkopf, B., Birbaumer, N.: Classifying event-related desynchronization in EEG, ECoG and MEG signals. In: Dornhege, G., d R Millán, J., Hinterberger, T., McFarland, D., Müller, K.R. (eds.) Toward Brain–Computer Interfacing, pp. 235–260. MIT press, Cambridge, MA (2007)
31. Hitchcock, D., Taylor, A.: Simulation for Inclusion–true user centred design. In: Proceedings of International Conference on Inclusive Design, Royal College of Art, London, Citeseer (2003)
32. Hitchcock, D., Lockyer, S., Cook, S., Quigley, C.: Third age usability and safety–an ergonomics contribution to design. Int. J. Hum. Comput. Stud. **55**(4), 635–643 (2001)
33. Höhne, J., Schreuder, M., Blankertz, B., Tangermann, M.: A novel 9-class auditory ERP paradigm driving a predictive text entry system. Front. Neuroprosthetics **99**(5), (2011)
34. Houde, S., Hill, C.: What do prototypes prototype? Handbook Hum. Comput. Interact. **2**, 367–381 (1997)
35. Johnson, J.: Designing with the Mind in Mind: a Simple Guide to Understanding User Interface Design Rules. Morgan Kaufmann, Burlington (2011)
36. Kübler, A., Kotchoubey, B., Kaiser, J., Wolpaw, J.P., Birbaumer, N.: Brain–computer communication: Unlocking the locked in. Psychol. Bull. **127**(3), 358–375 (2001)
37. Kübler, A., Furdea, A., Halder, S., Hammer, E.M., Nijboer, F., Kotchoubey, B.: A brain–computer interface controlled auditory event-related potential (p300) spelling system for locked-in patients. Ann. N. Y. Acad. Sci. **1157**, 90–100 (2009)

38. Langer, E.J., Roth, J.: Heads I win, tails it's chance: The illusion of control as a function of the sequence of outcomes in a purely chance task. J. Pers. Soc. Psychol. **32**(6), 951–955 (1975)
39. Leeb, R., Lee, F., Keinrath, C., Scherer, R., Bischof, H., Pfurtscheller, G.: Brain–computer communication: Motivation, aim, and impact of exploring a virtual apartment. IEEE Trans. Neural Syst. Rehabil. Eng. **15**(4), 473–482 (2007)
40. Lemm, S., Blankertz, B., Curio, G., Müller, K.R.: Spatio-spectral filters for improving classification of single trial EEG. IEEE Trans. Biomed. Eng. **52**(9), 1541–1548 (2005)
41. Millán, J.: On the need for on-line learning in brain–computer interfaces. In: Proceedings of the International Joint Conference on Neural Networks, Budapest, Hungary, iDIAP-RR 03-30 (2004)
42. Millán JdR., Rupp, R., Mueller-Putz, G., Murray-Smith, R., Giugliemma, C., Tangermann, M., Vidaurre, C., Cincotti, F., Kübler, A., Leeb, R., Neuper, C., Müller, K.R., Mattia, D.: Combining brain–computer interfaces and assistive technologies: State-of-the-art and challenges. Front. Neurosci. **4** 161 (2010)
43. Mönßinger, J., Halder, S., Kleih, S., Furdea, A., Raco, V., Hösle, A., Kübler, A.: Brain painting: first evaluation of a new brain–computer interface application with als-patients and healthy volunteers. Front. Neurosci. **4**, 1–11 (2010)
44. Müller, K.R., Krauledat, M., Dornhege, G., Curio, G., Blankertz, B.: Machine learning and applications for brain–computer interfacing. In: Smith, M.J., Salvendy, G. (eds.) Human Interface, Part I, HCII 2007, LNCS, vol. 4557, pp. 705–714. Springer, Berlin, Heidelberg (2007)
45. Nielsen, J.: Usability Engineering. Academic Press, Morgan Kaufmann, San Francisco (1993)
46. Norman, D.: Emotional design. Basic Books, New York (2005)
47. Norman, K.L.: The Psychology of Menu Selection: Designing Cognitive Control at the Human/Computer Interface. Greenwood, Westport, CT, USA (1991)
48. Obermaier, B., Neuper, C., Guger, C., Pfurtscheller, G.: Information transfer rate in a five-classes brain–computer interface. IEEE Trans. Neural Syst. Rehabil. Eng. **9**(3), 283–288 (2001). DOI 10.1109/7333.948456
49. Patel, S.H., Azzam, P.N.: Characterization of N200 and P300: selected studies of the Event-Related Potential. Int. J. Med. Sci. **2**, 147–154 (2005)
50. Plass-Oude Bos, D., Poel, M., Nijholt, A.: A study in user centered design and evaluation of mental tasks for BCI. In: Lee, K.T., Tsai, W.H., Liao, H.Y.M., Chen, T., Hsieh, J.W., Tseng, C.C. (eds.) Proc. of the 17th International Multimedia Modeling Conference, MMM 2011, Taipei, Taiwan, Lecture Notes in Computer Science, vol. 6524, pp. 122–134. Springer, Berlin, (2011)
51. Polich, J.: Updating P300: an integrative theory of P3a and P3b. Clin. Neurophysiol. **118**, 2128–2148 (2007)
52. Polich, J., Ellerson, P.C., Cohen, J.: P300, stimulus intensity, modality, and probability. Int. J. Psychophysiol. **23**, 55–62 (1996)
53. Quek, M., Boland, D., Williamson, J., Murray-Smith, R., Tavella, M., Perdikis, S., Schreuder, M., Tangermann, M.: Simulating the feel of brain–computer interfaces for design, development and social interaction. In: Proc. of the 2011 annual conference on Human factors in computing systems, CHI '11, pp. 25–28. ACM, New York, NY, USA (2011)
54. Ravden, D., Polich, J.: Habituation of P300 from visual stimuli. Int. J. Psychophysiol. **30**(3), 359–365 (1998)
55. Schreuder, M., Tangermann, M., Blankertz, B.: Initial results of a high-speed spatial auditory BCI. Int. J. Bioelectromagn. **11**(2), 105–109 (2009)
56. Schreuder, M., Blankertz, B., Tangermann, M.: A new auditory multi-class brain–computer interface paradigm: Spatial hearing as an informative cue. Plos One **5**(4), e9813 (2010)
57. Schreuder, M.; Höhne, J.; Treder, M.; Blankertz, B.; Tangermann, M.; , Performance optimization of ERP-based BCIs using dynamic stopping. IEEE Eng. Med. Biol. Soc. (2011) pp.4580–4583
58. Shneiderman, B., Plaisant, C., Cohen, M., Jacobs, S.: Designing the User Interface: Strategies for Effective Human–Computer Interaction (5th edn.). Addison Wesley, Boston (2009)

59. Simpson, R.C., Koester, H.H.: Adaptive one-switch row-column scanning. IEEE Trans. Rehabil. Eng. **7**(4), 464–473 (1999)
60. Strachan, S., Murray-Smith, R., O'Modhrain, S.: Bodyspace: inferring body pose for natural control of a music player. In: CHI '07 extended abstracts on Human factors in computing systems, CHI EA '07, pp. 2001–2006. ACM, New York, NY, USA (2007)
61. Szekely, P.: User interface prototyping: Tools and techniques. In: Software Engineering and Human–Computer Interaction, pp. 76–92. Springer, Heidelberg (1995)
62. Thimbleby, H.: Understanding User Centred Design (UCD) for People with Special Needs. Computers Helping People with Special Needs, pp. 1–17 (2008)
63. Tonin, L., Leeb, R., Tavella, M., Perdikis, S., del R Millán, J.: The role of shared-control in BCI-based telepresence. In: Proc. of 2010 IEEE International Conference on Systems, Man and Cybernetics (2010)
64. Townsend, G., Graimann, B., Pfurtscheller, G.: Continuous EEG classification during motor imagery-simulation of an asynchronous BCI. IEEE Trans. Neural Syst. Rehabil. Eng. **12**(2), 258–265 (2004)
65. Tractinsky, N., Hassenzahl, M.: Arguing for aesthetics in human–computer interaction. I-com **4**(3/2005), 66–68 (2005)
66. Tractinsky, N., Katz, A., Ikar, D.: What is beautiful is usable. Interact. Comput. **13**(2), 127–145 (2000)
67. Treder, M.S., Blankertz, B.: (C)overt attention and visual speller design in an ERP-based brain–computer interface. Behav. Brain Funct. **6**, 28 (2010)
68. Vidaurre, C., Kawanabe, M., von Bünau, P., Blankertz, B., Müller, K.R.: Toward unsupervised adaptation of lda for brain–computer interfaces. IEEE Trans. Biomed. Eng. **58**(3), 587–597 (2011a)
69. Vidaurre, C., Sannelli, C., Müller, K.R., Blankertz, B.: Machine-learning based co-adaptive calibration. Neural Comput. **23**(3), 791–816 (2011b)
70. Virzi, R.A., Sokolov, J.L., Karis, D.: Usability problem identification using both low- and high-fidelity prototypes. In: Proc. of the SIGCHI conference on Human factors in computing systems: common ground, CHI '96, pp. 236–243. ACM, New York, NY, USA (1996)
71. Williamson, J., Murray-Smith, R., Blankertz, B., Krauledat, M., Müller, K.R.: Designing for uncertain, asymmetric control: Interaction design for brain–computer interfaces. Int. J. Hum. Comput. Stud. **67**(10), 827–841 (2009)
72. Wilson, A.: Sensor-and recognition-based input for interaction. In: Sears, A., Jacko, J. (eds.) The Human Computer Interaction Handbook, pp. 177–200. Lawrence Erlbaum Associates (2007)
73. Wolpaw, J.R., Birbaumer, N., McFarland, D.J., Pfurtscheller, G., Vaughan, T.M.: Brain–computer interfaces for communication and control. Clin. Neurophysiol. **113**(6), 767–791 (2002)
74. Zhang, B., Wang, J., Fuhlbrigge, T.: A review of the commercial brain–computer interface technology from perspective of industrial robotics. In: 2010 IEEE International Conference on Automation and Logistics ICAL 2010, pp. 379–384 (2010)

Chapter 10
Combining BCI with Virtual Reality: Towards New Applications and Improved BCI

Fabien Lotte, Josef Faller, Christoph Guger, Yann Renard, Gert Pfurtscheller, Anatole Lécuyer, and Robert Leeb

10.1 Introduction

Historically, the main goal of Brain–Computer Interface (BCI) research was, and still is, to design communication, control and motor substitution applications for patients with severe disabilities [75]. These last years have indeed seen tremendous advances in these areas with a number of groups having achieved BCI control of prosthetics, wheelchairs and spellers, among other [49]. More recently, new applications of BCI have emerged that can be of benefit to both patients and healthy

F. Lotte (✉)
INRIA Bordeaux Sud-Ouest, 351 cours de la libération, F-33405, Talence, France
e-mail: fabien.lotte@inria.fr

J. Faller · G. Pfurtscheller
Institute for Knowledge Discovery, Laboratory of Brain–Computer Interfaces, Graz University of Technology, Krenngasse 37, A-8010 Graz, Austria
e-mail: josef.faller@tugraz.at; pfurtscheller@tugraz.at

C. Guger
g.tec medical engineering GmbH, Sierningstrasse 14, A-4521 Schiedlberg, Austria
e-mail: guger@gtec.at

Y. Renard
Independant Brain–Computer Interfaces Consultant, Rennes Area, France
e-mail: yann.renard@aliceadsl.fr

A. Lécuyer
INRIA Rennes Bretagne-Atlantique, Campus Universitaire de Beaulieu, F-35042 Rennes Cedex, France
e-mail: anatole.lecuyer@inria.fr

R. Leeb
Chair in Non-Invasive Brain-Machine Interface, École Polytechnique Fédérale de Lausanne, Station 11, CH-1015 Lausanne, Switzerland
e-mail: robert.leeb@epfl.ch

B. Z. Allison et al. (eds.), *Towards Practical Brain-Computer Interfaces*, Biological and Medical Physics, Biomedical Engineering, DOI 10.1007/978-3-642-29746-5_10,
© Springer-Verlag Berlin Heidelberg 2012

users alike, notably in the areas of multimedia and entertainment [52]. In this context, combining BCI with Virtual Reality (VR) technologies has rapidly been envisioned as very promising [37,39]. Such a combination is generally achieved by designing a system that provides the user with immersive 3D graphics and feedback with which it can interact in real-time by using the BCI. The promising potential of this BCI-VR combination is visible at two levels. On one hand, BCI is seen by the VR community as a new input device that may completely change the way to interact with Virtual Environments (VE) [37]. Moreover, BCI might also be more intuitive to use than traditional devices. In this sense, BCI can be seen as following a path similar to that of haptic devices a few years ago [7], that led to new ways of conceiving VR interaction. On the other hand, VR technologies also appear as useful tools for BCI research. VE can indeed be a richer and more motivating feedback for BCI users than traditional feedbacks that are usually in the form of a simple 2D bar displayed on screen. Therefore a VR feedback could enhance the learnability of the system, i.e., reduce the amount of time needed to learn the BCI skill as well as increase the mental state classification performance [39,64]. VE can also be used as a safe, cost-effective and flexible training and testing ground for prototypes of BCI applications. For instance, it could be used to train a patient to control a wheelchair with a BCI [40] and to test various designs for the wheelchair control, all of this without any physical risk and with a very limited cost. As such, VR can be used as an intermediary step before using BCI applications in real-life. Finally, VR could be the basis of new applications of BCI, such as 3D video games and artistic creation for both patients and healthy users, as well as virtual visits (cities, museums ...) and virtual online communities for patients, in order to address their social needs.[1]

Designing a system combining BCI and VR comes with several important challenges. First, the BCI being used as an input device, it should be, ideally, as convenient and intuitive to use as other VR input devices. This means that (1) the BCI should provide the user with several commands for the application, (2) the user should be able to send these commands at anytime, at will, i.e., the BCI should be self-paced (a.k.a. asynchronous), (3) the mapping between the mental states used and the commands (i.e., the interaction technique) should be intuitive, efficient, and not lead to too much fatigue for the user. This last point is particularly challenging since current BCI are usually based on a very small number of mental states, typically only two or three, whereas the number of interaction tasks that can be performed on a typical VE is very large, usually much larger than three. From the point of view of the VE design and rendering, the challenges include (1) to provide a meaningful VR feedback to the user, in order to enable him to control the BCI, (2) to integrate the stimuli needed for BCI based on evoked potentials as tightly and seamlessly as possible in order not to deteriorate the credibility and thus the immersiveness of the VE, and (3) to design a VR application that is useful and usable despite the huge differences between a typical VE and the standard BCI training protocols.

---

[1]See, for instance, the work achieved as part of the BrainAble project: http://www.brainable.org/

This chapter presents an overview of the research works that have combined BCI and VR and addressed these challenges. As such, (1) it surveys recent works that use BCI to interact with VE, (2) it highlights the critical aspects and solutions for the design of BCI-based VR applications, and (3) it discusses the related perspectives. It is organized as follows: Sect. 10.2 provides some introductory material on VR and the way to interact with VE using a BCI. Then, Sect. 10.3 reviews existing BCI-based VR applications according to the different neurophysiological signals used to drive the BCI. More particularly, Sect. 10.3.1 discusses VR applications controlled with a motor imagery (MI)-based BCI, Sect. 10.3.2 those based on Steady State Visual Evoked Potentials (SSVEP) and Sect. 10.3.3 those exploiting a P300-based BCI. Then, Sect. 10.4 elaborates on the impact of VR on BCI use, notably in terms of BCI performance and user experience. Finally, Sect. 10.5 concludes the chapter.

10.2 Basic Principles Behind VR and BCI Control

This section gives some insights about how VE can be controlled with a BCI. In the first subsection, VR is defined and the typical interaction tasks are described. The suitability of the different BCI neurophysiological signals (MI, P300, SSVEP) for each interaction task is also briefly mentioned. In the second subsection, a general architecture for BCI-based VR applications is proposed. This architecture is illustrated with examples of existing VR applications using BCI as input device.

10.2.1 Definition of Virtual Reality

A VR environment can be defined as an immersive system that provides the user with a sense of presence (the feeling of "being there" in the virtual world [8]) by means of plausible interactions with a real-time simulated synthetic world [36]. Such plausible interaction is made possible thanks to two categories of devices: input and output devices. First, the user must be able to interact with the virtual world in real time. This is achieved by using input devices such as game pads, data gloves, motion tracking systems or, as described in this chapter, BCI. Second, the user must be provided with real time feedback about the virtual world state. To this end, various output devices are generally used to render the virtual world content, such as visual displays, spatial sound systems or haptic devices.

According to Bowman et al. [6] typical interaction tasks with a 3D-VE can be described as belonging to one of the following categories:

- *Object selection*: It consists in selecting an object among those available in the virtual world, typically in order to subsequently manipulate it.

- *Object manipulation*: It consists in changing attributes of an object in the virtual world, typically its position and orientation or other properties such as appearance and size.
- *Navigation*: It consists in modifying the user's own position and orientation in the virtual world in order to explore it. In other words, navigation can be defined as moving around the VE and changing the current point of view.
- *Application control*: It consists in issuing commands to the application, to change the system mode or to activate various functionalities, for instance.

All these categories of interaction tasks can be performed with a BCI. However, each BCI paradigm is more or less suitable for each category of interaction task. For instance, MI and SSVEP-based BCI are more suitable for navigation tasks and possibly object manipulation because they can issue commands continuously and potentially in a self-paced way. On the other hand, P300-based BCI let the user pick one item among a list of usually at least four, such command being issued in a discrete and synchronous way. For this reason, they are more suitable for object selection tasks. The suitability of each BCI paradigm is discussed more in details and illustrated in Sects. 10.3.1–10.3.3 respectively.

10.2.2 General Architecture of BCI-Based VR Applications

Implementing a BCI control for a VR system can be seen as using the BCI as an input device to interact with the VE. Therefore, it consists in providing the user with a way to act on the virtual world only by means of brain activity, and using the available output devices to provide a meaningful feedback to the user. So far, only visual feedback has been deeply investigated in the context of BCI-based VR applications, but other modalities, in particular audio and haptics, would also be worth studying in the future. A BCI-based VR setup typically involves two independent softwares: (1) a BCI software to record brain signals, process them to extract relevant features and classify mental states in real-time in order to generate commands, and (2) a VR software to simulate and render a virtual world, provide feedback to user and process the received commands. Therefore, these two softwares must be able to communicate in order to exchange information and commands. Figure 10.1 provides a schematic representation of BCI control of a VR application.

In addition to these software considerations, there are also several hardware related issues that must be considered when using a BCI system in a VE: (1) the biosignal amplifiers must be able to work in such a noisy environment, (2) the recordings should ideally be done without wires to avoid collisions and irritations within the environment, (3) the BCI system must be coupled with the VR system to exchange information fast enough for real-time experiments, and (4) in the case of CAVE systems, users mostly want to move around and therefore active EEG electrodes should be used to avoid movement artifacts.

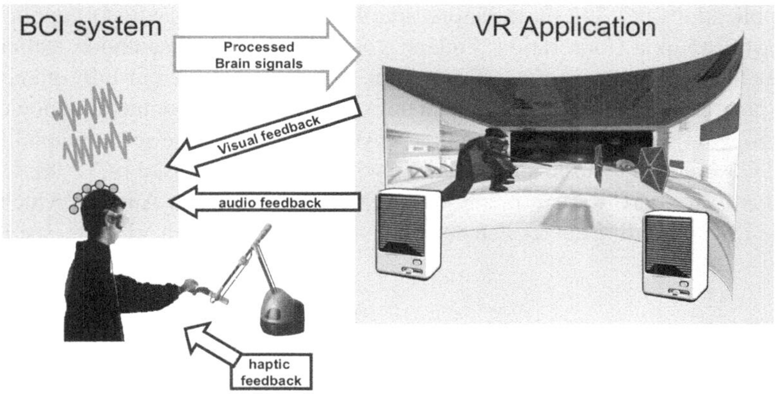

Fig. 10.1 General architecture of a BCI-based VR application: the user generates specific brain activity patterns that are processed by the BCI system and sent as command to the VR application. In return, the VR application provides meaningful feedback to the user, this feedback being potentially any combination of visual, audio or haptic feedback forms. This combination of "control on the VE" and "feedback from the VE" can elicit the sense of presence

In order to illustrate this general architecture implementation and propose a complete setup, we can mention two softwares which are devoted to BCI and VR as an example: OpenViBE and Ogre3D. OpenViBE[2] is a free software platform to design, test and use BCI [63]. OpenViBE has been successfully used for the three major families of BCI: Motor Imagery [47], P300 [10] and SSVEP [44]. Ogre3D[3] is a scene-oriented, flexible 3D engine that is capable of producing realistic representations of virtual worlds in real time. Ogre3D also includes extensions for spatial sound, physics simulation, etc. Moreover, it has been successfully used to simulate VE on equipments ranging from basic laptops to fully immersive systems such as CAVE systems [11]. These two softwares can communicate and exchange information, commands and responses using the Virtual Reality Peripheral Network (VRPN), a widely used library proposing an abstraction of VR devices [69]. Since both OpenViBE and Ogre3D have VRPN support, either natively or through contributions, they are able to communicate efficiently in order to design BCI-based VR applications. Those softwares have been used to design BCI-based VR applications such as those described in [44, 46, 47]. Generally not only VRPN, but any other interface (like proprietary TCP, UDP connections) can be used to communicate with existing VR systems.

Naturally, various other software and hardware can also be used to design BCI-based VR applications, such as Matlab/Simulink for real-time EEG signal processing and XVR (eXtremeVR 3D software, VRMedia, Italy) for VE design

---

[2]http://openvibe.inria.fr/

[3]http://www.ogre3d.org

and applications [27,30]. Furthermore, simple projection walls with Qt[4] application framework (Nokia Corporation, Finland), or stereoscopic presentation techniques such as head-mounted display (HMD) with VRjuggler, or even in fully-immersive multi-projection stereo-based and head tracked VE systems (commonly known as a "CAVE" [11] using DIVE software, or "DAVE" [21]) with the scene graph library OpenSG were already used and combined with a MATLAB based BCI [38]. On the EEG hardware part, we can mention the gMOBIlab+[5] (g.tec, Austria) which is a mobile EEG recording device that has been successfully used in VE (e.g., see [24]).

10.3 Review of BCI-Controlled VR Applications

This section reviews works that have used BCI to interact with VR applications. These works are arranged according to the neurophysiological signal used to drive the BCI: Motor Imagery (Sect. 10.3.1), SSVEP (Sect. 10.3.2) and P300 (Sect. 10.3.3). It should be mentioned that Sect. 10.3.1 describes more works than the other two sections, since more groups have used MI as the input signal to BCI-based VR applications. This is probably due to the fact that MI is a popular and well-studied neurophysiological signal for BCI [55], and that, contrary to SSVEP and P300, MI does not require any external stimulus which could be more convenient and natural for the user of a VR application.

10.3.1 Motor Imagery Controlled VR Environments

In this section we will focus on BCI based on MI, meaning on the analysis and classification of sensorimotor electroencephalographic (EEG) patterns generated during the imagination of specific movements (MI of left and right hand) [55,58]. The imagination of different types of movements results in a characteristic change of the EEG over the sensorimotor cortex which is called event-related de-/synchronisation [54]. After the computer learned the user-specific patterns, they can be used to control the movement of a bar to the right or left, just by imagining right or left hand movements. The same principle can be used to control simple movements in VEs.

The progress and comparison of MI-BCI controlled VR was first shown by Leeb and Pfurtscheller, by increasing the complexity of their studies from controlling a simple bar feedback in a synchronous manner till a self-paced (asynchronous) BCI in highly immersive VE [38]. In their first work, users perceived the feeling of rotating with constant speed to the right and left depending on the imagined hand

[4]http://qt.nokia.com/

[5]http://www.gtec.at

movement (see Fig. 10.2a), while the rotation information was integrated over one trial [42]. Interestingly, no differences between HMD and CAVE feedback could be found, but all users performed better compared to standard bar feedback. The reason for the same VE performance was that users lost the spatial orientation while rotating, which disturbed them. In a similar experiment the imagination of foot movement was used to walk forward in a virtual street [39, 56]. Correct classification of foot motor imagery was accompanied by forward movement at constant speed, whereas a correct classification of hand motor imagery stopped the motion. Incorrect classification of hand motor imagery resulted in backward motion (same speed) and incorrect foot in halting. The walking distance was scored as a "cumulative achieved mileage" (CAM, [39]), which was the integrated forward/backward distance covered during foot movement imagination and was used as performance measurement. All users achieved their best results within the CAVE and the worst in the standard BCI condition, so we can assume that the use of VR as feedback stimulated the participant's performances. The results indicate that foot motor imagery is a suitable mental strategy to control events within the VEs, because the imagination of feet movement is a mental task which comes very close to that of natural walking. It was observed that in the CAVE condition (highest immersion) the performance variation were stronger than in the control condition. One possible interpretation is that VR feedback amplifies both positive and negative feedback effects on the performance. The wrong behaving rich visual feedback can modify the EEG activity and thereby results in a further deterioration of performance [39].

The next important step was to overcome the cue-based interactions and to incorporate free will decisions (intentional control). Thereby users could navigate freely through a virtual apartment (see Fig. 10.2b), whereby at every junction the users could decide by their own, how they wanted to explore the VE [41]. The apartment (maze like) was designed similar to a real world application, with a goal-oriented task (predefined target room), a high mental workload and a variable decision period for the user. For comparison reasons, synchronous BCI sessions with a standard BCI bar feedback have been performed before and after the sessions with the virtual apartment, whereby the experiments with the virtual apartment were performed both in front of a normal TFT monitor and in an immersive VE. The users noted that the task in the apartment was much harder compared to the prior feedback training, because it was necessary not only to perform the "correct" imagination, but also the shortest way through the apartment had to be found. Therefore the cognitive load was much higher compared to the standard BCI paradigm. According to the hypothesis, it was found that the performance improves (decrease of error) over the sessions and the statistically significant lowest error could be found during the sessions with virtual feedback [41].

Giving the user the full control over timing and speed, was demonstrated in a study where users had to explore the Austrian National Library (see Fig. 10.2c). The participants were navigating down the library hall at their own pace but had to stop at several specific points (e.g. statue, column) [43]. After a variable pause time (between 20 and 95 s) the experimenter gave the command to restart moving.

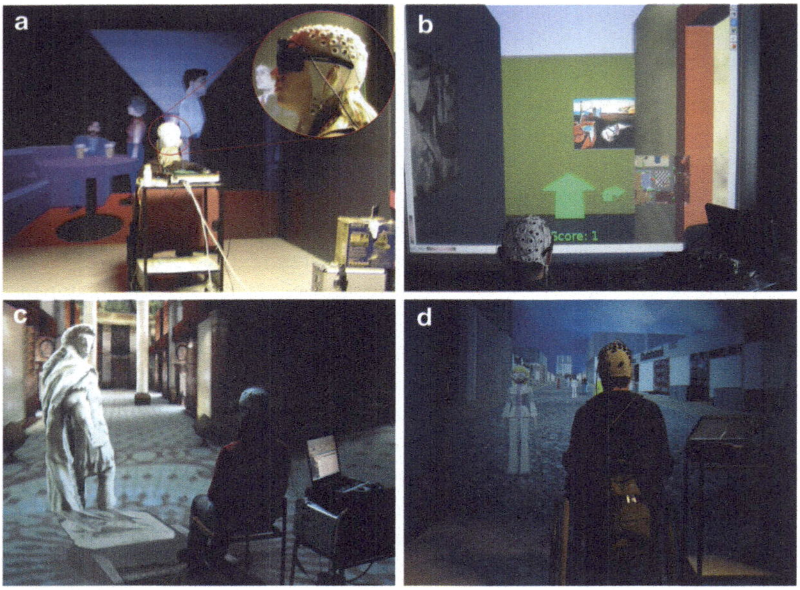

Fig. 10.2 Pictures of different MI-controlled VE: (**a**) exploring a virtual pub or (**b**) an apartment, (**c**) visiting the national library, and (**d**) walking by thoughts in case of a wheelchair person

Navigating always happened when the users performed foot motor imagery. Seven users accomplished the study with a very small number of false positive. Most interestingly in this study are the extremely long periods (up to 1.5 min) of pause times, where the user intentionally delivered no commands.

In their final study a tetraplegic patient used the imagination of his paralyzed feet to control forward movement of his wheelchair in VR [40]. The task was to go / move down a virtual street and stop at every avatar which was lined up along the street (see Fig. 10.2d). The patient achieved in some runs 100 % performance and in average 90 %. This work demonstrated for the first time that a tetraplegic user, sitting in a wheel chair, could control his movements in a VE by the usage of a self-paced BCI based on one single EEG recording. It has to be mentioned that VEs are especially attractive for a person who is wheelchair-bound. First, simply using a VE can give such persons access to experiences that may be long forgotten (or which they have never had). The fact that the user could still perform feet motor imagery, years after an injury that rendered him unable to use his feet, is a testament to the plasticity of the human brain (similar to [34]).

Another BCI controlled wheelchair study was performed by Grychtol et al. [25] with healthy users. Their results confirmed how voluntary behavioral modification brought about by VR feedback can help to improve the performance of a BCI system. VR feedback played an important role in the users' ability to learn and perform the activity well.

While these applications are already very impressive and innovative, one could argue that most of them provide the user with only a single or two commands. This could be inconvenient for the user and restrict the range of possible VE the user could interact with. This has motivated some researchers to explore BCI-based VR applications providing a larger number of commands to the user. For instance, Scherer et al. [67] proposed a 3-class self-paced BCI to freely navigate in a VE. With this BCI, the user could turn left, turn right or move forward by imagining a left hand, right hand or foot movement, respectively. While this proves to work and to be convenient for the user, this also highlighted some limitations of BCI-based interaction with VR. First, it stressed the well known performance problem of BCI, the performance being generally modest and decreasing when the number of classes to be identified increases [32]. Second, it suggested that performing navigation tasks in VR with a BCI can be tiring, especially when the user has to perform mental tasks continuously to go for one point to another or to keep the imagination over very long periods [43].

Some groups have recently proposed solutions to alleviate these issues, based on the use of specific interaction techniques. To address the limited classification performances of BCI system, the DIANA[6] group proposed to navigate VE using a self-paced BCI based on one or two motor imagery tasks only [65, 72]. Indeed, with a number of classes as small as possible, the classification performances of the BCI are much more likely to be high. In order to still provide the user with three or more commands (in order to go forward, turn left or turn right) despite the BCI recognizing only one or two motor imagery states, they proposed specific interaction techniques. These techniques are based on a scanning principle (similar to the hex-o-spell [74]). This means that to select a given interaction command, the user had to perform a motor imagery task during a given time frame, each frame being associated to a different command. Their evaluations showed that, with this approach, users can actually freely navigate in a VE with a simple brain-switch [72].

In order to alleviate the fatigue caused by BCI-based navigation in VR, as well as to efficiently use the small number of MI tasks recognized by a BCI, INRIA[7] also proposed a new interaction technique for BCI-based VR applications [47]. This technique, based on a 3-class self-paced BCI, provides the user with high-level commands, thus leaving the application in charge of performing the complex and tedious details (low-level aspects) of the interaction task. Thus, this can be seen as a form of shared-control [51]. The user can explore the VE by selecting points of interest such as navigation points (e.g., junctions, room entrances, etc.) or artworks. Interestingly enough, these navigation points can be generated completely automatically from the geometry of the VE. The user can select these points due to a sequence of binary choices. In addition to the two commands used to perform these binary choices, the user can use a third command to cancel any of his/her choice. Once a navigation point has been selected, the application takes all the

[6]http://www.diana.uma.es

[7]http://www.inria.fr/en/

Fig. 10.3 Exploring a virtual museum using a BCI and high-level commands [47]

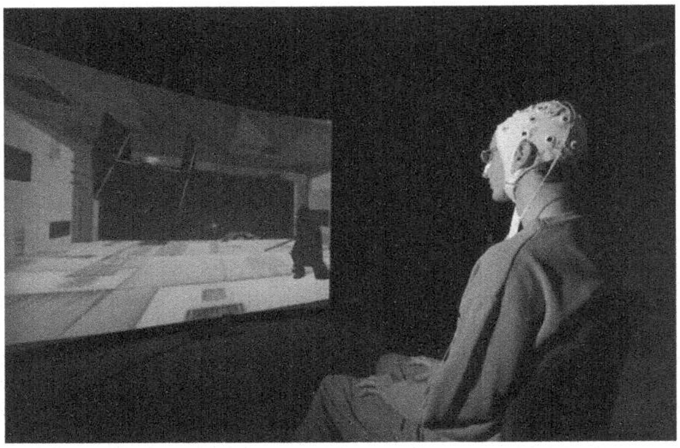

Fig. 10.4 The "Use-the-force" entertaining application [46], which enables its users to lift a virtual spaceship by using the BCI (© Hubert Raguet/Phototheque CNRS)

necessary actions to perform the interaction task such as moving from the current navigation point to the next selected one. Evaluations, performed in the context of the exploration of a virtual museum (see Fig. 10.3), showed that with this approach, users can navigate from one room to the other nearly twice as fast as with low-level commands, and with less fatigue.

Due to the huge potential of BCI-based VR applications, not only for patients but also for healthy users, it quickly became necessary to evaluate them outside laboratories, in close to real-life conditions. Such an evaluation was performed with the "use-the-force" application [46], a BCI-based VR game inspired by the Star Wars™ movie. With this game, users were asked to control the takeoff of a virtual spaceship by using real or imagined foot movements (see Fig. 10.4). The system relied on a simple brain switch that detects the beta rebound posterior to the real or imagined foot movement, at electrode Cz. The game was evaluated with 21 naïve users, during a public exhibition dedicated to VR. Despite the simplicity of the BCI design and the noisy environment, results showed that, without training, half the users could control the virtual object's motion by using real foot movements. A quarter of them could do so by using imagined foot movements. Furthermore, the whole application appeared enjoyable and motivating to the users.

Another instance of BCI to interact with complex applications was shown by Scherer et al. [66]. The Brainloop interface provides a new way to interact with complex programs like Google Earth™. Thereby through remapping of commands and options the interface can be customized. In this study, a multi-level selection process and the use of mental tasks in parallel enabled the user to send multiple commands to the application.

Although not based on motor imagery, another BCI-based VR application deserves to be mentioned here since it was also evaluated in public places, outside the lab: the "AlphaWoW" application [52]. In this game, based on World Of Warcraft®, the player controlled his avatar using a classical keyboard but can turn it from a fragile elf to a powerful bear by using a BCI. More particularly, the avatar shape (bear or elf) depended on the band power in the alpha band (8–12 Hz), the alpha rhythm power being related to the player's state of relaxation. In other words, when the player was stressed the avatar changed into a bear, and he/she has to relax to turn back the avatar into an elf. The evaluations showed that the game was received very positively despite the modest BCI performances, which were considered by the players more as a challenge than as a shortcoming. These different close-to-real-life-studies thus highlight the potential of BCI-based VR application and the need to push research efforts in these directions [45, 52, 66]

Table 10.1 summarizes the numerous studies presented in this section, by describing some key characteristics of these BCI-based VR applications using MI. Several interesting points come out of this table. First, this table highlights the importance of self-paced BCI design for VR application in general and for navigation tasks in particular. Indeed, navigation is inherently a self-paced task. Moreover, despite the fact that self-paced BCIs are a challenging research topic that is not well explored [48], most of the aforementioned studies have designed and used such a BCI. Another point to notice is that, although most BCI-based VR applications provide the user with as many commands as MI tasks used, it is possible to provide more commands than MI tasks by using appropriate interaction techniques. Finally, this table stressed that MI has been almost exclusively used to perform navigation tasks in VR. Indeed, MI appears as particularly suitable for such a task since it enables spontaneous and self-paced control, which navigation should be. On the contrary, it is not convenient to perform selection tasks with MI, since MI provides only a few mental states and thus a few commands whereas virtual objects to be selected can be potentially numerous. As such, and as it will be highlighted in subsequent sections, BCI based on Evoked Potentials (SSVEP, P300) are more suitable for selection tasks since they can use numerous stimuli and corresponding brain responses.

10.3.2 SSVEP Based VR/AR Environments

A SSVEP is an electroencephalographic response occurring when a user perceives a visual stimulus flickering at a constant frequency [73]. This response is observed over the visual cortex (occipital electrodes), and consists of an EEG pattern oscillating at the same frequencies as the flickering stimulus and its harmonics.

Table 10.1 Summary of BCI-based VR applications using Motor Imagery

Interaction task	Number of MI tasks	Number of commands	Synchronous or self-paced	VE	Reference
Navigation	2	2	Synchronous	Exploring a virtual pub	[38]
Navigation	2	2	Synchronous	Navigating along a virtual street	[39, 56]
Navigation	2	2	Synchronous	Navigating	[25]
Navigation	2	2	Semi synchronous	Exploring a virtual apartment	[41]
Navigation	1	1	Self-paced	Exploring a virtual library	[43]
Navigation	1	1	Self-paced	Moving along a virtual street	[40]
Navigation	3	3	Self-paced	Exploring the "free-space"	[67]
Navigation	1–2	4	Self-paced	Exploring a maze or park	[65, 72]
Navigation + selection	3	More than 3 (depends on the VE)	Self-paced	Exploring a virtual museum	[47]
manipulation	1	1	Self-paced	Lifting a virtual spaceship	[46]
Navigation + selection	3	More than 3	Self-paced	Controlling Google Earth™	[66]

Interestingly enough, SSVEP can be modulated by attention, which means that the SSVEP response to a given stimulus will be stronger (i.e., with a larger amplitude) when the user focuses his/her attention on this stimulus.

Lalor et al. [35] were the first to use an SSVEP-based BCI to control a character in a 3D gaming environment. In this game, a monster went from platform to platform by moving along a tight rope. From time to time, the monster lost its balance, and the user had to restore it by using the BCI. To do so, two flickering checkerboard were placed on each side of the VE, in order to elicit SSVEP at different frequencies. When the system detected that the user was focusing on the left or right checkerboard, it restored the monster's balance towards the left or right respectively. Later, Touyama worked towards more immersive applications based on SSVEP and showed that they could be used to change the point of view towards the left or right in a VE displayed in a CAVE-like system [70].

The works mentioned above proved that SSVEP-based BCI is a suitable and efficient way to interact with VE. One of its limitations though, is that it requires flickering stimuli in order to be used. In the context of VR applications, this has been mainly achieved by relying on flickering squares or checkerboards statically overlayed over the screen. As a consequence, the VE may look unnatural and is unlikely to elicit a strong sense of presence for the user. In order to address these limitations, Faller et al. [20] presented a desktop-based virtual environment, where

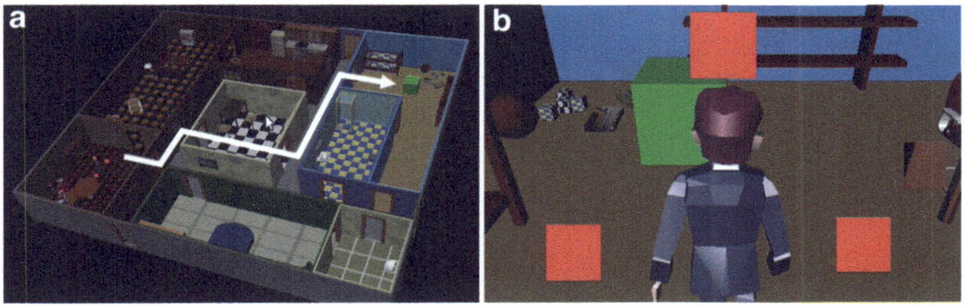

Fig. 10.5 Panel (**a**) shows an overview of the apartment scenario. The users were instructed to navigate the avatar along the path that is depicted by the *white arrow*. Panel (**b**) is a screenshot from an actual online scene, where the user navigates the avatar in third person perspective [20]

the stimuli were tightly integrated within 3D scenarios that allowed controlling avatar interaction and navigation. In one of the scenarios, seven healthy volunteers successfully controlled an avatar to alternately push one of two buttons in an asynchronous paradigm. The stimuli were fixed to the hands and hence dynamically following every avatar movement.

In another of the presented scenarios [20], five out of seven users successfully navigated an avatar in third person perspective through the same apartment scenario (see Fig. 10.5a) as presented in Leeb et al. [41]. They could guide the avatar in discrete steps and turns by visually fixating one of three SSVEP stimuli that were fixed to the avatars back. Each successful classification would then trigger one of three associated navigation commands, go one step ahead, turn left 45 ° or turn right 45 ° (see Fig. 10.5b).

Still dealing with the integration of SSVEP-stimulus within VE, Legeny et al [44]. worked towards an even more natural and ecological approach. In their work, which aimed at navigating in a virtual forest, the flickering stimuli necessary for SSVEP generation were displayed on butterfly wings. Three of these butterflies were displayed on screen, flying up and down in front of the user (see Fig. 10.6). The user had to focus his/her attention on the left, center or right butterfly in order to go left, forward or right, respectively. The butterflies' antennas were also used to provide feedback to the user. Indeed, the further apart the two antennas of a butterfly were, the more likely this butterfly will be selected by the classifier as the one the user pays attention to. Such stimuli are therefore more naturally incorporated into the VE, and formal evaluations suggested that it indeed increased the subjective preferences and feeling of presence of the users.

Finally, moving beyond traditional VE, Faller et al. [18, 19] extended their previous work into a SSVEP BCI system that relies on stimuli that are presented within immersive virtual and more interestingly, in an Augmented Reality (AR) environments. In a pilot study, three healthy volunteers were able to successfully navigate an avatar through an immersive VR slalom scenario based on embedded SSVEP stimuli. The complete scene was presented using a head-mounted display (HMD). Two of these three volunteers also succeeded in the immersive AR

Fig. 10.6 Mimetic integration of SSVEP stimulation and feedback in a Virtual Environment [44]. The butterfly wings are flickering at different frequencies in order to enable SSVEP-based BCI control, while their antenna positions represent the real-time feedback, i.e., the butterfly the most likely selected by the user according to the classifier

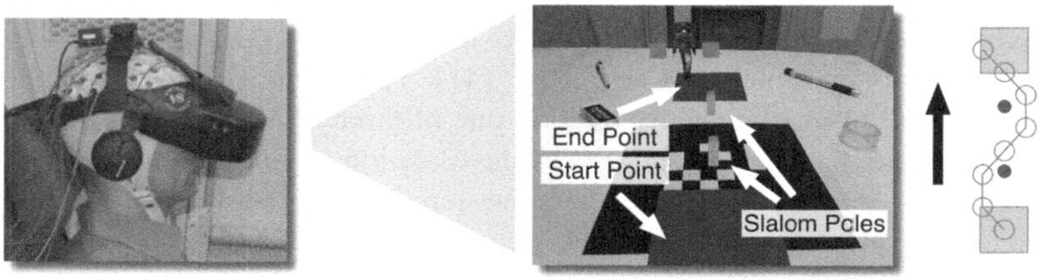

Fig. 10.7 The middle picture shows a screenshot of how the user saw the scene through the HMD seen in the *left picture*. The 3D graphics were tracked to the underlying fiducial marker. The participants were instructed to navigate the avatar through the slalom like in the picture on the *right* [18, 19]

condition, where a camera was mounted on the HMD and the slalom scenario 3D graphics were injected into the live, real-world video by tracking fiducial markers (see Fig. 10.7).

The positive results from this first feasibility study suggest that AR SSVEP BCIs have the potential to vastly improve real-world practicality and usability of BCI systems by compensating for some of their traditional shortcomings such as the low bandwidth, by offering a richer, more direct, and intuitive interface. This would allow for a more goal-directed and seamless real-world interaction. In an AR SSVEP BCI system, stimuli targets can be spatially associated to distinct points of interest in the physical world. These may be abstract or may overlap physical objects such as devices, people or controls, which is an elegant and intuitive way of presenting the user with all possible interaction options. These systems could provide patients with a higher degree of self autonomy and functional independence by introducing more intuitive and effective smart home control. Apart from that, AR SSVEP BCIs can further introduce a valuable, additional communication or control channel for user groups that require or benefit from hands free operation like pilots, car-drivers or office workers.

10.3.3 P300 Based VR Control

One of the first efforts to combine VR and BCI technologies was achieved by Bayliss and Ballard [3, 4], and made use of the P300 evoked potential. The P300 is a positive waveform occurring roughly 300 ms after a rare and relevant stimulus [17, 75]. In order to use a P300-based BCI, users have to focus their attention on a given stimulus randomly appearing among many others, each stimulus corresponding to a given command [14]. The appearance of the desired stimulus being rare and relevant, it is expected to trigger a P300 in the user's brain activity. In their study, Bayliss introduced a simple virtual smart home in which users could control different appliances (e.g., a TV or a light) using the P300-based BCI. 3D spheres were randomly appearing over the objects that can be manipulated and the user could turn them on or off simply by counting the number of times a sphere appears over the desired object.

More recently, a more interactive and richer virtual version of a smart home was implemented [24]. This smart home consists of a living room, a kitchen, a sleeping room, a bathroom, a floor and a patio as shown in the right side of Fig. 10.8. Each room has several devices that can be controlled: TV, MP3 player, telephone, lights, doors, etc. Therefore, all the different commands were summarized in seven control masks: a light mask, a music mask, a phone mask, a temperature mask, a TV mask, a move mask and a go to mask. The left side of Fig. 10.8 shows the TV mask and as an example the corresponding XVR image of the living room [15]. The user can e.g. switch on the TV by looking first at the TV symbol. Then, the station and the volume can be regulated. The bottom row of Fig. 10.8 shows the go to mask with an underlying plan of the smart home. Inside the mask, there are letters indicating the different accessible spots in the smart home which flash during the experiment. Therefore, the user has to focus on the spot where he wants to go. After the decision of the BCI system, the VR program moves to a bird's eye view of the apartment and zooms to the spot that was selected by the user. This is a goal oriented BCI control approach, in contrast to MI navigation task, where each small navigational step is controlled. Experiments with three users yielded accuracies of the BCI system between 83 % and 100 % and showed that such a BCI system can be used for the smart home control [29]. For comparison a group study with healthy people with the standard P300 speller gave an average accuracy of 91 % [28]. The Virtual Reality approach is a very cost effective way for testing the smart home environment together with the BCI system. Currently the BCI technology is interfaced to real smart home environments within the EC project

Fig. 10.8 *Top left*: Smart home control icons for TV, Telephone,... *Top right*: VR representation of the living room [15,29]. *Bottom*: Control icons to move to a certain position inside the apartment and corresponding bird eyes view of the apartment

SM4all.[8] The project aims at studying and developing an innovative middleware platform for inter-working of smart embedded services in immersive and person-centric environments [30].

These different experiments yielded two important new facts for P300-based BCIs: (1) instead of displaying characters and numbers to the user, it appears that different icons can be used as well, (2) the BCI system does not have to be trained on each individual character. The BCI system was trained with EEG data of the spelling experiment and the user specific information was used also for the smart home control. This allows using icons for many different tasks without prior time consuming and boring training of the user on each individual icon. This reduces the training time in contrast to other BCI implementations were hours or even weeks of training are needed [5,26,71], which might be important for locked-in and ALS patients who have problems with the concentration over longer time periods. The P300 concept works also better if more items are presented in the control mask as the P300 response is more pronounced if the likelihood that the target character is highlighted drops down [33]. This results of course in a lower information transfer rate, but enables to control almost any device with such a BCI system. Especially applications which require reliable decisions are highly supported. Therefore the P300 based BCI system enables an optimal way for the smart home control. In a further study the P300 smart home control was combined with a head tracker to switch on and off the BCI system. This means if the person was looking at the BCI system then it was switched on and a selection could be done. If the person turned

[8] http://www.sm4all-project.eu

to the VR projected the BCI system was switched off. Recently a hybrid version of a P300 and SSVEP BCI was used for controlling the smart home environment. The P300 BCI was used to select the command and SSVEP was used to switch on and off the BCI system [16]. These hybrid BCI systems hence demonstrated that BCI could be used practically to interact with a virtual smart home, hence potentially offering new and promising applications for patients, at home. It is worth mentioning that Groenegress compared the P300-based BCI control with a gaze-based selection method coupled with wand navigation [24]. Results suggested that the P300 BCI gives lower presence scores which might be due to the lack of motor actions which are relevant for semantic tasks and more breaks in presence.

10.4 Impact of Virtual Reality on BCI

In contrast to traditional interfaces like mouse or keypad, BCI systems could potentially promise a more direct and intuitive way of interacting and thereby overcome some limitations of navigating within VEs [68]. This is especially obvious for stimulus-dependent BCI-VR systems, where users can control appliances in the VE by simply directing their eye gaze and/or focus of attention towards the desired element (e.g., looking at the TV to switch it on, looking at the door to open it [1]). On the other hand motor imagery offers an intuitive way of VE control, for example, imagining foot movements for moving forward in a VE [43, 56]. This could overcome the problem of the contradictory stimuli while navigating VEs using a hand-held device and the reduced sense of being present in the VE [68]. On the other hand it is well known that feedback is one of the key components of a BCI, as it provides the user with information about the efficiency of his/her strategy and enables learning. The studies mentioned above show realistic and engaging VR feedback scenarios, which are closely related to the specific target application. However, the processing of such a realistic feedback stimulus may also interfere with the motor imagery task, and thus might impair the development of BCI control [50]. Furthermore, characteristic EEG changes during VE conditions were reported in [56], where a dominant ERS pattern which was permanently present in the CAVE, was less pronounced in the HMD and not existing at all in the normal feedback. Nevertheless, it was presented in Sect. 10.3.1 that VR improves the BCI performance; either the users achieved their best results within VR [25, 39, 56] compared to normal feedback or even 100 % performance result in VR [40] or achieved the lowest error with virtual feedback [41]. Generally it can be stated that VR feedback amplifies both positive and negative feedback effects on the performance.

Besides BCI performances, other data can also be used to investigate the influence and impact of VR on the BCI. For most of the MI studies mentioned in the beginning of Sect. 10.3.1, the electrocardiogram was recorded in addition and questionnaires were conducted. An interesting aspect is that mental simulation of a movement (motor imagery) results in cardiovascular changes explained by two factors: anticipation of movement and central preparation of movement [12,53]. The heart rate (HR) generally decreases during motor imagery in normal BCI conditions (without VR feedback) [38,57] which is similar to that observed during preparation for a voluntary movement. In case of VR feedback, the HR can be increased during effortful imagery [38,57,59]. The heart rate acceleration in the VE is interpreted as effect of an increased mental effort [13] to move as far as possible in VE. This underlines the importance of VR feedback in modifying emotional experiences and enhances autonomic and visceral responses. The HR changes can be in the order of several beats-per-minute (bpm) and therefore could be used to increase the classification accuracy of an ERD-based BCI when both the EEG and the HR are analyzed simultaneously [60].

These heart rate changes were found in most studies of Sect. 10.3.1: (1) In the "walking from thought" study [56], instead of the normal decrease of 3–5 %, an increase of up to 5 % was found. Furthermore, the results provide provisional evidence that moving backwards (negative feedback) resulted in a stronger and longer-lasting HR increase than forward moving (positive one) [57]. (2) In the virtual apartment study [41] the analysis of the heart rate showed that during the BCI condition a preparatory HR deceleration could be found, which is in line with the study [57] but not in the VE conditions since no preparation cue was provided. Generally, the visible HR deceleration is stronger for good trials than for bad trials in all conditions (with removed preparatory phase). Furthermore, a better classification accuracy was accompanied with a stronger deceleration [38]. (3) In contrast to these results, HR increases are observed for two users during VE feedback in study [57]. Interestingly, the slight HR increase (0.5–1 %) before the cue in the VR feedback conditions could be the effect of the anxiety of the user to find the best and correct way for the next decision. Moreover this increase is more dominant in the immersive VE condition, which correlates with the reported higher motivation. (4) In the case of the self-paced navigating study inside the virtual library [43], a phase relationship between the HR and the EEG could be identified. Movement onset occurred during periods of increasing HR, only one user showed a not statistically significant decreasing HR [38]. Although the users were participating in a self-paced experiment, the performance was not completely self-paced but aligned with the underlying cardio-vascular pace. (5) Finally, in the study with the tetraplegic patient [40], the analysis revealed that the induced beta oscillations were accompanied by a characteristic heart rate (HR) change in form of a preparatory HR acceleration followed by a short-lasting deceleration in the order of 10–20 bpm [59]. This provides evidence that mental practice of motor performance is accompanied not only by activation of cortical structures but also by central commands into the cardiovascular system with its nuclei in the brain stem. Another reason for the observed preparatory HR increase could be that the tetraplegic patient was highly motivated and therefore directed increased attention to "walk" in the immersive environment.

Summing up, the use of VR enhanced the user's BCI and application performances and provided motivation (see [38, 39, 56, 61, 64]). These findings are supported by the outcome of the questionnaires and heart rate analysis, where the users self-rated their success stronger than their failure and a stronger HR decrease could be found as well for good classification results. Especially the HR outcome, in the case of an asynchronous (self-paced) BCI, was interesting and it can be speculated that the "free will" of the users was affected by processes operating automatically and unconsciously [31]. Similar influences on self-paced hand movements without awareness of the participants can be caused by transcranial magnetic stimulation [2].

10.5 Conclusion

In this chapter, we have presented and discussed how BCI and VR could be combined and surveyed the related works. As a summary, recent works have shown that BCI could be used to navigate virtual worlds, mostly thanks to motor imagery and SSVEP-based BCI, since these signals enable continuous and self-paced control. BCI could be used to select and manipulate virtual objects as well, for which evoked potentials (P300, SSVEP) seem to be the most used and probably the most appropriate neurophysiological signals. Indeed, such signals enable to select objects simply by paying attention to the corresponding stimulus, and a BCI can deal with numerous such stimuli. On the contrary, MI-based BCI can use only a limited number of mental tasks and are thus less suitable for tasks involving the selection and/or manipulation of numerous virtual objects. These works have also highlighted the challenge in designing BCI-based VR applications, BCI control being usually slow, error-prone and with limited degrees of freedom whereas a VE can be highly interactive and complex. In this context, the design of appropriate interaction techniques and paradigms has shown to be a suitable way to alleviate these limitations and should thus be further studied. Finally, this chapter has highlighted that not only BCI can be a useful interaction device for VE, but that VR could also be a useful technology for BCI. In particular, VR being a rich and motivating environment for the BCI user, it has been shown that this could lead to improved BCI performances, higher motivation and engagement, and reduced human training time in comparison to classical feedback forms. Therefore, BCI and VR can certainly be seen as complementary tools, BCI being useful as an interaction device to enhance the VR experience, and VR being an environment that benefits BCI research and performances.

The various works described in this chapter have also opened the doors to exciting and promising new research topics to further develop the connection between BCI and VR. Indeed, it would be interesting to study how BCI could be used more naturally, transparently and ecologically with virtual environments, in order to make the interactive experience even more immersive. In addition to the classical need for BCI with higher recognition performances, it would be interesting to study whether new mental states and neurophysiological signals could be used to drive a BCI more naturally within a VE. For instance, a study by Plass-Oude Bos et al. suggested that visual spatial attention could be detected, to some extent, from

EEG signals and could thus be used in the future to naturally look around in a VE [62]. Such kind of research efforts should be encouraged in order to develop the repertoire of mental states that could be used to interact mentally with VE. Similarly, further research in the area of passive BCI [23, 76] could help to monitor different mental states of the user (e.g., flow, presence, emotions, attention, etc.) and dynamically adapt the content of the VE accordingly, thus providing an enhanced experience for the user. Since it has been shown that VR could lead to enhanced BCI performances, it would also be interesting to further study the impact of various VR feedback forms (e.g., visual, tactile or audio) on BCI, in order to identify how VR technologies can best optimize the performance and learnability of the system. The specificity of BCI, which do not rely on peripheral nerves and muscles contrary to traditional interfaces, also raise some interesting questions (and maybe answers) related to embodiment and embodied cognition. As such, a system combining BCI and VR might prove a worthy tool and research topic for philosophy and cognitive sciences [9]. Finally, using BCI to interact with VE has the potential to lead to several practical and useful applications. For patients, BCI-based VR applications could enable them to have access to entertainment (e.g., 3D video games), art and culture (e.g., digital creation of paintings, virtual visits of museums and cities) as well as a better social life (e.g., with virtual online communities), which their disabilities might prevent them from doing. This will enable BCI to be useful beyond restoring mobility and basic communication by addressing other important needs of patients [77]. For healthy users, BCI-based VR applications could also be useful, in areas such as entertainment as well [52]—although this may require more improvements in BCI design [45]—and artistic expression [22]. In short, it appears that combining BCI and VR is a promising research topic that is worth being further explored.

Acknowledgements This work was supported by the European Union projects PRESENCIA (IST-2001-37927) and PRESENCCIA (IST-2006-27731), furthermore by the French National Research Agency projects OpenViBE (ANR-05-RNTL01601) and OpenViBE2 (ANR-09-CORD-017).

References

1. Allison, B.Z., McFarland, D.J., Schalk, G., Zheng, S.D., Jackson, M.M., Wolpaw, J.R.: Towards an independent brain–computer interface using steady state visual evoked potentials. Clin. Neurophysiol. **119**(2), 399–408 (2008)
2. Ammon, K., Gandevia, S.C.: Transcranial magnetic stimulation can influence the selection of motor programmes. J. Neurol. Neurosurg. Psychiatry **53**(8), 705–707 (1990)
3. Bayliss, J.D.: Use of the evoked potential P3 component for control in a virtual apartment. IEEE Trans. Neural Syst. Rehabil. Eng. **11**(2), 113–116 (2003)
4. Bayliss, J.D., Ballard, D.H.: A virtual reality testbed for brain–computer interface research. IEEE Trans. Rehabil. Eng. **8**(2), 188–90 (2000)
5. Birbaumer, N., Ghanayim, N., Hinterberger, T., Iversen, I., Kotchoubey, B., Kübler, A., Perelmouter, J., Taub, E., Flor, H.: A spelling device for the paralysed. Nature **398**, 297–298 (1999)

6. Bowman, D., Kruijff, E., Jr, J.L., Poupyrev, I.: 3D User Interfaces: Theory and Practice. Addison-Wesley/Pearson Education, Redwood, USA (2005)
7. Burdea, G.: Force and touch feedback for virtual reality. Wiley, New York, USA (1996)
8. Burdea, G., Coiffet, P.: Virtual Reality Technology. Wiley, New York, USA (2003)
9. Clark, A.: Supersizing the mind: Embodiment, action, and cognitive extension. Oxford University Press, USA (2008)
10. Congedo, M., Goyat, M., Tarrin, N., Varnet, L., Rivet, B., Ionescu, G., Jrad, N., Phlypo, R., Acquadro, M., Jutten, C.: "Brain Invaders": a prototype of an open-source P300-based video game working with the OpenViBE platform. In: 5th International BCI Conference (2011)
11. Cruz-Neira, C., Sandin, D., Defanti, T., Kentyon, R., Hart, J.: The CAVE : audio visual experience automatic virtual environment. Commun. ACM **35**(6), 64–72 (1992)
12. Damen, E.J., Brunia, C.H.: Changes in heart rate and slow brain potentials related to motor preparation and stimulus anticipation in a time estimation task. Psychophysiology **24**(6), 700–713 (1987)
13. Decety, J., Jeannerod, M., Germain, M., Pastene, J.: Vegetative response during imagined movement is proportional to mental effort. Behav. Brain Res. **42**, 1–5 (1991)
14. Donchin, E., Spencer, K.M., Wijesinghe, R.: The mental prosthesis: assessing the speed of a P300-based brain–computer interface. IEEE Trans. Neural Syst. Rehabil. Eng. **8**, 174–179 (2000)
15. Edlinger, G., Holzner, C., Groenegress, C., Guger, C., Slater, M.: Goal-oriented control with brain–computer interface. In: Lecture Notes in Computer Science, Springer, Berlin/Heidelberg, vol. 5638, pp. 732–740 (2009)
16. Edlinger, G., Holzner, C., Guger, C.: A hybrid brain–computer interface for smart home control. In: Human Computer Interface Conference, Springer, Berlin/Heidelberg, 417–425 (2011)
17. Elshout, J., Molina, G.G.: Review of brain–computer interfaces based on the P300 evoked potential. Tech. Rep. PR-TN 2009/00066, Koninklijke Philips Electronics (2009)
18. Faller, J., Allison, B., Brunner, C., Schmalstieg, D., Pfurtscheller, G.: A software SSVEP BCI integrating stimuli within motivating and immersive virtual and augmented reality environments. In: Real Actions in Virtual Environments (RAVE) conference, Barcelona, Spain (2010a)
19. Faller, J., Leeb, R., Pfurtscheller, G., Scherer, R.: Avatar navigation in virtual and augmented reality environments using an SSVEP BCI. In: International Conference on Applied Bionics and Biomechanics (ICABB) 2010, Venice, Italy (2010b)
20. Faller, J., Müller-Putz, G.R., Schmalstieg, D., Pfurtscheller, G.: An application framework for controlling an avatar in a desktop based virtual environment via a software SSVEP brain–computer interface. Presence (Camb.) **19**(1), 25–34 (2010c)
21. Fellner, D., Havemann, S., Hopp, A.: Dave - eine neue technologie zur preiswerten und hochqualitativen immersiven 3d-darstellung. In: Möller, R. (ed.) Proc. 8. Workshop: Sichtsysteme - Visualisierung in der Simulationstechnik, pp. 77–83. Shaker Verlag, Bremen (2003)
22. Friedman, D., Donenfeld, A., Zafran, E.: Neurophysiology-based art in immersive virtual reality. Int. J. Arts Technol. **2**(4), 331–347 (2009)
23. George, L., Lécuyer, A.: An overview of research on passive brain–computer interfaces for implicit human–computer interaction. In: International Conference on Applied Bionics and Biomechanics (2010)
24. Groenegress, C., Holzner, C., Guger, C., Slater, M.: Effects of P300-based BCI use on reported presence in a virtual environment. Presence (Camb.) **19**(1), 1–11 (2010)
25. Grychtol, B., Lakany, H., Valsan, G., Conway, B.A.: Human behavior integration improves classification rates in real-time BCI. IEEE Trans. Neural Syst. Rehabil. Eng. **18**(4), 362–368 (2010)
26. Guger, C., Schlögl, A., Neuper, C., Walterspacher, D., Strein, T., Pfurtscheller, G.: Rapid prototyping of an EEG-based brain–computer interface (BCI). IEEE Trans. Rehab. Eng. **9**(1), 49–58 (2001)

27. Guger, C., Holzner, C., Groenegress, C., Edlinger, G., Slater, M.: Control of a smart home with a brain–computer interface. In: 4th International Brain–Computer Interface Workshop, pp. 339–342 (2008)
28. Guger, C., Daban, S., Sellers, E., Holzner, C., Krausz, G., Carabalona, R., Gramatica, F., Edlinger, G.: How many people are able to control a P300-based brain–computer interface (BCI)? Neurosci. Lett. **462**(1), 94–98 (2009a)
29. Guger, C., Holzner, C., Groenegress, C., Edlinger, G., Slater, M.: Brain–computer interface for virtual reality control. In: Proceedings of ESANN 2009, pp. 443–448 (2009b)
30. Guger, C., Edlinger, G., Krausz, G.: Recent Advances in Brain–Computer Interface Systems, InTech, chap. Hardware/Software Components and Applications of BCIs, Rijeka, Croatia, 1–24 (2011)
31. Haggard, P.: Conscious intention and motor cognition. Trends Cogn. Sci. **9**(6), 290–295 (2005)
32. Kronegg, J., Chanel, G., Voloshynovskiy, S., Pun, T.: EEG-based synchronized brain–computer interfaces: A model for optimizing the number of mental tasks. IEEE Trans. Neural Syst. Rehabil. Eng. **15**(1), 50–58 (2007)
33. Krusienski, D., Sellers, E., Cabestaing, F., Bayoudh, S., McFarland, D., Vaughan, T., Wolpaw, J.: A comparison of classification techniques for the P300 speller. J. Neural Eng. **3**, 299–305 (2006)
34. Kuebler, A., Nijboer, F., Mellinger, J., Vaughan, T.M., Pawelzik, H., Schalk, G., McFarland, D.J., Birbaumer, N., Wolpaw, J.R.: Patients with ALS can use sensorimotor rhythms to operate a brain–computer interface. Neurology **64**(10), 1775–1777 (2005)
35. Lalor, E., Kelly, S., Finucane, C., Burke, R., Smith, R., Reilly, R.B., McDarby, G.: Steady-state vep-based brain computer interface control in an immersive 3-d gaming environment. EURASIP J. Appl. Signal Process. **19**, 3156–3164 (2005)
36. Lécuyer, A.: Using eyes, hands, and brain for 3D interaction with virtual environments: A perception-based approach. Tech. rep., Habilitation thesis (2010)
37. Lécuyer, A., Lotte, F., Reilly, R., Leeb, R., Hirose, M., Slater, M.: Brain–computer interfaces, virtual reality and videogames. IEEE Computer **41**(10), 66–72 (2008)
38. Leeb, R.: Brain-computer communication: the motivation, aim, and impact of virtual feedback. PhD thesis, Graz University of Technology (2008)
39. Leeb, R., Keinrath, C., Friedman, D., Guger, C., Scherer, R., Neuper, C., Garau, M., Antley, A., Steed, A., Slater, M., Pfurtscheller, G.: Walking by thinking: the brainwaves are crucial, not the muscles! Presence (Camb.) **15**, 500–514 (2006)
40. Leeb, R., Friedman, D., Müller-Putz, G.R., Scherer, R., Slater, M., Pfurtscheller, G.: Self-paced (asynchronous) BCI control of a wheelchair in virtual environments: a case study with a tetraplegics. Comput. Intell. Neurosci. **2007**, 79,642 (2007a)
41. Leeb, R., Lee, F., Keinrath, C., Scherer, R., Bischof, H., Pfurtscheller, G.: Brain-computer communication: motivation, aim and impact of exploring a virtual apartment. IEEE Trans. Neural Syst. Rehabil. Eng. **15**, 473–482 (2007b)
42. Leeb, R., Scherer, R., Friedman, D., Lee, F.Y., Keinrath, C., Bischof, H., Slater, M., Pfurtscheller, G.: Combining BCI and virtual reality: scouting virtual worlds. In: Dornhege, G., Millán, J., Hinterberger, T., McFarland, D.J., Müller, K.R. (eds.) Toward brain–computer interfacing, chap 23, pp. 393–408. MIT Press, Cambridge/London (2007c)
43. Leeb, R., Settgast, V., Fellner, D.W., Pfurtscheller, G.: Self-paced exploring of the Austrian National Library through thoughts. Int. J. Bioelectromagn. **9**, 237–244 (2007d)
44. Legény, J., Viciana-Abad, R., Lécuyer, A.: Navigating in virtual worlds using a self-paced SSVEP-based brain–computer interface with integrated stimulation and real-time feedback. Presence – Teleoperators and Virtual Environments, vol 20(6), 529–544, 2011
45. Lotte, F.: Brain–computer interfaces for 3D games: Hype or hope? In: Foundations of Digital Games, pp. 325–327 (2011)
46. Lotte, F., Renard, Y., Lécuyer, A.: Self-paced brain–computer interaction with virtual worlds: a qualitative and quantitative study "out-of-the-lab." In: 4th International Brain–Computer Interface Workshop and Training Course, pp. 373–378 (2008)

47. Lotte, F., Langhenhove, A.V., Lamarche, F., Ernest, T., Renard, Y., Arnaldi, B., Lécuyer, A.: Exploring large virtual environments by thoughts using a brain–computer interface based on motor imagery and high-level commands. Presence (Camb.) **19**(1), 54–70 (2010)
48. Mason, S., Kronegg, J., Huggins, J., Fatourechi, M., Schloegl, A.: Evaluating the performance of self-paced BCI technology. Tech. rep., Neil Squire Society (2006)
49. Millán JdR., Rupp, R., Müller-Putz, G., Murray-Smith, R., Giugliemma, C., Tangermann, M., Kübler, A., Leeb, R., Neuper, C., Müller, K.R., Mattia, D.: Combining brain-computer interfaces and assistive technologies: state-of-the-art and challenges. Front. Neuroprosthetics, 4(161), 1–15 (2010)
50. Neuper, C., Scherer, R., Wriessnegger, S., Pfurtscheller, G.: Motor imagery and action observation: modulation of sensorimotor brain rhythms during mental control of a brain–computer interface. Clin. Neurophysiol. **120**(2), 239–247 (2009)
51. Nijholt, A., Tan, D., Pfurtscheller, G., Brunner, C., del R Millán, J., Allison, B., Graimann, B., Popescu, F., Blankertz, B., Müller, K.R.: Brain–computer interfacing for intelligent systems. IEEE Intell. Syst. **23**, 72–79 (2008)
52. Nijholt, A., Bos, D.P.O., Reuderink, B.: Turning shortcomings into challenges: Brain–computer interfaces for games. Entertain. Comput. **1**(2), 85–94 (2009)
53. Oishi, K., Kasai, T., Maeshima, T.: Autonomic response specificity during motor imagery. J. Physiol. Anthropol. Appl. Human Sci. **19**(6), 255–261 (2000)
54. Pfurtscheller, G., Lopes da Silva, F.H.: Event-related EEG/MEG synchronization and desynchronization: basic principles. Clin. Neurophysiol. **110**, 1842–1857 (1999)
55. Pfurtscheller, G., Neuper, C.: Motor imagery and direct brain–computer communication. Proc. IEEE **89**, 1123–1134 (2001)
56. Pfurtscheller, G., Leeb, R., Keinrath, C., Friedman, D., Neuper, C., Guger, C., Slater, M.: Walking from thought. Brain Res. **1071**(1), 145–152 (2006a)
57. Pfurtscheller, G., Leeb, R., Slater, M.: Cardiac responses induced during thought-based control of a virtual environment. Int. J. Psychophysiol. **62**, 134–140 (2006b)
58. Pfurtscheller, G., Müller-Putz, G.R., Schlögl, A., Graimann, B., Scherer, R., Leeb, R., Brunner, C., Keinrath, C., Lee, F., Townsend, G., Vidaurre, C., Neuper, C.: 15 years of BCI research at Graz University of Technology: current projects. IEEE Trans. Neural Syst. Rehabil. Eng. **14**, 205–210 (2006c)
59. Pfurtscheller, G., Leeb, R., Friedman, D., Slater, M.: Centrally controlled heart rate changes during mental practice in immersive virtual environment: a case study with a tetraplegic. Int. J. Psychophysiol. **68**, 1–5 (2008)
60. Pfurtscheller, G., Allison, B., Bauernfeind, G., Brunner, C., Solis Escalante, T., Scherer, R., Zander, T., Müller-Putz, G., Neuper, C., Birbaumer, N.: The hybrid BCI. Front. Neurosci. **4**, 42 (2010)
61. Pineda, J.A., Silverman, D.S., Vankov, A., Hestenes, J.: Learning to control brain rhythms: making a brain–computer interface possible. IEEE Trans. Neural Syst. Rehabil. Eng. **11**, 181–184 (2003)
62. Plass-Oude Bos, D., Duvinage, M., Oktay, O., Delgado Saa, J., Guruler, H., Istanbullu, A., Van Vliet, M., Van de Laar, B., Poel, M., Roijendijk, L., Tonin, L., Bahramisharif, A., Reuderink, B.: Looking around with your brain in a virtual world. In: IEEE Symposium on Computational Intelligence, Cognitive Algorithms, Mind, and Brain (SSCI'2011 CCMB) (2011)
63. Renard, Y., Lotte, F., Gibert, G., Congedo, M., Maby, E., Delannoy, V., Bertrand, O., Lécuyer, A.: OpenViBE: An open-source software platform to design, test and use brain–computer interfaces in real and virtual environments. Presence (Camb.) **19**(1), 35–53 (2010)
64. Ron-Angevin, R., Diaz-Estrella, A.: Brain–computer interface: Changes in performance using virtual reality technique. Neurosci. Lett. **449**(2), 123–127 (2009)
65. Ron-Angevin, R., Diaz-Estrella, A., Velasco-Alvarez, F.: A two-class brain computer interface to freely navigate through virtual worlds. Biomedizinische Biomed. Tech. (Berl.) **54**(3), 126–133 (2009)
66. Scherer, R., Schlögl, A., Lee, F., Bischof, H., Jansa, J., Pfurtscheller, G.: The self-paced graz brain–computer interface: methods and applications. Comput. Intell. Neurosci. **2007**, 1–9 (2007) (Article ID 79826)

67. Scherer, R., Lee, F., Schlögl, A., Leeb, R., Bischof, H., Pfurtscheller, G.: Towards self-paced brain–computer communication: Navigation through virtual worlds. IEEE Trans. Biomed. Eng. **55**(2), 675–682 (2008)
68. Slater, M., Usoh, M., Steed, A.: Taking steps: the influence of a walking technique on presence in virtual reality. ACM Trans. Comput. Hum. Interact. **2**(3), 201–219 (1995)
69. Taylor, R.M., Hudson, T.C., Seeger, A., Weber, H., Juliano, J., Helser, A.: VRPN: A device-independent, network-transparent VR peripheral system. In: VRST '01, Proceedings of the ACM symposium on Virtual reality software and technology, pp. 55–61. ACM, New York, NY, USA (2001)
70. Touyama, H.: Advances in Human Computer Interaction, InTech Education and Publishing, chap Brain-CAVE Interface Based on Steady-State Visual Evoked Potential, pp. 437–450 (2008). No. 26 in ISBN 978-953-7619-15-2
71. Vaughan, T.M., Wolpaw, J.R., Donchin, E.: EEG-based communication: prospects and problems. IEEE Trans. Rehabil. Eng. **4**, 425–430 (1996)
72. Velasco-Álvarez, F., Ron-Angevin, R.: Free virtual navigation using motor imagery through an asynchronous brain–computer interface. Presence (Camb.) **19**(1), 71–81 (2010)
73. Vialatte, F., Maurice, M., Dauwels, J., Cichocki, A.: Steady-state visually evoked potentials: Focus on essential paradigms and future perspectives. Prog. Neurobiol. **90**, 418–438 (2010)
74. Williamson, J., Murray-Smith, R., Blankertz, B., Krauledat, M., Müller, K.: Designing for uncertain, asymmetric control: Interaction design for brain–computer Int. J. Hum. Comput. Stud. **67**(10), 827–841 (2009)
75. Wolpaw, J., Birbaumer, N., McFarland, D., Pfurtscheller, G., Vaughan, T.: Brain–computer interfaces for communication and control. Clin. Neurophysiol. **113**(6), 767–791 (2002)
76. Zander, T., Kothe, C.: Towards passive brain–computer interfaces: applying brain–computer interface technology to human-machine systems in general. J. Neural Eng. **8**(2), 025005 (2011)
77. Zickler, C., Donna, V.D., Kaiser, V., Al-Khodairy, A., Kleih, S., Kuebler, A., Malavasi, M., Mattia, D., Mongardi, S., Neuper, C., Rohm, M., Rupp, R.: Brain computer interaction applications for people with disabilities: Defining user needs and user requirements. In: AAATE (2009)

Part III
Application Interfaces and Environments

Chapter 11
Brain–Computer Interfaces and User Experience Evaluation

Bram van de Laar, Hayrettin Gürkök, Danny Plass-Oude Bos, Femke Nijboer, and Anton Nijholt

11.1 Introduction

Brain–computer interfaces (BCIs) aim to provide a reliable control signal for assistive technology for disabled persons. With the merge of the fields of human–computer interaction (HCI) and BCI new applications are being developed for entertainment and education which may be interesting for users with and without disabilities. BCIs will be integrated into existing interactive applications. The aim of such applications is to create positive experiences that enrich our lives rather than only providing reliable control. Recently, it was suggested at several keynote presentations at large BCI conferences that reliability is the most important issue to be addressed to achieve technology transfer to the market and the society. However, perfectly reliable systems are not necessarily usable. Even reliable assistive technologies may get abandoned by users when usability is not warranted [35]. Making interactive systems usable is the core expertise of the field of HCI (see Chap. 9. The process of designing interactive systems in the field of HCI consists of analysis of requirements, design and implementation of the system and user evaluation. To evaluate such systems, the user experience (UX) needs a more important role in BCI studies. Researchers should not only focus on the reliability of the control signal, so that we can better understand how such a system can satisfy the needs of the user (see also Chap. 8 for user centered design).

At this point we should make clear that the concept of usability is not the same as user experience, although they are related. The most widely accepted model of measuring user-oriented quality assessment of interactive systems consists of three elements: functionality, usability and the user experience [24]. Functionality

B. van de Laar (✉) · H. Gürkök · D.P.-O. Bos · F. Nijboer · A. Nijholt
Human Media Interaction, University of Twente, Enschede, The Netherlands
e-mail: b.l.a.vandelaar@utwente.nl; h.gurkok@utwente.nl; d.plass@utwente.nl; femke.nijboer@utwente.nl; a.nijholt@utwente.nl

B. Z. Allison et al. (eds.), *Towards Practical Brain-Computer Interfaces*, Biological and Medical Physics, Biomedical Engineering, DOI 10.1007/978-3-642-29746-5_11,
© Springer-Verlag Berlin Heidelberg 2012

is about what one can do with the system, id est, what role does it fulfill? Technical aspects such as performance, maintainability, reliability and durability are important. Usability contains higher level concepts such as satisfaction, efficiency, effectiveness, learnability and usefulness. These can partly result from the functionality but are mainly defined by the interaction of the user with the system. Hence, these concepts cannot be tested without real users. User experience is about what the user feels and experiences using the system. User experience therefore contains higher level concepts as immersion (the user is involved and/or lost track of time), fun, engagement, presence (in case of a game, users experience being "in" the virtual world) et cetera. Even though usability and user experience evaluation is not common in current BCI studies, the user's experience may influence objective performance measures, such as BCI classifier accuracies, and has a big impact on whether users are actually willing to use a specific system.

In this paper, we review studies that investigate user experience in BCI research and the benefits of including such evaluations. Then, we will argue how the use of various techniques from the field of HCI can be advantageous for evaluating BCIs. In the last part of this paper we will elaborate on some case studies and provide recommendations for evaluating user experience with BCIs.

11.2 Current State of User Experience Evaluation of BCI

11.2.1 User Experience Affects BCI

User-centered approaches can increase usability and user acceptance, which is why some BCI groups involve users in the design process. They assess user needs, develop user requirements, and evaluate the usability [14, 21, 33, 41] and Chap. 8. What is often ignored, however, is the importance of assessing the UX and user acceptance in a structured way *during* or *directly after* interaction with the system. The BCI studies that do include UX evaluations indicate three main reasons: its potential to increase user acceptance, to improve performance of the system, and to increase enjoyment. Each of these are discussed in more detail next.

In a study by Münßinger et al., the mood and motivation of users of a BCI painting application was evaluated using a visual analogue scale (VAS) [26]. Patients with amyotrophic lateral sclerosis (ALS) were more motivated to train with the application than healthy users. While the healthy users also had other options for creative expression, this BCI application provided a unique opportunity to the paralyzed patients. Several BCI studies suggest a relation between motivation and BCI task performance [20, 27], and small but significant effects have been found [17] using an adapted version of the Current Motivation Questionnaire. This questionnaire assesses the current motivation in learning and performance situations [34]. Similarly, the users' belief of how accurately they can control a BCI has an influence on their actual performance. Barbero and Grosse-Wentrup observed that participants who normally perform around chance level, perform better when they think they are doing better than they actually are (positive bias). Capable participants, however, performed worse when given inaccurate feedback, whether the bias was positive or negative [1].

Motivation may be only one of the performance-related factors that are influenced by the UX. By evaluating and improving the UX, other relations between the user and BCI recognition performance could be exploited to improve performance measures. There could also be mechanisms with indirect influences. For example, a system that is perceived as more beautiful is also perceived as more usable [38]. This perception could influence motivation which in turn could influence performance. Similarly, a more positive experience may cause users to be more indulgent towards minor usability problems, increasing the user acceptance [29].

Most current BCI applications still serve only as a proof of concept [25], which may be why the entertainment value is often not evaluated. An exception is the BCI game BrainBasher, which was evaluated for the influence of different graphical interfaces and different user tasks [18,32]. The Game Experience Questionnaire was used to assess immersion, tension, competence, flow, negative affect, positive affect, and challenge [15]. In the first study, the UX and performance were determined for a clinical setup with minimal information on the screen. This was compared to a game-like setup of exactly the same task. The game version resulted in higher immersion. The second study compared the UX for imaginary and actual movement. Imagined movement was perceived as more challenging, but when using actual movement the participants stayed more alert.

While more research is still needed, the few studies so far suggest that UX can affect a BCI system in important ways. Therefore it is vital that the UX of BCI systems is properly evaluated.

11.2.2 BCI Affects User Experience

UX can influence the performance of BCIs, but BCIs can affect the UX as well, in two ways: (1) through the effects of using this particular input modality, and (2) by using information about the user's mental state to adapt the interface or the interaction itself, with as goal to improve usability and UX. Here are some examples to illustrate this.

Using BCI for input can in itself influence the UX (see also Chap. 10). Friedman et al. [9] investigated whether the use of imaginary movement to walk in a virtual world would increase the sense of being present there, using the Slater–Usoh–Steed presence questionnaire combined with a non-structured interview [9,36]. In a follow-up experiment, Groenegress et al. compared the presence experienced with a P300 interface to eye gaze and wand navigation [10]. Both experiments concluded that the BCI did not have a positive influence on presence. In a study by Vilimek and Zander [39], an eye gaze system was augmented with a BCI to simulate the mouse click. The resulting workload of the BCI method was compared to the standard method of using dwell times for activation, using the NASA TLX [13]. There was no significant difference between the workload for either activation method, so the BCI did not result in a higher cognitive demand. A more recent study by Hakvoort et al. [12] compared a BCI selection method with a non-BCI selection method, made equivalent in terms of time and effort necessary for selection [12]. The comparison was based on affect, evaluated with the self-assessment manikin [4], and on immersion, which was determined with the questionnaire developed by Jennett et al. [16]. In this case, the BCI did turn out to be more immersive and to result in a more positive experience.

With the help of BCI, users can also be supported in the tasks they are trying to accomplish, which in turn should increase user satisfaction. For example, error-related brain activity can be detected and used to fix user or system errors for improved error handling [40]. The amount of information presented on screen can be adjusted according to the user's workload [37]. BCI could also be used to create or maintain specific user experiences. As an example, brain activity indicators of stress or boredom can be used to keep the user in the optimal state of flow, where the challenge of the task is matched to the skill of the user [6, 7].

But the influence of BCI on UX may extend even further. Obbink et al. [30] investigated the influence of using a BCI on social interaction in a cooperative game [30]. The social interaction was assessed in terms of the amount of speech, number of utterances, and gestures. Additionally, a custom questionnaire at the end of the experiment was provided to evaluate the participants' self-reported, subjective experience. Because of the higher difficulty of the BCI-based selection, compared to point-and-click with a traditional mouse, there were more utterances and empathic gestures (see also Sect. 11.4).

All in all, whether BCI is used to affect the UX purposefully or whether this happens by simply using this input modality, in both cases it is important to evaluate and be aware of the effects. The next section will show different methods to do this, and discuss the implications for using them to evaluate BCIs specifically.

11.3 Applying HCI User Experience Evaluation to BCIs

Although evaluating the usability and UX of BCI systems is not common practice, in HCI research and development, especially for entertainment technologies which simply aim to improve the well-being of users, UX is a major concern. Therefore, the HCI community designs for UX and develops methods to evaluate it. Current methods for evaluating UX in entertainment technologies can be classified into quadrants of a plane which has an objective versus subjective axis and a qualitative versus quantitative axis [22] (see Fig. 11.1). The objective methods are based on overt and covert user responses during interaction while the subjective methods rely on user expressions after the interaction. The quantitative methods employ statistical analysis on collected data whereas the qualitative methods are interpretations of user responses by researchers. Below, we describe the methods corresponding to

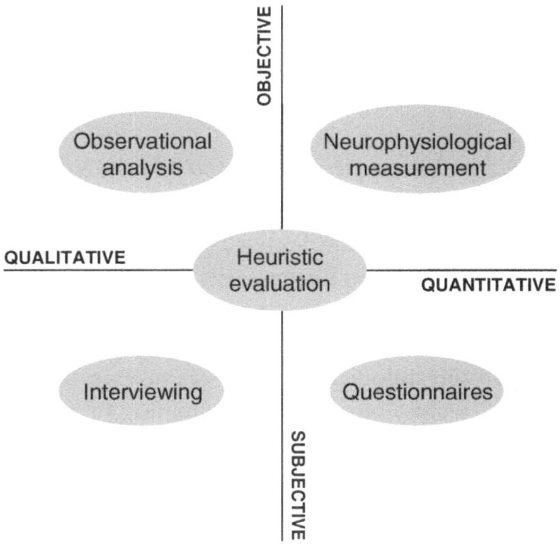

Fig. 11.1 A classification of current user experience evaluation methods used in human–computer interaction for entertainment technologies (adapted from [22])

the quadrants formed by these two axes and discuss their contribution in evaluating BCI systems.

11.3.1 Observational Analysis

Observational analysis is a qualitative–objective method which relies on overt user response. The classical way of observing overt user behaviour is through audiovisual recorders which provide qualitative data for gestures, facial expressions and verbalisations. There are some difficulties associated with annotating and analysing such rich data though. Firstly, while analysing the data, the researchers should acknowledge their biases, address inter-rater reliability and not read inferences where none are present. Secondly, there is an enormous time commitment associated with observational analysis. The ratio of analysis time to data sequence time ranges from 5:1 to 100:1 [23]. Thirdly, the operation of audiovisual recorders impose restrictions such as a noise-free environment during audio recording or consistent illumination during video capturing. Some restrictions are also imposed by brain activity recording devices. For example, the electroencephalogram (EEG, measuring electrical brain activity) is affected by the user's movement [8], so users are usually asked to keep their bodies and faces motionless. Thus, overt behaviour of users of BCIs will be minimal and observational analysis may not obtain sufficient data to analyze UX. Moreover, severely disabled people, such as patients with locked-in syndrome (LiS) who lose all their muscle control except for vertical eye movements [3] and who constitute a non-negligible user group for BCIs, are not able to show any overt behaviour at all. Consequently, in clinical experiments observational analysis is not a strong method for evaluating UX, although for studies in natural environments it might prove useful.

11.3.2 Neurophysiological Measurement

Task performance metrics have been suggested as quantitative-objective measures of UX but these are not necessarily the indicators of UX. Especially in entertainment applications, there might not be a clear task or users might prefer navigating in the virtual environment without any urge to complete tasks. More recently, use of neurophysiological signals was proposed to model the emotional state of users in play technologies [23]. Examples of psychophysiological signals are EEG, galvanic skin response (GSR, measuring skin conductivity) and electrocardiogram (ECG, measuring electrical heart activity). Measured emotions capture usability and playability through metrics relevant to play experience so they provide objective data. They account for user emotion and they are represented continuously over a session. While interacting with a BCI, at least one neurophysiological signal, the EEG, can already be recorded as it is used as an input signal. It is a golden opportunity to extract UX-related features from the brain signals using the same signals. Several problematic issues can be identified when recording psychophysiological signals. First of all, the research on using neurophysiological sensors to measure UX is in its infancy. The neurophysiological correlates of UX or its components are not well-defined which makes this method rather a questionable one. Secondly, the sensors attached to the user might induce discomfort to the user, restrict movements or influence the experience. So, the researchers should limit the number of sensors applied on the user. Thirdly, while measuring the UX through the same neurophysiological sensor that is used for controlling the application, UX-related responses should be differentiated from task-related activity.

11.3.3 Interviewing and Questionnaires

Interviews and questionnaires provide subjective data for assessing UX. They take place after interacting with a system thus are unobtrusive but then not able to extract instantaneous experiences during interaction. One way to converge capturing short-term UX might be to conduct questionnaires and interviews incrementally, id est, in multiple sessions, rather than conducting a single questionnaire/interview after the interaction has taken place. For disabled users, especially those with LiS, using subjective methods might not seem to be the easiest way to assess UX as these people might not be able to talk or write. However, if the interviews and questionnaires are prepared in such a way that they can be answered using a small number of choices, such as yes, no and maybe, then they can be completed by these users as well.

Interviewing is a qualitative–subjective technique. During interviews, researchers should be careful to pose the right questions during the interview, if necessary, by monitoring the interaction and detecting unexpected events. The interviewers should remain neutral and refrain from asking leading questions. An example demonstrating the use of interviews in BCI UX evaluation is the study by Gürkök et al. [11]. In their study, the authors conducted interviews with participants to find out the reasons why people switched between BCI and speech control in a multimodal game.

Questionnaires are designed to provide quantitative–subjective data. Users rate the items in a questionnaire on a Likert-scale or a Visual Analogue scale, which yields a number of how much they agreed with a statement. Development of UX questionnaires for entertainment applications has received attention from researchers, especially those who are interested in games. The recently developed Game Engagement Questionnaire [5] includes items related to absorption, flow, presence and immersion. There are also questionnaires focusing exclusively on the components that contribute to UX such as presence [2] and immersion [16].

11.3.4 Other Methods

Another concept that is often related to UX is the usability of the interface. Many heuristics have been proposed for evaluating the usability of video games [31]. However heuristic evaluation does not involve actual users and the usability of an interface alone does not represent the UX. Before questionnaires are used to evaluate BCIs, they may require adaptation taking into account that state-of-the-art BCI applications are relatively simple thus modest in providing rich UX. BCI recognition performance should also be taken into account, as a relatively low performance might influence the UX.

Analysing logged software data is also considered as a quantitative–objective method for UX evaluation in some studies. Logs are not direct correlates of UX but they might be helpful in understanding the course of interaction, identifying problems or certain preferences, and thus in designing for better UX. For example, by analysing the frequency of key presses in a game, one can derive a cluster of events to which the player was more reactive and can use this new information to design better interaction.

The important factors in selecting the right UX assessment method for BCIs can be listed as the ease of deployment and analysis for the researcher, the comfort of deployment on the user, the strength and reliability in representing the actual UX, and the width of the user spectrum. As seen within this section, all the methods partially fulfill these criteria. Nevertheless, questionnaires stand as strong candidates as they are easy and comfortable to apply, suitable for extracting statistical analyses quickly, strong and reliable when validated and applicable to the majority of the BCI users.

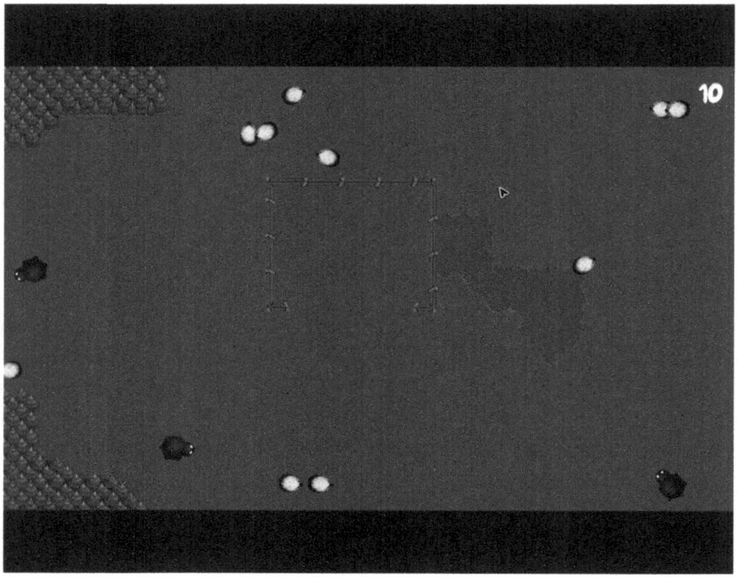

Fig. 11.2 A screenshot from the game Mind the Sheep! depicting the game world with ten sheep, three dogs and the pen

11.4 Case Studies

In this section we will elaborate on two case studies in which we applied various methods of UX evaluation and will try to explain why we chose a certain method and how it answers our research questions (see other experiences in Chap. 8).

11.4.1 Case Study: Mind the Sheep!

We did a series of UX evaluation studies using the multimodal game we developed, called Mind the Sheep!. The game world (see Fig. 11.2) is a meadow on which a number of (white) sheep move autonomously and the (black) shepherd dogs can be commanded by the player. When a dog approaches some sheep, the sheep will tend to flock and move away from the dog. This way the sheep are herded in a desired direction. The goal of the game is to gather the sheep in a pen as quickly as possible.

The game can be played using different modalities in different ways. In the BCI controlled version of the game, to command a dog, the player positions the cursor at the point to which the dog is supposed to move. The player holds the mouse button pressed to provide the command to select the dog. Meanwhile, the dog images are replaced by circles flickering at different frequencies and the player concentrates on the circle replacing the dog they want to select (so as to obtain an SSVEP). The stimulation persists and EEG data is accumulated as long as the mouse button is held. When the player releases the mouse button, the signal is analysed and a dog is selected based on this analysis. The selected dog immediately moves to the location where the cursor was located at the time of mouse button release.

In the first study we describe here, we compared BCI control to simple mouse control to study the social interaction between players in the cooperative multimodal version of the game [30]. To control the dogs using the mouse, the player first clicks on the dog they want to select and then on the location they want the dog to move to. In the cooperative multiplayer version of the game, co-located players work together to pen the sheep so they need to interact while playing to develop a strategy. However, interaction means such as speech and bodily movements might impair the accuracy of the BCI due to the noise they impose on the EEG. So there is a trade-off between maintaining a strategy during the game and maintaining a certain accuracy level. We did an experiment with ten pairs playing the game with both controllers. We performed an observational analysis of the audio–visual data recorded during the play. Though non-significantly, during mouse control the participants produced more speech and instrumental gestures, which are overt messaging channels. This implies that they interacted more freely with mouse control. On the other hand, during BCI control participants produced, again non-significantly, more utterances and empathic gestures, which are emotion signalling channels. This finding suggests that the participants were affected more by the events during the BCI game; perhaps they were surprised, or things went wrong more often.

In another study with the single-player version of Mind the Sheep!, we evaluated UX in terms of immersion and affect through questionnaires [12]. Again, we compared BCI control to mouse control but this time the way the mouse was used was different. Now, the player had to hold the mouse button pressed when they wanted to make a selection. The dogs were highlighted one at a time with an increasing highlight period. When the player released the mouse button, the currently highlighted dog was selected. To make an accurate selection, the player needs to react in the time when the dog they want to select is highlighted. This way, mouse control becomes similar to BCI control so that they both offer some challenge to the player. In our experiment we let 17 participants play the game with BCI and mouse control and after each game we evaluated UX using the self assessment manikin [4] for affect and the immersion questionnaire [16]. Evaluation results showed that BCI control was found to be more immersive ($p = 0.031$) and positively affective ($p = 0.044$) than mouse control. Furthermore, analysis of the logged game data revealed that participants appeared to have more patience with BCI control than mouse control, which could have been caused by the curiosity of participants for BCI control or by their self overestimation during mouse control.

11.4.2 Case Study: Hamster Lab

The aim of the study was to investigate the effect of level of control on user experience. We conducted an experiment, with 200 participants, in which they played a game with a varying amount of control: the game is called Hamster Lab [19]. The user controls a hamster situated in maze like game levels, set in a laboratory setting (see Fig. 11.3). The user controls the hamster by pressing the arrow keys on a keyboard. This way, the user has five possible options for control: up, right, down, left and do not move. The 15 controlled conditions in the game are specified with a certain amount of control. From perfect control, where every press of the button is directly translated to the corresponding action in the game, to 20 % control, id est, chance level, where there is an even chance for a certain action to be translated to any of the five possible actions. Using this manipulation of control we can simulate the feeling of unreliable input such as is the case with a BCI. Thus, whereas the relation between control and user experience can only be investigated through correlational analysis in conventional BCI experiments, through this simulation we can study the effects using controlled conditions. Hamster Lab is an online game suitable for playing in a web browser. This made it easier to gather participants for the study we will next describe.

This study was conducted to find the relationship between fun and control. After the user commanded the hamster to the exits of the four mazes a short questionnaire was presented to the user. The questionnaire was kept to a bare minimum of what we wanted to know. There were nine questions in total: six were Visual Analogue Scale (VAS) items and three questioned for basic demographics (age, gender and a field where user could give feedback. As it was an online game probably not all participants were motivated to answer the number of questions one would normally ask after an experiment. Furthermore, we wanted the participants to play multiple sessions to gather more data. Based on IP addresses 200 unique participants started a round. Three hundred and fifty-one rounds in total were started. Two hundred and twelve (60.4 %) of these runs were continued through the four levels and filled in the questionnaire completely. By most participants the short questionnaire was appreciated (12 people played five rounds of four levels). Though for 39.6 % of the runs, even nine questions were too much for the participant to bother, as some entries in our database showed with comments like “Why these questions, I want to play!” and items that remained unanswered. This number includes the runs that were started and terminated halfway, for example, participants that did not like the game and closed their browser window.

The six VAS items provided us with information on the UX of the participants on the following concepts: fun, frustration, control, dominance and empowerment. First we had to assert that our way of influencing the amount of control was also perceived by the user as such. Regression analysis on the amount of control and the perceived control showed a very significant linear trend, hence, the higher the amount of control “given” to the user, the higher the amount of control they

Fig. 11.3 A screenshot from the game Hamster Lab depicting the game world with the hamster in the first level of the game

perceived. For the relationship between fun and control we hypothesized that fun is positively influenced by control (users are able to do the things they intend) but an optimum exists before the maximum amount of control (the game is too easy and/or nothing surprising happens). A regression analysis showed that more variance was explained using a third order polynomial (34.9 %) than using a linear (29.1 %) trend would. The third order polynomial showed an optimum before the 100 % control mark which was also shown by comparing the medians of the conditions with a high amount of control. This supports our hypothesis and is also in line with the idea that users playing a game want to have some kind of challenge in a game. In this way, the unreliable input from a BCI can be used as a challenge in a game [28]. The result is validated only for this specific game, other games might show a curve looking slightly different but with the same characteristics. Games that are in itself a big challenge for the user might require 100 % control. An interesting conclusion from this is that one wants to use a BCI for control, the game difficulty needs to be adjusted to balance the user skills and game challenge for the optimal experience of flow [6].

11.5 Discussion and Conclusion

In this paper we stressed the need for UX evaluation of BCI applications. While some research has been done on this, it remains largely an uncultivated area of research. However, we can learn from methods developed in the field of human–computer interaction.

To evaluate a BCI system several methods are available: observational analysis can be used in settings where the interaction of the user(s) with the BCI system as a whole is important. For example in the case study of Mind the Sheep! we showed that observational analysis is a useful method when evaluating systems in a realistic setting especially when users can also interact with each other. When overt user response to the system is limited, id est, in case of a clinical experiment, when the user is disabled, or in non observable settings such as for example a web-based experiment, observational analysis is less useful.

Neurophysiological measurements are a quantitative–objective method to assess UX. However, these techniques are still topic of research and most are not very reliable at the present. If a reliable neurophysiological method is used however, this provides a worthful source of information as the signal is continuous of nature, as opposed to for example a questionnaire.

Interviews are especially useful in explorative studies. Asking open (non-leading) questions can lead to the reason *why* a user does or does not like a certain aspect of the system or why the users did what they did. This information is hard to capture through other methods, as it is quite detailed in nature.

Questionnaires are quantitative of nature and answers to the questionnaire can easily be quantified to prove effects over groups of participants. This makes it a frequently used method to evaluate systems. Standardized questionnaires exist on various aspects of UX. However, if one wants to evaluate a system on all these aspects the user has to fill in hundreds of questions, with the risk of "questionnaire fatigue" (filling in the questionnaire at random, or the same answer for each item) and users choosing for the safe middle option, because at some point, all questions seem to be the same. In the case study Hamster Lab, we showed that when a high number of participants is involved or data is gathered over multiple sessions it is possible to limit the items in the questionnaire to exactly what is needed to answer your research question.

Acknowledgements The authors acknowledge the support of the BrainGain Smart Mix Programme of the Netherlands Ministry of Economic Affairs and the Netherlands Ministry of Education, Culture and Science. This work was partially supported by the ITEA2 Metaverse1 Project (http://www.metaverse1.org).

References

1. Barbero, A., Grosse-Wentrup, M.: Biased feedback in brain–computer interfaces. J. Neuroeng. Rehabil. **7**(1), 34 (2010)
2. van Baren, J., IJsselsteijn, W.: Measuring presence: A guide to current measurement approaches. Deliverable 5 for OmniPres project (2004)
3. Bauer, G., Gerstenbrand, F., Rumpl, E.: Varieties of the locked-in syndrome. J. Neurol. **221**(2), 77–91 (1979)
4. Bradley, M.M., Lang, P.J.: Measuring emotion: The self-assessment manikin and the semantic differential. J. Behav. Ther. Exp. Psychiatry **25**(1), 49–59 (1994)
5. Brockmyer, J.H., Fox, C.M., Curtiss, K.A., McBroom, E., Burkhart, K.M., Pidruzny, J.N.: The development of the game engagement questionnaire: A measure of engagement in video game-playing. J. Exp. Soc. Psychol. **45**(4), 624–634 (2009)
6. Csikszentmihalyi, M.: Flow: the psychology of optimal experience. Harper & Row, New York (1990)
7. Fairclough, S.: Fundamentals of physiological computing. Interact. Comput. **21**(1-2), 133–145 (2009)
8. Fatourechi, M., Bashashati, A., Ward, R.K., Birch, G.E.: EMG and EOG artifacts in brain computer interface systems: A survey. Clin. Neurophysiol. **118**(3), 480–494 (2007)
9. Friedman, D., Leeb, R., Guger, C., Steed, A., Pfurtscheller, G., Slater, M.: Navigating virtual reality by thought: What is it like? Presence (Camb.) **16**(1), 100–110 (2007)
10. Groenegress, C., Holzner, C., Guger, C., Slater, M.: Effects of P300-based BCI use on reported presence in a virtual environment. Presence (Camb.) **19**(1), 1–11 (2010)
11. Gürkök, H., Hakvoort, G., Poel, M.: Modality switching and performance in a thought and speech controlled computer game. In: Proceedings of ICMI 2011, ACM, New York, NY, USA (2011)
12. Hakvoort, G., Gürkök, H., Plass-Oude Bos, D., Obbink, M., Poel, M.: Measuring immersion and affect in a brain-computer interface game. In: Campos, P., Graham, N., Jorge, J., Nunes, N., Palanque, P., Winckler, M. (eds.) 13th IFIP TC 13 International Conference on Human-Computer Interaction, INTERACT 2011, Lisbon, Portugal. Lecture Notes in Computer Science, vol. 6946, pp. 115–128. Springer, Berlin (2011)
13. Hart, S., Staveland, L.: Development of NASA-TLX (Task Load Index): Results of empirical and theoretical research. Adv. Psychol. **52**, 139–183 (1988)
14. Huggins, J., Wren, P., Gruis, K.: What would brain–computer interface users want? opinions and priorities of potential users with amyotrophic lateral sclerosis. Amyotroph. Lateral Scler. **5**, 1–8 (2011)
15. Ijsselsteijn, W.A., de Kort, Y.A.W., Poels, K.: The Game Experience Questionnaire: Development of a self-report measure to assess the psychological impact of digital games. (Manuscript in Preparation)
16. Jennett, C., Cox, A.L., Cairns, P., Dhoparee, S., Epps, A., Tijs, T., Walton, A.: Measuring and defining the experience of immersion in games. Int. J. Hum. Comput. Stud. **66**, 641–661 (2008)
17. Kleih, S., Nijboer, F., Halder, S., Kübler, A.: Motivation modulates the p300 amplitude during brain–computer interface use. Clin. Neurophysiol. **121**(7), 1023–1031 (2010)
18. van de Laar, B., Reuderink, B., Plass-Oude Bos, D., Heylen, D.: Evaluating user experience of actual and imagined movement in BCI gaming. Int. J. Gaming Comput. Mediated Simulat. **2**(4), 33–47 (2010)
19. van de Laar, B., Plass-Oude Bos, D., Reuderink, B., Nijholt, A.: Optimizing fun with unreliable input. Internal report CTIT (2011)
20. Leeb, R., Lee, F., Keinrath, C., Scherer, R., Bischof, H., Pfurtscheller, G.: Brain–computer communication: motivation, aim, and impact of exploring a virtual apartment. IEEE Trans. Neural Syst. Rehabil. Eng. **15**(4), 473–482 (2007)
21. Lightbody, G., Ware, M., McCullagh, P., Mulvenna, M., Thomson, E., Martin, S., Todd, D., Medina, V., Martinez, S.: A user centred approach for developing brain–computer interfaces.

In: 2010 4th International Conference on Pervasive Computing Technologies for Healthcare, IEEE, Piscataway, NJ, USA, pp. 1–8 (2010)
22. Mandryk, R.L., Atkins, M.S., Inkpen, K.M.: A continuous and objective evaluation of emotional experience with interactive play environments. In: CHI '06: Proceedings of the SIGCHI Conference on Human Factors in Computing Systems, ACM, New York, NY, USA, pp. 1027–1036 (2006)
23. Mandryk, R.L., Inkpen, K.M., Calvert, T.W.: Using psychophysiological techniques to measure user experience with entertainment technologies. Behav. Infor. Technol. **25**, 141–158 (2006)
24. McNamara, N., Kirakowski, J.: Functionality, usability, and user experience: three areas of concern. Interactions **13**, 26–28 (2006). DOI http://doi.acm.org/10.1145/1167948.1167972, URL http://doi.acm.org/10.1145/1167948.1167972
25. Moore Jackson, M., Mappus, R.: Applications for Brain–Computer Interfaces. In: Nijholt, A., Tan, D. (eds.) Brain–Computer Interfaces: Applying our Minds to Human–Computer Interaction, pp. 89–103. Springer, London, UK (2010)
26. Münßinger, J., Halder, S., Kleih, S., Furdea, A., Raco, V., Hösle, A., Kübler, A.: Brain painting: First evaluation of a new brain–computer interface application with als-patients and healthy volunteers. Front. Neuroprosthetics **4**, 182 (2010)
27. Nijboer, F., Birbaumer, N., Kübler, A.: The influence of psychological state and motivation on brain–computer interface performance in patients with amyotrophic lateral sclerosis – A longitudinal study. Front. Neurosci. **4**, 55 (2010)
28. Nijholt, A., Plass-Oude Bos, D., Reuderink, B.: Turning shortcomings into challenges: Brain–computer interfaces for games. Entertain. Comput. **1**(2), 85–94 (2009)
29. Norman, D.: Emotion & design: attractive things work better. Interactions **9**(4), 36–42 (2002)
30. Obbink, M., Gürkök, H., Plass-Oude Bos, D., Hakvoort, G., Poel, M., Nijholt, A.: Social interaction in a cooperative brain–computer interface game. In: Proceedings 4th International ICST Conference on Intelligent Technologies for Interactive Entertainment (INTETAIN 2011). Springer, Heidelberg, Germany (2011)
31. Omar, H., Jaafar, A.: Heuristics evaluation in computer games. In: Bakar, Z.A., Sembok, T.M.T., Zaman, H.B., Bruza, P., Crestani, F., Urs, S.R., Awang, Z. (eds.) 2010 International Conference on Information Retrieval Knowledge Management, IEEE, Piscataway, NJ, USA, pp. 188–193 (2010)
32. Oude Bos D., Reuderink, B.: BrainBasher: a BCI Game. In: Extended Abstracts of the International Conference on Fun and Games, Eindhoven TU/e (2008)
33. Pasqualotto, E., Simonetta, A., Gnisci, V., Federici, S., Belardinelli, M.O.: Toward a usability evaluation of BCIs. International Journal of Bioelectromagnetism. **13**(3), 121–122, (2011)
34. Rheinberg, F., Vollmeyer, R., Burns, B.: FAM: Ein Fragebogen zur Erfassung aktuller Motivation in Lern-und Leistungssituationen. Diagnostica **47**(2), 57–66 (2001)
35. Scherer, M.: Living in the State of Stuck : How Assistive Technology Impacts the Lives of People with Disabilities, 3rd edn. Brookline Books, Cambridge, Massachusetts, USA (2000)
36. Slater, M., Steed, A.: A virtual presence counter. Presence (Camb.) **9**(5), 413–434 (2000)
37. Solovey, E., Girouard, A., Chauncey, K., Hirshfield, L., Sassaroli, A., Zheng, F., Fantini, S., Jacob, R.: Using fNIRS brain sensing in realistic HCI settings: experiments and guidelines. In: Proc. of the 22nd ACM Symposium on User Interface Software and Technology, ACM, New York, NY, USA, pp. 157–166 (2009)
38. Tractinsky, N., Katz, A., Ikar, D.: What is beautiful is usable. Interact. Comput. **13**(2), 127–145 (2000)
39. Vilimek, R., Zander, T.: Bc (eye): Combining eye-gaze input with brain–computer interaction. Universal Access in Human–Computer Interaction Intelligent and Ubiquitous Interaction Environments, pp. 593–602 (2009)
40. Zander, T., Kothe, C., Jatzev, S., Gaertner, M.: Enhancing human–computer interaction with input from active and passive brain–computer interfaces. In: Nijholt, A., Tan, D. (eds.) Brain–Computer Interfaces: Applying our Minds to Human–Computer Interaction, pp. 181–199. Springer, London, UK (2010)

41. Zickler, C., Donna, V.D., Kaiser, V., Al-Khodairy, A., Kleih, S., Küebler, A., Malavasi, M., Mattia, D., Mongardi, S., Neuper, C., Rohm, M., Rupp, R.: BCI applications for people with disabilities: Defining user needs and user requirements. In: Assistive Technology from Adapted Equipment to Inclusive Environments: AAATE2009, pp. 185–189. IOS Press, Amsterdam, the Netherlands (2009)

Chapter 12
Framework for BCIs in Multimodal Interaction and Multitask Environments

Jan B.F. van Erp, Anne-Marie Brouwer, Marieke E. Thurlings, and Peter J. Werkhoven

12.1 Introduction

The initial development of Brain Computer Interfaces (BCIs) focused on providing users with special needs a way to communicate when other interaction means failed. However, there are also several good reasons to consider BCIs for healthy users, for instance to make control or communication more intuitive or reduce the risk of overloading sensory modalities or the motor system [29]. As a result, the scope of BCI applications under investigation expands rapidly and starts to include applications for gaming and adaptive automation.

For users with special needs, a BCI is often developed as the only interaction device and used for a specific communication task performed in isolation. In more recent applications, a BCI is part of a multimodal user interface and may be used in a multitask situation where the user performs different tasks sequentially or even in parallel. This introduces relatively new user–system interaction issues and here we aim to have a closer look at for instance the (human information processing) models relevant for these situations. We consider appropriate integration of BCIs in multimodal interaction and multitask environments as a prerequisite for the development of successful BCI applications for healthy users [33]. Important questions concern usability, ruggedized and comfortable sensors, and robust signal processing.

The expanding scope of BCI applications also requires reconsidering common BCI definitions. The assistive technology community often uses the strict definition provided by [36]: A BCI is a communication and control system that does not depend in any way on the brain's normal neuromuscular output channels, that

J.B.F. van Erp (✉) · A.-M. Brouwer · M.E. Thurlings · P.J. Werkhoven
TNO, Kampweg 5, 3769DE Soesterberg, The Netherlands
e-mail: jan.vanerp@tno.nl; anne-marie.brouwer@tno.nl; marieke.thurlings@tno.nl; peter.werkhoven@tno.nl

B. Z. Allison et al. (eds.), *Towards Practical Brain-Computer Interfaces*, Biological and Medical Physics, Biomedical Engineering, DOI 10.1007/978-3-642-29746-5_12,
© Springer-Verlag Berlin Heidelberg 2012

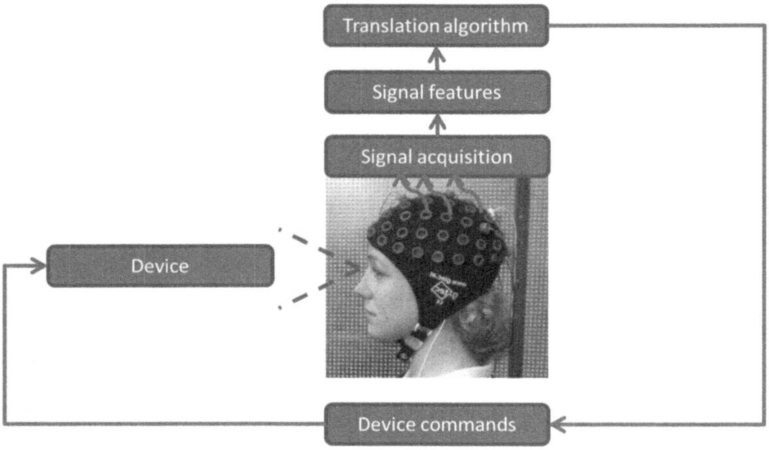

Fig. 12.1 Classic view of a BCI system from the assistive technology approach

provides real-time interaction and includes feedback of the outcome to the user. We propose a broader definition more adjusted to the HCI community: "*a BCI uses signals from the brain to control a device or the interaction between the user and a device (near) real time, and/or provides signals directly to the brain to either communicate information or alter brain activity*." This definition includes systems that use brain signals to assess the user state, for instance to adjust the task allocation or interaction modality between user and system. As such, brain signals can be considered an expansion of the set of physiological measures already used in user–system interaction such as heart rate variability. Also, BCIs can either refer to communication from the brain to a system, or vice versa (sometimes referred to as Computer Brain Interface, CBI), or both. However, the vast majority of current BCIs for healthy users uses communication from the brain to a device only.

Zander and colleagues [37] made a useful distinction between active, reactive and passive BCIs, based on the user's effort and task to control the BCI. In active BCIs, users actively generate specific brain signals to give a specific command, for instance by performing mental calculation or imaginary limb movement. Reactive BCIs do not require active generation of brain signals but interpret the brain's automatic reaction to so-called probe stimuli. The user can modulate this reaction pattern by modulating attention, which can be used to select a specific probe stimulus. Finally, a passive BCI analyses brain signals without the user needing to perform specific mental tasks or to process probe stimuli, but uses neural correlates of constructs such as engagement, mental workload and drowsiness [34].

Figure 12.1 depicts the classic view of a BCI. The user actively generates a specific brain pattern (e.g., motor imagery). A sensor system (e.g., EEG) acquires and processes the brain signals followed by extraction and classification of the signal features by computer algorithms (see [3] for a review). The results are translated into device commands and executed by the device (e.g., a wheelchair) and the user

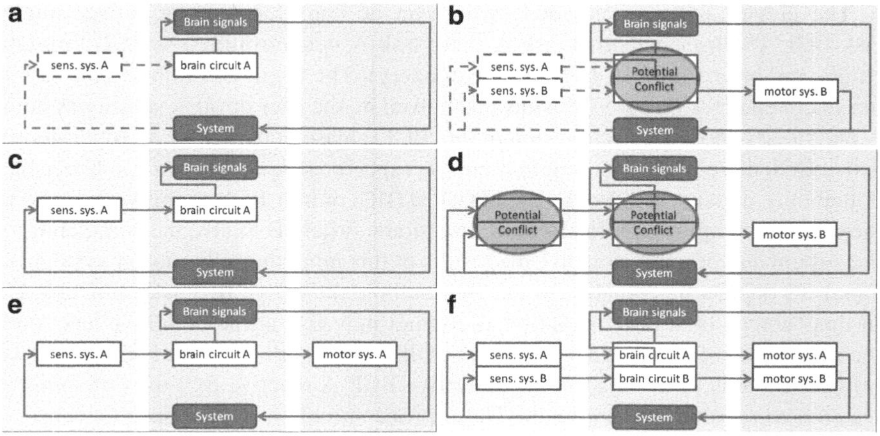

Fig. 12.2 Extending the three BCI classes from a single task situation (*left column*) to a dual task situation (*right column*). The ovals indicate a potential conflict that may arise in BCI use in dual-task situations. (**a**) active BCI, single task, (**b**) active BCI, dual task (**c**) reactive BCI, single task, (**d**) reactive BCI, dual task, (**e**) passive BCI, single task, (**f**) passive BCI, dual task

can perceive the result. In this classic setup, there is only one task and no other interaction channels between user and device. We will discuss the extensions: to a dual-task situation, to combining two BCIs and to the integration with other user–system interaction modalities.

12.2 Challenges for the Use of BCIs in a Dual Task Environment

Figure 12.2 uses a model of a (closed-loop) user–system interaction in which the user is simply modeled with a perception, cognition, and action step. The user perceives system information (this phase—arbitrarily—includes bottom-up processing in the brain, i.e., by the sensory cortices), further processes this information in the brain (i.e., higher order cognitive processes) and performs an action affecting the system. With this latter phase we exclusively refer to the peripheral motor system. We include planning the action—again arbitrarily—in the cognition phase. The panels of Fig. 12.2 illustrate the extension from a single task situation (left column) to a dual task situation (right column) for an active, reactive and passive BCI. Please note that the division in perception, cognition and action is useful in the current context because dual task situations can affect these phases separately, while the effects from one phase to the next are rather independent, for instance error rates or the distribution of errors are unaffected by earlier phases [13, 15, 23].

The classic BCI as described earlier can be considered as an active, single task BCI. The user performs task A (here task A is controlling the BCI) through employing brain circuit A (e.g., motor imagery). The acquired brain signals result in system changes that are (possibly) perceived by the user through sensory system A and there is no motor action step in this BCI (please note that for an (open-loop) active BCI, there is no strict requirement for a specific sensory system A—hence the dotted lines in panel (a)). For a second task B (BCI or non-BCI, cognitive or motor), the user may employ a brain circuit B, use motor system B to give the commands to the system and possibly perceive the results of this input through sensory system B. Panel (b) depicts the situation when we combine the active BCI with this task B. In this "active, dual task" situation, a conflict may rise at the cognitive level and the acquisition of brain signals which we will describe in more detail below. Panels (c) and (d) depict the situation for a reactive BCI. A reactive BCI uses the brain's reaction to specific probe stimuli. These probe stimuli rely on a specific sensory system A and the reaction to this probe stimulus in brain circuit A. As in panel (a), there is no motor action involved in a single task, reactive BCI. Panel (d) depicts the situation for a task B added to the reactive BCI and shows that potential conflicts may arise at both the sensory system and the brain (e.g., because both may use the same sensory channel (e.g., visual) or brain process (e.g., attention)). Finally, panels (e) and (f) depict the situation for a single and dual task passive BCI. Here, the BCI uses naturally occurring brain patterns when the user performs task A, and the same when the user performs tasks A and B. Of course, a conflict may occur when the user performs both tasks, but this will not affect the workings of the passive BCI. On the contrary, the goal of the BCI may even be the detection of such a conflict.

12.2.1 Psychological Models for Dual Task Situations and Coping with Conflicts

Here we briefly introduce Wickens' Multiple Resource Theory (MRT, e.g., see [35] for an overview) because this model provides relevant guidance on how to reduce dual task interference. Figure 12.3 depicts the information processing loop used by Wickens and many other authors.

The basic version of the MRT knows three independent dimensions, here given with their associated brain circuitry: (a) stage of processing: perceptual and cognitive (posterior to the central sulcus) vs. selection and execution of action (anterior to the central sulcus), (b) code of processing: spatial (right hemisphere) vs. verbal/linguistic (left hemisphere), and (c) modality: auditory (auditory cortex), visual (visual cortex), and possibly tactile (somatosensory cortex). A large body of evidence confirms the assertion that the degree to which two tasks use different levels along each of the three dimensions reduces interference between the tasks. Several variants (e.g., [32]) and extensions to this basic MRT have been suggested in recent years. For instance Boles et al. extended the number of perceptual resources

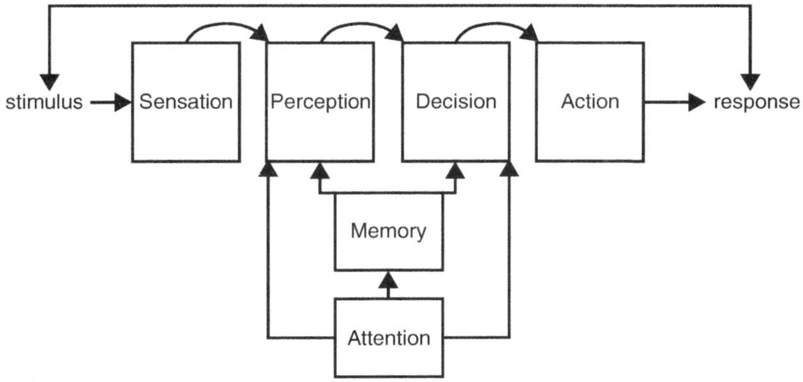

Fig. 12.3 The basic human information processing loop often used in the HCI domain

by distinguishing spatial positional, spatial quantitative and other resources (for recent work here, see [4]).

Applying a reactive BCI in a dual task environment can potentially lead to a conflict at the stage of the sensory system (perceptual processes) and the brain (higher order cognitive processes). At the sensory system, the probe stimuli required by a reactive BCI may interfere with sensory processing required for task B. This risk of sensory overload is relatively common in user–system interactions and several information processing models further detail the risks and possible solutions. A way to reduce the effects of a potential conflict is to employ different sensory systems for the probe stimuli of task A and the system feedback for task B (but see [25] for interference of concurrent stimulus processing). Although both may often be visual, the use of auditory and haptic displays increases [30]. Recent examples of using tactile stimuli as probes in a reactive BCI (see Fig. 12.4) show that this is feasible and performance is comparable to that with visual stimuli [5, 6]. Interestingly, the use of multisensory stimuli in the context of BCIs is not widely used while this is a proven solution in other domains.

A more complicated and challenging issue is the potential conflict that can arise at the cognitive level for active and reactive BCIs. There are actually two issues here. The first is similar to the sensory conflict described above: the tasks may use the same resources (brain circuits) and thus result in an overload situation (this is not different from two non-BCI tasks that use the same cognitive resources). The second is that even when tasks A and B use different and not-interfering brain circuits, the brain signals acquired by the BCI may still be affected by those of task B. This is inherent to most signal acquisition systems currently in use. Invasive signal acquisition systems will suffer much less from this issue, but are currently not considered a viable option for use outside strict medical conditions. The electrical signals acquired by for instance EEG sensors on the outside of the user's scalp have a low spatial specificity and not only represent activity of brain areas directly underlying the sensor but also areas centimetres away. Solving this

Fig. 12.4 Example of a tactile BCI developed by TNO in the Netherlands [5, 30]. The *white boxes* inside the jacket contain small vibrators that can deliver localized somatosensory probe cues in a reactive BCI paradigm

issue is outside the scope of this paper and progress made in both sensor technology and computational algorithms may reduce this issue.

Coping with dual tasks that use the same cognitive resources (brain circuits) is an important challenge. Within the BCI domain, this challenge has also been tackled from a single task perspective, e.g., [8] provides a good overview. Dual task situations will further complicate the challenge. First, we must state that people are not very good at executing two tasks at the same time or in close succession [22], even though the brain seems to adjust to dual tasks situations by dividing tasks among the left and right anterior prefrontal cortex compared to using both in a single task situation [7, 26], which will only work for two tasks, but not more.

Apart from possible distractions, some tasks interfere less with each other than others. A rule of thumb is that the more the two tasks share (or compete for) the same resources or brain circuits, the more they interfere. Although there is a large set of possible task combinations that has not been investigated yet, data indicating such competition are available for the more common combinations. For instance, working memory and visual search compete for the inferior and middle frontal cortex [1], manual tracking and visual detection seem to compete for the primary motor and somatosensory cortices involved with controlling the tracking-hand [9], manual tracking (driving) and listening affects the parietal lobe [12], and two motor tasks compete for the primary motor cortex [10]. Unfortunately, the literature is more keen on reporting tasks that do interfere than those that don't. Identifying two tasks that do not or only to a certain degree interfere and of which the brain circuits are spatially separated is an important challenge and must be based on neuroscientific as well as behavioural studies. A good point of departure are the dimensions of the MRT.

Another relevant aspect is that simply trying harder cannot overcome the limitations of a central cognitive bottleneck [20], but training may reduce the amount of interference. This training effect is not only visible in increased performance but also in the reduced overlap in employed brain circuits. For instance, Rémy et al. [19] investigated the combination of a bimanual task with a visual search task. After training of the manual task, the overlap between regions involved in both tasks was reduced, possibly due to automaticity of the manual task.

12.3 Combining BCIs

In this section, we will have a closer look at the consequences of combining different BCI classes. Please note that in the assistive technology domain, the term hybrid BCI was introduced to refer to combinations of BCIs or of a BCI with other control devices (e.g., [18]). In the user–system interaction domain, the common term is multimodal interface. Earlier work on this topic was restricted to either the serial use of two BCIs (e.g., one BCI serving as an on-off switch of a second BCI) or as redundant input channels in the same task [18, Fig. 1]. Here we focus on multi-task situations in which two BCIs are used (simultaneously) for two different tasks. Again, conflicts may be expected at the sensory system (when combining two reactive BCIs) and/or the involved brain circuits (when combining two active BCIs or an active and a reactive BCI). For combinations with a passive BCI, no conflicts are involved. Recent examples are the multimodal BCI described by [18] using event-related desynchronization and steady-state evoked potentials in a redundant set-up (i.e., both provide input to the same task and are integrated to provide the input to this single task) and [17] using (actively generated) alpha rhythm and Steady State Visually Evoked Potentials (SSVEPs).

Combining two active BCIs or an active and a reactive BCI may result in a conflict between the involved brain circuits. Combining two active BCIs is essentially a dual task and as such may suffer from the same effects described in the previous section, and choosing two appropriate tasks is essential. For instance, Sangals and Sommer [21] showed that a simple choice task with foot responses interferes with response preparation for a manual choice task. This indicates that two active BCIs based on motor (imagery) tasks may not be a good choice. The situation is even more complicated than a single active BCI in a dual task situation in the sense that the two brain circuits involved should not only be "independent" of each other, but also spatially distributed or the sensor system may have difficulties in classifying the two tasks. The brain circuit conflict for combining an active and a reactive BCI is potentially less severe. In principle, performing a mental task and paying attention to probe stimuli can be combined. Since the "attention wave" required by the reactive BCI will be located centrally, it is recommended to use a mental task for the active

BCI that involves brain circuits located more lateral or frontal, or which signal is clearly distinct from the attention wave (e.g., based on spectral features).

When combining two reactive BCIs (i.e., each connected to a different task), it is strongly recommended to use two different sensory modalities to present the probe stimuli. But even then, it is doubtful whether the user is able to pay sufficient attention to targets in the two modalities to obtain a unique and measurable brain pattern. This is related to the fact that both BCIs will also compete for the same central brain circuits or resources as for instance shown for auditory and visual stimuli [2, 14]. Or in other words: the location of the relevant brain signal indicating whether attention was paid to a stimulus is more or less independent from the stimulus modality. For instance, Brouwer and colleagues [5, 6] investigated visual, tactile and bimodal visual/tactile probe stimuli and found only small differences in location of the P3 (i.e., the "attention wave") as function of sensory modality. This means that exact time-locking of probe stimuli and EEG is critical and probe stimuli in both modalities should be out-of-phase. Another problem that may arise is the cost involved in switching attention between sensory modalities (e.g., in terms of required time [24, 31]). This means that for instance, using the two BCIs consecutively and not parallel may introduce a new bottleneck.

12.4 Integrating BCIs in a Multimodal User Interface: Relevant Issues

Especially in applications for healthy users, a BCI will likely not be a stand-alone interface between user and system but part of a multimodal user interface. Like other input and output modalities, integrating a BCI in a multimodal interface requires careful consideration of several aspects. Up till now, little or no attention in the design of BCI applications is given to usability aspects such as comfort and ease of use. Here we will list several issues that are of particular relevance for BCIs, but we do like to stress that general guidelines with respect to interaction design should also be taken into account such as adjusting the interaction to the user, task and context of use characteristics (see ISO 9241 series on international usability standards):

- The BCI dialogue design should take into account the user's conceptual model and the task sequence.
- BCI as a control device should be combined with a compatible display modality. Known compatible combinations in multimodal interfaces are for instance manual control—visual display, and voice control—auditory display [27]. This is an important research topic for BCIs.
- The choice for a specific BCI category should be based on the task requirements and the strength and weaknesses of specific BCIs and should minimize mapping between interface and task semantics. As we coined the term modality appropriateness for the choice of display modality, we propose to use BCI appropriateness for this choice.

- Minimize memory load and stimulate recognition over recall.
- The BCI should be consistent in the use of labels, menus, shortcuts etc.
- Always provide an exit (escape) option.
- A topic of specific interest is how to combine a (active or reactive) BCI with other control devices and prohibit interference [16].
- For passive BCIs, policies related to the fusion of BCI results with other physiological data must be developed.
- A general BCI issue is the question of how to switch a BCI on and off. Since users cannot simply switch their brain activity on and off, specific solutions are required. One should also ensure that the current system interaction state is communicated to the user and that the system appropriately provides feedback when it initiates a modality change.
- In case the classification accuracy is limited, the system should confirm its interpretations of the user input (when appropriate after fusion and not for each modality in isolation). In this situation, users should be allowed to switch to a different modality.
- Include the end-user in the design and his or her specific abilities (see Chaps. 8 and 11).
- The BCI system should provide ample feedback of its state and decisions.

12.5 Discussion and Conclusion

We started by making an inventory of relevant issues when extending the use of different types of BCI from a single task environment to a multitask environment. Although we can build on the lessons learned from the user–system interaction domains and relevant information processing models such as the MRT, there is a need to get better insight in how to choose BCI modalities and tasks that only minimally interfere, i.e., they should be functionally and spatially separated (i.e., with respect to brain region). We expect that the identification of non-conflicting tasks will benefit from studies in high resolution brain imaging. A relevant addition of the MRT is linked to the sharing of task goals. For instance, tasks that would normally interfere like driving and listening (to a cell phone) will do so to a lesser degree when they share the same task goal, i.e., driving and listening to navigation instructions. The same may hold for BCI tasks in a multitask environment. We also looked into the situation where BCI feedback or probe stimuli may lead to sensory overload or interference. The use of alternative sensory modalities or multisensory stimuli may reduce this risk, but the sensory modality should also be compatible with the BCI task. An important next step here is to have a quantitative evaluation of the identified conflicts.

Now that BCI technology is maturing and the range of possible applications expands, it is necessary to look more closely at general usability aspects. So far, usability does not seem to play the role it should when preparing BCIs for operational use outside the lab and for a growing range of users. ISO 9241-11 [11] defines usability as the extent to which a product can be used by specified users to achieve specified goals with effectiveness, efficiency and satisfaction in a specified context of use. Usability guidelines can help in improving effectiveness and avoiding errors, improving performance, and enhancing the comfort and well-being of users.

The above also means that the range of BCI paradigms should be stretched beyond the commonly used motor imagery for active BCIs and P3 matrix speller for reactive BCIs. A relevant issue is the design and application of multipurpose BCIs. A multipurpose BCI can be used for different applications like communication, control of domestic appliances and a videogame. This requires a BCI design that is either independent of task requirements or that can be easily adjusted.

Systematically looking into the task requirements and context of use can result in a better match of BCI as control device and other user–system interaction components [28]. Having said that, we would also like to stress that BCI applications can still be considered embryonic and that many technological issues should be solved in hardware development, signal processing and system integration. Based on the issues we discussed and general user–system guidelines we formulate the following preliminary guidelines for BCI design and multimodal interaction:

- A BCI should be used if it improves satisfaction, efficiency, or other aspects of performance for a given user, task and context of use.
- The BCI category should match the task requirement.
- The BCI coding should match the task requirements (e.g., letters for a spelling device and directions for a navigation task).
- The feedback to the user or the BCI probe stimuli should match the BCI coding and presented in the appropriate sensory modality (e.g., letters visually, directions through spatial audio).
- The feedback to the user should match with the strengths, weaknesses and possibilities of the BCI and not go beyond its capabilities.
- Ensure that the display modalities are well synchronized temporally as well as spatially.
- Minimise possible interference between the BCI task and other tasks, both functionally and in relation to the spatial specificity of the BCI signal acquisition system.
- Aim to combine tasks that share the same task goals.
- Minimise possible interference between the sensory system involved in the BCI and other user–system interaction components.
- Performing tasks sequentially instead of simultaneously may reduce sensory or cognitive conflicts but may also involve costs of task and/or modality switching.

Acknowledgements The authors gratefully acknowledge the support of the BrainGain Smart Mix Programme of the Netherlands Ministry of Economic Affairs and the Netherlands Ministry of Education, Culture and Science. This research has been supported by the GATE project, funded by the Netherlands Organization for Scientific Research (NWO) and the Netherlands ICT Research and Innovation Authority (ICT Regie).

References

1. Anderson, E.J., Mannan, S.K., Rees, G., Sumner, P., Kennard, C.: Overlapping functional anatomy for working memory and visual search. Exp. Brain Res. **200**(1), 91–107 (2010)
2. Arnell, K.M.: Visual, auditory, and cross-modality dual-task costs: Electrophysiological evidence for an amodal bottleneck on working memory consolidation. Percept. Psychophys. **68**(3), 447–457 (2006)
3. Bashashati, A., Fatourechi, M., Ward, R.K., Birch, G.E.: A survey of signal processing algorithms in brain–computer interfaces based on electrical brain signals. J. Neural Eng. **4**(2), R32–R57 (2007)
4. Boles, D.B., Bursk, J.H., Phillips, J.B., Perdelwitz, J.R.: Predicting dual-task performance with the multiple resource questionnaire. Hum. Factors **49**, 32–45 (2007)
5. Brouwer, A.-M., Van Erp, J.B.F.: A tactile P300 brain–computer interface. Front. Neurosci. **4**(19), 1–12 (2010)
6. Brouwer, A.-M., Van Erp, J.B.F., Aloise, F., Cincotti, F.: Tactile, visual and bimodal P300s: Could bimodal P300s boost BCI performance? SRX Neurosci. **2010**, 1–9 (2010)
7. Charron, S., Koechlin, E.: Divided representation of concurrent goals in the human frontal lobes. Science **328**(5976), 360–363 (2010)
8. Curran, E.A., Stokes, M.J.: Learning to control brain activity: A review of the production and control of EEG components for driving brain–computer interface (BCI) systems. Brain Cogn. **51**(3), 326–336 (2003)
9. Gazes, Y., Rakitin, B.C., Steffener, J., Habeck, C., Butterfield, B., Ghez, C., Stern, Y.: Performance degradation and altered cerebral activation during dual performance: Evidence for a bottom-up attentional system. Behav. Brain Res. **210**(2), 229–239 (2010)
10. Hiraga, C.Y., Garry, M.I., Carson, R.G., Summers, J.J.: Dual-task interference: Attentional and neurophysiological influences. Behav. Brain Res. **205**(1), 10–18 (2009)
11. ISO ISO 2941-11.: Ergonomic requirements for office work with visual display terminals (VDTs) Guidance on usability. International Organization for Standardization (1998)
12. Just, M.A., Keller, T.A., Cynkar, J.: A decrease in brain activation associated with driving when listening to someone speak. Brain Res. **1205**(C), 70–80 (2008)
13. Kamienkowski, J.E., Sigman, M.: Delays without mistakes: Response time and error distributions in dual-task. PLoS ONE **3**(9), (2008). art. no. e3196
14. Low, K.A., Leaver, E.E., Kramer, A.F., Fabiani, M., Gratton, G.: Share or compete? Load-dependent recruitment of prefrontal cortex during dual-task performance. Psychophysiology **46**(5), 1069–1079 (2009)
15. Marois, R., Larson, J.M., Chun, M.M., Shima, D.: Response-specific sources of dual-task interference in human pre-motor cortex. Psychol Res. **70**(6), 436–447 (2006)
16. Mochizuki, H., Tashiro, M., Gyoba, J., Suzuki, M., Okamura, N., Itoh, M., Yanai, K.: Brain activity associated with dual-task management differs depending on the combinations of response modalities. Brain Res. **1172**(1), 82–92 (2007)
17. Mühl, C., Gürkök, H., Plass-Oude Bos, D., Thurlings, M.E., Scherffig, L., Duvinage, M., Elbakyan, A.A., Kang, S., Poel, M., Heylen, D.: Bacteria Hunt: Evaluating multi-paradigm BCI interaction. J. Multimodal User Interfaces **4**(1), 11–25 (2010)
18. Pfurtscheller, G., Allison, B.Z., Bauernfeind, G., Brunner, C., Solis-Escalante, T., Scherer, R., Zander, T.O., Müller Putz, G., Neuper, C., Birbaumer, N.: The hybrid BCI. Front. Neurosci. **4**, 30, 1–12 (2010)
19. Rémy, F., Wenderoth, N., Lipkens, K., Swinnen, S.P.: Dual-task interference during initial learning of a new motor task results from competition for the same brain areas. Neuropsychologia **48**(9), 2517–2527 (2010)
20. Ruthruff, E., Johnston, J.C., Remington, R.W.: How strategic is the central bottleneck: Can it be overcome by trying harder? J. Exp. Psychol. Hum. Percept. Perform. **35**(5), 1368–1384 (2009)

21. Sangals, J., Sommer, W.: The impact of intervening tasks on response preparation. J. Exp. Psychol. Hum. Percept. Perform. **36**(2), 415–429 (2010)
22. Sigman, M., Dehaene, S.: Dynamics of the central bottleneck: Dual-task and task uncertainty. PLoS Biol. **4**(7), 1227–1238 (2006)
23. Sigman, M., Dehaene, S.: Brain mechanisms of serial and parallel processing during dual-task performance. J. Neurosci. **28**(30), 7585–7598 (2008)
24. Spence, C., Driver, J.: Cross-modal links in attention between audition, vision, and touch: implications for interface design. Int. J. Cogn. Ergon. **1**(4), 351–373 (1997)
25. Stelzel, C., Brandt, S.A., Schubert, T.: Neural mechanisms of concurrent stimulus processing in dual tasks. NeuroImage **48**(1), 237–248 (2009)
26. Stelzel, C., Kraft, A., Brandt, S.A., Schubert, T.: Dissociable neural effects of task order control and task set maintenance during dual-task processing. J. Cogn. Neurosci. **20**(4), 613–628 (2008)
27. Stelzel, C., Schumacher, E.H., Schubert, T., D'Esposito, M.: The neural effect of stimulus-response modality compatibility on dual-task performance: An fMRI study. Psychol. Res. **70**(6), 514–525 (2006)
28. Thurlings, M.E., Van Erp, J.B.F., Brouwer, A.-M., Blankertz, B., Werkhoven, P.J.: Control-display mapping in brain–computer interfaces. Ergonomics, **55**(5), (2012).
29. Thurlings, M.E., Van Erp, J.B.F., Brouwer, A.-M., Werkhoven, P.J.: EEG-Based navigation from a Human Factors perspective. In: Tan, D.S., Nijholt, A. (eds.) Brain–Computer Interfaces, Human–Computer Interaction series, pp. 117–132. Spinger, London (2010)
30. Van Erp, J.B.F., Van Veen, H.A.H.C.: Vibrotactile in-vehicle navigation system. Transp. Res. Part F Traffic Psychol. Behav. **7**(4–5), 247–256 (2004)
31. Van Erp, J.B.F., Werkhoven, P.J.: Vibro-tactile and visual asynchronies: Sensitivity and consistency. Perception **33**, 103–111 (2004)
32. Van Erp, J.B.F., Werkhoven, P.J.: Validation of Principles for Tactile Navigation Displays. HFES 50th annual meeting, 1687–1691. Santa Monica: Human Factors and Ergonomics Society. (2006)
33. Van Erp, J.B.F., Tangermann, M., Lotte, F.: Brain-Computer Interfaces: Beyond Medical Applications. IEEE computer, **45**(4), 26–34 (2012).
34. Van Erp, J.B.F., Veltman, J.A., Grootjen, M.: Brain-Based Indices for User System Symbiosis. In: Tan, D.S., Nijholt, A. (eds.) Brain–Computer Interfaces, Human–Computer Interaction series, pp. 201–219. Spinger, London (2010)
35. Wickens, C.D.: Multiple resources and mental workload. Hum. Factors **50**(3), 449–455 (2008)
36. Wolpaw, J., Birbaumer, N.: Brain–computer interfaces for communication and control. In: Selzer, M., Clarke, S., Cohen, L., Duncan, P., Gage, F. (eds.) Textbook of neural repair and rehabilitation; neural repair and plasticity, pp. 602–614. Cambridge University Press (2006)
37. Zander, T.O., Kothe, C., Jatzev, S., Gaertner, M.: Enhancing Human–Computer Interaction with Input from Active and Passive Brain–Computer Interfaces. In: Tan, D.S., Nijholt, A., (eds.) Brain–Computer Interfaces, Human–Computer Interaction series, pp. 181–200. Spinger, London (2010)

Chapter 13
EEG-Enabled Human–Computer Interaction and Applications

Olga Sourina, Qiang Wang, Yisi Liu, and Minh Khoa Nguyen

13.1 Introduction

Human brain is still a mystery of twenty-first century. Recent advances in development of EEG devices made possible to add EEG-enabled dimension to human–computer interfaces. Real-time EEG-enabled systems could be developed for medical applications, e-learning, entertainment, marketing or even for human performance training in high risk environments. Neurofeedback systems could monitor EEG signals of the user and give the user real-time visual, audio or haptics feedback of his/her efforts to voluntarily change the brain state. Traditionally, neurofeedback systems were used in medical applications to help patients with psychological disorders such as Attention Deficit Hyperactivity Disorder (ADHD), Autism Spectrum Disorder (ASD), Central Pain, etc. Recently, neurofeedback training systems were started to be used in non-medical applications to enhance human performance in mathematics, motor skills, creativity, driving, etc. Human brain could be trained as other parts of our body. The effectiveness of such neurofeedback systems has been demonstrated in research and clinical publications. With recent advances in EEG devices such as easy installation, portability, mobility, low cost, etc., EEG technology can be used not only in the laboratories with doctors' help but also at home. The brain state recognition algorithms could be implemented and integrated in different applications to provide optimal performance workload, to enhance personnel short-term and long-term performance, to provide psychological support, to enable game characters with the user's emotions, to interact with objects and subjects in the games, etc.

O. Sourina (✉) · Q. Wang · Y. Liu · M.K. Nguyen
Nanyang Technological University, Nanyang Ave, Singapore
e-mail: eosourina@ntu.edu.sg; wang0586@ntu.edu.sg; liuy0053@ntu.edu.sg; raymondkhoa@ntu.edu.sg

B. Z. Allison et al. (eds.), *Towards Practical Brain-Computer Interfaces*, Biological and Medical Physics, Biomedical Engineering, DOI 10.1007/978-3-642-29746-5_13,
© Springer-Verlag Berlin Heidelberg 2012

In this work, we describe a spatio-temporal fractal based approach to real-time brain state recognition. We proposed and implemented fractal based algorithms of real-time brain state recognition from EEG including the user's concentration, stress levels and human emotions. Traditionally, the signal processing algorithms used in neurofeedback systems were based on frequency analysis, and event related potential analysis. Frequency training is the most prevalent method in the clinic application together with the Quantitative EEG (QEEG) [17] protocol. In our work, we study non-linear Fractal Dimension (FD) [51] features of EEG signals for the brain state classification. Fractal dimension allows quantify complexity of the EEG signal. Our hypothesis is that changes of FD values over time and over 3D spatial brain model correspond to changes in the brain states. For emotion recognition, we use two-dimensional Arousal-Valence emotion model where all emotions could be defined as 2D ellipsoids in the 2D space. We could map FD values to the 2D emotion model. With such approach, for example, "happy" is defined as high arousal and positive emotion, and "sad" is defined as low arousal and negative one. EEG-based approach in brain state recognition provides high temporal resolution that could be employed in real time application systems. Even 128 Hz device could provide 128 samples per second which is sufficient for brain state assessment in our algorithm. To improve spatial representation of the EEG-based approach we proposed real-time 3D visualization of EEG signal samples. With such system we could assess 3D EEG distribution patterns corresponding to different brain states. For example, we could visually assess what parts of the brain (brain lobes) are involved in any cognitive process. We also confirmed hypotheses that emotions finally could be recognized just from the frontal lobe, and negative and positive emotions have lateralization pattern.

This paper is organized as follows. In Sect. 13.2, neurofeedback systems for medical and non-medical applications, neurofeedback algorithms and emotion recognition algorithms are reviewed. General fractal based approach to brain state recognition is described in Sect. 13.3. EEG-enabled applications for human performance enhancement and emotion-enabled applications are proposed and described in Sect. 13.4. Conclusion is given in Sect. 13.5.

13.2 Brain State Recognition Algorithms and Systems

13.2.1 Neurofeedback Systems for Medical Application

Let us give a traditional definition of neurofeedback from [24]: "Like other forms of biofeedback, neurofeedback uses monitoring devices to provide moment-to-moment information to an individual on the state of their physiological functioning. The characteristic that distinguishes neurofeedback from other biofeedback is a focus on the central nervous system and the brain. Neurofeedback training (NFT) has its foundations in basic and applied neuroscience as well as data-based clinical practice. It takes into account behavioral, cognitive, and subjective aspects as well as brain activity. NFT is preceded by an objective assessment of brain activity and psychological status." The above definition is cited on the most sites of neurofeedback research and clinical applications. Another short definition of the neurofeedback is as follows: neurofeedback is the technique that presents the real-time feedback to the user based on the EEG signals sampling from the scalp of the user, with the form of video display and/or sound [18].

Traditionally, neurofeedback systems were used in medical applications. Neurofeedback research reveals that the EEG power and ERP (Event Related Potential) distortions always accompany psychological disorders such as Attention Deficit Hyperactivity Disorder (ADHD) [15, 36], Autistic spectrum Disorders (ASD) [10, 28], Substance Use Disorders (SUD) that could include alcoholics and drug abuse [41, 44], etc. Similar to other parts of our body, the brain function can be trained as well. Neurofeedback (NF) is an alternative choice as a treatment to these disorders besides the medicine treatment. Many neurofeedback games were assessed by the healing effect of the ADHD, one of the most known psychological disorders with significant EEG distortion. The θ/β ratio abnormal behavior was reported in [9]. Besides the ratio, the distortion in Slow Cortical Potential (SCP) was also notified by [16]. Both the frequency neurofeedback training and the SCP neurofeedback training can achieve a good healing effect for ADHD [16]. Current treatments for pain syndrome employ multidiscipline approaches such as chemical (drugs), physical (therapeutic exercise, acupuncture), psychological approach (relaxation with music, biofeedback, hypnosis) or a combination of the above-mentioned approaches. There are reports on virtual game applications for pain management [11, 55]. 3D virtual games were used during burn dressing of children with pain, during treatments of wounds [54, 55], etc. Recently, it was reported successful application of EEG-based games for Central Pain Syndrome (CPS) pain management, and migraine management as well.

13.2.2 Signal Processing Algorithms for Neurofeedback Systems

Traditionally, signal processing algorithms used in neurofeedback systems could be generally concluded into two main methods, i.e. frequency analysis, and event related potential analysis. Frequency training is the most prevalent method in the clinic application together with the QEEG protocol.

EEG signal can be divided into several different frequency bands, i.e. δ band (<4 Hz), ϕ band (4–7 Hz), α band (8–12 Hz), β band (12–30 Hz) and γ band (>30 Hz). Specially, the Sensorimotor rhythm activity (12–15 Hz) is also used in several neurofeedback systems. Each frequency band is related to different brain functions. Generally, δ band is prevalent in infant's EEG or EEG when the subject is sleeping; ϕ band is prevalent in EEG when the subject feels drowsiness; α band is significant when the subject is relax; β band is associated with fast activities and γ band is related to the problem solving and memory work [12]. The power over different bands were assessed and extracted from the patient EEG signals and then compared to the QEEG database (QEEG protocol) or statistical analysis was run to generate the pathology and corresponding recovery protocols. The frequency training method is the most prevalent method used in the neurofeedback training systems and other EEG applications because the frequency band power is easy to obtain and analyze with the existing signal processing tools.

ERP analysis is the process analyzing the EEG signal synchronized with an event. SCP and P300 are the important event related potential approaches used in the neurofeedback treatment. SCP reflects the changes in cortical polarization, i.e. negative and positive trends, of EEG last for 300 ms to several seconds after event stimulus [4]. Abnormalities in SCP of ADHD patient were studied in [16], and the corresponding neurofeedback protocol could enhance the continuous performance. The P300 component of ERP occurs during 300–600 ms after event stimulus which is obtained by oddball paradigm. Researches indicated that amplitude of P300 component is related to allocation of attention resource and the latency reflects the stimulus evaluation and classification time. The pathology of P300 component in drug abuse patients was reported in [44], and neurofeedback based on P300 component training was proposed.

Although the signal processing algorithms embedded in neurofeedback games are well applied in clinic treatment, the linear features (power spectral density or amplitude) extracted from EEG cannot represent the brain activities perfectly due to the nonlinearity of EEG signal itself. Nonlinear methods, e.g., entropy analysis and fractal dimension analysis, become popular in many EEG processing for medical applications and could be applied to real-time neurofeedback systems to model brain activities [13, 47, 56]. Our hypothesis is that non-linear fractal dimension approach allows quantify different brain states with better accuracy, allow to use less electrodes and could recognize different levels of brain states such as attention, concentration, stress levels, etc.

13.2.3 Neurofeedback Systems for Performance Enhancement

Recently, neurofeedback systems were started to be used for performance enhancement in healthy people. As it was mentioned in Sect. 13.2.2, the signal processing algorithms used in neurofeedback systems are mostly based on frequency analysis, and event related potential analysis. The following approaches were proposed to improve performance in healthy people: enhance focused attention process with SMR/theta training, enhance mental rotation ability with upper alpha and theta training, and improve memory and attention process with SCPs training. In the Table 13.1, some examples of works on human performance enhancement are described. In this work, we proposed algorithms based on fractal dimension for human performance enhancement systems that have better accuracy than frequency based ones. Then, we designed and implemented stress management game "Shooting," and concentration games "Breaking Wall."

Table 13.1 Examples of human performance approaches

Training results	EEG bands and location	Details of the protocols
Golf Performance improved 25 % [3]	Personalized Profile (QEEG method). All bands are used as feature	In the assessment, a personal event-locked EEG profile at FPz was determined for successful versus unsuccessful putts. Target frequency bands and amplitudes marking optimal prefrontal brain state were derived from the profile by two raters.
Visual Attention, Performance in shooting enhanced [26]	Gamma Power (>30 Hz) and alpha power in T3	An expert shooter has higher alpha power (8–12 Hz) and lower level in beta and gamma power in T3.
Focused attention process [53]	With SMR (8–13 Hz) theta (4–8 Hz) training	After eight sessions of neurofeedback the SMR-group (Enhance SMR band power training) was able to selectively enhance their SMR activity, cued recall performance, using a semantic working memory task, and to a lesser extent showed improved accuracy of focused attention.
Enhance mental rotation ability [19]	Upper alpha and theta (4–8 Hz) training	As expected, those subjects who were able to enhance their upper alpha power performed better on the cube rotations after upper alpha neurofeedback.
Improve memory and attention process [21]	Slow cortex Potential Training (ERP)	ERP training.
ADHD patient [16]	Theta (4–8 Hz)/beta (13–30 Hz) ratio in Cz,C3	Frequency training decrease of theta activity and increase of beta to increase arousal.
Working memory workload [32]	Theta (4–8 Hz) in frontal lobe	The results indicate that enhanced working memory load induces an increase in theta power and decrease in alpha power in frontal lobe.
Drive workload [32]	In parietal alpha (8–12 Hz) activity	Increased driving task load led to the decrease in alpha power in parietal lobe.

13.2.4 Emotion Recognition Algorithms

Real-time emotion recognition from EEG could reveal the "inner" feeling of the user, and then, it could be used in emotional state monitoring, workload optimization, post-traumatic therapies, human performance enhancement, etc. Generally, emotion recognition algorithms consist from two parts: feature extraction and classification. For real-time applications, an objective is to develop fast algorithms recognizing more emotions with fewer electrodes used. Currently, mostly off-line recognition algorithms were proposed which are shown in the Table 13.2. EEG-based emotion recognition algorithms could be divided into two groups: a subject-dependent and a subject-independent one. In the Table 13.2, the algorithms are compared by feature extraction types algorithms and classification methods used, by emotion types recognized and by the algorithms accuracy. The best accuracy is given if there are more than one type of feature extraction methods or classifiers. The algorithms are also differed by the number of the electrodes used in the emotion recognition. In the Table 13.2, in works [25,38] two to five electrodes were used. All other works employed more than 32 electrodes to collect EEG data and recognize emotions.

Although, in the past few years, the field of EEG-based emotion recognition has been receiving more and more attention, it is still a relatively new area. There are the following limitations: most of the works are off-line implementation of the algorithms; the number of electrodes used is usually large; the types of emotions recognized are limited, etc. In [34], we proposed a real-time algorithm only using three channels (AF3, F4 and FC6) in total. Fractal dimension algorithms were applied to compute fractal based features, and the real-time EEG-based emotion recognition algorithm was implemented with predefined thresholds based on the training session analysis.

13.3 Spatio-Temporal Fractal Approach

In this work, we describe spatio-temporal fractal based approach to brain study that was first proposed in [35]. The spatio-temporal approach combines two methods: a spatio-temporal analysis and fractal based analysis. The spatio-temporal analysis includes real-time 3D mapping of EEG signal amplitude or other parameters, for example, fractal dimension values, with Blobby model defined by implicit functions and applying set-theoretic operations over the moving shapes to isolate activities common for the signal during the time interval, as well as those that are unique ones. The proposed fractal based method would allow us to estimate the signal complexity changing over time and recognize the brain state.

13.3.1 3D Mapping of EEG for Visual Analytics

We proposed a method of EEG analysis based on 3D mapping of EEG data for visual analytics. We employed a concept of a dynamic 3D volumetric Blobby shape to visualize the EEG signal changes over time [30, 48]. A time-dependent Blobby object is defined using implicit functions that allow us to propose and implement set-theoretic operations over the time changing shapes for further analysis. Besides applying the set-theoretic ("Boolean") operations to the moving shapes to isolate

Table 13.2 Off-line emotion recognition algorithms

Author	Feature and classifier	Emotion	Classification accuracy
Subject-dependent emotion recognition works			
Ishino and Hagiwara [25]	**Feature:** FFT; Wavelet transform; Variance, mean **Classifier:** Neural Network	Joy, sad, angry, relaxed	Joy: 54.5 % Anger: 67.7 % Sorrow: 59 % Relaxation: 62.9 %
Zhang and Lee [58]	**Feature:** PCA **Classifier:** Linear Kernel SVM; RBF Kernel SVM	Negative and Positive	73 %
Chanel et al. [8]	**Feature:** Six frequency bands from different locations **Classifier:** Naïve Bayes; Fisher Discriminant Analysis	Three degree of arousal	58 %
Ansari-Asl et al. [2]	**Feature:** Synchronization Likelihood **Classifier:** LDA	Exciting/negative, Exiting/positive Calm/neutral	55.3 %
Chanel et al. [7]	**Feature:** Short Time Fourier Transform; Mutual Information **Classifier:** Discriminant Analysis; SVM; Relevance Vector Machine	Positive/arousal, neutral/calm, negative/arousal	63 %
Lin et al. [33]	**Feature:** ASM 12 **Classifier:** SVM	Joy, anger, sadness, pleasure	90.72 %
Schaaff and Schultz [42]	**Feature:** Peak alpha frequency, alpha power and cross-correlation features **Classifier:** SVM	Pleasant, neutral, unpleasant	66.7 %
Subject-independent emotion recognition works			
Khalili and Moradi [27]	**Feature:** Statistical feature combined with correlation dimension **Classifier:** Quadratic Discriminant Analysis	Calm, positive arroused, negative aroused	76.66 %
Takahashi [50]	**Feature:** Statistical features **Classifier:** SVM; Neural Networks	Joy, anger, sadness, fear and realization	41.7 % for five emotions
Petrantonakis and Hadjileontiadis [38]	**Feature:** Statistical features, wavelet based features, higher order crossings. **Classifier:** SVM; QDA; KNN; Mahalanobis Distance	Happy, surprised, angry, fear, disgust, sad	62.3 % for single channel case, 83.33 % for combined channel case

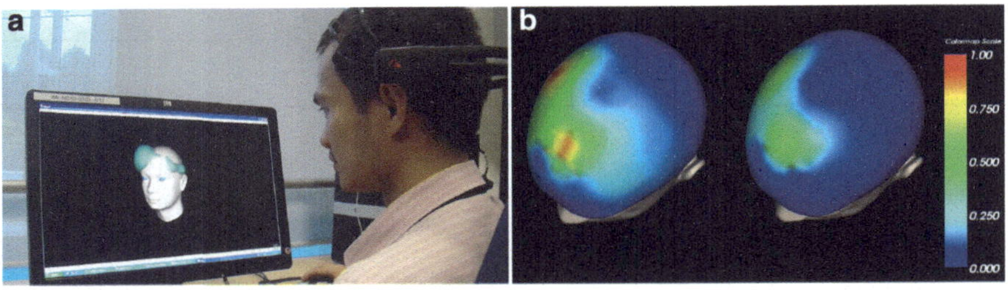

Fig. 13.1 Visual analytics. (**a**) 3D Blobby functions, (**b**) Colour mapping

activities common for both of them per time point, as well as those that are unique for either one, for a better visual impression, the blobby shape is superimposed on a 3D head model. On one data set, we can do intersection of all shapes to show constant activity on the time interval, and union of all shapes to show the overall maximum activity. On two data sets, we could apply an intersection to show common activity, union to show overall maximum activity and subtraction to show activities which are characteristic to one set. Its size and appearance visually reflect the brain activity. In Fig. 13.1a, the user looks in real-time on his brain parameters visualized with the Blobby functions and changing over time. He could "look inside his brain" in real time. In Fig. 13.1b, two snapshots of visualization of parameters with colour mapping corresponding to different time are shown. The advantage of the proposed and implemented "VisBrain" software is that it uses novel Blobby model changing over time and could work with any EEG device after accessing raw data in real-time. As it was mentioned above, with this proposed system we improve spatial resolution of the brain study based on EEG.

13.3.2 Fractal-Based Approach

Fractal dimension (FD) is a measurement of complexity of the object based on an entropy analysis. Entropy is a measure of the disorder in physical systems, or an amount of information that may be gained by observations of the disordered systems. A common practice to distinguish among possible classes of time series is to determine their so-called correlation dimension. The correlation dimension, however, belongs to an infinite family of fractal dimensions [22]. Hence, there is a hope that the use of the whole family of fractal dimensions may be advantageous in comparison to using only some of these dimensions. The concept of generalized entropy of a probability distribution was introduced by Alfred Renyi [39]. Based on the moments of order of the probability Renyi obtained the following expression for entropy

$$S_q = \frac{1}{q-1} \log \sum_{i-1}^{n} p_i^q \tag{13.1}$$

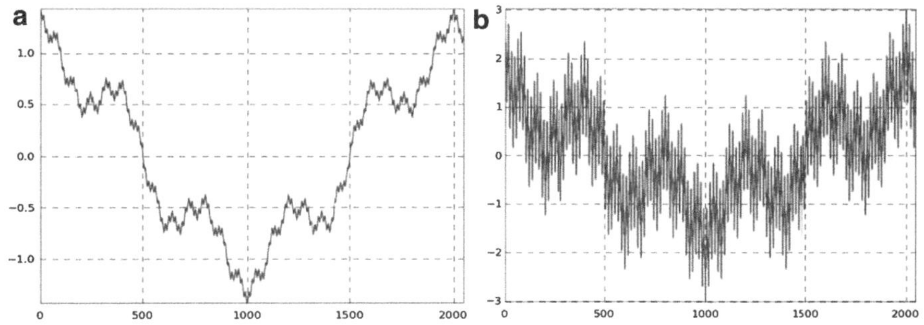

Fig. 13.2 Example of Weierstrass functions with (**a**) $FD = 1.25$ and (**b**) $FD = 1.75$

where q is not necessarily an integer and log denotes log 2. Note that for $q \to 1$, Eq. (13.1) yields the well-known entropy of a discrete probability distribution [43]:

$$S_1 = -\sum_{i-1}^{n} p_i \tag{13.2}$$

There are various methods to calculate fractal dimensions. In works [29, 30], the generalized Renyi approach based on Renyi entropy and calculation of the whole spectra of fractal dimensions to quantify brain states were studied. In this project, we are going to study whole spectra of fractal dimensions and propose novel algorithms for fractal dimension estimation for real-time applications. So far, we applied Hausdorff dimension when in Eq. (13.1). We implemented two well-known Higuchi [23] and Box-counting [5] algorithms for fractal dimension calculation. Both of them were evaluated using fractional Brownian and Weierstrass functions where theoretical FD values are known [37]. Higuchi algorithm gave better accuracy as FD values were closer to the theoretical FD ones. In Fig. 13.2, Weierstrass functions with FD = 1.25 and FD = 1.75 are presented to asses visually correlation between the signal complexity and FD values.

13.3.3 Real-Time Brain State Recognition

We proposed the spatio-temporal approach to the implementation of real-time systems. A diagram of the system is shown in Fig. 13.3. The user receives stimuli from the computer system such as visual, audio, haptic, etc. Then, the mental process is recognized from his/her EEG that is acquired by the EEG device. An overall recognition algorithm used in the real-time applications consists from the following steps: data sampling and pre-processing including data filtering, feature extraction, and machine learning algorithm. Then, the command to the feedback system is formed based on the recognition results. Depending on the application, the command could be the recognized emotion, concentration level, etc.

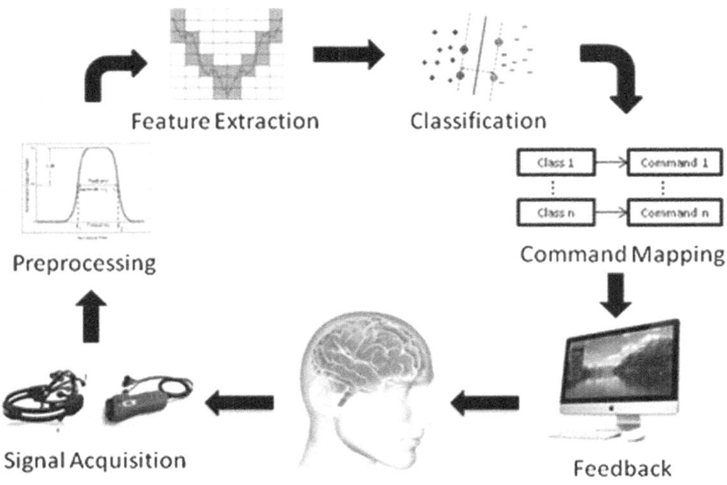

Fig. 13.3 Diagram for non-invasive real-time EEG-based system [49]

13.3.3.1 Pre-processing

The collected data are filtered by a 2–42 Hz bandpass filter since major waves of EEG lie in this band.

13.3.4 Features Extraction

The next step after the data pre-processing is the features extraction. We will apply a sliding window and calculate one FD value per sample per channel. Number of channels used in the recognition algorithm defines a size of the feature vector as follows:

$$F = \{FD_1, FD_2, \ldots FD_m\} \tag{13.3}$$

where m is number of channels.

In our preliminary study on the concentration level recognition algorithm, we have one feature as only one channel is used [57]. In the emotion recognition algorithm, there are three features in the vector as three channels—AF3, F4 and FC6 are used [34]. To improve accuracy of the algorithms we propose to use FD values of different order per one channel.

13.3.4.1 Classification Algorithms

Currently, we use a simple real-time subject-dependent classification algorithm based on threshold FD values that are calculated during a short training session.

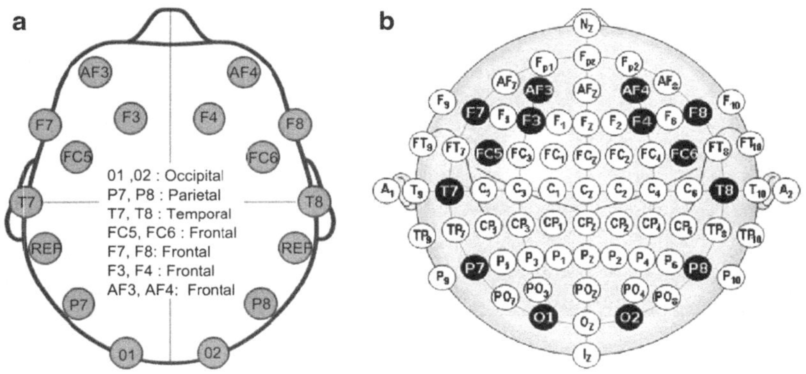

Fig. 13.4 Diagram electrodes location. (**a**) Emotiv device [14], (**b**) Standardized system [40]

Note that off-line processing with SVM classifier of the EEG labelled with emotions and concentration levels gave us similar accuracy as the real time implementation algorithm used thresholds [46].

13.3.4.2 EEG Data

EEG data are collected by Emotiv device [14] with 14 electrodes located at AF3, F7, F3, FC5, T7, P7, O1, O2, P8, T8, FC6, F4, F8, AF4 standardized by the American Electroencephalographic Society [1]. In Fig. 13.4, the electrodes location is shown. The sampling rate of Emotiv is 128 Hz. To be able to use any EEG device a program reading raw EEG signals is needed to be implemented. Currently, our applications could also work with Pet 2 and Mindset 24. All electrodes can be active in the system. The steps of an overall algorithm of the real-time application are as follows. First, raw data are read from the EEG device, filtered with band pass filter 2–42 Hz, and processed by the corresponding brain state recognition algorithm. Then, the results of the recognition are fed to the developed game, web site, or any other real-time software.

13.4 Real-Time EEG-Enabled Applications

Electroencephalogram (EEG) is a non-invasive technique recording the electrical potential over the scalp which is produced by the activities of brain cortex and reflects the state of the brain. Different from other mental state interpreters, e.g. the analysis based on fMIR, EEG technique gives us an easy and portable way to monitor brain activities with the help of suitable signal processing and classification

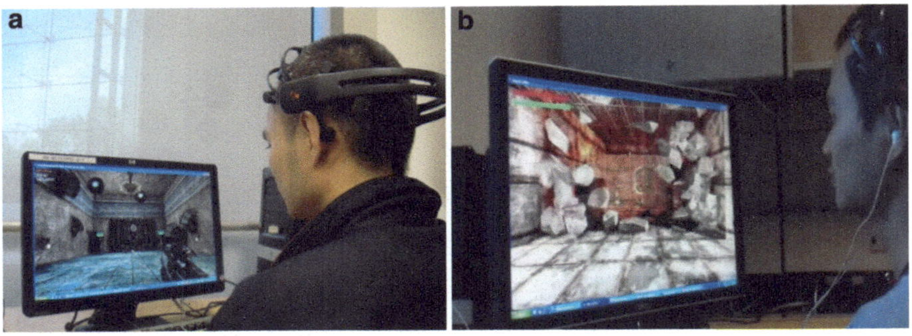

Fig. 13.5 (**a**) EEG-enabled stress management "Shooting" game (**b**) EEG-enabled concentration "Breaking Wall" game

algorithms. The main advantages of EEG-based approach is high temporal resolution of brain states recognition, portability and mobility of the modern EEG devices that makes it possible to develop real-time, portable and mobile applications.

13.4.1 Neurofeedback Training Systems

We proposed and developed concentration and stress management training systems based on the fractal model. The EEG-based training system design includes two parts: signal processing algorithms and a 2D/3D or virtual reality game part. Raw EEG signals collected by the device from the user brain are filtered and analyzed by the subject-dependent fractal based algorithms in real-time, and the resulting values are interpreted in the game as an additional game control using just the "brain power" or power of "thinking." A training effect of such systems consists from combination of the distraction effect of the game and the user's ability to learn how to control the game by changing voluntarily his brain state. For example, the user could learn how to improve his/her concentration ability. Currently, the proposed subject-dependent fractal-based algorithms recognize two levels of concentration or stress brain state from EEG [57]. The proposed algorithms give better accuracy comparing to traditional neurofeedback algorithms. Based on the proposed algorithms we designed and implemented "Shooting" and "Breaking Wall" games. In Fig. 13.5a, the proposed and implemented stress management "Shooting" game is shown. The user is able to shoot flying objects only if the stress level is less than the user's predefined threshold. Thus, the user could voluntarily change his/her stress level. In Fig. 13.5b, EEG-enabled concentration training "Breaking Wall" system is shown. In the "Breaking Wall" system, the wall is broken when the level of the user concentration recognized in real time from EEG reaches the threshold level.

13.4.2 Real-Time EEG-Based Emotion Recognition and Monitoring

There is no easily available benchmark database of EEG data labeled with emotions. But there are labeled databases of audio stimuli for emotion induction—International Affective Digitized Sounds (IADS) [6] and visual stimuli—International Affective Picture System (IAPS) [31]. In work [34], we proposed and carried out an experiment on emotion induction using IADS database of the labeled audio stimuli. We also proposed and implemented an experiment with music stimuli to induce emotions by playing music pieces and prepared questionnaire for the participants to label the recorded EEG data with corresponding emotions. To avoid the EEG signal being contaminated by artifacts such as facial expression, eye blinking, etc., the subjects were required to keep still and close eyes during the experiments. By analyzing the obtained EEG data based on the fractal dimension algorithm, in work [35], we proposed and implemented a real-time subject-dependent fractal-based emotion recognition algorithm where we mapped fractal dimension values to 2D Arousal-Valence emotion model. Since frontal lobe is believed to play an important role in emotion [52], we partially confirmed the hypotheses that emotions could be recognized using frontal lobe channels. As a result, in work [34, 35], only three channels which include AF3, F4 and FC6 were used. The arousal level was recognized with the accuracy of 84.9% and the valence level was recognized with the accuracy of 90%. By combining the recognized arousal and valence level, we could recognize in real-time the following emotions: satisfied, pleasant, happy, frustrated, sad, and fear. In Fig. 13.6a, mapping of six emotions to 2D Arousal-Valence space is shown. In Fig. 13.6b, six emotions are visualised with Haptek system [20] on the user's avatar.

Based on the fractal-based emotion recognition algorithm we implemented music therapy site for stress and depression management, music player, and emotion-enabled games. An emotion enabled game "Dancing Penguin" was designed and implemented. In Fig. 13.7a, an emotion is induced to the user by audio stimuli through earphones, recognized in real time from EEG and visualized on the user's 3D Penguin avatar. The user's emotion is interpreted as the Penguin motions.

Currently, the Penguin Avatar has dance movements corresponding to six recognised emotions: satisfied, pleasant, happy, frustrated, sad, and fear. In Fig. 13.7b, example of "happy" emotion movement is shown. Videos of the implemented real-time EEG-enabled applications are presented in [45].

13.5 Conclusion

Real-time brain states recognition and EEG-enabled applications need inter-disciplinary approach. It includes research on signal and bio-signal processing, pattern recognition and cognitive informatics, human–computer interfaces, game

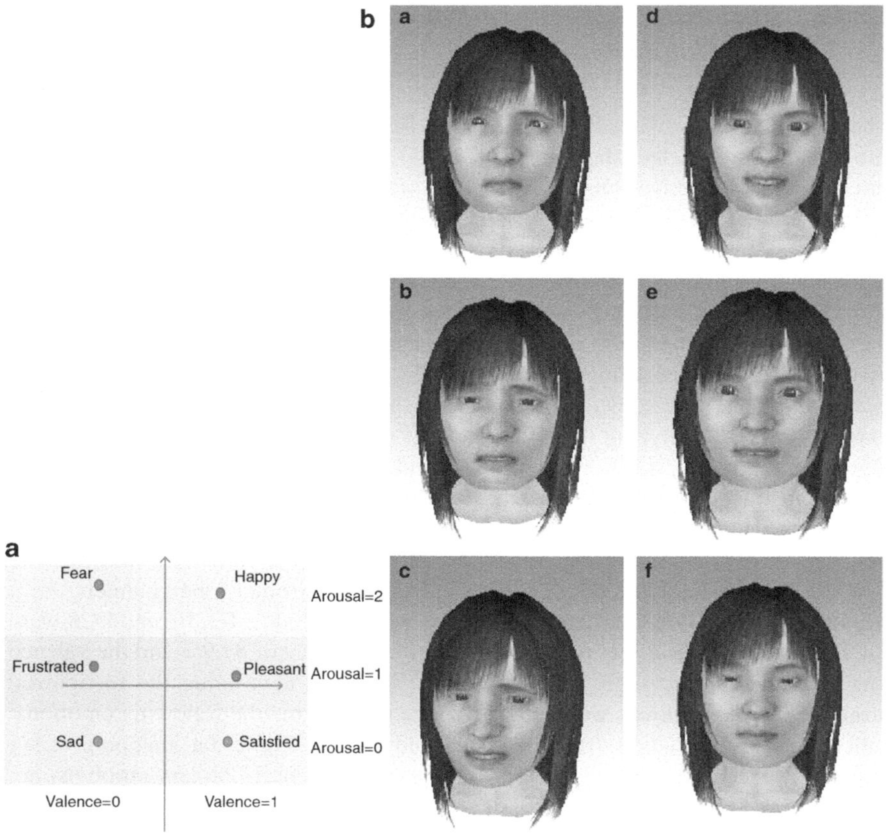

Fig. 13.6 (**a**) Six emotions mapping in Valence-Arousal emotion model (**b**) Real-time recognition from EEG and visualization of happy, pleasant, satisfied, fear, frustrated, sad emotions on the user's Avatar with Haptek system [35]

design, etc. Currently, algorithms embedded in neurofeedback systems are mainly based on EEG frequency band power assessment. Those linear methods may not represent nonlinear brain process. In our work, fractal dimension (FD) algorithms to extract non-linear FD features changing with time that describe complexity of the signal over time were implemented. The experiments on EEG recoding of different brain states induced by external stimuli were proposed and carried out. Algorithms of brain states recognition including stress, concentration levels recognition and emotion recognition were implemented. Use of the nonlinear FD features in the brain state recognition algorithms could increase brain state classification accuracy. Emotion-enabled game "dancing Penguin," stress management game "Shooting," and concentration training game "Breaking Wall" were proposed and implemented. The proposed spatio-temporal fractal based approach is cost effective as the

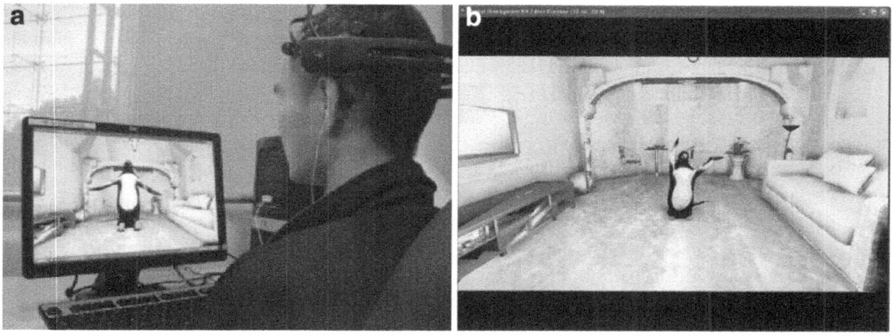

Fig. 13.7 (**a**) Emotion-enabled "Dancing Penguin" game (**b**) Snapshot of "happy" emotion animation

implemented training systems based on it are self-contained and do not require any staff to conduct the training. As each system has an entertainment element, it is easy to use especially if motivation of enhancing the human capabilities is explained before the training. The proposed brain states quantification algorithms including stress, concentration, and emotion recognition would advance research on human cognition and human–computer interaction by adding one more dimension to human computer interfaces.

Acknowledgements This project is supported by the Media Digital Authority grant NRF2008IDM-IDM004-020 "Emotion-based personalized digital media experience in Co-Spaces" of National Research Fund of Singapore.

References

1. American electroencephalographic society.: American electroencephalographic society guidelines for standard electrode position nomenclature. J. Clin. Neurophysiol. **8**(2), 200–202 (1991)
2. Ansari-Asl, K., Chanel, G., Pun, T.: A channel selection method for EEG classification in emotion assessment based on synchronization likelihood. In: Proc. 15th European Signal Processing Conference, Poznan 2007, pp. 1241–1245
3. Arns, M., Kleinnijenhuis, M., Fallahpour, K., Breteler, R.: Golf performance enhancement and real-life neurofeedback training using personalized event-locked EEG profiles. J. Neurother. **11**(4), 11–18 (2007)
4. Birbaumer, N.: Slow cortical potentials: Plasticity, operant control, and behavioral effects. Neuroscientist **5**(2), 74–78 (1999)
5. Block, A., Von Bloh, W., Schellnhuber, H.J.: Efficient box-counting determination of generalized fractal dimensions. Phys. Rev. A **42**(4), 1869–1874 (1990)
6. Bradley, M.M., Lang, P.J.: The International Affective Digitized Sounds (2nd edn. IADS-2): Affective ratings of sounds and instruction manual. University of Florida, Gainesville, (2007)
7. Chanel, G., Kierkels, J.J.M., Soleymani, M., Pun, T.: Short-term emotion assessment in a recall paradigm. Int. J. Hum. Comput. Stud. **67**(8), 607–627 (2009)

8. Chanel, G., Kronegg, J., Grandjean, D., Pun, T.: Emotion Assessment: Arousal Evaluation Using EEG's and Peripheral Physiological Signals. In: Multimedia Content Representation, Classification and Security, vol. 4105. Lecture Notes in Computer Science, pp. 530–537. Springer, Berlin/Heidelberg (2006)
9. Clarke, A.R., Barry, R.J., McCarthy, R., Selikowitz, M.: Electroencephalogram differences in two subtypes of Attention-Deficit/Hyperactivity Disorder. Psychophysiology **38**(2), 212–221 (2001)
10. Coben, R., Linden, M., Myers, T.E.: Neurofeedback for autistic spectrum disorder: A review of the literature. Appl. Psychophysiol. Biofeedback **35**(1), 83–105 (2010)
11. Current Statistics on Chronic Pain. http://www.beyondchronicpain.com/site/media/currentStatisticsOnChronicPain.php
12. Demos, J.N.: Getting Started with Neurofeedback. WW Norton and Company, New York (2005)
13. Easwaramoorthy, D., Uthayakumar, R.: Improved generalized fractal dimensions in the discrimination between Healthy and Epileptic EEG Signals. J. Comput. Sci. **2**(1), 31–38 (2011)
14. Emotiv. http://www.emotiv.com
15. Fuchs, T., Birbaumer, N., Lutzenberger, W., Gruzelier, J.H., Kaiser, J.: Neurofeedback treatment for attention-deficit/hyperactivity disorder in children: A comparison with methylphenidate. Appl. Psychophysiol. Biofeedback **28**(1), 1–12 (2003)
16. Gevensleben, H., Holl, B., Albrecht, B., Schlamp, D., Kratz, O., Studer, P., Wangler, S., Rothenberger, A., Moll, G.H., Heinrich, H.: Distinct EEG effects related to neurofeedback training in children with ADHD: A randomized controlled trial. Int. J. Psychophysiol. **74**(2), 149–157 (2009)
17. Hammond, D.C.: QEEG-guided neurofeedback in the treatment of obsessive compulsive disorder. J. Neurother. **7**(2), 25–52 (2003)
18. Hammond, D.C.: What is neurofeedback? J. Neurother. **10**(4), 25–36 (2006)
19. Hanslmayr, S., Sauseng, P., Doppelmayr, M., Schabus, M., Klimesch, W.: Increasing individual upper alpha power by neurofeedback improves cognitive performance in human subjects. Appl. Psychophysiol. Biofeedback **30**(1), 1–10 (2005)
20. Haptek. http://www.haptek.com
21. Heinrich, H., Gevensleben, H., Strehl, U.: Annotation: Neurofeedback - Train your brain to train behaviour. J. Child. Psychol. Psychiatry **48**(1), 3–16 (2007)
22. Hentschel, H.G.E., Procaccia, I.: The infinite number of generalized dimensions of fractals and strange attractors. Physica D **8**(3), 435–444 (1983)
23. Higuchi, T.: Approach to an irregular time series on the basis of the fractal theory. Physica D **31**(2), 277–283 (1988)
24. International Society of Neurofeedback & Research. http://www.isnr.org/information/index.cfm
25. Ishino, K., Hagiwara, M.: A feeling estimation system using a simple electroencephalograph. In: Proc. IEEE International Conference onSystems, Man and Cybernetics, 2003, pp. 4204–4209, vol. 4205, 5–8 Oct. 2003
26. Janelle, C.M., Hatfield, B.D.: Visual attention and brain processes that underlie expert performance: Implications for sport and military psychology. Military Psychol. **20**(suppl. 1), S39–S69 (2008)
27. Khalili, Z., Moradi, M.H.: Emotion recognition system using brain and peripheral signals: Using correlation dimension to improve the results of EEG. In: Proc. International Joint Conference on Neural Networks 2009, pp. 1571–1575
28. Kouijzer, M.E.J., van Schie, H.T., de Moor, J.M.H., Gerrits, B.J.L., Buitelaar, J.K.: Neurofeedback treatment in autism. Preliminary findings in behavioral, cognitive, and neurophysiological functioning. Res. Autism Spectr. Disord. **4**(3), 386–399 (2010)
29. Kulish, V., Sourin, A., Sourina, O.: Analysis and visualization of human electroencephalograms seen as fractal time series. J. Mech. Med. Biol. **6**(2), 175–188 (2006a)
30. Kulish, V., Sourin, A., Sourina, O.: Human electroencephalograms seen as fractal time series: Mathematical analysis and visualization. Comput. Biol. Med. **36**(3), 291–302 (2006b)

31. Lang, P.J., Bradley, M.M., Cuthbert, B.N.: International affective picture system (IAPS): Affective ratings of pictures and instruction manual. Technical Report A-8. University of Florida, Gainesville, FL, (2008)
32. Lei, S., Roetting, M.: Influence of task combination on EEG spectrum modulation for driver workload estimation. Hum. Factors **53**(2), 168–179 (2011)
33. Lin, Y.P., Wang, C.H., Wu, T.L., Jeng, S.K., Chen, J.H.: EEG-based emotion recognition in music listening: A comparison of schemes for multiclass support vector machine. In: Proc ICASSP, IEEE International Conference on Acoustics, Speech and Signal Processing, Taipei 2009, pp. 489–492
34. Liu, Y., Sourina, O., Nguyen, M.K.: Real-time EEG-based Human Emotion Recognition and Visualization In: Proc. 2010 Int. Conf. on Cyberworlds, Singapore, pp. 262–269, 20–22 Oct. 2010
35. Liu, Y., Sourina, O., Nguyen, M.K.: Real-Time EEG-based Emotion Recognition and Applications. Trans. Comput. Sci. XII, LNCS 6670 TOC, 256–278 (2011)
36. Lubar, J.F., Swartwood, M.O., Swartwood, J.N., O'Donnell, P.H.: Evaluation of the effectiveness of EEG neurofeedback training for ADHD in a clinical setting as measured by changes in T.O.V.A. scores, behavioral ratings, and WISC-R performance. Biofeedback Self Regul. **20**(1), 83–99 (1995)
37. Maragos, P., Sun F-K.: Measuring the fractal dimension of signals: morphological covers and iterative optimization. IEEE Trans. Signal Process. **41**(1), 108–121 (1993)
38. Petrantonakis PC., Hadjileontiadis, L.J.: Emotion recognition from EEG using higher order crossings. IEEE Trans. Inf. Technol. Biomed. **14**(2), 186–197 (2010)
39. Renyi, A.: Probability Theory. Dover, Mineola, NY (2007)
40. Sanei, S., Chambers, J.A.: EEG Signal Processing. WILEY, San Francisco (2007)
41. Saxby, E., Peniston, E.G.: Alpha-theta brainwave neurofeedback training: An effective treatment for male and female alcoholics with depressive symptoms. J. Clin. Psychol. **51**(5), 685–693 (1995)
42. Schaaff, K., Schultz, T.: Towards emotion recognition from electroencephalographic signals. In: IEEE International Workshop on Robot and Human Interactive Communication 2009, pp. 792–796
43. Shannon, C.E.: A mathematical theory of communication. Bell System Tech. J. **27**(4), 623–656 (1948)
44. Sokhadze, T.M., Cannon, R.L., Trudeau, D.L.: EEG biofeedback as a treatment for substance use disorders: Review, rating of efficacy, and recommendations for further research. Appl. Psychophysiol. Biofeedback **33**(1), 1–28 (2008)
45. Sourina, O.: IDM-Project. (2008), Emotion-based personalized digital media experience in Co-Spaces. http://www3.ntu.edu.sg/home/eosourina/CHCILab/projects.html
46. Sourina, O., Liu, Y.: A Fractal-Based Algorithm of Emotion Recognition from EEG Using Arousal-Valence Model. In: Proc. Biosignals 2011, Rome, Italy, pp. 209–214, 26–29 Jan 2011
47. Sourina, O., Liu, Y., Wang, Q., Nguyen, M.K.: EEG-based Personalized Digital Experience. In: Stephanidis C. (ed.) Universal Access in HCI, Part II, HCII 2011, Heidelberg 2011, pp. 591–599. Springer, Heidelberg (2011)
48. Sourina, O., Sourin, A., Kulish, V.: EEG Data Driven Animation and Its Application. In: Proc. Computer Vision/Computer Graphics Collaboration Techniques 2009. Lecture Notes in Computer Science, pp. 380–388
49. Sourina, O., Wang, Q., Liu, Y., Nguyen, M.K.: A Real-time Fracal-based Brain State Recognition from EEG and Its Application In: Proc. Biosignals 2011, Rome Italy, pp. 82–91, 26–29 Jan. 2011,
50. Takahashi, K.: Remarks on emotion recognition from multi-modal bio-potential signals. In: Proc. IEEE ICIT '04, vol. 1133, pp. 1138–1143, 8–10 Dec. 2004
51. Theiler, J.: Estimating fractal dimension. J. Opt. Soc. Am. A **7**, 1055–1073 (1990)
52. Train, B.: Introduction to psychology. Pearson Education, South Africa (2007)

53. Vernon, D., Egner, T., Cooper, N., Compton, T., Neilands, C., Sheri, A., Gruzelier, J.: The effect of training distinct neurofeedback protocols on aspects of cognitive performance. Int. J. Psychophysiol. **47**(1), 75–85 (2003)
54. Video Game Therapy Helping Soldiers. http://www.myfoxaustin.com/dpp/news/local/111909-Video-Game-Therapy-Helping-Soldiers
55. Virtual Pain Relief. http://videos.howstuffworks.com/sciencentral/2888-virtual-pain-relief-video.htm
56. Wang, Q., Sourina, O., Nguyen, M.K.: EEG-based "Serious" Games Design for Medical Applications. In: Proc. 2010 Int. Conf. on Cyberworlds, Singapore 2010, pp. 270–276
57. Wang, Q., Sourina, O., Nguyen, M.K.: Fractal dimension based neurofeedback. Vis. Computer **27**, 299–309 (2011)
58. Zhang, Q., Lee, M.: Analysis of positive and negative emotions in natural scene using brain activity and GIST. Neurocomputing **72**(4-6), 1302–1306 (2009)

Chapter 14
Phase Detection of Visual Evoked Potentials Applied to Brain Computer Interfacing

Gary Garcia-Molina and Danhua Zhu

14.1 Introduction

The steady state visual evoked potential (SSVEP) refers to the activity of the cerebral cortex that results from attending to a repetitive visual stimulus (RVS) oscillating at a constant *stimulation frequency*. The SSVEP manifests in the scalp recorded electroencephalogram (EEG) as oscillatory components at the stimulation frequency and/or harmonics. The SSVEP is more prominent at parietal and occipital locations due to their relative proximity to the primary visual cortex [18]. In Fig. 14.1, the signal recorded between EEG positions Oz-Cz clearly shows the SSVEP that appears in response to an RVS at 15 Hz.

Among non-invasive EEG based brain computer interfaces (BCI), BCIs based on the SSVEP have the advantages of providing higher information transfer rates (ITR) and of requiring shorter calibration [4]. SSVEP based BCIs operate by presenting the user with a set of repetitive visual stimuli (RVSi) which, in general, have different stimulation frequencies from each other [7, 9, 14]. The SSVEP corresponding to the RVS receiving the user's overt or covert [23] attention is more prominent and can be detected from the ongoing EEG. Each RVS is associated with an action or command which is executed by the BCI system when the corresponding SSVEP is detected.

G. Garcia-Molina (✉)
Philips Research North-America, Madison WI, United States
e-mail: gary.garcia@philips.com

D. Zhu
College of Biomedical Engineering and Instrument Science, Zhejiang University, Hangzhou, China
e-mail: danzhu.zhu@gmail.com

B. Z. Allison et al. (eds.), *Towards Practical Brain-Computer Interfaces*, Biological and Medical Physics, Biomedical Engineering, DOI 10.1007/978-3-642-29746-5_14,
© Springer-Verlag Berlin Heidelberg 2012

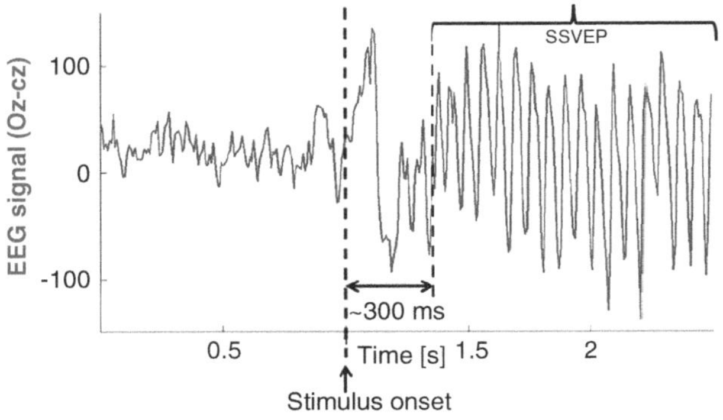

Fig. 14.1 Steady state visual evoked potential (SSVEP) elicited by an RVS at 15 Hz. The *top signal* shows the light intensity of the RVS across time. The EEG signal (Oz-Cz) shows the SSVEP evoked by such stimulation. It is important to notice that the appearance of the SSVEP does not immediately follow the onset of stimulation but manifests few hundred milliseconds after

Most of the current SSVEP-based BCIs use stimulation frequencies between 4 and 30 Hz because the SSVEP is more prominent in this frequency range [24]. The visual stimuli at these frequencies as compared to higher frequencies, entail the following disadvantages: (1) visual fatigue occurs faster and this decreases the SSVEP strength, and (2) a higher risk of photic or pattern-induced epileptic seizure exists [6]. High frequency SSVEPs are thus preferable for the sake of safety and comfort of the BCI application.

Only a limited number of frequencies above 30 Hz can elicit a sufficiently strong SSVEP for BCI purposes [25]. In the classical SSVEP based BCI operation where each RVS has a unique stimulation frequency, this limitation implies a reduction in the number of possible commands which can also limit the bitrate.

A possible way to tackle this limitation is to combine two or more frequencies to drive a single RVS [3, 16]. If N frequencies can be used and k of them are combined to drive a particular RVS, the total number of distinct RVS is $\binom{N}{k}$ which is larger than N if $N > k + 1$ and $k > 1$.

An alternative way is to use the same stimulation frequency but different phases [12, 15, 20]. Detecting the phase of the stimulus that receives the user's attention is possible because the SSVEP is phased-locked to the visual stimulus [18] (see also Fig. 14.1).

The SSVEP phase can be estimated using Fourier analysis [13,20,21]. To ensure reasonable accuracy, these methods need relatively long signal segments having a duration that is a multiple of the stimulation period. The phase locking analysis necessary for accurate detection of the attended phase, imposes synchronous BCI operation. This means that the BCI operation is paced by the system because the user is notified whenever the system is ready to receive commands. Asynchronous operation where the operation is paced by the user, can be achieved with phase SSVEP if the *stimulation signal* (i.e. the one that drives the RVS) is jointly recorded

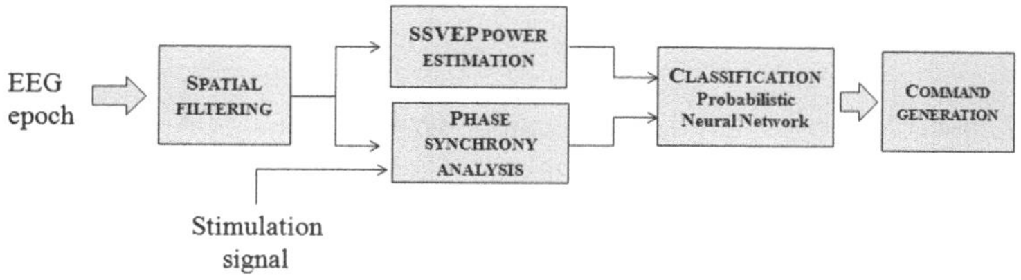

Fig. 14.2 After spatial filtering of the EEG epoch, a two-dimensional feature vector is obtained having as components the SSVEP power and phase. The resulting vector is processed by a neural network whose output estimates the RVS that receives the user's attention

with the brain signals. The stimulation signal serves then as a reference for phase detection of the stimulus that receives the user's attention.

This chapter presents an implementation of a high frequency SSVEP BCI using phase coding which allows for asynchronous operation. The signal processing methods are presented in Sect. 14.2. Section 14.3 presents the experimental evidence and Sect. 14.4 summarizes this chapter and suggests options for further research.

14.2 Signal Processing and Pattern Recognition Methods

Signal processing methods are proposed here to obtain a two-dimensional *feature vector* from a multi-channel EEG recording. The name *EEG epoch* is used here to refer to a set of EEG signals of a given duration.

Pattern recognition methods are applied to the feature vectors to recognize the user's intended command. The basic diagram illustrating this process is presented in Fig. 14.2.

The EEG epoch is first spatially filtered by linearly combining the signals from all EEG channels into a univariate signal. The algorithm to estimate the linear combination coefficients is detailed in Sect. 14.2.1.

The first component of the feature vector is the SSVEP power. This is estimated by applying a peak filter centered at the stimulation frequency, on the univariate signal resulting from spatial filtering, squaring the resulting signal, and averaging over the duration of the EEG epoch. The SSVEP power is used as feature to ensure that the user's attention was sufficiently high in order to elicit an SSVEP.

The second component of the feature vector is the SSVEP phase which is estimated by computing the average phase difference between the instantaneous phases of the stimulation signal and the spatially filtered signal. The instantaneous phases are estimated at the stimulation frequency. We refer to this process as to phase synchrony analysis [25].

The feature vectors are classified using a probabilistic neural network to estimate the RVS which received the user's attention.

14.2.1 Spatial Filtering

An EEG epoch $\mathbf{X}$ is considered. which can be written as a $T \times N$ matrix having N columns which are T-sample long signals $\mathbf{x}_i \;\; i = 1, \ldots, N$. The components of $\mathbf{x}_i$ correspond to the samples of the signal recorded at electrode site i.

A spatially filtered signal $\mathbf{x_w}$ is considered, which corresponds to a linear combination of the $\{\mathbf{x}_i\}$, i.e. $\mathbf{x_w} = \sum_i w_i \mathbf{x}_i = \mathbf{Xw}$, where $\mathbf{w} = [w_1, \ldots, w_N]'$.

The spatial filter coefficients w_i can be selected in such a way that the ratio between the power, at the stimulation frequency, in $\mathbf{x_w}$ due to SSVEP and that due to background brain activity is maximized. The coefficients selected in this manner lead to the so-called *maximum contrast combination* [8] where $\mathbf{w}$ is estimated on a per epoch basis by solving the following optimization problem.

$$\mathbf{w} = \arg\max_{\tilde{\mathbf{w}}} \frac{\tilde{\mathbf{w}}'\mathbf{X}'\mathbf{X}\tilde{\mathbf{w}}}{\tilde{\mathbf{w}}'(\mathbf{X} - \mathbf{QX})'(\mathbf{X} - \mathbf{QX})\tilde{\mathbf{w}}}, \tag{14.1}$$

where Q is the projection matrix on the vector space spanned by sinusoidal signals at the stimulation frequency and up to H harmonics.

Let $\boldsymbol{\Phi} = \{\sin(2\pi hf\mathbf{t}), \cos(2\pi hf\mathbf{t}) | h = 1, ..., H\}$ with $\mathbf{t} = [0, \ldots, T-1]'$ be the set of sinusoidal signals at the stimulation frequency and up to H harmonics. Then, Q can be written as: $Q = \mathbf{S}\left(\mathbf{S}^{\mathrm{T}}\mathbf{S}\right)^{-1}\mathbf{S}'$ where $\mathbf{S}$ is a matrix which has as columns the signals in the set $\boldsymbol{\Phi}$.

Under this modeling, $\mathbf{Xw} - Q\mathbf{Xw}$ is orthogonal to the vector space spanned by the components in $\boldsymbol{\Phi}$ and its Euclidean norm can be used as an estimate of the power, at the stimulation frequency, of the background brain activity.

As mentioned in Sect. 14.1, only high stimulation frequencies (larger than 30 Hz) are considered in this chapter. Since for most practical purposes the EEG spectral content is restricted to frequencies lower than 60 Hz, the number of harmonics H to be considered is set to one.

In (14.1), the per-epoch covariance matrices $\mathbf{X}'\mathbf{X}$ and $(\mathbf{X} - Q)'(X - QX)$ are used to estimate the SSVEP and the background activities respectively. A better and more stable estimate of the covariance matrix can be obtained through averaging of the covariance matrices over several EEG epochs. Thus, we propose to estimate the optimum spatial filter by solving the following optimization problem.

$$\mathbf{w} = \arg\max_{\tilde{\mathbf{w}}} \frac{\tilde{\mathbf{w}}' \sum_{k=1}^{K} \mathbf{X}_k'\mathbf{X}_k\tilde{\mathbf{w}}}{\tilde{\mathbf{w}}' \sum_{k=1}^{K} (\mathbf{X}_k - \mathbf{QX}_k)'(\mathbf{X}_k - \mathbf{QX}_k)\tilde{\mathbf{w}}}, \tag{14.2}$$

where $\mathbf{X}_k$ is the kth EEG epoch and K is the total number of epochs that are considered.

The SSVEP power (first component of the feature vector) is estimated by applying to the signal $x_{\mathbf{w}}(t)$, a 1 Hz narrow band FIR filter centered around the stimulation frequency (*peak-filter*). This results in the narrow band signal $z(t)$ from which the SSVEP power E can be estimated in a time window $[t, t + \Delta t]$ as: $E = \frac{1}{\Delta t} \int_{t}^{t+\Delta t} |z(t)|^2 dt$.

14.2.2 Phase Synchrony Analysis

The phase (second component of the feature vector) is estimated through a process which we refer to as phase synchrony analysis. The Hilbert transforms of $z(t)$ and the stimulation signal $l(t)$ are first calculated to obtain the analytical signals $A_z(t)$ and $A_l(t)$ as follows.

$$\begin{aligned} A_z(t) &= z(t) + jH_z(t) = \rho_z(t)e^{j\theta_z(t)}, \\ A_l(t) &= l(t) + jH_l(t) = \rho_l(t)e^{j\theta_l(t)}, \end{aligned} \quad (14.3)$$

where $H_z(t)$ and $H_l(t)$ are the Hilbert transforms of $z(t)$ and $l(t)$. The instantaneous amplitude and phase are respectively $\rho_{(\cdot)}$ and $\theta_{(\cdot)}$.

The SSVEP phase of the corresponding EEG epoch is estimated as the median value of the instantaneous phase difference $\delta_f(t) = \theta_z(t) - \theta_l(t)$ across the EEG epoch.

14.3 Experimental Evidence

The BCI implementation used in this chapter to illustrate the principles presented in Sect. 14.2, builds on the BCI2000 platform [19]. The application is a cursor navigation task along a computer rendered 2D maze. The allowed movements in this task follow four possible directions, namely upper-left, upper-right, lower-left, and lower-right (see Fig. 14.3a).

These directions are associated to the visual stimuli arranged around the computer screen as illustrated in Fig. 14.3b. Each stimulus was rendered through a 10×10 cm box containing a (green) power LED shining through a diffusion screen. The stimulation signal was a square wave with 50 % duty cycle at the stimulation frequency. Four phases $(\phi, \phi + \frac{\pi}{2}, \phi + \pi, \phi + \frac{3\pi}{2})$ were used to command the RVSi (see Fig. 14.3d) where ϕ is the initial phase at the onset of the stimulation signal. The corresponding stimulation signals were generated using four synchronized function generators (from Agilent technologies, model 33220A).

The EEG signals were collected using a BioSemi Active-two acquisition system [2] under normal office illumination. The signals from the 32 electrodes shown in Fig. 14.3c were recorded. The signal were re-referenced during pre-processing to the average of all recorded signals (i.e. Common average referencing). The sampling frequency was 2,048 Hz. During pre-processing the signals were downsampled to 256 Hz.

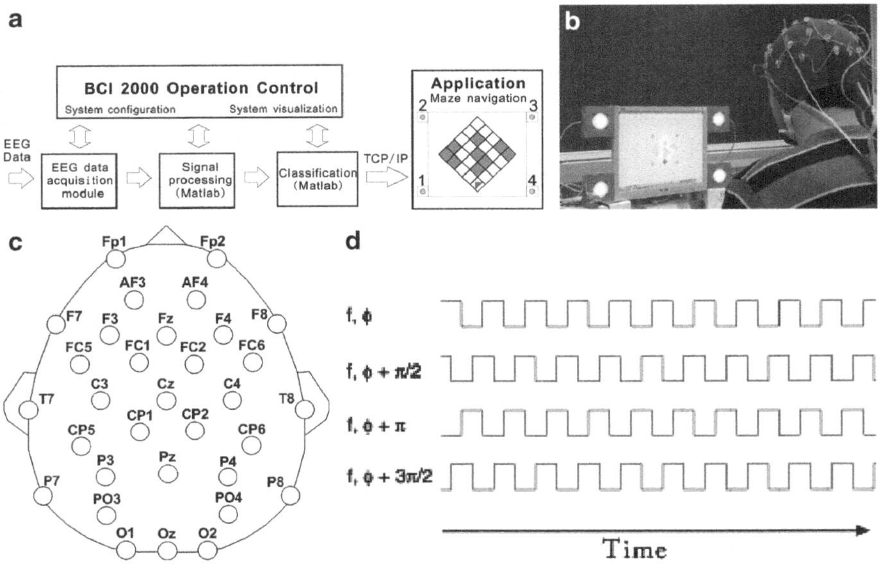

Fig. 14.3 (**a**) Software architecture of the BCI implementation. The application consists of a 2D maze in which the cursor moves in the direction of the RVS that receives the user's attention. In this particular configuration the numbers 1, 2, 3, and 4 represent the directions *bottom left*, *top left*, *top right*, and *bottom right* respectively. The command sequence to successfully complete this maze configuration is "2232323344144111." (**b**) RVSi arrangement around the computer screen. (**c**) Measured EEG sites. (**d**) Stimulation signals illustrating the phase difference

The stimulation signal was measured using a photodiode located near the RVS with phase ϕ. The signal from the photodiode was recorded simultaneously to the EEG signals to allow for the phase synchrony analysis (see Sect. 14.2.2).

14.3.1 Optimal Stimulation Frequency

The SSVEP responses for different stimulation frequencies depend on individual factors [10]. To determine the range of stimulation frequencies having sufficient comfort level, the participants in this study were presented with 10-s long flicker stimulation at frequencies ranging from 15 to 60 Hz (in steps of 5 Hz) and they were requested to subjectively rate their visual comfort in a scale from 1 to 5 (five is the most comfortable). The presentation order was randomized.

The average comfort level for different frequencies is reported in Fig. 14.4. It can be seen that for stimulation frequencies higher than 30 Hz, the comfort level exceeds three. Thus, we implemented a procedure aiming at determining the individual optimal stimulation frequency in the range from 30 to 40 Hz. The upper limit was decided upon considerations related to the SSVEP detectability.

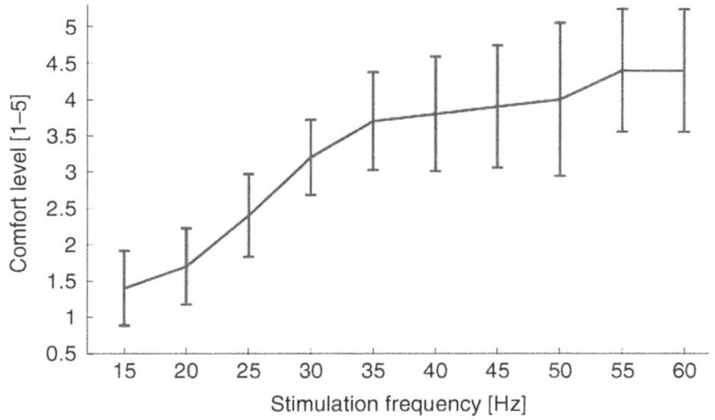

Fig. 14.4 Average visual comfort level for different stimulation frequencies. The vertical bars correspond to the standard deviation at each stimulation frequency

For a given user, the procedure to estimate the optimal stimulation frequency consisted in presenting RVSi at all the integer stimulation frequencies between 30 and 40 Hz. The presentation order was randomized.

For each stimulation frequency, the stimulation was presented in a sequence of four intervals. Each interval was composed of a 4-s long period of stimulation followed by a 4-s long break. To determine the stimulation frequency that elicits the strongest SSVEP response, the following procedure was followed:

1. The stimulation period in the first interval was used to estimate the spatial filter as described in Sect. 14.2.1.
2. This filter was applied to the EEG signals in the four intervals. This resulted in a one-dimensional spatially filtered signal.
3. The spatially filtered signal was then temporally filtered through a peak filter centered at the stimulation frequency. The result of this operation was squared.
4. The SSVEP power was then estimated in each 1-s long window using the result of the previous point. This resulted in 32 values, i.e. 16 SSVEP power values corresponding to the stimulation periods and 16 values corresponding to the break periods.
5. Using the set of 32 values, a threshold based detection of the SSVEP during stimulation was performed. Since this constitutes a detection problem with a single threshold, a receiving-operator curve (ROC) [5] can be determined by progressively varying the threshold from the lowest SSVEP power during the stimulation periods to the highest SSVEP power during the break periods. The area under the ROC (AUC) is a good indicator of the detectability of the SSVEP at the stimulation frequency. The optimal stimulation frequency corresponded to the one which resulted in the highest AUC.

The spatial filter of the optimal stimulation frequency was then used in the BCI calibration phase described in next section.

14.3.2 Calibration of the BCI Operation

The goal of the calibration is to determine the optimal classifier parameters to detect the phase of the stimulus on which the user's attention is focussed. Similarly to the protocol in Sect. 14.3.1, the stimulation was presented in a sequence of 16 intervals each of them composed of a 4-s long stimulation period followed by a 4-s long break. During the stimulation period all four RVSi flickered at the optimal stimulation frequency but having different phases from each other as depicted in Fig. 14.3d. At the onset of each interval, the user was instructed to pay attention to a specific RVS. The sequence of attention was randomized.

Using the optimal spatial filter obtained in the frequency selection phase (Sect. 14.3.1), the two parameters: (a) SSVEP phase, and (b) SSVEP power were estimated for each 1-s long segment during stimulation. The feature vectors for the four considered phases were then used to set the parameters of the probabilistic neural network classifier.

14.3.3 BCI Operation and Information Transfer Rate

During operation participants were instructed to move a cursor within the 2D maze along a pre-specified path and as fast as possible. In this research, participants were allowed to overtly focus their attention (i.e. by moving their eyes) on the repetitive visual stimuli.

On detection of the estimated user's intended direction of movement, the cursor moved along this direction only if this movement was following the pre-specified path. This restriction facilitated the estimation of the information transfer rate.

Cursor moves were accompanied by a low pitched tone. Detections corresponding to non-allowed directions were accompanied by a high pitched tone to notify the user of the error.

As shown in Fig. 14.3b, the command sequence to successfully navigate through this maze was "2232323344144111," where 1, 2, 3 and 4 are associated with the directions: bottom-left, top-left, top-right, and the bottom-right respectively. This sequence is balanced so that each direction is represented four times. This avoids biasing the results due to preferred directions.

The bitrate was estimated based on the user's proficiency in moving the cursor through the maze and along the specified path. Each user was requested to go through the maze two times. We used as an estimate for the accuracy, the ratio between the number of correct moves and the total number of moves.

To estimate the bitrate the most popular approach consists in using the information transfer rate (ITR) based on the definition in [22].

This definition suggests the following formula to obtain bitrate and ITR for C classes and classification accuracy p.

Table 14.1 Experimental results

Subject	Freq. (Hz)	Accuracy	Time-per-command (seconds)	ITR (bits/minute)
S1	40	1.00	4.17	28.78
S2	40	1.00	2.95	40.68
S3	39	0.89	2.29	39.71
S4	40	1.00	3.05	39.34
S5	39	0.84	2.41	34.06
S6	40	1.00	3.70	32.43
S7	40	0.80	4.30	17.45
S8	38	0.96	2.40	44.55
S9	39	1.00	4.01	29.12
S10	40	0.88	3.52	25.14
S11	37	0.82	2.91	26.90
S12	33	0.86	2.48	34.17
S13	31	0.95	3.02	34.96
S14	35	0.92	3.15	30.66
S15	39	0.90	2.32	39.93
Mean	37.9	0.92	3.11	33.19
S.D.	2.88	0.07	0.69	7.13

$$\begin{aligned} R(bits/symbol) = \log_2(C) + p\log(p) \\ +(1-p)\log_2[(1-p)/(C-1)], \end{aligned} \tag{14.4}$$

$$ITR(bits/minute) = R \times 60/\tau, \tag{14.5}$$

where τ is the average time (in seconds) necessary to detect a symbol or to execute a command, and C is equal to four.

During operation, a 1.5-s long window was used to take a decision about the direction of movement. This window was subdivided into three 1-s long windows having 75 % overlapping. For each sub-window, the two-dimensional feature vector was extracted and classified. A decision was taken by majority vote among the classification of all the sub-windows. In case of a tie, no decision was taken and consequently the cursor did not move.

Fifteen volunteers (S1–S15) participated in the experiment. All of them were able to complete the maze navigation task. Their performance is shown in Table 14.1. For each participant, the optimal stimulation frequency is reported in the second column of Table 14.1. Six out of the 15 participants had a optimum stimulation frequency of 40 Hz. The prevalence of this frequency may be due to resonant processes of the alpha-peak frequency [11].

The accuracy for each participant is reported in the third column of Table 14.1. Five participants were able to complete the navigation task without any error. The mean accuracy across all participants was 0.92. Such a high accuracy is a desirable feature in several BCI applications dealing with patient locomotion.

The average time-per-command in the fourth column results from dividing the total time it took the subject to complete the maze by the total number of commands. This term was used as τ in (14.5) to estimate the information transfer rate in bits-per-minute. The ITR is reported in the fifth column of Table 14.1. The across-subject averages and corresponding standard deviations (SD) are reported in the last rows of Table 14.1. Our results show an average ITR of 33.2 bits-per-minute which shows the potential of our approach especially because high-frequency repetitive visual stimulation was used together with phase coding. Because of the high frequency stimulation, all the participants reported low visual discomfort due to flickering (see also Fig. 14.4).

14.4 Discussion and Conclusion

This chapter has presented an approach to use high frequency repetitive visual stimulation and phase coding in an SSVEP based BCI. Using high frequencies is desirable for reasons of comfort and safety however only few frequencies in the high frequency range can evoke a sufficiently strong SSVEP. The solution consisting in selecting a single frequency and different phases efficiently overcomes the frequency limitation.

The higher the frequency is, the lower the visual discomfort becomes. While it is desirable to increase the stimulation frequency beyond the 40 Hz limit applied in this chapter, SSVEP detectability needs to be considered. Promising directions in detecting SSVEPs even beyond the perceptual threshold are reported in [17].

The signal processing methods proposed in this chapter rely on the use of optimal spatial filters and phase synchrony analysis. The experimental evidence to support our approach, shows an average accuracy of 0.92 and an ITR of 33.2 bits-per-minute across 15 volunteers. In the low frequency range (around 15 Hz), ITRs of up to 92 bits-per-minute are reported [1]. This however comes at a price of visual discomfort which limits the usability of the system.

The optimal stimulation frequency that was selected more often was 40 Hz. This suggests a resonant process of the alpha-peak frequency. Further research is clearly needed to determine whether it could be possible to envision the use of an universal stimulation frequency. This can be of great interest in view of improving the usability of SSVEP based BCIs.

Frequency and phase modulation can be mixed in order to increase even further the number of possible stimuli in a BCI application. This approach is proposed in [12] for low frequency flicker stimulation. Similar principles could be applied to high frequency stimulation. Yet, one needs to take into account that in a practical application only few different commands may be necessary.

Acknowledgements The authors would like to express their gratitude to Dr. Vojkan Mihajlovic for his valuable suggestions to improve the quality of this chapter.

The research leading to these results has received funding from the European Community's Seventh Framework Programme under grant agreement number 224156.

References

1. Bin, G., Gao, X., Wang, Y., Hong, B., Gao, S.: VEP-based brain–computer interfaces: time, frequency, and code modulations [Research Frontier]. IEEE Comput. Intell. Mag. **4**(4), 22–26 (2009)
2. Biosemi system. http://www.biosemi.com
3. Cheng, M., Gao, X., Gao, S., Xu, D.: Multiple color stimulus induced steady state visual evoked potentials. In: Proceedings of the 23rd Annual International Conference of the IEEE Engineering in Medicine and Biology Society, vol. 2, pp. 1012–1014 (2001)
4. Cheng, M., Gao, X., Gao, S., Xu, D.: Design and implementation of a brain–computer interface with high transfer rates. IEEE Trans. Biomed. Eng. **49**, 1181–1186 (2002)
5. Fawcett, T.: An introduction to ROC analysis. Pattern Recogn. Lett. **27**, 861—874 (2006)
6. Fisher, R.S., Harding, G., Erba, G., Barkley, G.L., Wilkins, A.: Photic-and pattern-induced seizures: a review for the Epilepsy Foundation of America Working Group. Epilepsia **46**, 1426–1441 (2005)
7. Friman, O., Lüth, T., Volosyak, I., Gräser, A.: Spelling with steady-state visual evoked potentials. In: Proceedings of the 3rd International IEEE EMBS Conference on Neural Engineering, pp. 354–357 (2007a)
8. Friman, O., Volosyak, I., Gräser, A.: Multiple channel detection of steady-state visual evoked potentials for brain–computer interfaces. IEEE Trans. Biomed. Eng. **54**, 742–750 (2007b)
9. Gao, X., Xu, D., Cheng, M., Gao, S.: A BCI-based environmental controller for the motion-disabled. IEEE Trans. Neural Syst. Rehabil. Eng. **11**, 137–140 (2003)
10. Garcia-Molina, G., Mihajlovic, V.: Spatial filters to detect steady state visual evoked potentials elicited by high frequency stimulation: BCI application. Elektromed Biomed. Tech. **55**(3), 173–182 (2010)
11. Herrmann, C.: Human EEG responses to 1–100 Hz flicker: resonance phenomena in visual cortex and their potential correlation to cognitive phenomena. Exp. Brain Res. **137**(3–4), 346–53 (2001)
12. Jia, C., Gao, X., Hong, B., Gao, S.: Frequency and phase mixed coding in SSVEP-based brain computer interface. IEEE Trans. Biomed. Eng. **58**(1), 200–206 (2010)
13. Kluge, T., Hartmann, M.: Phase coherent detection of steady-state evoked potentials: Experimental results and application to brain–computer interfaces. In: Proceedings of the 3rd International IEEE EMBS Conference on Neural Engineering, pp. 425–429 (2007)
14. Lalor, E.C., Kelly, S.P., Finucane, C., Burke, R., Smith, R., Reilly, R.B., McDarby, G.: Steady-state VEP-based brain–computer interface control in an immersive 3D gaming environment. EURASIP J. Appl. Signal Process. **2005**, 3156–3164 (2005)
15. Lee, P., Sie, J., Liu, Y., Wu, C., Lee, M., Shu, C., Li, P., Sun, C., Shyu, K.: An SSVEP-actuated brain computer interface using phase-tagged flickering sequences: A cursor system. Ann. Biomed. Eng. **38**(7), 2383–2397 (2010)
16. Mukesh, T.M.S., Jaganathan, V., Reddy, M.R.: A novel multiple frequency stimulation method for steady state VEP based brain computer interfaces. Physiol. Meas. **27**, 61–71 (2006)
17. Porbadnigk, A., Scholler, S., Blankertz, B., Ritz, A., Born, M., Scholl, R., Müller, K., Curio, G., Treder, M.: Revealing the Neural Response to Imperceptible Peripheral Flicker with Machine Learning. In: Conf. Proc. IEEE. Eng. Med. Biol. Soc. (2011)
18. Regan, D.: Human brain electrophysiology: evoked potentials and evoked magnetic fields in science and medicine. New York, NY, United States Elsevier (1989)
19. Schalk, G., McFarland, D., Hinterberger, T., Birbaumer, N., Wolpaw, J.: BCI2000: A General-Purpose Brain–Computer Interface (BCI) System. IEEE Trans. Biomed. Eng. **51**(6), 1034–1043 (2004)
20. Wang, Y., Gao, X., Hong, B., Jia, C., Gao, S.: Brain–computer interfaces based on visual evoked potentials. IEEE Eng. Med. Biol. Mag. **27**, 64–71 (2008)

21. Wilson, J.J., Palaniappan, R.: Augmenting a SSVEP BCI through single cycle analysis and phase weighting. In: Proceedings of the 4th International IEEE EMBS Conference on Neural Engineering, pp. 371–374 (2009)
22. Wolpaw, J.R., Birbaumer, N., McFarland, D.J., Pfurtscheller, G., Vaughan, T.M.: Brain–computer interfaces for communication and control. Clin. Neurophysiol. **113**, 767–791 (2002)
23. Zhang, D., Maye, A., Gao, X., Hong, B., Engel, A., Gao, S.: An independent brain–computer interface using covert non-spatial visual selective attention. J. Neural Eng. **7**(1) 16010 (11 pp) (2010)
24. Zhu, D., Bieger, J., Garcia-Molina, G., Aarts, R.: A survey of stimulation methods used in SSVEP-based BCIs. Comput. Intell. Neurosci. (2010a)
25. Zhu, D., Garcia-Molina, G., Mihajlovic, V., Aarts, R.: Phase synchrony analysis in SSVEP-based BCIs. In: The 2nd International Conference on Computer Engineering and Technology (ICCET-2010), vol. 2, pp. 329–333 (2010b)

Chapter 15
Can Dry EEG Sensors Improve the Usability of SMR, P300 and SSVEP Based BCIs?

Günter Edlinger and Christoph Guger

15.1 Motivation of BCI Research

The basic idea of a brain–computer interface (BCI) is to enable a new communication channel that bypasses the standard neural pathways and output channels and in order to control an external device [28]. One major goal for BCI technology from the very beginning of the research was to enable lost body or communication functions in handicapped persons. Persons suffering from, e.g., amyotrophic lateral sclerosis (ALS), stroke or spinal cord injuries might lose the ability to fully control (peripheral) muscle activity. Depending on the disease either the neural pathway might be affected or the muscle itself. In a first attempt one can substitute the neural pathways or the affected muscles with still functional pathways or muscles. This approach might be very beneficial to the subjects, though the approach might also put limitations. Subjects can use for example eye movements for communication or control. In the BCI approach body functions are restored by detecting the proper neural or muscle activity above the level of injury. These signals can serve as input to the BCI which properly encodes the patterns and converts the activity into control commands. After a certain time of training the BCI can predict the users' intentions and the user can operate, e.g., the closing/opening a robotic hand or control a wheelchair. However, recently BCI technology has been utilized in non medical applications as well to, e.g., control computer games, control other devices like mobile phones or controlling smart homes and avatars in virtual reality environments.

G. Edlinger (✉) · C. Guger
g.tec medical engineering GmbH, Sierningstraße 14, A-4521 Schiedlberg, Austria

Guger Technologies OG, Herbersteinstraße 60, 8020 Graz, Austria
e-mail: edlinger@gtec.at; guger@gtec.at

B. Z. Allison et al. (eds.), *Towards Practical Brain-Computer Interfaces*, Biological and Medical Physics, Biomedical Engineering, DOI 10.1007/978-3-642-29746-5_15,
© Springer-Verlag Berlin Heidelberg 2012

Brain activity can be observed by various methods like functional magnetic resonance imaging, functional near infrared spectroscopy, positron emission tomography, magnetoencephalography or by more invasive methods like electrocorticography or single neural cell recordings and others. For practical applications, availability, costs and probably home usage for end users the noninvasive measurement of brain electrical activity the electroencephalogram (EEG) is still the method of choice for many research groups. BCI systems have been successfully realized based on different EEG phenomena whereby most of the research up to now has been focused on two major groups of BCI:

- *Endogenous BCIs*: In this type of BCIs subjects learn and train to perform specific mental tasks to change willingly brain activity. This type of BCI includes slow cortical potentials (SCP) and SMR (sensorimotor rhythmic) or event related desynchronization/synchronization (ERD/ERS) based BCIs.

  1. SCP based BCIs: Very early approaches of BCIs include the use of SCP [2]. This approach required months of training. Today the SCP approach is not widely used anymore for BCI control.
  2. SMR and ERD/ERS based BCIs: BCI systems based on the oscillatory brain electrical activity use motor imagery strategies that generate ERD and ERS in the alpha and beta frequency ranges of the EEG. In ECoG also gamma band activities have been used to construct BCI control [22]. More specifically, changes in sensorimotor rhythms associated with imagined hand or feet movements are mostly used to realize this type of BCI. However, even less specific movement imagery developed via training can be used [3, 19, 28]. Applications of this so called SMR BCI are found for cursor control on computer screens, for navigation of wheelchairs or controlling virtual environments (see [6, 15, 16, 19] and Chaps. 6 and 10).

- *Exogenous BCIs*: In this type of BCIs the presentation of external stimuli evokes a specific change in the brain activity. Typical evoked potentials that are found in the ongoing EEG, depending on focused or selective attention to an external stimulus, are the P300 response and Steady-State Visually Evoked (SSVEP) potentials.

  1. P300 based BCIs: the P300 BCI approach requires the user to focus on a visual or tactile stimulus, whereby the brainwaves differ for stimuli that the user attends versus ignores. Such a system uses the effect that an unlike event induces a P300 component in the EEG, i.e., a positive deflection in the EEG signal occurring around 300 ms after the event [7,23]. In spelling applications typically several letters are displayed on a computer screen in a row-column format. All the letters are flashed transiently. The user selects and attends the letter she/he wants to select by simply counting the number of times it is flashed. Then the BCI system can determine which of several visual targets the user attends. Applications so far comprise mostly spelling devices as P300 or environmental control [5,6,14,24]. In a similar way recently P300 BCIs are realized on tactile stimulation. Several tactors are mounted to different parts of

the body and transiently switched on. The BCI system can determine which of several tactile targets contains the desired information. Applications here are supposed to aid in situations where tactile stimulations are more suitable than, e.g., visual cues [4].
2. SSVEP based BCI: steady-state visual evoked potentials approaches uses the fact that flickering light sources with flickering frequencies in the range of 5–20 Hz induce brain oscillations of the same flickering frequency. Similarly to the P300 BCI here the brainwaves differ again for stimuli that the user attends versus ignores. Applications so far comprise, e.g., robot control or mobile phone control [8, 17].

Hence BCI systems are used for communication purposes, to control robotic devices or wheelchairs, to control games or for rehabilitation. This means BCI systems are not only built for special user groups but also for healthy people. One limiting factor in the wide-spread application is the usage of abrasive gel and conductive paste to mount EEG electrodes. From the authors personal experience subjects report discomfort participating in EEG experiments or even rejected participation as hair washing after the experiments is necessary.

Therefore many research groups are now working on the practical usability of dry electrodes to completely avoid the usage of electrode gel. Before EEG electrodes are mounted to the head the skin is typically cleaned with an abrasive gel to remove the outer dry layer of the skin to obtain a lower skin-electrode impedance ensuring high quality EEG recordings. This procedure is performed as the outer layer of the skin can contribute up to several mV of DC potentials because of small electrode position shifts. Tam and Webster showed that with 20 strokes of fine sand-paper the impedance of the epidermis layer can be reduced and this reduces motion artifacts produced, e.g., by stretching the skin [26]. The abrasion does not reach the capillary layers, but removes a barrier layer that protects the skin. Furthermore the skin electrode impedance is influenced by temperature and humidity. The lower the impedance the less artifacts and noise are picked up. The disadvantage of the skin preparation is that the cleaning is time consuming and can even cause some pain especially if it is done every day over the same electrode location. Also the drying of the gel and the skin regrowing after the abrasion degrades the performance of EEG electrodes in long term recordings [13]. Therefore so called active electrodes were developed that have an amplifier already in the electrode itself and can therefore accept much higher electrode skin impedances. As a result active electrodes pick up less artifacts and for the electrode montage no prior skin preparation using abrasive gel is necessary. However, the usage of conductive gel ensuring a galvanic connection between the skin and the electrode is still needed. Active electrodes are also larger in its size, have additional electronic components and are more expensive.

Dry electrodes use either micro-needles to penetrate the first layer of the skin and to get in contact with the conducting layers, use capacitive sensors or are penetrating the skin with mechanical springs that press the electrodes into the skin [20, 25]. Early work focused on the usage of active and dry electrodes for the recording of electrocardiogram signals [21] which is easier to do because of the larger signal to noise ratio and easier montage on the thorax.

All these EEG based BCIs use electrodes mounted over specific positions on the human scalp to pick up brain activity in order to extract control signals to operate external devices. Several different external noise sources can influence EEG measurements: (a) electromagnetic interference produced by currents flowing through nearby wires to supply other devices, (b) triboelectrical noise produced by friction between conductors and insulators, or (c) skin-electrode potentials shifts produced by movements of the electrodes because the skin has different ionic charges in its layers [25]. The ongoing EEG displays amplitudes of about $\pm 50\,\mu V$ and comprises a frequency range between DC and about 40 Hz. Because of the rather small amplitudes it is important to have a low noise biosignal amplifier input stage ($<0.3\,\mu$Vrms in the interesting frequency range) before the EEG is converted from analog to digital and processed in a computer system. For noninvasive BCI systems the EEG signals must be measured from about 0.1 Hz up to 40 Hz to enable using ERD/ERS, SSVEP or P300 as control signals. Furthermore the EEG data should be noise and artifact free as much as possible.

Single trial classification of motor imagination using six dry electrodes was already shown by the Berlin BCI group [20] and resulted in about 30 % lower information transfer rate than with gel electrodes. Gargiulo [9] constructed a dry electrode system with conductive rubber showing a high correlation between gel based and dry electrodes. A stainless steel disk with 3 mm was used to prove the usefulness of it for spontaneous EEG and evoked potentials (EP) [25]. In another approach avoiding electrode gel, Volosyak [27] showed the successful usage of water-based electrodes for an SSVEP based BCI.

In this chapter results for endogenous and exogenous BCI approaches are presented and discussed based on a dry electrode sensor concept. Therefore raw EEG data, the power spectrum, the time course of evoked potentials, ERD/ERS values and BCI accuracy are compared for three BCI setups based on SSVEP, P300 and SMR BCIs. The focus in this study was set to P300 evoked potentials as it can be expected that the signal to noise ratio is small in the low EEG frequency ranges for a dry electrode system. However the feasibility of the dry sensor concept for SMR BCI and SSVEP BCI was also evaluated.

15.2 Methods

15.2.1 g.SAHARA Dry Electrode Sensor Concept

Another obstacle found in BCI literature is the fact that a certain percentage of the population cannot operate a specific type of BCI due to various reasons. Inter-subject as well as intra-subject variability often leads to a so-called BCI illiteracy [1]. Across the different BCI approaches around 20–25% of subject are

Fig. 15.1 (**a**) Seven pin 8 mm gold coated sensor elements attached to an elastic head cap; (**b**) detachable dry sensor element in combination with active clip lead connector; (**c**) reconfigurable dry sensors attached to the occipital/parietal areas

unable to control one type of BCI in a satisfactory way [11]. Therefore, the usage of "hybrid" BCIs has been introduced into the literature to overcome these problems using the output of somato-sensory rhythm BCI as well as P300 or steady state visually evoked potentials based BCIs enabling subjects to choose between these different approaches for optimal BCI control [18].

Hence a dry sensor concept should support all major BCI approaches. Therefore following issues have to be addressed. Among them:

a. The sensor setup must be reconfigurable, i.e., the positioning of sensor should be possible on an arbitrary location according to the extended 10/20 electrode system.
b. The number of utilized sensors should be changeable.
c. Depending on the type of BCI approach electrodes in the central areas, parietal areas as well as occipital areas are attached; the sensor concept shall work on the various head shapes as well as hair thickness and haircut.
d. The usage of the electrode should impose no risks for the subject as this cannot be ruled out for certain suggested sensor concepts like micro tip electrodes which could break off [10].

In order to fulfill the requirements the following concept has been investigated and utilized to realize the dry sensors: an elastic textile has been selected as basis for an appropriate head cap; At predefined locations according to the extended 10/20 system holes are foreseen for attaching the sensor; The elastic material can guarantee the proper fixation of the sensors while still arbitrary positions can be chosen for optimal recordings depending on the experimental paradigm; furthermore the elastic material provides the necessary pressure to ensure physical contact of the sensor to the skin. The electrode itself was constructed inserting seven golden coated pins on a plate mounted in a circular arrangement (d = 10 mm). A small preamplifier is located in the electrode itself ensuring that the electrode can work even with very high skin to electrode impedances. The length of the pins can be varied in order to fulfill requirement (c), but was fixed for the experiments in this case to 8 mm. Figure 15.1 displays the dry sensors setup.

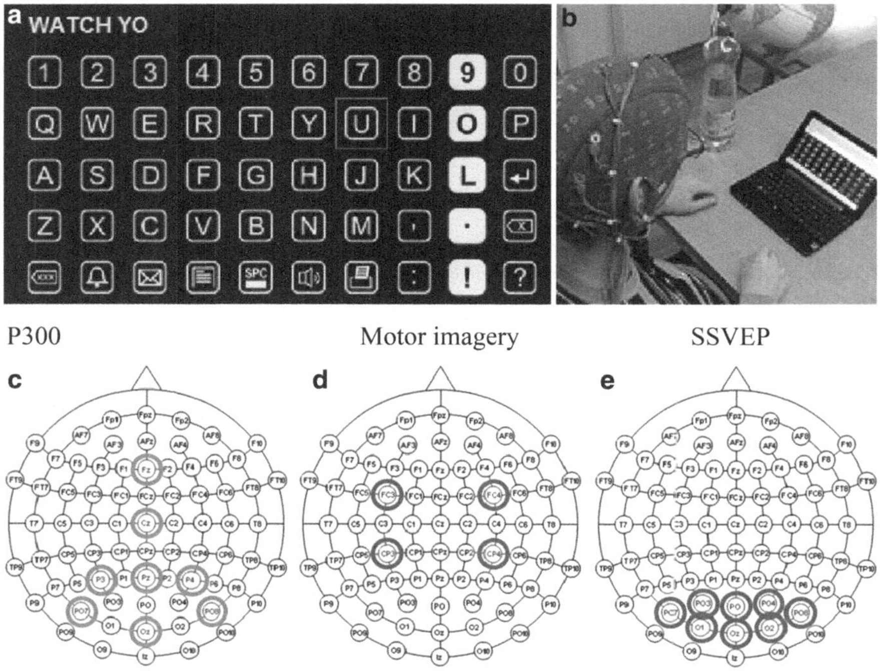

Fig. 15.2 (**a**) intendiX spelling matrix; (**b**) measurement setup with active electrodes. (**c**) electrode montage for P300, (**d**) motor imagery, and (**e**) SSVEP experiments. Ground on left mastoid, reference on right mastoid

15.3 Experimental Setup

A total of 13 subjects (three female, ten male, age: 19–41, P300: 11, motor imagery: 1, SSVEP: 1) participated in the study conducted over four weeks. The subjects were free of medication, had normal vision or vision corrected to normal and no history of central nervous system abnormalities.

Subjects sat in front of a laptop computer and were instructed to relax and remain as still as possible. Figure 15.2 shows the electrode configuration and the electrode locations used for the studies. EEG data were acquired using a g.USBamp (24 Bit biosignal amplification unit, g.tec medical engineering GmbH, Austria) with a sampling frequency set to 256 Hz. The ground electrode was mounted over the left mastoid and the reference was mounted over the right mastoid; for both positions disposable pre-gelled electrode pads were used. EEG electrodes were fixed to an EEG electrode cap (g.GAMMAcap) according to the extended international 10/20 electrode system. EEG recordings based on gel electrodes were done with the g.BUTTERfly electrode (golden ring electrode type with a hole in the middle to inject the gel) and EEG recordings based on dry electrodes were done with the g.SAHARA electrode (seven golden coated pins with 8 mm length mounted in a circular arrangement, diameter 10 mm). Both type of electrodes are active EEG electrodes with a small preamplifier located in the electrode itself.

15.4 P300 BCI

Subjects participating in the P300 study used the intendiX row/column (RC) speller as shown in Fig. 15.1. All 11 subjects used the dry electrodes and additionally one of the subjects used also the gel based electrodes for comparison. One subject has to quit the experiment due to too bad signal quality as there was hardly electrode skin contact possible due to thick hair. The RC speller shows 50 characters (A, B,. . . Z; 0, 1,. . . 9; and special characters) on the computer screen and highlights a whole column or row for 100 ms. Between the flashes there is a short time while only the grey matrix items are visible (60 ms). The subject's task is to attend to (or look at) the character he/she is prompted to spell and to count how many times the character is highlighted. The counting task helps the subject to remain focused on the task. After 15 times highlighting each row and column the signal processing unit calculates the evoked potential for each character and performs a linear discriminant analysis (LDA) classification to determine which matrix item the subject was attending to [11]. Then the highlighting sequence starts again and the subject is prompted to attend to the next character. The BCI system must be calibrated in a first step on individual EEG data. Therefore the subject was asked to "select" (or attend to) the word WATER, one letter at a time. This training procedure took about 5 min. After training the BCI system using the calibration data, the subject was asked to write the word LUCAS, one character at a time, taking about five more minutes.

15.5 Motor Imagery

As the goal of the study was to compare SMR BCI results obtained via dry and gel based electrodes a well trained subject performed the motor imagery experiments. First gel based and dry electrodes were mounted beside each other to record EEG data (run 1 with 80 trials of left and right hand movement imagination) almost from the same region (1.5 cm apart). Second the subject performed 160 trials with dry electrodes (run 2) and then 160 trials with gel electrodes (run 3). The first experiment (run 1) lasted about 30 min, the second one about 2 h (runs 2 and 3).

The motor imagery experiment started with the display of a fixation cross in the center of a screen. After 2 s, a warning stimulus was given in the form of a "beep." After 3 s, an arrow (cue stimulus) pointing to the left or right was shown for 1.25 s. The subjects were instructed to imagine a right-hand movement or left-hand movement until the end of the trial, depending on the direction of the arrow. One trial lasted 8 s and the time between two trials was randomized in a range of

0.5–2.5 s to avoid adaptation. No feedback was given to the subject to prevent distraction of the participant.

The motor imagery BCI estimated the bandpower in two different frequency bands of the EEG data. The reactive frequency bands in the alpha and beta range were identified from the power spectrum and a time-frequency evaluation of the ERD/ERS activity (ERDmaps). The bandpower features were classified with a linear discriminant analysis resulting in a subject specific weight vector [12].

15.6 SSVEP BCI

One subject performed the SSVEP experiment. In the first run dry electrodes, in the second run gel based electrodes were used. The task of the subject was to attend for 14 s to one of four LEDS flickering with a certain frequency (10, 11, 12, 13 Hz) and then to rest for 6 s. The task was repeated for the remaining three LEDS and the whole loop was repeated three times. The four LEDS were arranged in a 12×12 cm box and were controlled by a microcontroller resulting in a frequency error <0.025 Hz.

The SSVEP analysis process works with a sliding window containing 512 samples (2 s of EEG) with an overlap of 448 samples and consists of four steps: pre-processing, feature extraction, classification and change rate/majority weight analysis. These three steps are executed every 250 ms.

Two different methods are used to calculate features of the EEG data. The first method is based on the minimum energy approach (ME) and requires no training [8]. This algorithm is fed with raw EEG-data channels since it selects the best combination of channels by itself. First of all the EEG-data gets cleaned of potential SSVEP-signals. This is done by projecting artificial oscillations with stimulation frequencies and harmonics onto the orthogonal complement of the EEG-signals. After that operation the signals contain just the unwanted noise. Now a weight vector is generated, which has the property of combining the channels in a way, that the outcome has minimal energy. Now SSVEP detection is done utilizing a test statistic which calculates the ratio between the signal with an estimated SSVEP-response and the signal where no visual stimulus is present. This is done for all stimulation frequencies and all EEG-channels. The output of this classifier is the index of the frequency with the highest signal/noise ratio.

The second method is based on the Fast Fourier Transformation (FFT) and linear discriminant analysis (LDA). In a pre-processing step for this method Laplacian derivations are calculated. First of all incoming data are transformed after Laplacian derivation to the frequency spectrum with a FFT. A feature vector is extracted by taking the values of the stimulation frequencies and their first and second harmonics. With these feature vectors a weight/bias vector must be generated for each user in a training procedure. When the training is completed successfully the LDA classifier can then be used to classify new feature vectors to one of the stimulation frequency indices. In the model used for the experiments described in this paper four ME

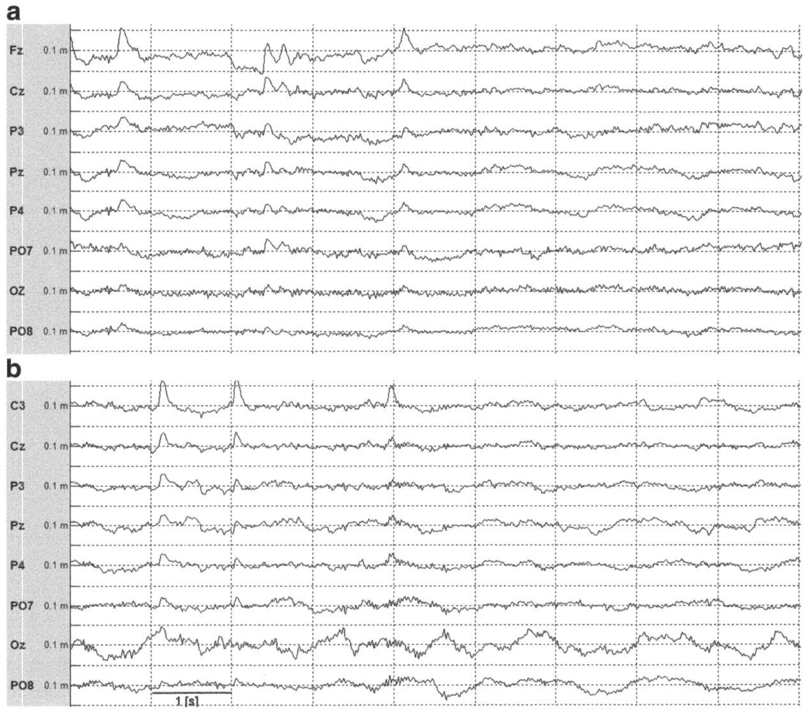

Fig. 15.3 Eight channel EEG data of the P300 experiment acquired with dry and gel electrodes on central, parietal and occipital sites. The EOG artifacts are mostly visible on Fz, Cz, P3, Pz and P4 in both cases. The y-axis is scaled with +/−100 μV, the x-axis in seconds. The data is bandpass filtered between 0.1–30 Hz and 50 Hz Notch filtered. (**a**) Dry, (**b**) Gel

classification units and four FFT+LDA classification units were used with different EEG channels as inputs.

The last step is a procedure called change rate/majority weight analysis. By having multiple classification units configured with slightly different input data there will be in general random classification results on noise input. This effect is used on one side to produce a zero class decision when the outputs of the classifiers are changing heavily and are very different. On the other side a low change rate and a high majority weight (the number of classifications of the different algorithms which are pointing in the same direction) can be used to strengthen the robustness of the decision.

15.7 Results

The first interesting step was to compare raw EEG data acquired with dry and gel based electrodes over central, parietal and occipital sites. Nine seconds EEG segments scaled with the sample amplitude are shown in Fig. 15.3. The data were

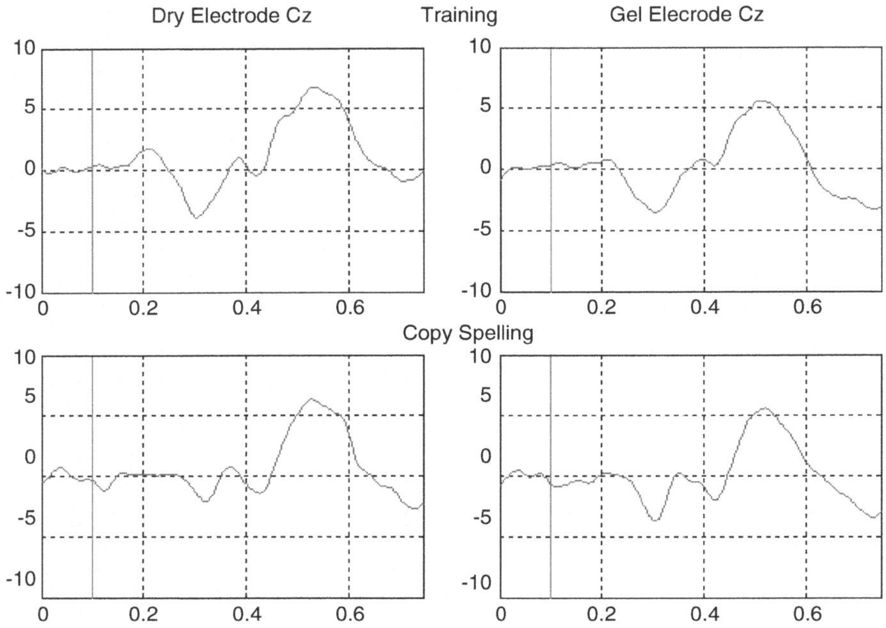

Fig. 15.4 P300 response for dry and gel electrodes in the training run and copy spelling run of one subject. Each run had five characters and each character flashed 30 times (15 rows, 15 columns). This give in total about 5 min/run. The y-axis is scaled with +/—10 μV, the x-axis in seconds. The P300 BCI accuracy increases with the number of flashes of the characters on the screen, but the communication rate drops down. Therefore for BCI control it is important to reduce the number of flashes

acquired in different session for the same subject and therefore are of course different. Important to note is that the dry electrodes are able to capture EEG data from all recording sites and that no visible noise differences can be noticed. In both segments the eye blinks are mostly visible on frontal and less in central and parietal sites.

15.8 P300 Paradigm

The P300 BCI system uses the evoked potential induced by the target character as control signal. Therefore the average of all target characters was calculated with a pre-stimulus interval of 100 ms and a post-stimulus interval of 700 ms. Beforehand a baseline correction was performed taking the first 100 ms as input. The EP for dry and gel based electrodes at electrode position Cz is shown in Fig. 15.4 for the training and copy spelling run for one subject. Data from electrode Cz are selected because it is one of the most important electrodes for the P300 speller. The EP

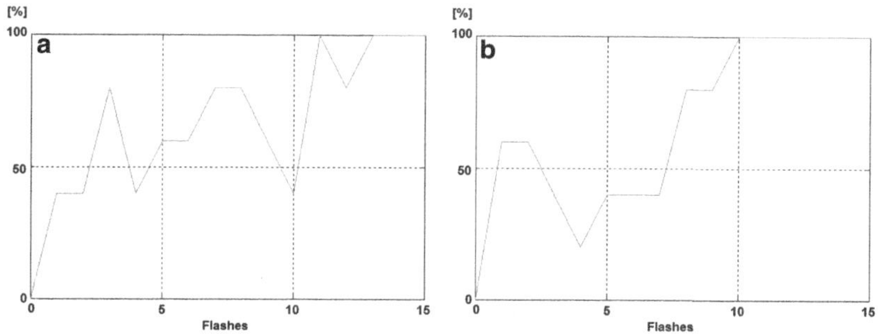

Fig. 15.5 P300 BCI accuracy over number of flashes for dry and gel electrodes for one subject. (**a**) Dry electrodes, (**b**) Gel electrodes

Table 15.1 Percentage of sessions which were classified with a certain accuracy. N specifies the number of subjects participating

Row–Column Speller classification accuracy in %	Gel electrodes (N = 81) [11]	Dry electrodes (N = 11)
100	72.8	63.6
80–100	88.9	90.9
60–79	6.2	0
40–59	3.7	9.1
20–39	0.0	0
0–19	1.2	0
Average accuracy of all subjects	91.0	89.1

reaches its maximum of about 6 μV after about 240 ms in both cases. The EP looks very similar for the dry and gel based electrodes and the comparison of the training and copy spelling run shows that the EP is very stable over time.

Figure 15.5 shows the accuracy over the number of flashes for dry and gel based electrodes. For both types of recordings an accuracy of 100 % is reached. However, for the gel based electrodes less flashes are needed.

Table 15.1 shows the results of a group study done with gel based electrodes (N = 81) [11] and the group study (N = 11) done with dry electrodes in this paper. The most important result is that the average accuracy for the gel based electrodes is 91 % and for the dry electrodes 89.1 %. The percentage of subjects that spelled with 100 % accuracy (i.e., all five characters of LUCAS were correctly selected by the LDA) is lower for dry electrodes (63.6 %) as for gel based electrodes (72.8 %). It must be noted that this is an on-line results and not a cross-validation result. Even 88.9% (gel) and 90.9 (dry) were able to make none or only one mistake. Moreover, only 1.2 % (gel) and 0 % (dry) were not able to spell a single character correctly.

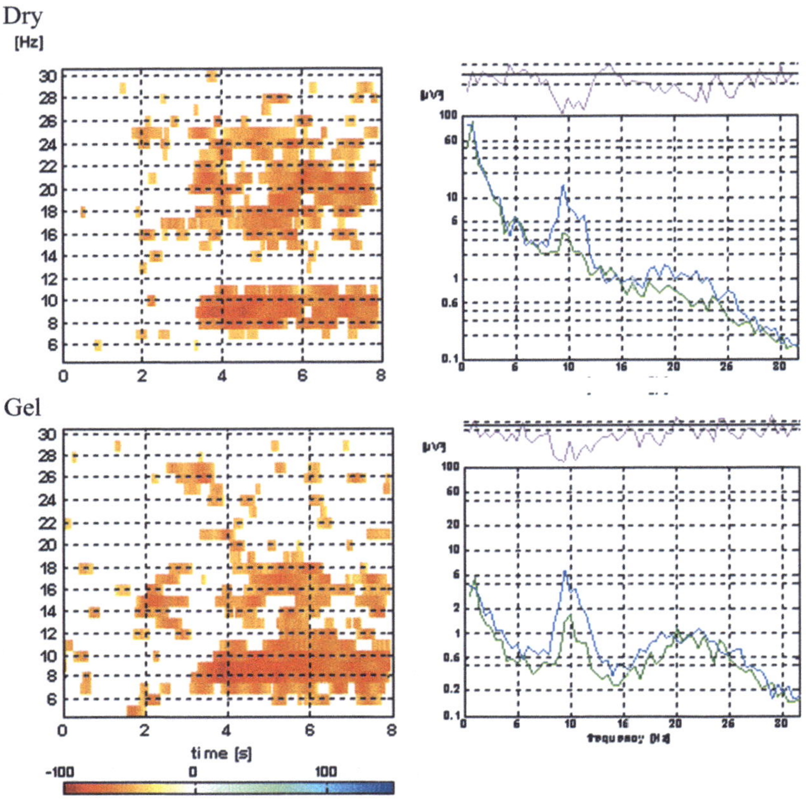

Fig. 15.6 *Left*: ERDmaps of electrode positions C3 during right hand movement imagination for dry (*top*) and gel electrodes (*bottom*). Both show a strong ERD in the alpha range from second 3.5 until 8 over C3. The dry electrodes show a broader beta ERD. Only pixels with significant ERD/ERS values are displayed (bootstrap, $p < 0.05$). *Right*: Reactive frequency components of the reference interval (0–2 s, *blue*) and active interval (6–8 s, *green*) of C3 of dry (*top*) and gel (*bottom*) electrodes. The graph above each power spectrum shows significant changes if the *line* crosses the *dashed line* (sign test, $p < 0.05$). (**a**) Dry, (**b**) Gel

15.9 Motor Imagery

ERDmaps and the power spectrum were calculated to compare the dry and gel based motor imagery based BCI system as shown in Fig. 15.6. First the EEG data was visually inspected and about 5 % of the trials containing artifacts were removed. In both cases an ERD in the alpha and beta ranges can be found. EEG measured with dry electrode recordings show a broader activity in the beta frequency range. The power spectrum allows to identify the reactive frequency components in the EEG data. In the baseline period (without movement imagination) two alpha peaks can be found for this subject in both derivations (dry and gel). It is known from previous

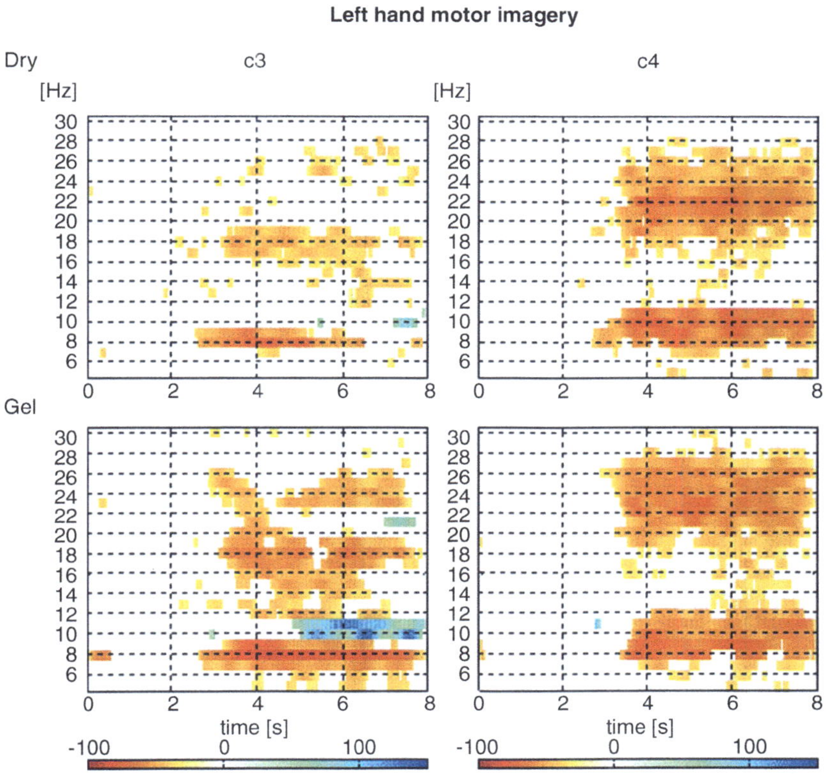

Fig. 15.7 (Continued)

experiments from this subject that the higher alpha activity is more suppressed during the hand movement imagination and therefore this frequency band is used for the BCI control. The significant difference between baseline and imagination is proven by the sign test ($p < 0.05$). EEG power spectra for the dry electrode show a higher difference in the beta region than the for the gel based electrodes. One reason could be that both measurements were done at nearby but still distinct locations. A clear difference comparing the two power spectra is the higher power found below 3 Hz for the dry electrode signal. However comparing power levels in alpha and beta ranges it can be stated that the ERDmaps and power spectra show very similar results for both types of electrodes (Figs. 15.7 and 15.8).

There will of course always be a difference between brain activity recording from different electrode locations even if the distance is small (1.5 cm). Therefore ERDmaps are compared for dry and gel based electrodes located at the same position but recorded after each other. The ERDmaps shown in Fig. 15.9 (page 293 and 294) display results for right and left hand movement imagination. The left hand imagination produces an ERD in the alpha and beta frequency range over electrode

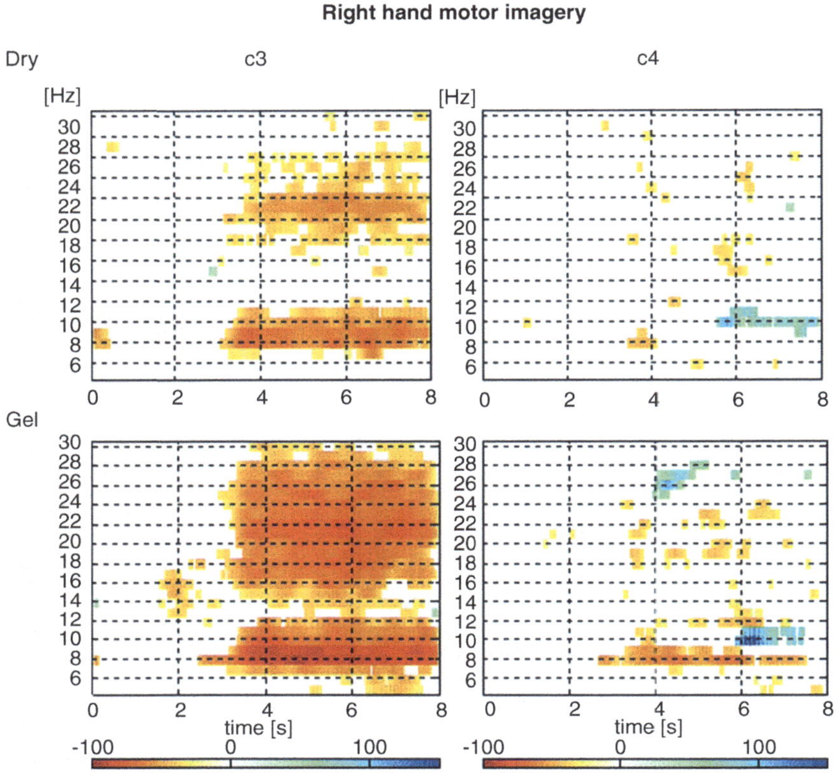

Fig. 15.7 (continued) (**a**) ERDmaps of left (see page 294) and right hand movement imaginations calculated over electrode positions C3 and C4. (**b**) ERDmaps of left [see page 293] and right hand movement imaginations over C3 and C4

position C4 that is very similar for both electrodes. On the ipsilateral side an ERD and also an ERS can be found under both conditions. The ERS is more pronounced for the gel based electrodes. For the right hand imagination an ERD in the alpha and beta frequency range can be found over electrode position C3. On the contralateral side over C4 an alpha ERD and ERS can be found that are very similar. The gel based electrodes show additionally a beta ERS over C4.

Figure 15.9 shows the power spectrum with the reactive frequency components for run 2 (dry) and 3 (gel) for right hand movement. Contralaterally to the recording site a significant difference between baseline and imagination can be found in the alpha and beta ranges. Important here is to note that the difference in the beta range is stronger for gel based electrodes and this is in contrast to Fig. 15.6 where dry electrodes showed a higher difference. Ipsilaterally the power spectrum looks very similar.

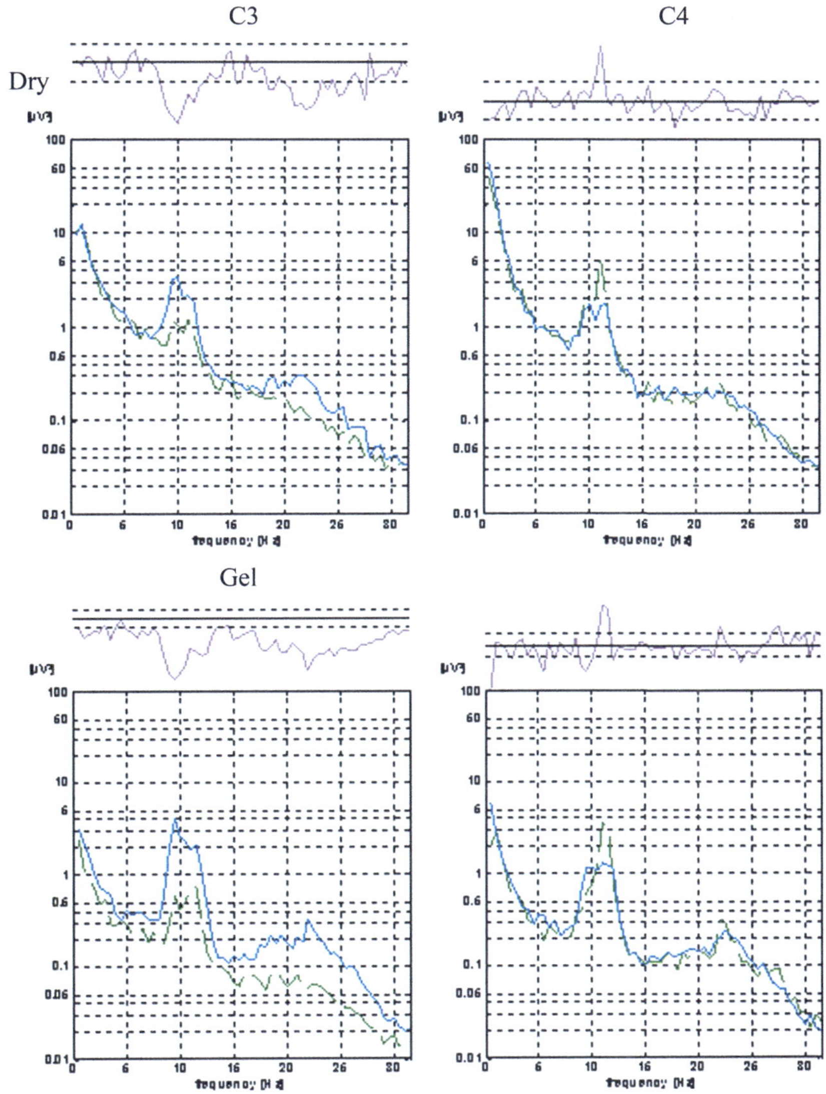

Fig. 15.8 Right hand movement imagination. Reactive frequency components of the reference interval (0–2 s, *blue*) and active interval (6–8 s, *green*, *dashed*) of C3/C4 of dry (*top*) and wet (*bottom*) electrodes. The graph above each power spectrum shows significant changes if the *line* crosses the *dashed line* (sign. test, $p < 0.05$). (**a**) Dry, (**b**) Gel

Table 15.2 Motor imagery error for dry and gel electrodes

Electrodes	Run	Error (%)	Time point (s)
Dry	1	15	7.5
Gel		18	8
Dry	2	14	7
Gel	3	5	7

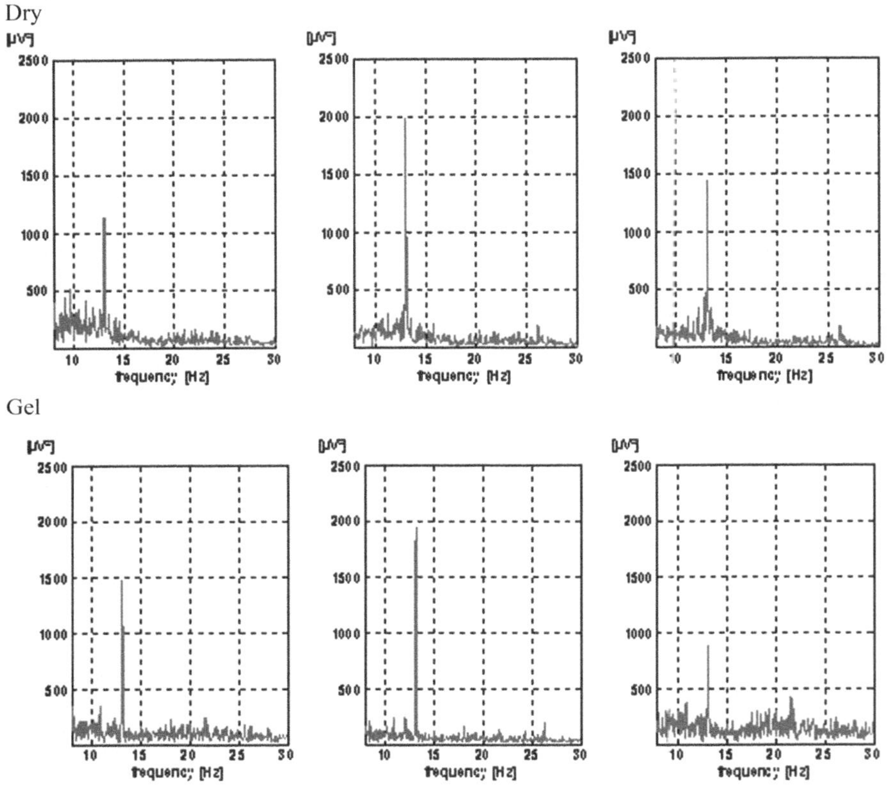

Fig. 15.9 Power spectrum of EEG data of electrodes O1, Oz and O3 during 13 Hz LED stimulation. (**a**) Dry, (**b**) Gel

Beside the ERDmaps and power spectrum it is also interesting to compare the BCI accuracy for dry and gel based electrodes as shown in Table 15.2. The motor imagery BCI accuracy was compared using a 10-times tenfold cross validation technique that mixes the data randomly to have separate training and testing data. The error of run 1 with gel based and dry electrodes mounted beside each other is 18 % versus 15 %. Dry electrodes performed slightly better and had an earlier best classification time point (7.5 s). In run 2, dry electrodes reached 14 % error compared to 5 % error in run 3 for gel based electrodes.

15.10 SSVEP Training

The SSVEP based BCI system is controlled with discrete frequency peaks showing if the subject is looking at a certain LED. Figure 15.9 displays the power spectrum of three important electrode positions (O1, Oz and O3) for dry and gel based electrodes when the subject is looking at the 13 Hz LED computed from the complete three 14 s segments. The highest and very similar peak can be found in both cases on electrode Oz at 13 Hz. The peak on O1 is smaller for dry and the peak on O2 is smaller for gel based electrodes.

Finally, we investigated the accuracy for the dry and gel based electrodes at each single time step of the experiment. Dry electrodes reached an accuracy of 53.7 % with only 3.9 % wrong decisions. This means that mostly the BCI system could not make a certain decision, but made just a few wrong decisions. The gel based electrodes reached an accuracy of 44.5 % with 3.0 % wrong decisions.

15.11 Discussion

We could show that the used dry electrode sensor concept can be used for motor imagery, SSVEP and P300 based BCI systems. For dry electrodes no conductive gel is used and therefore a much higher skin-electrode impedance than for gel based electrodes can be expected. Electrodes with higher impedance can pick up more artifacts and are mostly sensitive for movements of the electrodes and cable swings which results in signal amplitudes much higher than for normal EEG. Electrodes with high impedance can also pick up electrostatic voltages in the surrounding and electro-magnetic noise. To solve these problems we reduced the impedance with multiple gold coated pins per electrode being in contact with the skin. Secondly we integrated an amplifier unit into the electrode itself to make it resistant against artifacts and to be able to record EEG with a high electrode impedance. Dry electrodes also show a higher polarization voltage than gel based electrodes and therefore the recording equipment must be able to accept DC voltages up to several mV. This was solved with an amplification unit with high input range in combination with a 24 Bit ADC (g.USBamp, g.tec medical engineering GmbH).

A big question beside the avoidance and minimizing of technical artifacts is the signal to noise ratio as a function of frequency of dry electrodes. The P300 BCI system uses the EEG data with a lower cut-off frequency of 0.1 Hz, the motor imagery based BCI system uses EEG data in the alpha (8–12 Hz) and beta (14–32 Hz) frequency ranges and the SSVEP based systems uses the EEG data at the stimulation frequencies (mostly between 6 and 30 Hz).

To test the usefulness of dry electrodes for the P300 BCI we conducted the group study with 11 subjects and compared the EPs (for one subject) and accuracies (for all subjects). The latencies and amplitudes of the P300 appeared to be similar for dry and gel based electrodes. In a group study with 81 subjects the grand average maximum P300 response across all subjects was calculated for electrode Cz and was 7.9 μV, in the current study it was about 6 μV. This can be explained by the inter and intra-subject variation. The mean accuracy was in the same range for dry and gel based electrodes. But we had to exclude one subject from the dry EEG electrode experiment because she had too dense hair and therefore the pins of the dry electrodes did not contact the skin properly. This is a clear limitation of the dry electrodes and can only be solved by using longer pins. If gel based electrodes are used just more gel is injected. This means that there will be more subjects were dry electrodes cannot be used because of the hair-cut. The dry electrodes showed also higher signal drifts below 3 Hz compared to gel based electrodes.

For the motor imagery BCI the recordings were done with dry and gel based electrodes at the same time by locating the electrodes close to each other. This resulted in ERDmaps and power spectra that are very similar, but not identical. Of course different EEG sites give different results, but must be similar. Motor imagery BCIs show a high inter and intra-subject variability and therefore the comparison of different sessions can be difficult. We took an experienced BCI subject to perform a dry electrode session and a gel based electrode session. The sessions were made without feedback to avoid any influence to the subject. The resulting ERDmaps and power spectra were again very similar and showed typical alpha and beta ERD and ERS. Nevertheless the BCI error rate of run 3 was 5 % compared to 14 % in run 2. The difference seems to be high but can also be explained by the training effect with the sessions made beforehand and is considered as normal variability.

A big advantage of SSVEP based BCI systems is that we know exactly the EEG frequency that is modulated by the stimulating LED. Therefore the electrode must be able to pick up the EEG data at this frequency. The SSVEP system was tested with four frequencies at 10, 11, 12 and 13 Hz and showed that the peak frequency and amplitude are similar for dry and gel based electrodes.

The montage of the dry EEG electrodes was solved with an EEG cap that allows to use electrodes on all frontal, central, parietal and occipital sites without limitations. The cap is made of elastic material that presses each electrode with a similar pressure to the skin. The cap can be mounted easily and fits many different subjects without user specific adaptations. Other mechanical electrode fixation systems are more difficult to adjust to individual head shapes, record mostly from central and frontal sites and use less electrodes. The number of electrodes that can be used with the cap is only limited by the mechanical size of the electrodes.

The montage time of the dry electrodes is shorter than for gel based electrodes. If a preconfigured cap with its electrodes already inserted at the desired position is utilized the P300 speller montage with passive electrodes is mounted with abrasive gel within about 10 min. If a preconfigured cap with active electrodes is used the montage time is reduced to about 1–3 min and using the same cap but with dry electrodes the time is about 1 min or below. But after the cap is attached, dry electrodes need a few minutes to adjust and therefore the time span needed for preparation is comparable to active electrodes. The biggest advantage of dry electrodes is that no abrasive and conductive gel remains in the hair and therefore especially the time consuming cleaning of patients hair is avoided. Beside that a big advantage is also that the electrodes do not get in contact with water for cleaning and therefore the lifetime is enhanced. None of the subjects reported discomfort of the dry electrodes.

The results of the study has important consequences. The usage of dry electrodes speeds up the montage, enhances the acceptance and brings therefore the technology closer to many people and increases the possible recording time. Nevertheless the dry electrodes show higher signal power below 3 Hz resulting from low frequency drifts. However, considering the results of the experiments the dry sensor concept with its interplaying components of the stretchable electrode cap, the arbitrary positioning of the active electrodes and adjustable pin length we conclude that the concept works very well for SMR, P300 and SSVEP based BCIs.

References

1. Allison, B., Luth, T., Valbuena, D., Teymourian, A., Volosyak, I., Graser, A.: BCI demographics: How many (and what kinds of) people can use an SSVEP BCI? IEEE Trans. Neural Syst. Rehabil. Eng. **18**(2), 107–116 (2010)
2. Birbaumer, N., Ghanayim, N., Hinterberger, T., Iversen, I., Kotchoubey, B., Kubler, A., Perelmouter, J., Taub, E., Flor, H.: A spelling device for the paralysed. Nature **398**, 297–298 (1999)
3. Blankertz, B., Losch, F., Krauledat, M., Dornhege, G., Curio, G., Muller, K.-R.: The Berlin brain–computer interface: Accurate performance from first-session in BCI-naive subjects. IEEE Trans. Biomed. Eng. **55**(10), 2452–2462 (2008)
4. Cincotti, F., Kauhanen, L., Aloise, F., Palomaki, T., Caporusso, N., Jylanki, P., Babiloni, F., Vanacker, G., Nuttin, M., Marciani, M.G., Del, R.M., Mattia, D.: Preliminary experimentation on vibrotactile feedback in the context of mu-rhythm based BCI. Conf. Proc. IEEE Eng. Med. Biol. Soc., 2007, pp. 4739–4742 (2007)
5. Donchin, E., Spencer, K.M., Wijesinghe, R.: The mental prosthesis: assessing the speed of a P300-based brain–computer interface. IEEE Trans. Rehabil. Eng. **8**, 174–179 (2000)
6. Edlinger, G., Holzner, C., Guger, C., Groenegress, C., Slater, M.: Brain–computer interfaces for goal orientated control of a virtual smart home environment. 4th International IEEE/EMBS conference on Neural Engineering. NER09, pp. 463–465 (2009)
7. Farwell, L.A., Donchin, E.: Talking off the top of your head: toward a mental prosthesis utilizing event-related brain potentials. Electroencephalogr. Clin. Neurophysiol. **70**, 510–523 (1988)
8. Friman, O.: Multiple channel detection of steady-state visual evoked potentials for brain–computer interfaces. IEEE Trans. Biomed. Eng. **54**(4), 742 (2007)
9. Gargiulo, G., Bifulco, P., Calvo, R.A., Cesarelli, M., Jin, C., van Schaik, A.A.: mobile EEG system with dry electrodes. IEEE Biomedical Circuits and Systems Conference, pp. 273–276 (2008)
10. Grozea, C., Voinescu, C.D., Fazli, S.: Bristle-sensors-low-cost flexible passive dry EEG electrodes for neurofeedback and BCI applications. J. Neural Eng. **8**(2), 025008 (2011)
11. Guger, C., Daban, S., Sellers, E., Holzner, C., Krausz, G., Carabalona, R., Gramatica, F., Edlinger, G.: How many people are able to control a P300-based brain–computer interface (BCI)? Neurosci. Lett. **462**(1), 94–98 (2009)
12. Guger, C., Edlinger, G., Harkam, W., Niedermayer, I., Pfurtscheller, G.: How many people are able to operate an EEG-based brain–computer interface (BCI)? IEEE Trans. Neural Syst. Rehabil. Eng. **11**(2), 145–147 (2003)

13. Ko, W.H., Hynecek, J.: Dry electrodes and electrode amplifiers, In: Miller, H.A., Harrison, D.C. (eds.) Biomedical Electrode Technology, pp. 169–181. Academic Press, New York (1974)
14. Krusienski, D.J., Sellers, E.W., McFarland, D.J., Vaughan, T.M., Wolpaw, J.R.: Toward enhanced P300 speller performance. J. Neurosci. Methods **167**, 15–21 (2008)
15. Leeb, R., Lee, F., Keinrath, C., Scherer, R., Bischof, H., Pfurtscheller, G.: Brain–computer communication: Motivation, aim, and impact of exploring a virtual apartment. IEEE Trans. Neural Syst. Rehabil. Eng. **15**(4), 473–482 (2007)
16. Millan, J.R., Carmena, J.M.: Invasive or noninvasive: understanding brain-machine interface technology. IEEE Eng. Med. Biol. Mag. **29**(1), 16–22 (2010)
17. Muller-Putz, G.R., Pfurtscheller, G.: Control of an electrical prosthesis with an SSVEP-based BCI. IEEE Trans. Biomed. Eng. **55**, 361–364 (2008)
18. Pfurtscheller, G., Allison, B.Z., Brunner, C., Bauernfeind, G., Solis-Escalante, T., Scherer, R., Zander, T.O., Mueller-Putz, G., Neuper, C., Birbaumer, N.: The Hybrid BCI. Front. Neurosci. **21**(4), 42 (2010)
19. Pfurtscheller, G., Neuper, C., Müller, G.R., Obermaier, B., Krausz, G., Schlögl, A., Scherer, R., Graimann, B., Keinrath, C., Skliris, D., Wörtz, M., Supp, G., Schrank, C.: Graz-BCI: state of the art and clinical applications. IEEE Trans. Neural Syst. Rehabil. Eng. **11**(2), 177–180 (2003)
20. Popescu, F., Fazli, S., Badower, Y., Blankertz, B., Muller, K.-R.: Single trial classification of motor imagination using 6 dry EEG electrodes. PLoS One **2**(7), e637 (2007)
21. Portnoy, W., David, R., Akers, L.: Insulated ECG Electrodes. In: Miller, H.A., Harrison, D.C. (eds.) Biomedical Electrode Technology. Academic Press, New York (1974)
22. Schalk, G., Kubanek, J., Miller, K.J., Anderson, N.R., Leuthardt, E.C., Ojemann, J.G., Limbrick, D., Moran, D., Gerhardt, L.A., Wolpaw, J.R.: Decoding two-dimensional movement trajectories using electrocorticographic signals in humans. J. Neural Eng. **4**(3), 264–275 (2007)
23. Sellers, E.W.: Brain–computer interface research at the University of South Florida cognitive psychophysiology laboratory: the P300 speller. IEEE Trans. Neural Syst. Rehabil. Eng. **14**(2), 221 (2006)
24. Sellers, E.W., Krusienski, D.J., McFarland, D.J., Vaughan, T.M., Wolpaw, J.R.: A P300 event-related potential brain–computer interface (BCI): The effects of matrix size and inter stimulus interval on performance. Biol. Psychol. **73**, 242–252 (2006)
25. Taheri, B.A., Knight, R.T., Smith, R.L.: A dry electrode for EEG recording. Electroenceph. Clin. Neurophysiol. **90**(5), 376–383 (1994)
26. Tam, H.W., Webster, J.G.: Minimizing electrode motion artifact by skin abrasion. IEEE Trans. Biomed. Eng. **24**(2), 134–139 (1977)
27. Volosyak, I., Valbuena, D., Malechka, T., Peuscher, J., Graser, A.: Brain–computer interface using water-based electrodes. J. Neural Eng. **7**(6), 066007 (2010)
28. Wolpaw, J.R., Birbaumer, N., McFarland, D.J., Pfurtscheller, G., Vaughan, T.M.: Brain–computer interfaces for communication and control. Clin. Neurophysiol. **113**(6), 767–791 (2002)

Part IV
A Practical BCI Infrastructure: Emerging Issues

Chapter 16
BCI Software Platforms

Clemens Brunner, Giuseppe Andreoni, Lugi Bianchi, Benjamin Blankertz, Christian Breitwieser, Shin'ichiro Kanoh, Christian A. Kothe, Anatole Lécuyer, Scott Makeig, Jürgen Mellinger, Paolo Perego, Yann Renard, Gerwin Schalk, I Putu Susila, Bastian Venthur, and Gernot R. Müller-Putz

C. Brunner (✉) · C. Breitwieser · G.R. Müller-Putz
Institute for Knowledge Discovery, Graz University of Technology, Austria
e-mail: clemens.brunner@tugraz.at; c.breitwieser@tugraz.at; gernot.mueller@tugraz.at

C. Brunner · C.A. Kothe · S. Makeig
Swartz Center for Computational Neuroscience, INC, UCSD, San Diego, CA, USA
e-mail: clbrunner@ucsd.edu; ckothe@ucsd.edu; smakeig@ucsd.edu

G. Andreoni · P. Perego
INDACO, Politecnico di Milano, Milan, Italy
e-mail: giuseppe.andreoni@polimi.it; paolo.perego@polimi.it

L. Bianchi
Neuroscience Department, Tor Vergata University of Rome, Rome, Italy
e-mail: luigi.bianci@uniroma2.it

B. Blankertz · B. Venthur
Machine Learning Laboratory, Berlin Institute of Technology, Berlin, Germany
e-mail: benjamin.blankertz@tu-berlin.de; bastian.venthur@tu-berlin.de

S. Kanoh
Department of Electronics and Intelligent Systems,
Tohoku Institute of Technology, Sendai, Japan
e-mail: kanoh@tohtech.ac.jp

J. Mellinger
Institute of Medical Psychology and Behavioral Neurobiology, University of Tübingen, Germany
e-mail: juergen.mellinger@uni-tuebingen.de

Y. Renard · A. Lécuyer
National Institute for Research in Computer Science and Control (INRIA), Rennes, France
e-mail: yann.renard@irisa.fr; anatole.lecuyer@irisa.fr

G. Schalk
Wadsworth Center, New York State Department of Health, Albany, NY, USA
e-mail: schalk@wadsworth.org

I.P. Susila
Nuclear Equipment Engineering Center, Tangerang Selatan, Indonesia
e-mail: putu@batan.go.id

B. Z. Allison et al. (eds.), *Towards Practical Brain-Computer Interfaces*, Biological and Medical Physics, Biomedical Engineering, DOI 10.1007/978-3-642-29746-5_16,
© Springer-Verlag Berlin Heidelberg 2012

16.1 Introduction

Research on brain–computer interfaces (BCIs) started as early as 1973, when Jacques J. Vidal presented the first concept of direct brain–computer communication [60] (interestingly, the first BCI can also be attributed to Dr. Grey Walter, who reported on a BCI experiment in a talk in 1964, but he did not publish his results [21]). Since then, many research groups have developed this first idea into functional prototypes. While there are still many open issues that need to be addressed, the first BCIs are already being used outside the labs, for example in hospitals or at homes [41, 54, 58].

With the advent of modern personal computers, computational power was more than sufficient for most BCI requirements. Moreover, more user-friendly development environments started to emerge, and applications started to rely heavily on graphical representations of objects and relationships between objects. For example, the combination of MATLAB and Simulink (The Mathworks, Inc.) is probably one of the most popular commercial general-purpose platforms for developing a great variety of different scientific applications.

Software platforms specifically targeted towards the development of BCIs should offer frequently used building blocks such as data acquisition, feature extraction, classification, and feedback presentation modules. Many labs have developed their own custom set of tools over many years, based on different requirements, programming languages, and prospective users. These tools are typically closed source and not available to the public, since they are primarily used for rapid prototyping and in-house testing. Moreover, such tools might lack extensive documentation and might not be readily useable for others outside the lab.

On the other hand, several publicly available BCI platforms have been released during the past few years. These frameworks are targeted either towards BCI developers, BCI users, or both. Some platforms are released under popular open source licenses (such as the GNU General Public License[1]), which allow everyone to inspect, modify, and redistribute the source code. Moreover, many frameworks are cross-platform, which means that they can be deployed on several different operating systems, whereas others are restricted to either a specific operating system and/or require commercial software.

This article provides an overview of currently available platforms and frameworks for developing and deploying BCIs. We have identified seven major BCI platforms and one platform specifically targeted towards feedback and stimulus presentation. These platforms are: (1) BCI2000, (2) OpenViBE, (3) TOBI Common Implementation Platform (CIP), (4) BCILAB, (5) BCI++, (6) xBCI, and (7) BF++. The framework for feedback and stimulus presentation is called Pyff and does not have any direct competitors at the moment. Among the seven platforms, TOBI CIP plays a special role, because it is not a full-fledged BCI platform. Instead,

[1] www.gnu.org/licenses/gpl.html

this platform defines standardized interfaces between different BCI components. This allows other BCI platforms that implement specific modules (such as data acquisition, feature extraction or classification blocks) which adhere to the TOBI CIP specifications to talk to and interact with each other.

16.2 BCI2000

BCI2000[2] is a general-purpose software platform for BCI research. It can also be used for a wide variety of data acquisition, stimulus presentation, and brain monitoring applications. BCI2000 has been in development since 2000 in a project led by the Brain–Computer Interface R&D Program at the Wadsworth Center of the New York State Department of Health in Albany, New York, USA, with substantial contributions by the Institute of Medical Psychology and Behavioral Neurobiology at the University of Tübingen, Germany. In addition, many laboratories around the world, most notably the BrainLab at Georgia Tech in Atlanta, Georgia, and Fondazione Santa Lucia in Rome, Italy, have also played an important role in the project's development. BCI2000 is currently maintained and further developed by a core team consisting of six scientists and programmers, and by a community of contributors that constantly expand the capabilities of the system, such as by adding support for new hardware devices. The BCI2000 core team consists of Gerwin Schalk (Project Director and Chief Evangelist), Jürgen Mellinger (Chief Software Engineer), Jeremy Hill (Project Coordinator), Griffin Milsap (Software and Test Engineer), Adam Wilson (User Management and Support), and Peter Brunner (Workshops and Tutorials).

Main Features BCI2000 comes with support for different data acquisition hardware, signal processing routines, and experimental paradigms. Specifically, BCI2000 supports 19 different data acquisition systems by different manufacturers, including all major digital EEG amplifiers. It supports appropriate processing of EEG oscillations, evoked potentials, ECoG activity, and single-unit action potentials. The resulting outputs can control cursor movement and provide spelling. BCI2000 can also provide highly customizable auditory/visual stimulation that is synchronized with acquisition of brain signals. In addition to brain signals, input from other devices, such as joysticks, keyboards, or eye trackers may be recorded.

Modularity The design of BCI2000 is modular on multiple levels. First, it separates a BCI system into a number of modules specializing in data acquisition, signal processing, user application, and system control. These modules are realized as separate programs, which communicate with each other over TCP/IP connections, and may be distributed across a network. Except for the control module ("Operator"), all modules come in a number of incorporations that may be freely

[2] www.bci2000.org

combined at run-time. There exists a source module for each of the amplifiers supported; signal processing modules for spectral estimation by different methods, and for ERP analysis; user application modules providing cursor feedback, stimulus presentation, and speller interfaces. No programming or recompilation is required to use these modules in BCI experiments. There is a GUI provided that allows to select which combination of modules should be started up for an experiment. All modules allow for a high degree of customization without recompilation by adapting parameters from an Operator GUI. Parameters, and module versioning information, are stored in recorded data files, such that all information about an experiment is available in data analysis. The Operator GUI itself may be configured and automated such that only a minimum of configuration steps is exposed to recording staff.

At a second level, BCI2000 is modularized into a chain of filters operating on signals. These filters share a common programming interface, and form a chain reaching across module boundaries. Each filter's output may be visualized in form of signal time courses, or as a color field suited to display spectra for multiple channels and across time. Within modules, the filter chain may be built from serial and parallel combinations of existing filters, such that processing of brain signals may be split up into an arbitrary number of parallel data streams. Changes to this configuration require recompilation of a module but no actual programming knowledge.

A third level of modularization exists in form of re-usable software building blocks. Such building blocks support the creation of new signal processing filters, or application modules implementing a feedback or stimulus presentation paradigm. Using these building blocks requires some programming knowledge, but is simplified by programming tutorials, and wizard-like tools that create filter and module projects containing template code with guiding comments.

Documentation BCI2000 provides comprehensive documentation for both researchers and programmers. Documentation for researchers describes how to operate and configure existing BCI2000 components. Documentation for programmers describes the data structures, data types, and internal events in the BCI2000 online system. It also describes how to extend BCI2000 with new acquisition modules, signal processing components, or application modules. For both researchers and programmers, information is available in the form of tutorials as well as detailed references. BCI2000 documentation is provided with each BCI2000 installation and is also available online.[3] In addition, there is a bulletin board[4] for questions about BCI2000 and BCI systems in general. Finally, there is a book on the BCI2000 system, which includes an introduction to all major aspects of BCI operation [49].

Programming Languages and Compatibility BCI2000 has been written in C++, and is thus very efficient in terms of resource utilization. It provides a

[3]doc.bci2000.org

[4]bbs.bci2000.org

programming interface that allows to access system parameters, data signals, and event information in a concise and intuitive way. In addition, BCI2000 provides support for writing online signal processing code in MATLAB, and includes an entire layer of Python compatibility.[5] This Python layer allows for writing complete BCI2000 modules that support data acquisition, signal processing, or application output. For compatibility with even more programming languages and external applications, BCI2000's core functionality comes as a loadable library, and may be wrapped into an application that accesses this library. Furthermore, BCI2000 exposes its internal state over a UDP socket interface, which can be read and written to by an external application. For code compilation, BCI2000 supports Visual Studio—including the freely available Express versions—and GCC[6]/MinGW[7] in addition to the Borland C++ compiler it was originally developed with. This set of compilers allows compilation of BCI2000 on multiple platforms, including Windows and Mac OS X, though it is currently fully tested and supported on Windows only. BCI2000 is freely available under the terms of the GNU General Public License.

Deployment The BCI2000 platform does not rely on third-party software components. A full BCI2000 installation is contained in a single directory tree. BCI2000 may be deployed simply by copying this tree, without the need of administrative rights, and without the need to install additional software. Maintenance of BCI2000 installations across multiple research sites is as easy as synchronizing a centrally maintained installation between sites.

Real-time Performance BCI2000 is usually executed on Microsoft Windows operating systems. Windows does not have dedicated support for real-time operation. However, BCI2000's timing behavior is well suited for BCI experiments. Generally, stimulus and feedback presentation is delivered with millisecond accuracy [62]. BCI2000 comes with a tool that comprehensively characterizes timing behavior for different configurations.

Impact BCI2000 has had a substantial impact on BCI and related research. As of April 2011, BCI2000 has been acquired by more than 900 users around the world. The original article that described the BCI2000 system [50] has been cited close to 400 times (Google Scholar, 4/29/11), and was awarded a Best Paper Award by IEEE Transactions on Biomedical Engineering. Furthermore, a review of the literature revealed that BCI2000 has been used in studies reported in more than 150 peer-reviewed publications. These publications include some of the most impressive BCI demonstrations and applications reported to date such as: the first online brain–computer interfaces using magnetoencephalographic (MEG) signals [37] or electrocorticographic (ECoG) signals [20, 28, 29, 61]; the first multi-dimensional

[5] bci2000.org/downloads/BCPy2000

[6] gcc.gnu.org

[7] www.mingw.org

BCI using ECoG signals [53]; the fastest BCI ever demonstrated in humans [13]; the first applications of BCI technology toward restoration of function in patients with chronic stroke [14, 63]; the use of BCI techniques to control assistive technologies [17]; the first real-time BCI use of high-resolution EEG techniques [16]; the first tactile P300 BCI [11]; demonstrations that non-invasive BCI systems can support multi-dimensional cursor movements without [36, 64, 65] and with [35] selection capabilities; control of a humanoid robot by a noninvasive BCI [3]; and the first demonstration that people severely paralyzed by amyotrophic lateral sclerosis (ALS) can operate a sensorimotor rhythm-based BCI [26]. BCI2000 is also supporting the only existing long-term in-home application of BCI technology for people who are severely disabled [54].

Many studies have used BCI2000 in fields related to BCI research. This includes the first large-scale motor mapping studies using ECoG signals [30, 39]; real-time mapping of cortical function using ECoG [12, 38, 52]; the optimization of BCI signal processing routines [15, 48, 66]; evaluation of steady-state visual evoked potentials (SSVEP) for BCI purposes [1]; the demonstration that two-dimensional hand movements and finger movements can be decoded from ECoG signals [25,51]; and determination of the electrical properties of the dura and its influence on ECoG recordings [57]. Facilitated by the easy exchange of data and experimental paradigms that BCI2000 enables, a number of these studies were performed as collaborations among several geographically widespread laboratories.

16.3 OpenViBE

OpenViBE[8] is a free and open-source software platform for designing, testing, and using brain–computer interfaces. The platform consists of a set of software modules that can be easily and efficiently integrated to develop fully functional BCIs. OpenViBE features an easy-to-use graphical user interface for non-programmers. Key aspects of the platform are described in the following paragraphs.

Development Team and Community OpenViBE is licensed under the GNU Lesser General Public License (version 2 or later).[9] It is officially available for Microsoft Windows (XP to 7) and Linux (Ubuntu and Fedora) platforms. Other operating systems have been addressed by the community. OpenViBE is released every three months by the French National Institute for Research in Computer Science and Control (INRIA). The core development team at INRIA works continuously on new features, integration of community contributions, and releases. People who contributed to OpenViBE include A. L'ecuyer, Y. Renard, F. Lotte, L. Bougrain, L. Bonnet, J. Leg'eny, V. Delannoy, B. Payan, M. Clerc,

[8]openvibe.inria.fr

[9]www.gnu.org/copyleft/lesser.html

T. Papadopoulo, and J. Fruitet (INRIA); O. Bertrand, J.-P. Lachaux, G. Gibert, E. Maby, and J. Mattout (INSERM); M. Congedo, G. Ionescu, M. Goyat, G. Lio, and N. Tarrin (GIPSA-LAB); A. Souloumiac and B. Rivet (CEA); and Dieter Devlaminck (Ghent University).

It is difficult to reliably estimate the number of OpenViBE users, because OpenViBE can be downloaded and used without any kind of registration. However, the OpenViBE Windows installer has been downloaded more than 300 times a month in 2010, and the OpenViBE website has been visited by more than 3,000 single visitors per month. The non-exhaustive list of identified users of OpenViBE is provided on the OpenViBE website and includes many universities, research institutes, and medical centers all around the world. OpenViBE is also used in a large variety of projects involving industrial or medical partners, for example in video games or assistance to disabled people.

Modularity and Reusability OpenViBE consists of a set of software modules devoted to the acquisition, preprocessing, processing, and visualization of cerebral data. The platform also has modules which handle the interaction with applications. OpenViBE is a general purpose platform and allows users to easily add new software modules specifically tailored towards their needs. This is largely made possible thanks to the OpenViBE box concept. A box is a graphical representation of an elementary component in the processing pipeline. Boxes can be connected and composed altogether in a complete BCI scenario. This design makes software components reusable at low cost, reduces development time, and helps to quickly extend functionality. Finally, there is no built-in limit for the number of boxes or connections in a scenario, allowing to merge existing state-of-the-art BCI scenarios in new BCI scenarios.

Different User Types OpenViBE is designed for different types of users, including researchers, developers, and clinicians. Their various needs are addressed and different tools are proposed for each user type, depending on their programming skills and their knowledge of brain physiology.

Portability The OpenViBE platform operates independently from different software targets and hardware devices. It includes an abstract layer of representation, which supports various acquisition devices such as EEG or MEG amplifiers. OpenViBE runs on Windows and Linux platforms. OpenViBE is based on free and portable software such as GTK+,[10] IT++,[11] VRPN,[12] and GCC.

Connection with External Applications OpenViBE can be easily integrated with high-level applications such as virtual reality applications. OpenViBE acts as an external peripheral device for any kind of real or virtual environment. It also takes advantage of virtual reality displays through a scenegraph management library,

---

[10]www.gtk.org

[11]sourceforge.net/apps/wordpress/itpp

[12]www.cs.unc.edu/Research/vrpn

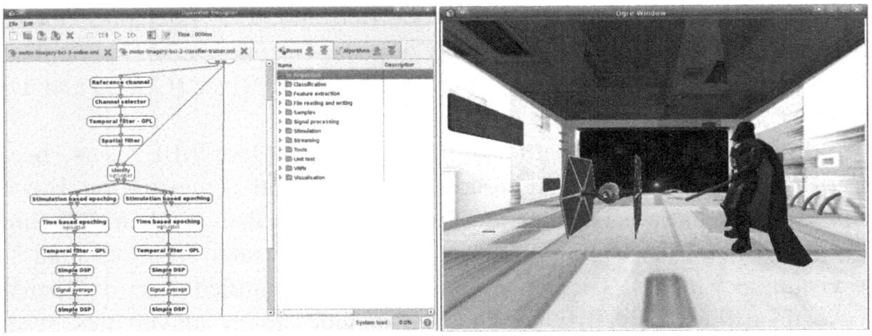

Fig. 16.1 *Left*: The OpenViBE designer supports intuitive graphical development of a BCI system. *Right*: Video game based on motor imagery using a self-paced BCI developed with OpenViBE [31]

allowing the visualization of cerebral activity in an intuitive way or the creation of incentive training environments for neurofeedback applications.

OpenViBE Tools The OpenViBE platform includes a large number of useful tools: the acquisition server, the designer, 2D visualization tools, and sample scenarios for BCIs or neurofeedback applications.

The acquisition server provides a generic interface to various kinds of acquisition devices. It allows an author to create hardware-independent scenarios with a generic acquisition box. This box receives data over the network from the acquisition server, which is connected to the hardware and transforms the recorded data in a generic way. The way the acquisition server is connected to the device mostly depends on the hardware manufacturer's tools to access the device. Some devices are shipped with a dedicated SDK, whereas others involve a communication protocol over the network, serial interface or a USB connection.

The designer makes it possible to create complete scenarios using a dedicated graphical language (see Fig. 16.1 *left*). The user can drag and drop existing modules from a panel to the scenario window. Each module appears as a rectangular box with inputs, outputs, and a dedicated configuration panel. Boxes can be connected through their inputs and outputs. The designer also allows to configure the arrangement of visualization windows. Finally, an embedded player engine supports testing and debugging the current scenario in real time.

The visualization features of OpenViBE are available as specific boxes and include 2D/3D brain activity plots. OpenViBE offers a wide range of visualization widgets such as raw signal display, gauges, power spectrum, time-frequency maps, and 2D/3D topography (where EEG activity is projected on the scalp surface). Virtually any data of a scenario can be visualized by these visualization boxes such as for instance: raw or filtered signals, extracted features or classifier outputs. OpenViBE also provides presentation widgets that display instructions to a user, for example, as used in typical BCI paradigms such as the classical cue-based motor imagery paradigm or the P300 speller.

Existing and pre-configured ready-to-use scenarios are provided to assist the user, such as:

- The motor imagery based BCI scenario uses OpenViBE as an interaction peripheral device with imagined movements of the left and right hands.
- The Self-paced BCI scenario implements a BCI based on real or imagined foot movements in a self-paced way (Fig. 16.1 *right*).
- The neurofeedback scenario displays the power of the brain activity in a specific frequency band for neurofeedback applications.
- The real time visualization scenario visualizes brain activity of a user in real time on a 2D or 3D head model. This scenario can be used together with inverse solution methods to visualize brain activity in the whole brain volume in addition to the scalp surface.
- The P300 speller scenario implements the famous P300 speller, a BCI used to spell letters by using the P300 component of visual event-related potentials.
- The SSVEP scenario allows a user to control a simple game by focusing on flickering targets on the screen. The scenario detects SSVEP at occipital sites to move a virtual object.

Extensive online documentation[13] is also available to help all types of users, either programmers or non-programmers, to start with the software.

OpenViBE Workflow Designing and operating an online BCI with OpenViBE follows a rather universal approach. Three distinct steps are required. In the first step, a training dataset must be recorded for a given subject, who performs specific mental tasks. The second step consists of an offline analysis of these recorded data to find the best calibration parameters (e.g. optimal features, relevant channels, etc.) for this subject. The last step involves using the BCI online in a closed loop scheme. Optionally, iterations can be done on data acquisition and offline training to refine the parameters. Recent BCI research has also focused on adaptative algorithms that automatically adapt the BCI to the subject's brain activity. Some of these algorithms do not perfectly fit in this workflow. Thus, future versions of OpenViBE will address new and specific software mechanisms adapted to these novel needs.

16.4 TOBI

The TOBI Common Implementation Platform (CIP)[14] is a cross-platform set of interfaces which connect parts of different BCI systems. These interfaces transmit raw data, extracted features, classifier outputs, and events over the network in a standardized way. Therefore, the TOBI CIP is not another BCI platform. In contrast,

---

[13]openvibe.inria.fr/documentation/latest

[14]www.tobi-project.org/download

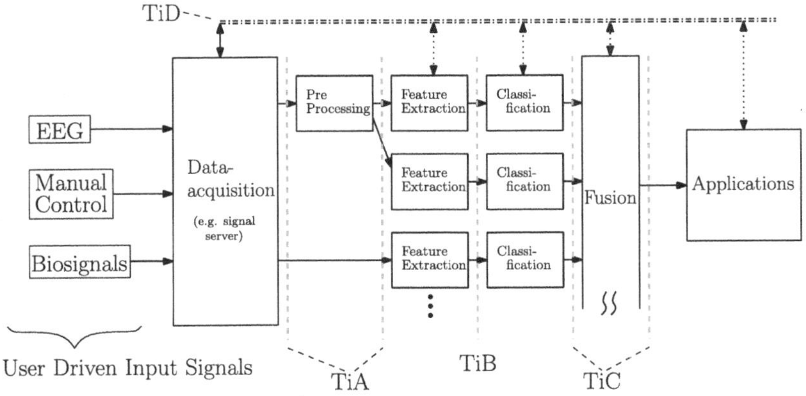

Fig. 16.2 Scheme of the TOBI Common Implementation Platform

it facilitates distributed BCI research and interoperability between different BCI systems and platforms. Therefore, the CIP mainly targets people who want to make their BCI compatible with others and potentially also want to use components and tools from other researchers. In addition, it attempts to introduce standardization into the BCI field, thereby bringing BCI technology one step further towards the end-user market.

Design The design of the CIP is based on the BCI model proposed by Mason and Birch [34]. As shown in Fig. 16.2, the CIP is based on a pipeline system. Data is acquired via a data acquisition system and forwarded to data processing modules. Different processing pipes are shown, because the TOBI CIP supports multiple (potentially distributed) processing streams. Modules are interconnected by different interfaces labeled as TiA, TiB, and TiC (TOBI interface A, B, and C). Each interface transmits specific types of signals used in BCI systems. A fourth interface (TiD) is used to transmit events and markers within the CIP. In case of multiple processing streams, a fusion module merges incoming information to one information stream. This merging process can be based on static or adaptive rules. The output of the fusion module can be used to control different types of applications or graphical user interfaces. The CIP synchronizes data streams by including the block number and time stamps of received data at each interface.

TiA TiA is an interface to transmit raw biosignals and information gathered from assistive devices or sensors [9, 10] in realtime. Data is transmitted via TiA by the data acquisition and the preprocessing modules. TiA assigns different signal types to acquired data (for example, EEG, EOG, buttons, joystick, and so on) and supports simultaneous multi-rate and multi-channel data transmission. Furthermore, multiple clients can attach to a TiA server during runtime. Meta information is exchanged via a defined handshaking procedure based on XML (extensible markup language). Raw data can be transmitted either using TCP (transmission control protocol) or UDP using TiA data packets. A detailed documentation of TiA is available online.[15]

TiB TiB is an interface for transmitting signal features such as band power. There is no further definition or implementation available yet.

TiC TiC is an interface to transmit detected classes and class labels within a BCI system. Information is encoded in XML messages. Each TiC message can consist of a different classifier and class, both with label and value fields. Therefore, the fusion module or an application module can interpret received TiC messages in a standardized way.

TiD TiD is an interface to transmit markers and events used in BCI systems. It is based on XML messages and is acting like a bus system using multiple TCP connections. A module can send an event to the bus, and this event is dispatched by a TiD server (must be integrated or attached to the data acquisition system) to all connected clients.

Implementation A cross-platform library for TiA (implemented in C++) is available online.[16,17] Libraries for TiB, TiC, and TiD are currently under development and will be released soon. Additionally, a cross-platform data acquisition system called signal server (which implements TiA) is also available for download. The signal server supports simultaneous multi-rate data acquisition of different kinds of signals from different devices. The signal server and the TiA library have successfully passed various timing and stability tests. In addition, both software products are very resource and memory efficient, as they are implemented using C++. For cross-platform compatibility, only established libraries such as Boost[18] or SDL[19] are used within the TiA library or the signal server. TiA was already successfully integrated into MATLAB and Simulink, BCI2000, and a Linux embedded board (FOX Board G20, ARM 400 MHz, 64 MB RAM). MATLAB clients are currently available for TiA and TiC. Although there are no official builds for Mac OS X (or related platforms such as iOS) at the moment, the library can be built on these platforms. For example, we have successfully implemented an iOS app (running on iPhone, iPod Touch, and iPad) using the TOBI library. The integration of the TiA library into an embedded board or iOS-based devices demonstrates its portability and low resource requirements. The different interfaces can either be used by re-implementing the protocol by oneself or by merely including the provided libraries into an existing BCI framework. The provided libraries are simple so that only minor programming experience (preferably in C++) is necessary to use them.

[15]http://arxiv.org/abs/1103.4717v1

[16]www.tobi-project.org/download

[17]bci.tugraz.at/downloads

[18]www.boost.org

[19]www.libsdl.org

Furthermore, Matlab and Matlab/Simulink clients are also provided to facilitate the distribution of the CIP.

Benefits By using the TOBI Common Implementation Platform, it is possible to interconnect different BCI systems with a minimum of additional work. Since the CIP uses network connections, building distributed BCI systems is also straightforward. The signal server acquires data from different devices at the same time, potentially also with different sampling rates. Therefore, BCIs and other assistive technology can be combined into an augmented assistive device, the so-called hybrid BCI [40]. Furthermore, as a result of the multiple data streams, it is easily possible to add additional processing modules such as mental state monitoring or passive BCI approaches to an existing system. Additional tools for monitoring the raw signal (scope) or the classifier output will be made available continuously on the project website.

16.5 BCILAB

BCILAB[20] is an open-source MATLAB-based toolbox for advanced BCI research. Its graphical and scripting user interfaces provide access to a large collection of well-established methods, such as Common Spatial Patterns [46] and shrinkage LDA [8], as well as more recent developments [24, 56]. Because of its MATLAB foundation, the major strengths of the toolbox are implementing rapid prototyping, real time testing, offline performance evaluation of new BCI applications, and comparative evaluation of BCI methods. The design of BCILAB is less focused on clinical or commercial deployment, although compiled versions of BCILAB are available to run standalone versions of BCI methods.

Workflow Most BCI methods depend on parameters that may vary dramatically across people and/or sessions. These parameters must be learned, often via machine learning methods, on pre-recorded training or calibration data. Thus, building and using a BCI typically involves recording a calibration session, performing offline analyses on this data to learn or refine a BCI model, and using the learned model to estimate (in real time) changes in the user's cognitive state, response, or intent. Offline analysis in BCILAB involves computing models from training data, but frequently also extends to post-hoc/simulated assessment of the performance of a BCI model on separate testing data, thereby avoiding the need for costly online method testing sessions when sufficient data are available. To this end, BCILAB automates rigorous cross-validation to assess test set performance, automatic parameter search, nested cross-validation, and online simulation. BCILAB also visualizes models, which facilitates psychophysiological interpretation of discriminating data features used by the model. For online processing, BCILAB provides a general-purpose real

[20]sccn.ucsd.edu/wiki/BCILAB

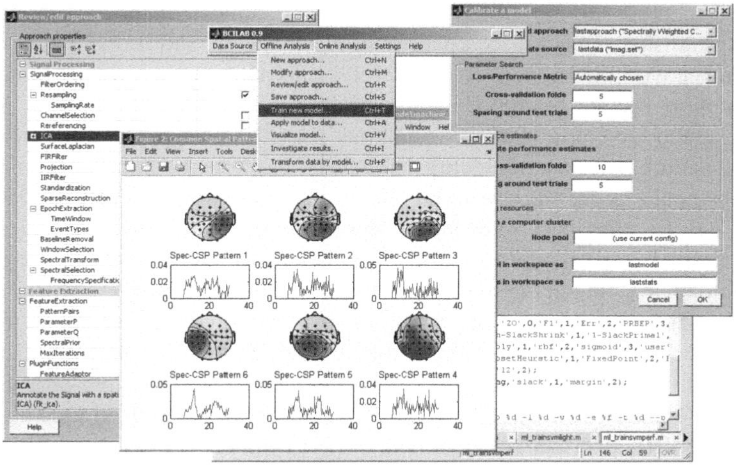

Fig. 16.3 The BCILAB graphical user interface showing the main menu (*top middle*), a model visualization window (*bottom middle*), a parameter settings dialog for a BCI approach (*left*), a method configuration window (*right*), as well as a MATLAB editor workspace (*bottom right*)

time data streaming and signal processing framework compatible with data collection and stimulus adaptation software (BCI2000, OpenViBE, ERICA), described below in more detail.

Features BCILAB puts emphasis on combining contemporary methods in machine learning, signal processing, statistical modeling, and electrophysiological imaging to facilitate methods-oriented research across disciplines. To this end, it provides several plug-in frameworks to speed up incorporation and testing of new BCI methods. Currently, BCILAB offers 15 machine learning methods, 20 signal processing methods (not counting variants), and ten feature extraction methods, all of which can be configured and combined freely both via a GUI (as shown in Fig. 16.3) and command line scripting. In addition to these dataflow-oriented components, BCI model building approaches can be realized that cut across several of these traditionally distinct processing stages, for example, methods involving joint optimization and/or probabilistic modeling. To shorten the time it takes to realize a particular BCI approach, the toolbox makes heavy use of default settings when possible, and provides a pre-configured palette of well-established and recently-proposed BCI approaches, many of which can be reused with little customization. Extensive documentation is available on the project website.

Through its linkage to EEGLAB [18], BCILAB makes available an extensive collection of neuroscience tools including the ability to operate on independent components found with Independent Component Analysis (ICA) methods [32], in particular Infomax [2] and Adaptive Mixture ICA (AMICA) [42]. Further capabilities include the use of prior information about relevant anatomical structures based on ICA-enabled source localization and probabilistic brain atlas look-up

[27], and methods to extract high-quality source time-frequency representations including transient inter-source coherence. Furthermore, and unlike many current neuroscience workflows, these steps run fully automatically in most settings.

To support mobile brain/body imaging (MoBI) research [33], BCILAB has been designed to work with classifications based on multiple data modalities collected simultaneously, including EEG, eye gaze, body motion capture, and other biosignals, as recorded with the DataRiver framework in ERICA [19]. This feature may be especially relevant for applications of BCI methods outside the clinical context, in particular for passive monitoring of cognitive state in cognition-aware human–system interaction applications including gaming [67]. BCILAB uses plug-ins to link to real time recording and stimulation environments. Currently, it can either be used in standalone mode (with current support for BioSemi, TCP and OSC[21] data protocols) or as a signal processing module in a general-purpose BCI platform. Currently, BCI2000 [50] and ERICA [19] are supported, with OpenViBE [47] support planned. For real-time operation, the number of simultaneous output streams is only limited by processing power (in versions 0.91+, up to 50–100 filter blocks can be executed concurrently on a 2007-era PC, e.g., configured as ten parallel output streams with 5–10 pipeline stages each). This number is further reduced when computationally expensive filters are used, such as time-frequency analysis on overlapped windows. The processing latency introduced by BCILAB when using Common Spatial Patterns on 32-channel EEG (sampled at 256 Hz) is approximately 5 ms on a desktop PC, plus latency of the involved device and presentation systems, although the strength of the platform lies in computationally more involved designs with correspondingly higher latencies.

Availability BCILAB has been developed at the Swartz Center for Computational Neuroscience, University of California San Diego.[22] Its design was inspired by an earlier PhyPA toolbox developed by C. Kothe and T. Zander at Technical University, Berlin. BCILAB is open source (GPL) and supports most versions of MATLAB (running on Windows/Linux/Mac OS X).

16.6 BCI++

BCI++[23] is an open source framework based on a sophisticated graphics engine. The platform provides a set of tools for the rapid development of brain–computer interfaces and human–computer interaction (HCI) in general.

[21] opensoundcontrol.org

[22] sccn.ucsd.edu

[23] www.sensibilab.campuspoint.polimi.it

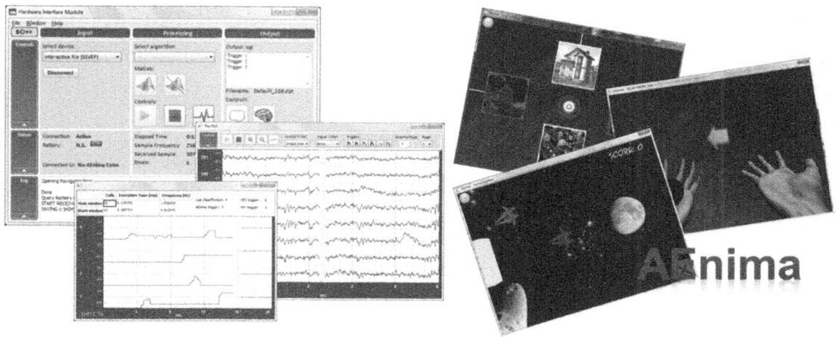

Fig. 16.4 Hardware Interface Module GUIs (*left*) and AEnima protocol examples (*right*)

Structure of the system The BCI++ framework is composed of two main modules, which communicate with each other via TCP/IP. The first module is called HIM (Hardware Interface Module) and handles signal acquisition, storage, visualization, and real-time processing. The second module is named AEnima and provides a Graphical User Interface (GUI). This module is dedicated to creating and managing different protocols based on a high-level 2D/3D graphics engine. This structure was devised to split the development of a real-time BCI system into two parts, namely into (1) signal processing algorithms, and (2) a graphical user interface (GUI).

Hardware Interface Module (HIM) HIM provides a reliable software solution for the acquisition, storage, visualization, and real-time processing of signals. HIM communicates with AEnima via TCP/IP, but both software modules can also run on the same machine. HIM is open source under the GNU GPL, the source code can be downloaded from the Sensibilab website or checked out from our Subversion repository (for the latest development version). HIM was written in C++ using the cross-platform wxWidgets library,[24] but the actual release build is for Microsoft Windows only. HIM has a core block, which handles all tasks common to all protocols and loads plug-ins. These plug-ins are encapsulated in dynamically linked libraries and contain algorithms that the user develops. Algorithms for real-time signal processing can be designed both in C/C++ and MATLAB. BCI++ provides a Visual C++ 2010 project wizard to assist developers during the creation of new algorithm classes. The framework also comes with some SSVEP and motor imagery tools to help researchers rapidly create new BCI systems. In summary, BCI++ comes with a solid set of tools, which simplify the development of updates without rebuilding everything, and which allow to share applications and algorithms without recompiling them. Figure 16.4 (*left*) shows the main window, the signal plot window, and the feedback window of HIM, respectively.

HIM supports several signal acquisition devices; some are real, others are virtual and are useful for debugging and simulation purposes. The list of compatible

[24] www.wxwidgets.org

devices includes: Brainproducts Brain Vision Recorder (supports most Brainproducts devices); Brainproducts Vamp; g.tec g.Mobilab; g.tec g.USBamp; Compumedics Neuroscan; Braintronics Brainbox (four different devices); SXT-telemed ProtheoII; and SXT-telemed Phedra.

BCI++ also provides compatibility with some of the devices developed in selected labs, including an Arduino-based amplifier that acquires up to 16 channels at 256 Hz (the electronic circuit and the Arduino sketch can be downloaded from the BCI++ website). It is also possible to add a new device by deriving a specific class from the base instrument class. In the source code, an example illustrates how to implement new devices. More instructions are also reported in the documentation.

Graphical User Interface (AEnima) AEnima is a flexible tool developed to simplify the implementation of new operating protocols for BCI-based applications. There are two version of AEnima: one is written in C++ using a multiplatform graphics engine (Irrlicht[25]), whereas the other one is written in C# using XNA Game Studio to use BCI++ on Xbox 360, Windows Phone or Windows 7 Tablet platforms (the latter version is still under development). Both versions are open source and can be downloaded from the Sensibilab website or checked out from the Subversion repository.

The user interface software is based on a sophisticated graphics engine to provide a more realistic and challenging experience to the BCI user, and to guarantee versatility and efficiency in application development. Just like HIM, AEnima has a core based on these graphics engines and a plug-in which contains the real GUI. The two different versions (Irrlicht and XNA) both support OpenGL and DirectX (versions 8, 9, and 10). Therefore, the engine runs on fast and slow computers alike (for example, the software was successfully tested on an old Pentium 3 machine with an embedded graphics card). AEnima includes an audio engine, which offers a set of high-level functions which allow the reproduction and management of sound effects and audio files in different formats (for example WAV, MP3, and OGG). This engine also supports positional and 3D audio, which can be a useful way to develop protocols and paradigms with auditory stimulation or feedback. Furthermore, AEnima features two stimulation modules; the first one is a stimulation module that sends messages via USB to control external stimuli like the ones usually used for SSVEP BCI paradigms [43]. The second stimulation module can send commands via TCP/IP to a FES (functional electrical stimulation) controller used in BCIs for rehabilitation purposes. A specific software module was also implemented to provide an application layer with a home automation system. In the latest release, AEnima includes augmented reality features based on ARtoolkit.[26] Figure 16.4 (*right*) shows some AEnima GUI examples.

Conclusion The BCI++ system simplifies interfacing a BCI with external devices (such as a BCI-based FES stimulator for rehabilitation). The advanced graphics

[25] irrlicht.sourceforge.net

[26] www.hitl.washington.edu/artoolkit

engine allows developers to focus on the design of the HCI aspect without having to spend a long time on developing a new system from scratch. BCI++ supports different kinds of acquisition devices, which could be used by both the end-user in their daily activities (for example, home automation control) and by the researcher to develop new protocols, algorithms, and software modules useful in a BCI laboratory. The framework is very flexible, and the large set of debugging tools dramatically simplifies debugging and testing of a new system.

However, the most relevant aspect of BCI++ is the possibility for unskilled developers to develop and test their own work and to actively help to increase the number of available instruments in the framework. All software modules and the source code are available on our web site along with some examples and documentation. The framework was also validated and tested on more than one-hundred users (healthy and disabled) with SSVEP and motor imagery BCI systems.

16.7 xBCI

xBCI[27] is a generic platform for developing online brain–computer interfaces [55]. This platform provides users with an easy-to-use system development tool and reduces the time needed to develop a BCI system. The main developers are I.P. Susila and S. Kanoh.

Features The main features of this platform are as follows:

- Extendable and modular system design: Functional modules can be added by users, and PCs or data acquisition devices (e. g. EEG or NIRS amplifiers) can be easily integrated into xBCI.
- GUI-based system development: A GUI-based editor for building and editing the BCI system is provided. Using the editor, even inexperienced users can easily build their own systems.
- Multi-threaded parallel processing: Users can build a multi-threaded parallel processing system without any detailed knowledge of the operating system or thread programming.
- Multi-OS support: The platform supports multiple operating systems, such as Microsoft Windows and GNU Linux.
- Open source: The xBCI platform was implemented with the GNU C/C++ compiler set, and only open source libraries were used to implement components and the platform itself. It does not depend on any commercial software products.

Workflow The platform consists of several functional modules (components), which can be used to realize a specific BCI system. Users can design and build

[27] xbci.sourceforge.net

various types of BCI systems by combining these components in a GUI-based editor. The ready-to-use components are listed below.

- Basic mathematical operations: Logical operation, arithmetic operation of scalar values and matrices, and basic mathematical functions such as trigonometric and logarithmic functions. Mathematical expressions are evaluated and calculated by these dedicated components.
- Data processing: Temporal and spatial filters, frequency analysis, averaging, pattern classifiers, data import and export, and so on.
- Data acquisition: Measured data or digital event marker signals are acquired by interface boards (e. g. A/D converter boards) or parallel ports.
- Network communications: Data transfer from/to other PCs or data acquisition devices over TCP/IP or UDP. These components allow users to easily build an experimental system with several PCs or data acquisition devices which are connected over a network.
- Data visualization: Real time data scopes for displaying and monitoring measured or processed data.
- Experiment control: Control of experimental protocols with a precise timing accuracy.
- Real time feedback presentation: Various ways to present the feedback information for neurofeedback experiments can be constructed.

Users can also add custom components to extend the functionality of the platform. A custom component can be added to the platform by either programming in C++ or by using a scripting language. Every component is completely independent as a plug-in, and components can be added or modified without rebuilding the whole platform. Plug-ins can then be distributed separately from the platform.

Each component is executed in its own thread and starts processing in parallel as soon as any incoming data becomes available. Data are transferred between components by means of a packet. A packet consists of a packet header and data to be processed. System parameters, such as sampling frequency and number of channels for measurement, are shared among components by the packet header.

Input and Output xBCI can transfer analog and digital data from/to external devices via interface boards. This means that generic data acquisition devices (e.g. biosignal amplifiers) with analog output can be used. On Linux, the interface was implemented with COMEDI, which supports many interface boards. On Windows, DAQ boards of National Instruments (Austin, Texas, USA) and Interface Corp. (Hiroshima, Japan) are currently supported.

The BCI platform can also communicate with external devices over TCP/IP or UDP.

Performance and Timing We evaluated the performance of xBCI during real-time processing and showed that (1) xBCI can acquire data of many input channels (tested on 16 channels) digitized at a sampling rate of 1 kHz and apply FFT to the acquired data in real time (the processing time divided by the number of processed

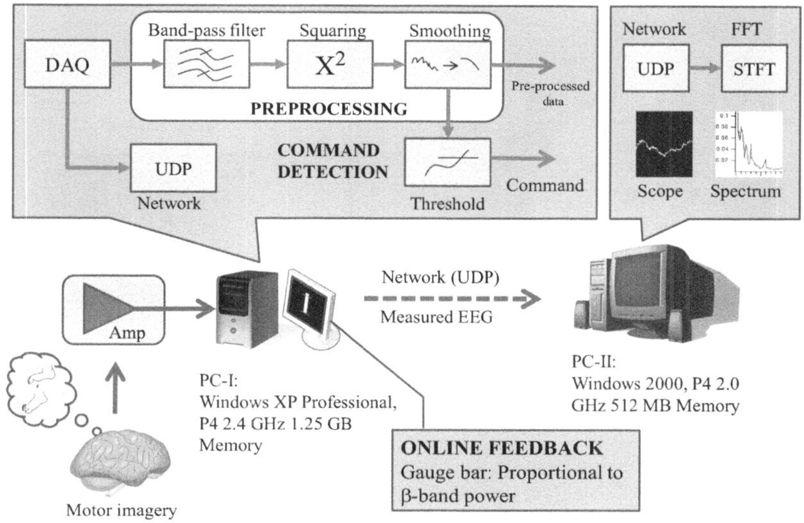

Fig. 16.5 Schematic diagram of the data processing chain in an example neurofeedback application

samples is about 1 μs), and (2) the processed data can be transferred to other PCs through the network with a jitter in the millisecond range (see [55] for more details). Since xBCI occupies only relow CPU and memory resources, the number of input and output streams is mainly limited by the performance of the interface boards or external equipment.

Applications Figures 16.5 and 16.6 show the application of xBCI to the online BCI neurofeedback training system based on a brain switch [22, 23], which detects a binary command (on/off) by an increase of EEG band power elicited during motor imagery recorded from a single bipolar EEG channel. Figure 16.5 shows a block diagram of the data processing chain. Data acquisition, online processing, and neurofeedback experiment control were carried out on PC-I, and the measured data were transmitted to PC-II and displayed for online monitoring. The realized system by xBCI is shown in Fig. 16.6. This data processing chain was implemented by connecting the components in the GUI editor (upper left), and the recorded EEG data (middle), the spectrum (lower left), as well as neurofeedback information (right) were displayed.

Conclusion In summary, the xBCI platform provides users with an easy-to-use system development tool and reduces the time needed to develop a BCI system. The complete platform along with documentation and example designs can be obtained from the project website and is freely available under the GNU General Public License.

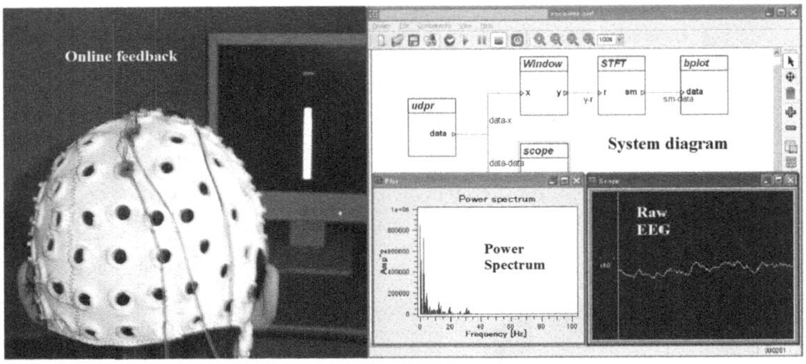

Fig. 16.6 An example neurofeedback application using xBCI

16.8 BF++

The aim of BF++[28] (Body Language Framework in C++) is to provide tools for the implementation, modeling and data analysis of BCI and HCI systems. The main objective of BF++ is to create unique methods, terminologies, and tools independent from the specific protocols such as P300, SSVEP, or SMR BCIs. BF++ is based on a well-defined abstract model, on top of which various methods and tools have been implemented. It is highly scalable, cross-platform, and programmed by adopting only well established technologies, such as C++ as the programming language, XML for storage, and UML (unified modeling language) for description and documentation. BF++ was one of the first cross-platform BCI platforms [4, 5], but it is mostly oriented towards data analysis and BCI system description and evaluation.

Comparing Performance Across Different BCI Systems Great effort has been made to allow a reliable comparison among different systems and the optimization of their performances. This was achieved by starting from a unique static functional model [34] as shown in Fig. 16.7. In this model, the two main elements are the transducer, which is responsible for the acquisition of neurophysiological signals and their classification, and the control interface, which processes the output of the classifier and controls external peripheral devices by feeding into the application control module.

This model was extended recently by adding dynamic behavior and a description of the model using UML sequence diagrams [45]. Following this model, the same actors (classes in object oriented programming terminology) have been successfully used in five different BCI protocols confirming its robustness and the high abstraction level achieved. The main advantage of this is that it is much easier to share software tools regardless of the BCI protocols and that it is much easier to compare them.

[28]www.brainterface.com

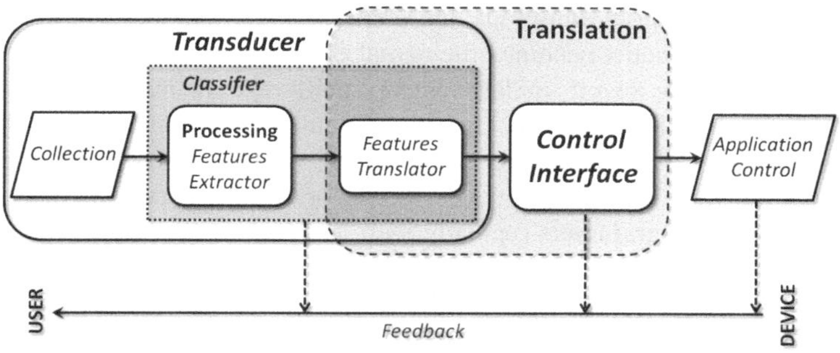

Fig. 16.7 Functional model of a BCI system as used in BF++

Another important aspect of BF++ is that it provides a unique and reliable performance metric, the efficiency [6] of the model. It is based on the characterization of either the transducer or the control interface and it is able to deal with their adaptation. Compared to other commonly used metrics (such as classification accuracy, information transfer rate, and so on), the efficiency is suitable for the description, simulation, and more importantly, optimization of the systems. For this reason, several software tools (the BF++ Toys) have been released. The advantage of using the same model and methods regardless of the specific protocol maximizes consistency among the tools and their (re)usability.

File Formats Moreover, specific file formats have been implemented using XML, which allows extensions by adding data without breaking the backward compatibility with already existing tools. To facilitate the exchange of data between different laboratories, support for several file formats are provided (for example, BCI2000, GDF, Brain Vision Analyzer, EDF, CTF MEG, and so on). However, only the BF++ native NPX file format (neurophysiological data in XML) is able to take advantage of all BF++ software analysis packages [7]. These packages allow to perform EEG and ERP analysis, spectral analysis, statistics, spatial filtering (for example, independent component analysis and common spatial patterns), classification, and 2D/3D mapping. All packages can be downloaded from the project website.

16.9 Pyff

Pyff[29] (Pythonic feedback framework) is a framework for the rapid development of experimental paradigms and a platform to run neuroscientific experiments. The foremost design goal was to make the development of BCI feedback and stimulus applications as fast and easy as possible. As stimulation and feedback paradigms

[29]bbci.de/pyff

are getting more and more ambitious and complex, one bottleneck in the process of conducting experiments becomes the actual development of the software. This problem is even more severe in labs where such software is not developed by computer scientists. Thus, we decided to implement the framework in Python. Python is a high level programming language and well known for its flat learning curve compared to low level languages like C or C++. Experience has shown us that non-expert programmers typically learn to program feedback and stimulus applications with Pyff within two days. Implementing equivalent applications in low level programming languages like C or C++ can easily take an order of magnitude more time, and even for experienced programmers usually a factor of two remains [44].

Pyff is completely written in Python and thus not tied to a special operating system. Pyff runs everywhere where Python runs, which in includes all major platforms such as Linux, Mac OS X, and Windows. Moreover, we tried our best to keep Pyff also independent from specific BCI systems. That is, our goal was to make it compatible with as many BCI systems as possible. We achieved that by coupling Pyff with the rest of the BCI system using UPD and XML. The network protocol is used to transport the data from the BCI system to Pyff, and XML is used to wrap arbitrary data in a format Pyff can handle. UDP is supported by almost all mainstream programming languages, and so is XML. A complete description of the interface can be found in [59]. Additionally, Pyff also supports the TOBI interface to communicate with the rest of the BCI system.

It is important to note that Pyff does not provide a complete BCI software stack. In a typical BCI environment, a BCI system consists of three parts: (1) data acquisition, (2) signal processing, and (3) feedback or stimulus presentation. Pyff provides only the third part of this stack. Moreover, it creates a layer above the BCI system and allows to implement feedback and stimuli without having to worry about the underlying BCI system. Therefore, Pyff is not only a framework for rapid development of feedback and stimulus applications, but also a platform to run neuroscientific experiments independent from BCI systems. Such a platform could foster a fruitful exchange of experimental paradigms between research groups, decrease the need of reprogramming standard paradigms, facilitate the reproducibility of published results, and promote standardization of feedback and stimulus presentation.

Pyff already comes with a variety of ready-to-use experimental paradigms, like the hex-o-speller or the matrix speller. Pyff is actively maintained by one developer and several others are regularly contributing code.

Overview of Pyff Pyff consists of four parts: (1) the feedback controller, (2) a graphical user interface, (3) a set of feedback paradigms and stimuli, and (4) a collection of base classes.

The feedback controller receives incoming signals from the BCI system and translates and forwards them to the feedback and stimulus application. The feedback controller is also responsible for controlling the execution of these applications, for example starting, pausing or stopping them.

The graphical user interface (GUI) controls the feedback controller remotely over the network. The experimenter can select, start, pause, and stop feedback and stimulus applications as well as inspect and modify their variables during runtime. Being able to modify all variables on the fly provides a great way to explore different settings in a pilot experiment, and this feature also makes the GUI an invaluable debugging tool. The GUI communicates with the feedback controller using the same UDP/XML protocol as the BCI system. This makes the GUI completely optional, every command can also be issued by the BCI system directly.

Pyff also provides a constantly growing set of ready-to-go feedback and stimulus applications, which can be used without or with only small modifications. Pyff supports loading and saving the feedback and stimulus application's parameters to a JSON[30] file, which is useful for providing supporting material in publications and facilitates the reproducibility of results.

The collection of feedback base classes provides methods and functionality shared by many feedback and stimulus applications. These methods can be used in derived classes, which reduces the overhead of developing new applications and minimizes code duplication. For example, Pygame[31] is often used for the graphical representation of stimuli. Applications using Pygame often share a huge amount code, for example for the initialization of the screen or polling Pygame's event queue. All this functionality is already available in the `PygameFeedback` base class and does not have to be rewritten in derived classes. All feedback base classes also provide the methods needed to communicate with the feedback controller. Therefore, every class derived from one of the feedback base classes is automatically a valid feedback (or stimulus) class.

Since Python can utilize existing libraries (e. g. shared objects or DLLs), it is straightforward to use special hardware within Pyff. Pyff already provides support for the IntelliGaze eye tracker by Alea Technologies and the g.STIMbox by g.tec.

License and Availability Pyff is completely open source and free software under the terms of the GNU General Public License. Pyff is available for download including documentation as well as other information and links on the project homepage. Furthermore, the source code is available from the public Git repository. The requirements to run Pyff are currently a working installation of Python 2.6.6[32] and PyQt 4.[33]

---

[30] www.json.org

[31] www.pygame.org

[32] www.python.org

[33] www.riverbankcomputing.co.uk/software/pyqt/intro

16.10 Conclusion

The number of user-friendly BCI software platforms has increased significantly over the past years. The days when a researcher had to start from scratch and develop all required BCI components are almost over, or at least there are viable alternatives available. Nowadays, people who want to use or develop BCIs can choose between many publicly available BCI platforms. We have described seven of the most popular BCI frameworks and one platform dedicated to feedback and stimulus presentation in this article. While some platforms have been available for many years and offer a great number of features (for example, BCI2000 and OpenViBE), each platform has its unique features and benefits. We addressed topics that might be important to potential users, such as licensing issues, availability for multiple platforms, supported hardware devices, interaction with other software applications,almost and so on. Table 16.1 compares all platforms with respect to supported operating systems, license, and requirements (see caption for more details). It is interesting to note that all platforms (except for BF++) have adopted either the GPL or LGPL as their license. Furthermore, most platforms run under more than one operating system, at least unofficially. However, Microsoft Windows remains the most popular target in officially supported versions. Most platforms are written in C/C++, which are very efficient programming languages. However, they are also more difficult to learn than MATLAB, which is a popular rapid prototyping environment for many researchers. To alleviate this potential problem for non-programmers, some platforms written in C/C++ offer a GUI and/or bindings to other simpler programming languages.

Future directions could include exploiting synergies and minimizing redundancies between platforms. The TOBI CIP could play an important role in reaching this goal, or in reaching less ambitious short term goals such as making the platforms talk to another. This would allow the data acquisition facility from one platform to be used with the feature extraction facility of another platform and the visualization capabilities of a third framework. It should be relatively straightforward to adapt existing platforms to support the TOBI interfaces in addition to their native data exchange formats. Even if platform-specific features had to be dropped because of a lack of support in the TOBI protocols, the possibility to use this standardized format opens up a wealth of opportunities. Work towards implementing TOBI interfaces (especially TiA) has already started in some platforms, and is planned for other frameworks. For example, Pyff has had built-in support for the TOBI interfaces for several months.

In summary, there probably is no best platform for everyone. With the information presented in this article, interested users should be able to identify platforms that might be suitable for their specific purposes.

Acknowledgements The views and the conclusions contained in this document are those of the authors and should not be interpreted as representing the official policies, either expressed or implied, of the corresponding funding agencies. The authors would like to acknowledge the following projects and funding sources:

Table 16.1 Feature comparison of BCI platforms. Columns 2–4 indicate if operating systems are officially supported. Support for Windows includes versions XP, Vista, and 7 unless otherwise noted. Support for Mac OS X includes versions 10.5 and 10.6 unless otherwise noted. If Linux is supported, the platform should run on any Linux distribution. The last column lists all required software components that are not open source or freely available

Platform	Windows	Mac OS X	Linux	License	Requirements
BCI2000	•	–[a]	–[a]	GPL	Windows[b]
OpenViBE	•[c]	–	•	LGPL[d]	–
TOBI	•	–[e]	•	GPL, LGPL[f]	–
BCILAB	•[g]	•[g]	•[g]	GPL	MATLAB[h]
BCI++	•	–[i]	–[i]	GPL	Windows[i]
xBCI	•	•	•	GPL	–
BF++	•[j]	–[i]	–[i]	Free[k]	Windows[i]
Pyff	•[l]	•[l]	•[l]	GPL	–

[a]Officially supported in next version, current version should run under Mac OS X and Linux.
[b]Next version will also run under Mac OS X and Linux.
[c]Also runs on Windows 2000.
[d]Version 2 or later.
[e]Unofficially, the TOBI library runs on Mac OS X and iOS platforms.
[f]TiA is licensed under the LGPL; the TOBI Signal Server is licensed under the GPL.
[g]All versions that run MATLAB R2006a or greater.
[h]MATLAB is not required to run BCILAB, only for making changes at the source code level.
[i]Unofficially, also runs and compiles on Mac OS X and Linux.
[j]Unofficially, BF++ also compiles on Windows CE.
[k]Free for non-commercial use.
[l]All versions that run Python 2.6.6.

BCI2000: This work was supported by grants from the US Army Research Office (W911NF-07-1-0415, W911NF-08-1-0216) and the NIH/NIBIB (EB006356 and EB000856).

OpenViBE: This work was partly supported by grants of the French National Research Agency under the OpenViBE (ANR-05-RNTL-016) and OpenViBE2 (ANR-09-CORD-017) projects.

TOBI: This work is supported by the European ICT Programme Project FP7-224631.

BCILAB: Research was sponsored by the Army Research Laboratory and was accomplished under Cooperative Agreement Number W911NF-10-2-0022. Initial development was supported by a gift from the Swartz Foundation (Oldfield, NY) and a basic research grant of the Office of Naval Research (ONR).

Pyff: This work was partly supported by grants of the Bundesministerium für Bildung und Forschung (BMBF) (FKZ 01IB001A, 01GQ0850) and by the FP7-ICT Programme of the European Community, under the PASCAL2 Network of Excellence, ICT-216886.

References

1. Allison, B.Z., McFarland, D.J., Schalk, G., Zheng, S.D., Jackson, M.M., Wolpaw, J.R.: Towards an independent brain–computer interface using steady state visual evoked potentials. Clin. Neurophysiol. **119**, 399–408 (2008)
2. Bell, A.J., Sejnowski, T.J.: An information-maximization approach to blind separation and blind deconvolution. Neural Comput. **7**, 1129–1159 (1995)

3. Bell, C.J., Shenoy, P., Chalodhorn, R., Rao, R.P.: Control of a humanoid robot by a noninvasive brain–computer interface in humans. J. Neural Eng. **5**, 214–220 (2008)
4. Bianchi, L., Babiloni, F., Cincotti, F., Salinari, S., Marciani, M.G.: An object oriented approach to biofeedback applications for disabled people. In: 3rd International Conference on BioElectroMagnetism, pp. 1–3. Bled, Slovenia (2000)
5. Bianchi, L., Babiloni, F., Cincotti, F., Mattia, D., Marciani, M.G.: Developing wearable bio-feedback systems: the BF++ framework approach. In: 1st International IEEE EMBS Conference on Neural Engineering, pp. 607–609. Capri, Italy (2003)
6. Bianchi, L., Quitadamo, L., Garreffa, G., Cardarilli, G., Marciani, M.: Performances evaluation and optimization of brain computer interface systems in a copy spelling task. IEEE Trans. Neural Syst. Rehabil. Eng. **15**, 207–216 (2007)
7. Bianchi, L., Quitadamo, L.R., Abbafati, M., Marciani, M.G., Saggio, G.: Introducing NPXLab 2010: a tool for the analysis and optimization of P300 based brain–computer interfaces. In: 2nd International Symposium on Applied Sciences in Biomedical and Communication Technologies, pp. 1–4 (2009)
8. Blankertz, B., Lemm, S., Treder, M., Haufe, S., Müller, K.R.: Single-trial analysis and classification of ERP components – a tutorial. NeuroImage **56**, 814–825 (2011)
9. Breitwieser, C., Daly, I., Neuper, C., Müller-Putz, G. R.: Proposing a standardized protocol for raw biosignal transmission. IEEE Trans. Biomed. Eng. **59**, 852–859 (2012)
10. Breitwieser, C., Neuper, C., Müller-Putz, G.R.: A concept to standardize raw biosignal transmission for brain–computer interfaces. In: Proceedings of the 33rd Annual International Conference of the IEEE Engineering in Medicine and Biology Society (2011b)
11. Brouwer, A.M., Van Erp, J.B.F.: A tactile P300 brain–computer interface. Front. Neurosci. **4** (2010)
12. Brunner, P., Ritaccio, A.L., Lynch, T.M., Emrich, J.F., Wilson, J.A., Williams, J.C., Aarnoutse, E.J., Ramsey, N.F., Leuthardt, E.C., Bischof, H., Schalk, G.: A practical procedure for real-time functional mapping of eloquent cortex using electrocorticographic signals in humans. Epilepsy Behav. **15**, 278–286 (2009)
13. Brunner, P., Ritaccio, A.L., Emrich, J.F., Bischof, H., Schalk, G.: Rapid communication with a "P300" matrix speller using electrocorticographic signals (ECoG). Front. Neurosci. **5** (2011)
14. Buch, E., Weber, C., Cohen, L.G., Braun, C., Dimyan, M.A., Ard, T., Mellinger, J., Caria, A., Soekadar, S., Fourkas, A., Birbaumer, N.: Think to move: a neuromagnetic brain–computer interface (BCI) system for chronic stroke. Stroke **39**, 910–917 (2008)
15. Cabrera, A.F., Dremstrup, K.: Auditory and spatial navigation imagery in brain–computer interface using optimized wavelets. J. Neurosci. Methods **174**, 135–146 (2008)
16. Cincotti, F., Mattia, D., Aloise, F., Bufalari, S., Astolfi, L., De Vico Fallani F., Tocci, A., Bianchi, L., Marciani, M.G., Gao, S., Millán, J., Babiloni, F.: High-resolution EEG techniques for brain–computer interface applications. J. Neurosci. Meth. **167**, 31–42 (2008a)
17. Cincotti, F., Mattia, D., Aloise, F., Bufalari, S., Schalk, G., Oriolo, G., Cherubini, A., Marciani, M.G., Babiloni, F.: Non-invasive brain–computer interface system: towards its application as assistive technology. Brain Res. Bull. **75**, 796–803 (2008b)
18. Delorme, A., Makeig, S.: EEGLAB: an open source toolbox for analysis of single-trial EEG dynamics including independent component analysis. J. Neurosci. Meth. **134**, 9–21 (2004)
19. Delorme, A., Mullen, T., Kothe, C., Acar, Z.A., Bigdely-Shamlo, N., Vankov, A., Makeig, S.: EEGLAB, SIFT, NFT, BCILAB, and ERICA: new tools for advanced EEG processing. Comput. Intell. Neurosci. **2011**, 130,714 (2011)
20. Felton, E.A., Wilson, J.A., Williams, J.C., Garell, P.C.: Electrocorticographically controlled brain–computer interfaces using motor and sensory imagery in patients with temporary subdural electrode implants – report of four cases. J. Neurosurg. **106**, 495–500 (2007)
21. Graimann, B., Allison, B., Pfurtscheller, G.: Brain-computer interfaces: a gentle introduction. In: Graimann, B., Allison, B., Pfurtscheller, G.: (eds.) Brain–Computer Interfaces: Revolutionizing Human–Computer Interaction, pp. 1–28. Springer Berlin Heidelberg, (2011)

22. Kanoh, S., Scherer, R., Yoshinobu, T., Hoshimiya, N., Pfurtscheller, G.: "Brain switch" BCI system based on EEG during foot movement imagery. In: Proceedings of the Third International Brain–Computer Interface Workshop and Training Course, pp. 64–65 (2006)
23. Kanoh, S., Scherer, R., Yoshinobu, T., Hoshimiya, N., Pfurtscheller, G.: Effects of long-term feedback training on oscillatory EEG components modulated by motor imagery. In: Proceedings of the Fourth International Brain–Computer Interface Workshop and Training Course, pp. 150–155 (2008)
24. Kothe, C., Makeig, S.: Estimation of task workload from EEG data: new and current tools and perspectives. In: Proceedings of the 33rd Annual International Conference of the IEEE Engineering in Medicine and Biology Society (2011)
25. Kubánek, J., Miller, K.J., Ojemann, J.G., Wolpaw, J.R., Schalk, G.: Decoding flexion of individual fingers using electrocorticographic signals in humans. J. Neural Eng. **6**, 066,001 (2009)
26. Kübler, A., Nijboer, F., Mellinger, J., Vaughan, T.M., Pawelzik, H., Schalk, G., McFarland, D.J., Birbaumer, N., Wolpaw, J.R.: Patients with ALS can use sensorimotor rhythms to operate a brain–computer interface. Neurology **64**, 1775–1777 (2005)
27. Lancaster, J.L., Woldorff, M.G., Parsons, L.M., Liotti, M., Freitas, C.S., Rainey, L., Kochunov, P.V., Nickerson, D., Mikiten, S.A., Fox, P.T.: Automated Talairach atlas labels for functional brain mapping. Hum. Brain Mapp. **10**, 120–131 (2000)
28. Leuthardt, E.C., Schalk, G., Wolpaw, J.R., Ojemann, J.G., Moran, D.W.: A brain–computer interface using electrocorticographic signals in humans. J. Neural Eng. **1**, 63–71 (2004)
29. Leuthardt, E.C., Miller, K.J., Schalk, G., Rao, R.P., Ojemann, J.G.: Electrocorticography-based brain computer interface – the Seattle experience. IEEE Trans. Neural Syst. Rehabil. Eng. **14**, 194–198 (2006)
30. Leuthardt, E.C., Miller, K.J., Anderson, N.R., Schalk, G., Dowling, J., Miller, J., Moran, D.W., Ojemann, J.G.: Electrocorticographic frequency alteration mapping: a clinical technique for mapping the motor cortex. Neurosurgery **60**, 260–270 (2007)
31. Lotte, F., Renard, Y., Lécuyer, A.: Self-paced brain–computer interaction with virtual worlds: a qualitative and quantitative study "out-of-the-lab." In: Proceedings of the Fourth International Brain–Computer Interface Workshop and Training Course, pp. 373–378 (2008)
32. Makeig, S., Bell, A.J., Jung, T.P., Sejnowski, T.J.: Independent component analysis of electroencephalographic data. In: Touretzky, D., Mozer, M., Hasselmo, M. (eds.) Advances in Neural Information Processing Systems, pp. 145–151. MIT Press (1996)
33. Makeig, S., Gramann, K., Jung, T.P., Sejnowski, T.J., Polzner, H.: Linking brain, mind and behavior. Int. J. Psychophysiol. **73**, 95–100 (2009)
34. Mason, S.G., Birch, G.E.: A general framework for brain–computer interface design. IEEE Trans. Neural Syst. Rehabil. Eng. **11**, 70–85 (2003)
35. McFarland, D.J., Krusienski, D.J., Sarnacki, W.A., Wolpaw, J.R.: Emulation of computer mouse control with a noninvasive brain–computer interface. J. Neural Eng. **5**, 101–110 (2008)
36. McFarland, D.J., Sarnacki, W.A., Wolpaw, J.R.: Electroencephalographic (EEG) control of three-dimensional movement. J. Neural Eng. **7**, 036,007 (2010)
37. Mellinger, J., Schalk, G., Braun, C., Preissl, H., Rosenstiel, W., Birbaumer, N., Kübler, A.: An MEG-based brain–computer interface (BCI). NeuroImage **36**, 581–593 (2007)
38. Miller, K.J., Dennijs, M., Shenoy, P., Miller, J.W., Rao, R.P., Ojemann, J.G.: Real-time functional brain mapping using electrocorticography. Neuroimage **37**, 504–507 (2007a)
39. Miller, K.J., Leuthardt, E.C., Schalk, G., Rao, R.P., Anderson, N.R., Moran, D.W., Miller, J.W., Ojemann, J.G.: Spectral changes in cortical surface potentials during motor movement. J. Neurosci. **27**, 2424–32 (2007b)
40. Millán, J., Rupp, R., Müller-Putz, G.R., Murray-Smith, R., Giugliemma, C., Tangermann, M., Vidaurre, C., Cincotti, F., Kübler, A., Leeb, R., Neuper, C., Müller, K.R., Mattia, D.: Combining brain–computer interfaces and assistive technologies: state-of-the-art and challenges. Front. Neurosci. **4** (2010)

41. Müller-Putz, G.R., Kaiser, V., Solis-Escalante, T., Pfurtscheller, G.: Fast set-up asynchronous brain-switch based on detection of foot motor imagery in 1-channel EEG. Med. Biol. Eng. Comput. **48**, 229–233 (2010)
42. Palmer, J.A., Makeig, S., Kreutz-Delgado, K., Rao, B.D.: Newton Method for the ICA Mixture Model. In: Proceedings of the 33rd IEEE International Conference on Acoustics and Signal Processing (ICASSP), pp. 1805–1808 (2008)
43. Parini, S., Maggi, L., Turconi, A.C., Andreoni, G.: A robust and self-paced BCI system based on a four class SSVEP paradigm: algorithms and protocols for a high-transfer-rate direct brain communication. Comput. Intell. Neurosci. **2009**, 864,564 (2009)
44. Prechelt, L.: An empirical comparison of seven programming languages. IEEE Comput. **33**, 23–29 (2000)
45. Quitadamo, L.R., Marciani, M.G., Cardarilli, G.C., Bianchi, L.: Describing different brain computer interface systems through a unique model: a UML implementation. Neuroinformatics **6**, 81–96 (2008)
46. Ramoser, H., Müller-Gerking, J., Pfurtscheller, G.: Optimal spatial filtering of single trial EEG during imagined hand movement. IEEE Trans. Rehabil. Eng. **8**, 441–446 (2000)
47. Renard, Y., Lotte, F., Gibert, G., Congedo, M., Maby, E., Delannoy, V., Bertrand, O., Lécuyer, A.: OpenViBE: an open-source software platform to design, test, and use brain–computer interfaces in real and virtual environments. Presence **19**, 35–53 (2010)
48. Royer, A.S., He, B.: Goal selection versus process control in a brain–computer interface based on sensorimotor rhythms. J. Neural Eng. **6**, 016,005 (2009)
49. Schalk, G., Mellinger, J.: A Practical Guide to Brain-Computer Interfacing with BCI2000: General-Purpose Software for Brain-Computer Interface Research, Data Acquisition, Stimulus Presentation, and Brain Monitoring. Springer London (2010)
50. Schalk, G., McFarland, D.J., Hinterberger, T., Birbaumer, N., Wolpaw, J.R.: BCI2000: A General-Purpose Brain-Computer Interface (BCI) System. IEEE Trans. Biomed. Eng. **51**, 1034–1043 (2004)
51. Schalk, G., Kubánek, J., Miller, K.J., Anderson, N.R., Leuthardt, E.C., Ojemann, J.G., Limbrick, D., Moran, D., Gerhardt, L.A., Wolpaw, J.R.: Decoding two-dimensional movement trajectories using electrocorticographic signals in humans. J. Neural Eng. **4**, 264–275 (2007)
52. Schalk, G., Leuthardt, E.C., Brunner, P., Ojemann, J.G., Gerhardt, L.A., Wolpaw, J.R.: Real-time detection of event-related brain activity. NeuroImage **43**, 245–249 (2008a)
53. Schalk, G., Miller, K.J., Anderson, N.R., Wilson, J.A., Smyth, M.D., Ojemann, J.G., Moran, D.W., Wolpaw, J.R., Leuthardt, E.C.: Two-dimensional movement control using electrocorticographic signals in humans. J. Neural Eng. **5**, 75–84 (2008b)
54. Sellers, E.W., Vaughan, T.M., Wolpaw, J.R.: A brain–computer interface for long-term independent home use. Amyotroph. Lateral Scler. **11**, 449–455 (2010)
55. Susila, I.P., Kanoh, S., Miyamoto, K., Yoshinobu, T.: xBCI: a generic platform for development of an online BCI system. IEEE Trans. Electr. Electron. Eng. **5**, 467–473 (2010)
56. Tomioka, R., Müller, K.R.: A regularized discriminative framework for EEG analysis with application to brain–computer interface. NeuroImage **49**, 415–432 (2010)
57. Valderrama, A.T., Oostenveld, R., Vansteensel, M.J., Huiskamp, G.M., Ramsey, N.F.: Gain of the human dura in vivo and its effect on invasive brain signals feature detection. J. Neurosci Methods **187**, 270–279 (2010)
58. Vaughan, T.M., McFarland, D.J., Schalk, G., Sarnacki, W.A., Krusienski, D.J., Sellers, E.W., Wolpaw, J.R.: The Wadsworth BCI Research and Development Program: at home with BCI. IEEE Trans. Neural Syst. Rehabil. Eng. **14**, 229–233 (2006)
59. Venthur, B., Scholler, S., Williamson, J., Dähne, S., Treder, M.S., Kramarek, M.T., Müller, K.R., Blankertz, B.: Pyff – a pythonic framework for feedback applications and stimulus presentation in neuroscience. Front. Neurosci. **4** (2010)
60. Vidal, J.J.: Toward direct brain–computer communication. Ann. Rev. Biophys. Bioeng. **2**, 157–180 (1973)

61. Wilson, J.A., Felton, E.A., Garell, P.C., Schalk, G., Williams, J.C.: ECoG factors underlying multimodal control of a brain–computer interface. IEEE Trans. Neural Syst. Rehabil. Eng. **14**, 246–250 (2006)
62. Wilson, J.A., Mellinger, J., Schalk, G., Williams, J.: A procedure for measuring latencies in brain-computer interfaces. IEEE Trans. Biomed. Eng. **7**, 1785–1797 (2010)
63. Wisneski, K.J., Anderson, N., Schalk, G., Smyth, M., Moran, D., Leuthardt, E.C.: Unique cortical physiology associated with ipsilateral hand movements and neuroprosthetic implications. Stroke **39**, 3351–3359 (2008)
64. Wolpaw, J.R., McFarland, D.J.: Multichannel EEG-based brain–computer communication. Clin. Neurophysiol. **90**, 444–449 (1994)
65. Wolpaw, J.R., McFarland, D.J.: Control of a two-dimensional movement signal by a noninvasive brain–computer interface in humans. Proc. Natl. Acad. Sci. USA **101**, 17,849–17,854 (2004)
66. Yamawaki, N., Wilke, C., Liu, Z., He, B.: An enhanced time-frequency-spatial approach for motor imagery classification. IEEE Trans. Neural Syst. Rehabil. Eng. **14**, 250–254 (2006)
67. Zander, T.O., Kothe, C.: Towards passive brain–computer interfaces: applying brain–computer interface technology to human-machine systems in general. J. Neural Eng. **8**, 025,005 (2011)

Chapter 17
Is It Significant? Guidelines for Reporting BCI Performance

Martin Billinger, Ian Daly, Vera Kaiser, Jing Jin, Brendan Z. Allison, Gernot R. Müller-Putz, and Clemens Brunner

Abstract Recent growth in brain-computer interface (BCI) research has increased pressure to report improved performance. However, different research groups report performance in different ways. Hence, it is essential that evaluation procedures are valid and reported in sufficient detail.

In this chapter we give an overview of available performance measures such as classification accuracy, cohen's kappa, information transfer rate (ITR), and written symbol rate (WSR). We show how to distinguish results from chance level using confidence intervals for accuracy or kappa. Furthermore, we point out common pitfalls when moving from offline to online analysis and provide a guide on how to conduct statistical tests on BCI results.

17.1 Introduction

Brain–computer interface (BCI) research is expanding in many ways. Within the academic research community, new articles, events, and research groups emerge increasingly quickly. Research labs have developed BCIs for communication [7, 19, 27, 34, 43–45, 67], for control of wheelchairs [24, 53] and neuroprosthetic devices

M. Billinger (✉) · I. Daly · V. Kaiser · B.Z. Allison · G.R.Müller-Putz · C. Brunner
Institute for Knowledge Discovery, Graz University of Technology, Austria
e-mail: martin.billinger@tugraz.at; ian.daly@tugraz.at; vera.kaiser@tugraz.at; allison@tugraz.at; gernot.mueller@tugraz.at; clemens.brunner@tugraz.at

C. Brunner
Swartz Center for Computational Neuroscience, INC, UCSD, San Diego, CA, USA
e-mail: clbrunner@ucsd.edu

J. Jin
Key Laboratory of Advanced Control and Optimization for Chemical Processes, Ministry of Education, East China University of Science and Technology, China
e-mail: jinjing@ecust.edu.cn

B. Z. Allison et al. (eds.), *Towards Practical Brain-Computer Interfaces*, Biological and Medical Physics, Biomedical Engineering, DOI 10.1007/978-3-642-29746-5_17,
© Springer-Verlag Berlin Heidelberg 2012

[31, 46]. Although BCI research has been conducted for more than 20 years now, only some research labs have successfully applied BCIs to patient use [30, 36–38, 47, 49, 51, 52, 65]. The popular media has also shown increased interest in BCIs, with BCIs featured prominently in science fiction as well as in the mainstream. Additionally, new businesses are gaining attention with various products sold as BCIs for entertainment.

Hence, there is growing attention in performance, and increased pressure to report improved performance. Recent articles that developed fast BCIs openly noted this feat [6, 12, 63]. Articles routinely highlight methods and results that improve accuracy or reduce illiteracy relative to earlier work [3, 4, 10, 11, 33, 55, 62]. However, different groups use different methods for reporting performance, and it is essential that (1) the evaluation procedure is valid from a statistical and machine learning point of view, and (2) this procedure is described in sufficient detail.

It is also important to distinguish any reported BCI performance from the chance level, the expected best performance obtainable by chance alone. Depending on the performance measure, the number of classes in the BCI task, and the number of available trials, the chance level varies and should be considered in every study [48].

In this chapter, we provide an introduction to common performance measures (such as classification accuracy, Cohen's kappa, and information transfer rate). Furthermore, we discuss confidence intervals of the classification accuracy and Cohen's kappa to estimate the associated chance level. We also summarize state of the art offline procedures to estimate performance on a pre-recorded data set and discuss common cross-validation pitfalls. In the last two sections, we describe statistical tests often used in BCI studies, such as t-tests, repeated measures ANOVA, and suitable post-hoc tests. We also mention the need to correct for multiple comparisons.

17.2 Performance Measures

17.2.1 Confusion Matrix

A number of metrics may be used to measure the performance of a BCI. These include the number of correct classifications and the number of mistakes made by the classifier. The most straightforward classification example is binary classification, in which the classifier need only differentiate two classes. For example, this might be the case in the popular P300 speller first presented by Farwell and Donchin [22]. The task of the classifier is to determine if there is a P300 event present in a particular time segment of the EEG. Therefore, the two classes are either "yes, there is a P300 present" or "no, there is no P300 present." When considering such binary classification problems, four classification results are possible:

(1) A trial is classified as containing a P300 when a P300 is present (true positive, TP).

Table 17.1 Confusion matrix for binary classification

		Predicted		
		Class 1	Class 2	
Actual	Class 1	TP	FN	TP+FN
	Class 2	FP	TN	FP+TN
		TP+FP	FN+TN	N

Table 17.2 Example of a confusion matrix for three classes. The diagonal contains all 257 correct classifications (86 + 45 + 126), whereas the 193 misclassifications are on the off-diagonal (45 + 19 + 32 + 73 + 10 + 14). The sum of all elements yields 450 and equals the total number of trials (shown in the lower right corner). The row sums reflect the relative frequencies of each class (rightmost column). In this example, the classes are balanced, because each class occurs 150 times. The column sums reveal how many trials were classified as the specific class. In this example, the classifier assigned 169 trials to left hand, 104 trials to right hand, and 177 trials to foot imagery. Since these numbers are not equal, and because the classes were equally distributed, the classifier is biased towards left hand and foot classes

		Predicted			
		Left hand	Right hand	Foot	
	Left hand	86	45	19	150
Actual	Right hand	73	45	32	150
	Foot	10	14	126	150
		169	104	177	450

(2) A trial is classified as containing a P300 when a P300 is not present (false positive, FP).
(3) A trial is classified as not containing a P300 when there is no P300 present (true negative, TN).
(4) A trial is classified as not containing a P300 when there is a P300 present (false negative, FN).

For two or more classes, it is useful to employ a so-called confusion matrix to present the results. A confusion matrix presents the results of the classifier over several trials against the actual known classes of items in the dataset. This allows for an evaluation of which classes are being correctly and incorrectly classified. For binary classification described above, the structure of the confusion matrix is illustrated in Table 17.1.

Consider the case of a motor imagery based BCI with three possible classes. The BCI user may imagine left hand movement, right hand movement or foot movement to control the BCI. In the classification example illustrated in Table 17.2, 450 trials were classified into three different possible classes. The columns list the output from the classifier, while the rows list the actual class that the trials corresponded to. For example, 86 trials were correctly classified as corresponding to left hand imagery, whereas 45 trials that corresponded to left hand imagery were misclassified as corresponding to right hand imagery. From this example, it is clear that the number of correct classifications for each class are found along the diagonal of the confusion matrix. The row sums reflect the a priori distribution of the classes, that is, the relative frequency of each class. Conversely, the column sums reveal a potential bias of the classifier towards one (or more) classes.

While the confusion matrix contains all information on the outcome of a classification procedure, it is difficult to compare two or more confusion matrices. Therefore, most studies usually report scalar performance measures, which can be derived from the confusion matrix. Metrics commonly used in reporting BCI results include classification accuracy, Cohen's kappa κ, sensitivity and specificity, positive and negative predictive value, the F-measure and the r^2 correlation coefficient [57].

17.2.2 Accuracy and Error Rate

The accuracy p is the probability of performing a correct classification. It can be estimated from dividing the number of correct classifications by the total number of trials

$$p = \frac{\sum C_{i,i}}{N}. \tag{17.1}$$

$C_{i,i}$ is the ith diagonal element of the confusion matrix, and N is the total number of trials. The error rate or misclassification rate $e = 1 - p$ is the probability of making an incorrect classification.

Accuracy and error rate do not take class balance into account. If one class occurs more frequently than the other, accuracy may be high even for classifiers that cannot discriminate between classes. See Tables 17.3 and 17.4 for examples.

17.2.3 Cohen's Kappa

Cohen's kappa (κ) is a measure for the agreement between nominal scales [15]. As such κ can be used to measure the agreement between true class labels and classifier output. It is scaled between 1 (perfect agreement) and 0 (pure chance agreement). Equation (17.2) shows how to obtain κ from accuracy p and chance level p_0.

$$\kappa = \frac{p - p_0}{1 - p_0} \tag{17.2}$$

The chance level p_0 is the accuracy under the assumption that all agreement occurred by chance (see Sect. 17.3.1). p_0 can be estimated from the confusion matrix by

$$p_0 = \frac{\sum C_{i,:} C_{:,i}}{N^2}. \tag{17.3}$$

Table 17.3 Confusion matrix for binary classification when the two classes are not balanced (class 1 occurs more often than class 2). *Left*: The classifier selected the classes with a probability of 50 %. *Right*: The classifier always selected the first class

		Predicted 1	Predicted 2	
Actual	1	45	45	90
	2	5	5	10
		50	50	100

$p = 0.5 \quad \kappa = 0$

		Predicted 1	Predicted 2	
Actual	1	90	0	90
	2	10	0	10
		100	0	100

$p = 0.9 \quad \kappa = 0$

Table 17.4 Confusion matrix for binary classification when the two classes are not balanced (class 1 occurs more often than class 2). *Left*: The classifier selected the first class with 90 % probability and the second class with 10 % probability. *Right*: The classifier classified all trials correctly

		Predicted 1	Predicted 2	
Actual	1	81	9	90
	2	9	1	10
		90	10	100

$p = 0.82 \quad \kappa = 0$

		Predicted 1	Predicted 2	
Actual	1	90	0	90
	2	0	10	10
		90	10	100

$p = 1 \quad \kappa = 1$

$C_{i,:}$ and $C_{:,i}$ are the ith row and column of the confusion matrix, and N is the total number of trials.

For both confusion matrices in Table 17.3 $\kappa = 0$, indicating classification at chance level. Neither of these confusion matrices represents a meaningful classifier, although accuracies are 0.5 and 0.9 respectively.

17.2.4 Sensitivity and Specificity

Alternative metrics reported in BCI studies include the sensitivity and specificity (see for example [5,21,25,60]), which measure the proportion of correctly identified positive results (true positives) and the proportion of correctly identified negative results (true negatives). Sensitivity is defined as

$$H = \mathrm{Se} = \frac{\mathrm{TP}}{\mathrm{TP} + \mathrm{FN}}. \tag{17.4}$$

Specificity is then defined as

$$\mathrm{Sc} = \frac{\mathrm{TN}}{\mathrm{TN} + \mathrm{FP}}. \tag{17.5}$$

The sensitivity is alternately referred to as the true positive rate (TPR) or the recall. The false positive rate (FPR) is then equal to 1—specificity.

The false detection rate F may be calculated as

$$F = \frac{\text{FP}}{\text{TP} + \text{FP}}. \tag{17.6}$$

From this, the positive predictive value (also referred to as the precision) may be calculated as $1 - F$. The HF difference $(H - F)$, as developed in [32], may then be derived.

These metrics may also be used to measure the receiver operator characteristic (ROC) curve [18, 29, 41]. This is a plot of how the true positive rate varies against the false positive rate for a binary classifier as the classification threshold is varied between its smallest and largest limit. The x axis of the ROC curve is the false positive rate (1—specificity), while the y axis is the true positive rate (sensitivity). The larger the area under the ROC curve, the larger the true positive rate and the smaller the false positive rate for a greater number of threshold values. Thus, an ROC curve that forms a diagonal from the bottom left corner of the plot to the top right is at theoretical chance level, whereas an ROC plot that reaches the top left corner is reporting perfect classification.

17.2.5 F-Measure

The terms precision and recall (sensitivity) may be used to describe the accuracy of classification results. Precision (also referred to as the positive predictive value) measures the fraction of classifications which are correct while recall measures the fraction of true positive classifications.

As the precision is increased, the recall decreases, and vice-versa. Therefore, for a given classifier, it is useful to have a measure of the harmonic mean of both measures. The F-measure is used to do this and is defined as

$$F_\alpha = \frac{(1 + \alpha) \cdot (1 - F) \cdot H}{\alpha \cdot (1 - F) + H}, \tag{17.7}$$

where α is the significance level of the measure and may be varied between 0 and 1. Thus, the F-measure may be analogous to the ROC curve, in that it provides a measure of the classifier performance across different significance levels.

17.2.6 Correlation Coefficient

The correlation coefficient may be used for either feature extraction or validation of classification results (see for example [13, 41, 60]). It is defined—via Pearson's correlation coefficient—as

$$r = \frac{\sum_i (y_i - \bar{y})(x_i - \bar{x})}{\sqrt{(\sum_i (y_i - \bar{y})^2)(\sum_i (x_i - \bar{x})^2)}}, \tag{17.8}$$

where x_i denotes output values, y_i the class labels, $\bar{x}$ the mean of x and $\bar{y}$ the mean of the labels y.

Pearson's correlation should be used for Gaussian data, while for non-Gaussian data the rank correlation is recommended. The rank correlation is defined as above with the difference that x_i and y_i values are replaced by rank(x_i) and rank(y_i).

The correlation varies between -1 and 1, with a 0 indicating no correlation between the classifier results and a 1 indicating perfect positive correlation. A correlation of -1 indicates perfect negative correlation and may be discounted if the squared correlation measure is chosen (as used in [13]).

17.3 Significance of Classification

Reporting classification results by providing performance measures alone is often not enough. Even accuracies as high as 90 % can be meaningless if the number of trials is too low or classes are not balanced (see Table 17.3).

The practical level of chance [48] provides a convenient tool to quickly verify if an accuracy value lies significantly above chance level. This practical level of chance is defined as the upper confidence interval of a random classifier's accuracy. Given the number of trials, the resulting accuracy of a BCI experiment must be higher than the practical level of chance. Then the BCI can be said to perform significantly better than chance.

The original publication assumes that classes are balanced [48]. In this section, we describe a more general approach that can handle arbitrary class distributions.

17.3.1 Theoretical Level of Random Classification

In order to test classification results for randomness, a sound definition of random classification is required: A random classifier's output is statistically independent from the true class labels.[1] More formally,

$$P(c_e = c \mid c_t) = P(c_e = c), \tag{17.9}$$

where c_e is the estimated class label and c_t is the true class label.

[1] Such randomness is not necessarily caused by the classifier alone. The BCI user failing at the task, electrode failures or inadequate features may all decrease the degree of agreement between the estimated and true class labels. The actual source of randomness is not relevant for this analysis.

The probability of such a random classifier correctly classifying a trial is

$$p_0 = \sum_{c \in C} P(c_e = c) \cdot P(c_t = c), \tag{17.10}$$

where C is the set of all available class labels.

While the probability $P(c_t = c)$ of a trial belonging to class c is determined by the experimental setup, the probability $P(c_e = c)$ of the classifier returning class c needs to be carefully considered. The most conservative approach is to find the highest possible p_0 for a given experiment. This is the case for a classifier that always returns the class that occurs most often. Intuitively, such a classifier would not be considered random since its output is purely deterministic, but the output is independent from the true class labels, thus (17.9) applies.

Alternatively, the values for $P(c_e = c)$ can be calculated from the experimental results using the confusion matrix (17.3). This yields the same p_0 that is used for the calculation of Cohen's κ, which is the theoretical chance level of an actual classifier. This approach is less conservative as the chance level no longer depends on the experimental setup alone, but also on the probability of each class to be selected by the classifier. However, this approach can only be applied after classification has been performed.

17.3.2 Confidence Intervals

Can a BCI identify the user's intended message or command more accurately than chance? This question can be formally defined with a statistical test, in which the null hypothesis H_0 represents the hypothesis that the BCI's classification is not more accurate than a random classifier. As discussed later, performing above chance is a necessary, but not sufficient, condition for an effective BCI. BCIs typically must perform well above chance to be useful. For example, a speller that identifies one of 36 targets with 50 % accuracy would perform much better than chance, but would not allow useful communication. Formally, the hypothesis test can be written as

$$H_0 : p \leq p_0$$
$$H_1 : p > p_0,$$

where p is the true classification accuracy, and p_0 is the classification accuracy of a random classifier. We compare the one-sided confidence interval of p against the theoretical level of chance, p_0. If p_0 lies outside the confidence interval of p, we can reject H_0 in favor of H_1, thereby indicating that the classifier performs significantly better than chance, at the chosen level of significance.

Regardless of the number of classes, classification can be reduced to either of two outcomes: correct or wrong classification. The correct classification of a trial is called "success." When the probability of success is p, then the probability of getting exactly K successes from N independent trials follows the binomial distribution:

$$f(K; N, p) = \binom{N}{K} p^K (1-p)^{N-K}. \tag{17.11}$$

In a BCI experiment, the classification accuracy is an estimate of p, the true probability of correctly classifying a trial. Given the observed classification accuracy $\hat{p}$, a confidence interval can be calculated that contains the true p with a probability of $1-\alpha$.

Different confidence intervals have been proposed in the literature. The Clopper–Pearson "exact" interval, as well as the Wald interval are too conservative and should not be used in favor of the adjusted Wald interval or the Wilson score interval [8]. We will focus on the adjusted Wald interval because of its simplicity.

17.3.2.1 Adjusted Wald Confidence Interval for Classification Accuracy

Consider the situation where we have N independent trials, of which K are correctly classified. Adding two successes and two failures to the experimental result leads to an unbiased estimator for the probability $\hat{p}$ of correct classification (17.12). Upper and lower confidence limits of $\hat{p}$ are given by (17.13) and (17.14) respectively.

$$\hat{p} = \frac{K+2}{N+4} \tag{17.12}$$

$$p_u = \hat{p} + z_{1-\alpha/2}\sqrt{\frac{\hat{p}(1-\hat{p})}{N+4}} \tag{17.13}$$

$$p_l = \hat{p} - z_{1-\alpha/2}\sqrt{\frac{\hat{p}(1-\hat{p})}{N+4}} \tag{17.14}$$

$z_{1-\alpha/2}$ is the $1-\alpha/2$ quantile of the standard normal distribution. For a one-sided confidence interval $z_{1-\alpha}$ can be used instead of $z_{1-\alpha/2}$.

17.3.2.2 Adjusted Wald Confidence Interval for Cohen's Kappa

Kappa is calculated by transforming accuracy values from the interval $[p_0, 1]$ to the interval $[0, 1]$ according to (17.2). Similarly, a confidence interval of the classification accuracy can be transformed, resulting in a confidence interval for κ

$$\kappa_{l/u} = \frac{p_{l/u} - p_0}{1 - p_0}. \tag{17.15}$$

This results in a modified null hypothesis that tests κ and associated confidence intervals against zero

$$H_0 : \kappa \leq 0$$
$$H_1 : \kappa > 0.$$

The original publication introducing κ [15] proposes a confidence interval that is derived from the Wald interval, which is too conservative according to [8]. Applying the adjusted Wald interval instead results in (17.16)–(17.19)

$$\hat{p} = \frac{K+2}{N+4} \tag{17.16}$$

$$\hat{\kappa} = \frac{\hat{p} - p_0}{1 - p_0} \tag{17.17}$$

$$\kappa_l = \hat{\kappa} - z_{1-\alpha/2} \frac{\sqrt{\hat{p}(1-\hat{p})}}{(N+4)(1-p_0)} \tag{17.18}$$

$$\kappa_u = \hat{\kappa} + z_{1-\alpha/2} \frac{\sqrt{\hat{p}(1-\hat{p})}}{(N+4)(1-p_0)}, \tag{17.19}$$

where $\hat{\kappa}$ is the value of kappa that follows from the unbiased estimator in (17.16).

17.3.3 Summary

It is important not only to consider point estimators of performance measures but also to use appropriate statistics to validate experimental results. In this section we showed how to test estimates of classification accuracy and Cohen's κ against results expected from random classification.

Care has to be taken to chose an appropriate model of random classification. Without knowledge of the classifier's behavior conservative assumptions have to be made about chance classification. When classification results are available a less conservative chance level can be estimated from the classifier output.

17.4 Performance Metrics Incorporating Time

Another critical factor in any communication system is speed—the time required to accomplish a goal, such as spelling a sentence or navigating a room. BCIs often report performance in terms of ITR or bit rate, a common metric for measuring the information sent within a given time [58, 66]. We will measure ITR in bits per minute, and bit rate in bits per trial, which can be calculated via

$$B = \log_2 C + \hat{p} \log_2 \hat{p} + (1 - \hat{p}) \log_2 \frac{1 - \hat{p}}{C - 1}, \tag{17.20}$$

where $\hat{p}$ is the estimated classification accuracy and C is the total number of classes (i.e. possible selections). This equation provides the amount of information (in bits) communicated with a single selection. Many BCI articles multiply B by the number of selections per unit time to attain the ITR, measured in bits per minute. In a trial based BCI this is accomplished by multiplying the ITR by the actual number of trials performed per minute.

However, in a typical BCI speller, users correct errors through a "backspace" function, which may be activated manually or automatically via detection of a neuronal error potential [56]. In contrast to the ITR, the WSR (17.21)–(17.22) incorporates such error correction functionality [23].

$$\mathrm{SR} = \frac{B}{\log_2 C} \tag{17.21}$$

$$\mathrm{WSR} = \begin{cases} (2\mathrm{SR} - 1)/T & \mathrm{SR} > 0.5 \\ 0 & \mathrm{SR} \leq 0.5 \end{cases}, \tag{17.22}$$

where SR is referred to as symbol rate, and T is the trial duration in minutes (including eventual delays).

The WSR incorporates correction of an error by two additional selections (backspace and new selection). However, another error may happen during the correction process. This has been addressed by the practical bit rate (PBR) [61], calculated via

$$\mathrm{PBR} = \begin{cases} B(2p - 1)/T & \hat{p} > 0.5 \\ 0 & \hat{p} \leq 0.5 \end{cases}. \tag{17.23}$$

However, WSR or PBR may not be suitable for systems that use other mechanisms to correct errors [1, 17], or if the user chooses to ignore some or all errors.

ITR calculation may seem to rest on a few simple formulae. However, ITR is often misreported, partly to exaggerate a BCI's performance and partly due to inadequate understanding of many assumptions underlying ITR calculation. Articles that only report the time required to convey a single message or command might ignore many delays that are inevitable in realworld BCI operation. BCIs often entail delays between selections for many reasons. A BCI system might need time to process data to reach a classification decision, present feedback to the user, clear the screen, allow the user to choose a new target, and/or provide a cue that the next trial will begin. Delays also occur if a user decides to correct errors.

Moreover, various factors could affect the effective information transfer rate [2], which incorporates advanced features that could help users attain goals more quickly without improving the raw bit rate. Some BCIs may feature automatic tools to correct errors or complete words or sentences. These tools may introduce some delays, which are presumably welcome because they avoid the greater delays that might be necessary to manually correct errors to complete their messages. Similarly, some BCIs may focus on goal-oriented selections rather than process-oriented selections [1, 64]. Consider two BCIs that allow a user to choose one of eight items with perfect accuracy every ten seconds. Each BCI has a raw ITR of 18 bits/min. However, the first BCI allows a user to move a wheelchair one meter in one of eight directions with each selection, and a second BCI might instead let users choose a target room (leaving the system to work out the details necessary to get there). Other BCIs might incorporate context in various ways. BCIs might change the mapping from brain signals to outcomes. For example, if a robot is in an open space, then imagining left hand movement could move the robot left, but if a wall is to the robot's left, then the same mental command could instruct the robot to follow the wall [42]. BCIs could also use context to change the options available to a user. For example, if a user turns a light off, or if the light bulb burns out, then the option of turning on a light might simply not be available [1].

Moreover, ITR has other limitations [4]. For example, ITR is only meaningful for some types of BCIs. ITR is best suited to synchronous BCIs. Self-paced BCIs, in which the user can freely choose when to make selections or refrain from communicating, are not well suited to ITR estimation. ITR also does not account for different types of errors, such as false positives vs. misses, which could influence the time needed for error correction. Reporting ITR might encourage developers to focus on maximizing ITR, even though some users may prefer higher accuracy, even if it reduces ITR.

In summary, ITR calculation is more complicated than it may seem. Articles that report ITR should include realworld delays, account for tools that might increase effective ITR, and consider whether ITR is the best metric. In some cases, articles present different ITR calculation methods such as practical bit rate or raw bit rate [33, 63]. In such cases, authors should clearly specify the differences in ITR calculation methods and explain why different methods were explored.

17.5 Estimating Performance Measures on Offline Data

BCI researchers often perform initial analysis on offline data to test out a new approach, e.g. a new signal processing method, a new control paradigm etc. For example, [4, 10] report on offline results of a hybrid feature set before they apply it in an online BCI [11].

Because the data is available offline it may be manipulated in a way that is not possible with online data. Common manipulations used in the analysis of offline data include, but are not limited to, cross validation, iteration over a parameter space and the use of machine learning techniques. When applying any offline analysis method, it is important to consider firstly the statistical significance of the reported results and secondly how well the results translate to online BCI operation.

Statistical significance must be reported on the results of classifying a dataset which is separate from the dataset used to train the classification function. The dataset the results are reported on is referred to as the verification (or testing) set while the dataset the classifier is trained on is referred to as the training set. Separating training and verification sets allows us to estimate the expected performance of the trained classifier on unseen data.

The ability to translate offline analysis results to online BCI operation depends on a number of factors including the effects of feedback in online analysis, any temporal drift effects in the signal and how well the offline analysis method is constructed to ensure that the results generalize well. These issues will be considered further in the subsequent sections.

17.5.1 Dataset Manipulations

In online BCI operation any parameters (e.g. classifier weights, feature indices) must be learned first before operation of the BCI begins. However with offline data the trials within the dataset may be manipulated freely.

The most straightforward approach is to simply split the dataset into a training and validation set. This could be done with or without re-sorting the trials. If no re-sorting is used and the trials at the beginning of the session are used for training, this is analogous to online analysis. On the other hand it may be desirable to remove serial regularities from the dataset via re-sorting the trial order prior to splitting into training and validation sets.

A common approach taken is to use either k-fold or leave 1 out cross validation. In k-fold cross validation the dataset is split into K subsets. One of these subsets (subset l) is omitted (this is denoted as the "hold out" set), the remainder are used to train the classifier function. The trained function is then used to classify trials in the lth hold out set. This operation is repeated K times with each set being omitted once. Leave 1 out cross validation is identical, except that each hold out set contains just one trial. Thus, every trial is omitted from the training set once.

Cross validation requires trials to be independent. In general this is not the case due to slowly varying baseline, background activity and noise influence. Trials recorded close to each other are likely to be more similar than trials recorded further apart in time. This issue is addressed through h-block cross validation [39]. h trials closest to each trial in the validation set are left out of the training set, in order to avoid overfitting due to temporal similarities in trials.

Another approach taken, particularly in situations where the size of the available dataset is small, is to use bootstrapping. The training (and possibly the validation) set is created from bootstrap replications of the original dataset. A bootstrap replication is a new trial created from the original dataset in such a way that it preserves some statistical or morphological properties of the original trial. For example, [40] describes a method to increase the training set for BCI operation by randomly swapping features between a small number of original trials to create a much larger set of bootstrap replications.

17.5.2 Considerations

Ultimately, the results reported from offline analysis should readily translate to online BCI operation. Therefore when deciding on any data manipulations and/or machine learning techniques the following considerations should be made:

1. Temporal drift in the dataset. During online BCI operation, factors such as fatigue, learning and motivation affect the ability of the BCI user to exert control. If trials are randomly re-sorted in offline analysis the effect of such temporally dependent changes in the signal are destroyed.
2. The effects of feedback. During online BCI operation the classifier results are fed back to the user via exerted control. This affects the users' motivation and hence the signals recorded from them.
3. Overlearning and stability. Classification methods should be stable when applied to large datasets recorded over prolonged periods of time. Thus, efforts must be made to ensure manipulations made to datasets during offline analysis do not lead to an overlearning effect resulting in poor generalisation and performance instability.

17.6 Hypothesis Testing

Statistical significance of the results obtained in a study is reported via testing against a null hypothesis (H_0), which corresponds to a general or default position (generally the opposite of an expected or desired outcome). For example, in studies reporting classification accuracies, the null hypothesis is that classification is random, i. e. the classification result is uncorrelated with the class labels (see Sect. 17.3). In Sect. 17.3 we discussed the use of confidence intervals in testing against the null hypothesis. This section will elaborate further on additional approaches to testing against the null hypothesis, issues that may arise, and how to properly report results.

Many BCI papers present new or improved methods such as new signal processing methods, new pattern recognition methods, or new paradigms, aiming to improve overall BCI performance. From a scientific point of view, the statement that one method is better than another method is only justifiable if it is based on a solid statistical analysis.

A prerequisite for all statistical tests described in the following subsections is a sufficiently large sample size. The optimal sample size depends on the level of the

α (type I) and β (type II) error and the effect size ϵ [9]. The effect size refers to the magnitude of the smallest effect in the population which is still of substantive significance (given the alternate hypothesis H_1 is valid). For smaller effect sizes, bigger sample sizes are needed and vice versa. Cohen suggested values for small, medium, and large effect sizes and their corresponding sample sizes [16].

The following guidelines are a rough summary of commonly used statistical tests for comparing different methods and should help in finding an appropriate method for the statistical analysis of BCI performance.

17.6.1 Student's t-Test vs. ANOVA

To find out if there is a significant difference in performance between two methods, a Student's t-test is the statistic of choice. However, this does not apply to the case where more than two methods should be compared. The reason for this is that every statistical test has a certain probability of producing an error of Type I—that is, incorrectly rejecting the null hypothesis. In the case of the t-test, this would mean that the test indicates a significant difference, although there is no difference in the population (this is referred to as the type I error, false positives, or α error). For t-test we establish an upper bound on the probability of producing an error of Type I. This is the significance level of the test, denoted by the p-value. For instance, a test with $p \leq 0.04$ indicates the probability of a Type I error is no greater than 4 %. If more than one t-test is calculated, this Type I (α) error probability accumulates over independent tests.

There are two ways to cope with this α-error accumulation. Firstly, a correction for multiple testing such as Bonferroni correction could be applied (see Sect. 17.6.3). Secondly, an analysis of variances (ANOVA) with an adequate post-hoc test avoids the problem of α-error accumulation. The advantage of an ANOVA is that it does not perform multiple tests, and in case of more than one factor or independent variable interactions between these variables can also be revealed (see Sect. 17.6.2).

17.6.2 Repeated Measures

There are different ways to study the effects of new methods. One way is to compare the methods by applying each method to a separate subgroup of one sample, meaning every participant is only tested with one method. Another way is to apply each method to every participant, meaning that each participant is tested repeatedly. For statistical analysis, the way the data has been collected must be considered. In case of repeated measures, different statistical tests must be used as compared to separated subgroups. For a regular t-test and an ANOVA, it is assumed that the samples are independent, which is not fulfilled if the same participants are measured repeatedly. A t-test for dependent samples and an ANOVA for repeated measures take the dependency of the subgroups into account and are therefore the methods of choice for repeated measurements.

In a repeated measures ANOVA design, the data must satisfy the sphericity assumption. This has to be verified (i.e. Mauchly's test for sphericity), and if the assumption is violated, a correction method such as Greenhouse Geisser correction must be applied. Most statistical software packages provide tests for sphericity and possible corrections.

In summary, comparing different methods on the same data set also requires repeated measures tests, which is the classical setting for most offline analyses.

17.6.3 Multiple Comparisons

Section 17.3.2 showed how to test a single classification result against the null hypothesis of random classification. This approach is adequate when reporting a single classification accuracy. However, consider the case when multiple classifications are attempted simultaneously. For example, if one has a dataset containing feature sets spanning a range of different time-frequency locations, one may train a classifier on each feature independently and report significantly better then chance performance, at a desired significance level (e.g. $p < 0.05$), if at least one of these classifiers perform better then chance. In this case, the probability of us falsely reporting better then chance performance for a single classifier is 5 % (the Type 1 error rate). However, if we have 100 classifiers each being trained on an independent feature, then we would expect on average five (5 %) of these classifiers to falsely appear to perform significantly better than chance. Thus, if fewer than six of our classifiers independently perform better than chance, we cannot reject the null hypothesis of random classification at the 5 % significance level.

To adjust for this multiple comparisons problem, Bonferroni correction is commonly applied. This is an attempt to determine the family-wise error rate (the probability of making a type 1 error when multiple hypotheses are tested simultaneously). For n comparisons, the significance level is adjusted by $1/n$. Thus, if 100 independent statistical tests are carried out simultaneously, the significance level for each test is multiplied by 1/100. In our previous example, our original significance level of 0.05 (5 %) would thus be reduced to $0.05/100 = 0.0005$. If we were then to select any single classifier which performs significantly better than chance at this adjusted significance level, we may be confident that in practical application it could be expected to perform better than chance at the 5 % significance level.

BCI studies often report features identified in biosignals which may be useful for BCI control. These signals produce very high-dimensional feature spaces due to the combinatorial explosion of temporal, spatial or spectral dimensions. Traditional analysis methods suggest that it is necessary to correct for multiple comparisons.

However, often in biomedical signal processing such corrections prove to be too conservative.

For example, in a plethora of studies from multiple labs, features derived from the event related desynchronization (ERD) have been successfully shown to reliably allow control of BCIs via imagined movement (see for example [20, 35, 41, 50, 54]).

However, if one attempts to report the statistical significance of the ERD effect in the time-frequency spectra—treating every time-frequency location as an independent univariate test—using Bonferroni correction, the effect may not pass the test of statistical significance. Say, for example, we observe an ERD effect in a set of time-frequency features spanning a 2 s interval (sampled as 250 Hz) and a frequency range of 1–40 Hz, in 1 Hz increments. Say also we have 100 trials, 50 of which contain the ERD effect and 50 of which do not. Our dataset contains $250 \times 2 \times 40$ features and we are interested in which of them contain a statistically significant difference between the 50 trials in which an ERD is observed and the 50 trials in which an ERD is not observed. We are making 20,000 comparisons, therefore the Bonferroni adjustment to our significance level is 1/20,000. With such a large adjustment, we find that classifiers trained on those time-frequency features encompassing the ERD do not exhibit performance surpassing this stringent threshold for significance. In fact, with this many comparisons, if we wished to continue using Bonferroni correction, we would need a much larger number of trials before we began to see a significant effect.

This highlights a fundamental issue with applying Bonferroni correction to BCI features. Namely, the Bonferroni correction assumes independence of the comparisons. This is an adequate assumption when considering coin tosses (and a number of other more interesting experimental paradigms). However, the biosignals used for BCI classification features, typically derived from co-dependent temporal, spatial, and spectral dimensions of the signal, cannot be assumed to be independent. This must be taken into account when correcting for multiple comparisons.

The false discovery rate (FDR) has been proposed to allow multiple comparison control that is less conservative than the Bonferroni correction, particularly in cases where the individual tests may not be independent. This comes at the risk of increased likelihood of Type 1 errors. The proportion of false positives is controlled instead of the probability of a single false positive. This approach is routinely used to control for Type I errors in functional magnetic resonance tomography (fMRI) maps, EEG/MEG, and functional near infrared spectroscopy (fNIRS) (see for example [14, 26, 28]). However, dependencies between time, frequency and spatial locations may not be adequately accounted for.

A new hierarchical significance testing approach proposed in [59] may provide a solution. The EEG is broken into a time-frequency hierarchy. For example, a family of EEG features at different time-frequency locations may be broken into frequency band sub-families (child hypothesis). Each of these frequency families may be further deconstructed into time sub-families. Hypothesis testing proceeds down the tree with pruning at each node of the tree if we fail to reject the null hypothesis at that node. Child hypotheses are recursively checked if their parents' null hypothesis is rejected. This pruning approach prevents the multiple comparisons correction from being overly conservative while accounting for time-frequency dependencies.

17.6.4 Reporting Results

To correctly report the results of a statistical analysis, the values of the test statistic (t-test: t-value, ANOVA: F-value), the degrees of freedom (subscripted to values of test statistic, e. g. t_{df}, $F_{\mathrm{df1,df2}}$, where df1 stands for in between degrees of freedom and df2 equals within degrees of freedom), and the significance level p (e.g. $p = 0.0008$, $p < 0.05$; $p < 0.01$; $p < 0.001$; or n. s. for not significant results) must be provided. If tests for the violation of assumptions (such as sphericity or normality) are applied, results of these tests and adequate corrections should be reported too.

17.7 Conclusion

A BCI is applied for online control of a computer or device. Yet, offline analysis, including preliminary analyses and parameter optimization, remains an important tool in successful development of online BCI technology. Special care must be taken so that offline analysis readily translates to accurate online BCI operation. Effects from temporal drift in the data, feedback which may not be available in training data, and the possibility of overfitting have to be considered.

A number of different metrics for reporting classification performance are available. From these, classification accuracy is probably the most comprehensible, as it directly corresponds to the probability of performing a correct classification. However, reporting only the accuracy is not sufficient. Depending on the number and distribution of classes, even bad performance can lead to high accuracy values. Therefore, the theoretical chance level and confidence interval should always be reported along with accuracy metrics. Additionally, confusion matrices or ROC curves may provide a more complete picture of classification performance.

When reporting performance metrics that incorporate time, one should always take into account the actual time required to reach a certain goal. This includes trial duration, repetitions, error correction, delays in processing or feedback, and even breaks between trials. Furthermore, this time may be reduced by application specific tools. For instance, consider a BCI spelling system. The time required to spell a complete sentence is likely to be the most important criteria for the BCI user. The bit rate measures the amount of information provided by a single trial, and bit rate multiplied by the rate at which trials are repeated allows one to determine the speed at which individual letters can be spelled. Finally, automatic word completion may reduce the time required to complete words and sentences.

Ultimately, as in almost every other applied science, results of a BCI study will need to be subject to a statistical test. Researchers often seek to demonstrate that a BCI can operate at a particular performance level. Or to demonstrate improved performance of a new method over a previously published method, or compare BCI performance in one population to that of a control group. An appropriate statistic, such as a t-test or ANOVA with or without repeated measures design must be chosen, and when necessary, care should be taken to account for multiple comparisons.

Acknowledgements The views and the conclusions contained in this document are those of the authors and should not be interpreted as representing the official policies, either expressed or implied, of the corresponding funding agencies.

The research leading to these results has received funding from the European Union Seventh Framework Programme FP7/2007-2013 under grant agreement 248320. In addition, the authors would like to acknowledge the following projects and funding sources:

- Coupling Measures for BCIs, FWF project P 20848-N15
- TOBI: Tools for Brain–Computer Interaction, EU project D-1709000020
- Grant National Natural Science Foundation of China, grant no. 61074113.

We would like to express our gratitude towards the reviewers, who provided invaluable thorough and constructive feedback to improve the quality of this chapter.

References

1. Allison, B.Z.: The I of BCIs: Next Generation Interfaces for Brain-Computer Interface Systems That Adapt to Individual Users. Human-Computer Interaction. Novel Interaction Methods and Techniques, vol. 5611, pp. 558–568. Springer Berlin/Heidelberg (2009)
2. Allison, B.Z.: Toward Ubiquitous BCIs. Brain-Computer Interfaces. The Frontiers Collection, pp. 357–387. Springer Berlin/Heidelberg (2010)
3. Allison, B.Z., Neuper, C.: Could Anyone Use a BCI? Brain-Computer Interfaces. Human-Computer Interaction Series, pp. 35–54. Springer London (2010)
4. Allison, B.Z., Brunner, C., Kaiser, V., Müller-Putz, G.R., Neuper, C., Pfurtscheller, G.: Toward a hybrid brain–computer interface based on imagined movement and visual attention. J. Neural Eng. **7**, 026,007 (2010). DOI 10.1088/1741-2560/7/2/026007
5. Atum, Y., Gareis, I., Gentiletti, G., Ruben, A., Rufiner, L.: Genetic feature selection to optimally detect P300 in brain computer interfaces. In: 32nd Annual International Conference of the IEEE EMBS (2010)
6. Bin, G., Gao, X., Wang, Y., Li, Y., Hong, B., Gao, S.: A high-speed BCI based on code modulation VEP. J. Neural Eng. **8**, 025,015 (2011). DOI 10.1088/1741-2560/8/2/025015
7. Birbaumer, N., Ghanayim, N., Hinterberger, T., Iversen, I., Kotchoubey, B., Kübler, A., Perelmouter, J., Taub, E., Flor, H.: A spelling device for the paralysed. Nature **398**, 297–298 (1999). DOI 10.1038/18581
8. Boomsma, A.: Confidence intervals for a binomial proportion. Unpublished manuscript, university of Groningen, Department of Statistics & Measurement Theory (2005)
9. Bortz, J.: Statistik für Sozialwissenschaftler. Springer, Berlin, Heidelberg, New York (1999)
10. Brunner, C., Allison, B.Z., Krusienski, D.J., Kaiser, V., Müller-Putz, G.R., Pfurtscheller, G., Neuper, C.: Improved signal processing approaches in an offline simulation of a hybrid brain–computer interface. J. Neurosci. Methods **188**, 165–173 (2010). DOI 10.1016/j.jneumeth.2010.02.002
11. Brunner, C., Allison, B.Z., Altstätter, C., Neuper, C.: A comparison of three brain–computer interfaces based on event-related desynchronization, steady state visual evoked potentials,

or a hybrid approach using both signals. J. Neural Eng. **8**, 025,010 (2011a). DOI 10.1088/1741-2560/8/2/025010
12. Brunner, P., Ritaccio, A.L., Emrich, J.F., Bischof, H., Schalk, G.: Rapid communication with a "P300" matrix speller using electrocorticographic signals (ECoG). Front. Neurosci. **5**, 5 (2011b)
13. Cabestaing, F., Vaughan, T.M., Mcfarland, D.J., Wolpaw, J.R.: Classification of evoked potentials by Pearson's correlation in a brain–computer interface. Matrix **67**, 156–166 (2007)
14. Chumbley, J.R., Friston, K.J.: False discovery rate revisited: FDR and topological inference using gaussian random fields. NeuroImage **44**(1), 62–70 (2009). DOI 10.1016/j.neuroimage.2008.05.021, http://www.ncbi.nlm.nih.gov/pubmed/18603449
15. Cohen, J.: A coefficient of agreement for nominal scales. Psychol. Meas. **20**, 37–46 (1960)
16. Cohen, J.: A power primer. Psychol. Bull. **112**(1), 155–159 (1992)
17. Dal Seno, B. Matteucci, M., Mainardi, L.: Online detection of P300 and error potentials in a BCI speller. Computational Intelligence and Neuroscience, pp. 1–5 (2010)
18. Daly, I., Nasuto, S., Warwick, K.: Single tap identification for fast BCI control. Cogn. Neurodyn. **5**, 21–30 (2011)
19. Dornhege, G., del R Millán, J., Hinterberger, T., McFarland, D.J., Müller, K.R.: (eds.) Towards Brain–Computer Interfacing. MIT Press (2007)
20. Eskandari, P., Erfanian, A.: Improving the performance of brain–computer interface through meditation practicing. In: Engineering in Medicine and Biology Society, 2008. EMBS 2008. 30th Annual International Conference of the IEEE, pp. 662–665 (2008). DOI 10.1109/IEMBS.2008.4649239
21. Falk, T., Paton, K., Power, S., Chau, T.: Improving the performance of NIRS-based brain–computer interfaces in the presence of background auditory distractions. In: Acoustics Speech and Signal Processing (ICASSP), 2010 IEEE International Conference on, pp. 517–520 (2010). DOI 10.1109/ICASSP.2010.5495643
22. Farwell, L.A., Donchin, E.: Talking off the top of your head: toward a mental prosthesis utilizing event-related brain potentials. Electroencephalogr. Clin. Neurophysiol. **70**, 510–523 (1988)
23. Furdea, A., Halder, S., Krusienski, D.J., Bross, D., Nijboer, F., Birbaumer, N., Kübler, A.: An auditory oddball (P300) spelling system for brain–computer interfaces. Psychophysiology **46**, 1–9 (2009). DOI 10.1111/j.1469-8986.2008.00783.x
24. Galán, F., Nuttin, M., Lew, E., Ferrez, P.W., Vanacker, G., Philips, J., del R Millán, J.: A brain-actuated wheelchair: asynchronous and non-invasive brain–computer interfaces for continuous control of robots. Clin. Neurophysiol. **119**, 2159–2169 (2008). DOI 10.1016/j.clinph.2008.06.001
25. Gareis, I., Gentiletti, G., Acevedo, R., Rufiner, L.: Feature extraction on brain computer interfaces using discrete dyadic wavelet transform: preliminary results. Journal of Physics: Conference Series (IOP) **313**, pp. 1–7 (2011)
26. Genovese, C., Wasserman, L.: Operating characteristics and extensions of the false discovery rate procedure. J. R. Stat. Soc. Series B Stat. Methodol. **64**(3), 499–517 (2002). DOI 10.1111/1467-9868.00347, http://doi.wiley.com/10.1111/1467-9868.00347
27. Guger, C., Ramoser, H., Pfurtscheller, G.: Real-time EEG analysis with subject-specific spatial patterns for a brain–computer interface (BCI). IEEE Trans. Neural Syst. Rehabil. Eng. **8**, 447–450 (2000). DOI 10.1109/86.895947
28. Hemmelmann, C., Horn, M., Süsse, T., Vollandt, R., Weiss, S.: New concepts of multiple tests and their use for evaluating high-dimensional EEG data. J. Neurosci. Methods **142**(2), 209–17 (2005). DOI 10.1016/j.jneumeth.2004.08.008, http://ukpmc.ac.uk/abstract/MED/15698661/reload=1
29. Hild II, K.E., Kurimo, M., Calhoun, V.D.: The sixth annual MLSP competition, 2010. Machine Learning for Signal Proc (MLSP '10) (2010)
30. Hoffmann, U., Vesin, J.M., Ebrahimi, T., Diserens, K.: An efficient P300-based brain–computer interface for disabled subjects. J. Neurosci. Methods **167**, 115–125 (2008). DOI 10.1016/j.jneumeth.2007.03.005

31. Horki, P., Solis-Escalante, T., Neuper, C., Müller-Putz, G.: Combined motor imagery and SSVEP based BCI control of a 2 DoF artificial upper limb. Med. Biol. Eng. Comput. (2011). DOI 10.1007/s11517-011-0750-2
32. Huggins, J.E., Levine, S.P., BeMent, S.L., Kushwaha, R.K., Schuh, L.A., Passaro, E.A., Rohde, M.M., Ross, D.A., Elisevich, K.V., Smith, B.J.: Detection of event-related potentials for development of a direct brain interface. J. Clin. Neurophysiol. **16**(5), 448 (1999)
33. Jin, J., Allison, B., Sellers, E., Brunner, C., Horki, P., Wang, X., Neuper, C.: Optimized stimulus presentation patterns for an event-related potential EEG-based brain–computer interface. Med. Biol. Eng. Comput. **49**, 181–191 (2011). doi:10.1007/s11517-010-0689-8
34. Kalcher, J., Flotzinger, D., Neuper, C., Gölly, S., Pfurtscheller, G.: Graz brain–computer interface II: towards communication between humans and computers based on online classification of three different EEG patterns. Med. Biol. Eng. Comput. **34**, 382–388 (1996). DOI 10.1007/BF02520010
35. Karrasch, M., Laine, M., Rapinoja, P., Krause, C.M.: Effects of normal aging on event-related desynchronization/synchronization during a memory task in humans. Neurosci. Lett. **366**(1), 18–23 (2004). DOI 10.1016/j.neulet.2004.05.010, http://dx.doi.org/10.1016/j.neulet.2004.05.010
36. Krausz, G., Ortner, R., Opisso, E.: Accuracy of a brain computer interface (p300 spelling device) used by people with motor impairments. Stud. Health Technol. Inform. **167**, 182–186 (2011)
37. Kübler, A., Birbaumer, N.: Brain-computer interfaces and communication in paralysis: extinction of goal directed thinking in completely paralysed patients? Clin. Neurophysiol. **119**, 2658–2666 (2008). DOI 10.1016/j.clinph.2008.06.019
38. Kübler, A., Nijboer, F., Mellinger, J., Vaughan, T.M., Pawelzik, H., Schalk, G., McFarland, D.J., Birbaumer, N., Wolpaw, J.R.: Patients with ALS can use sensorimotor rhythms to operate a braincomputer interface. Neurology **64**, 1775–1777 (2005)
39. Lemm, S., Blankertz, B. Dickhaus, T., Müller, K.R.: Introduction to machine learning for brain imaging. NeuroImage **56**(2), pp. 387–399 (2011)
40. Lotte, F.: Generating artificial EEG signals to reduce BCI calibration time. In: Proceedings of the 5th International Brain–Computer Interface Conference 2011, pp. 176–179 (2011)
41. Mason, S.G., Birch, G.E.: A brain-controlled switch for asynchronous control applications. IEEE Trans. Biomed. Eng. **47**, 1297–1307 (2000)
42. Millán, J., Mouriño, J.: Asynchronous BCI and local neural classifiers: an overview of the adaptive brain interface project. IEEE Trans. Neural Syst. Rehabil. Eng. **11**, 159–161 (2003)
43. Millán, J., Mouriño, J., Franzé M., Cincotti, F., Varsta, M., Heikkonen, J., Babiloni, F.: A local neural classifier for the recognition of EEG patterns associated to mental tasks. IEEE Trans. Neural Netw. **13**, 678–686 (2002)
44. Müller, K.R., Anderson, C.W., Birch, G.E.: Linear and nonlinear methods for brain–computer interfaces. IEEE Trans. Neural Syst. Rehabil. Eng. **11**, 165–169 (2003)
45. Müller, K.R., Tangermann, M., Dornhege, G., Krauledat, M., Curio, G., Blankertz, B.: Machine learning for real-time single-trial EEG analysis: from brain–computer interfacing to mental state monitoring. J. Neurosci. Meth. **167**, 82–90 (2008). DOI 10.1016/j.jneumeth.2007.09.022
46. Müller-Putz, G.R., Pfurtscheller, G.: Control of an electrical prosthesis with an SSVEP-based BCI. IEEE Trans. Biomed. Eng. **55**, 361–364 (2008). DOI 10.1109/TBME.2007.897815
47. Müller-Putz, G.R., Scherer, R., Pfurtscheller, G., Rupp, R.: EEG-based neuroprosthesis control: a step towards clinical practice. Neurosci. Lett. **382**, 169–174 (2005)
48. Müller-Putz, G.R., Scherer, R., Brunner, C., Leeb, R., Pfurtscheller, G.: Better than random? A closer look on BCI results. Int. J. Bioelectromagn. **10**, 52–55 (2008)
49. Neuper, C., Müller, G.R., Kübler, A., Birbaumer, N., Pfurtscheller, G.: Clinical application of an EEG-based brain–computer interface: a case study in a patient with severe motor impairment. Clin. Neurophysiol. **114**, 399–409 (2003)
50. Pfurtscheller, G., Neuper, C.: Motor imagery and direct brain–computer communication. Proc. IEEE **89**, 1123–1134 (2001). DOI 10.1109/5.939829

51. Pfurtscheller, G., Müller, G.R., Pfurtscheller, J., Gerner, H.J., Rupp, R.: "Thought"-control of functional electrical stimulation to restore handgrasp in a patient with tetraplegia. Neurosci. Lett. **351**, 33–36 (2003). DOI 10.1016/S0304-3940(03)00947-9
52. Piccione, F., Giorgi, F., Tonin, P., Priftis, K., Giove, S., Silvoni, S., Palmas, G., Beverina, F.: P300-based brain computer interface: reliability and performance in healthy and paralysed participants. Clin. Neurophysiol. **117**, 531–537 (2006). DOI 10.1016/j.clinph.2005.07.024
53. Rebsamen, B., Guan, C., Zhang, H., Wang, C., Teo, C., Ang, M.H., Burdet, E.: A brain controlled wheelchair to navigate in familiar environments. IEEE Trans. Neural Syst. Rehabil. Eng. **18**(6), 590–598 (2010). DOI 10.1109/TNSRE.2010.2049862, http://dx.doi.org/10.1109/TNSRE.2010.2049862
54. Roberts, S., Penny, W., Rezek, I.: Temporal and spatial complexity measures for electroencephalogram based brain–computer interfacing. Med. Biol. Eng. Comput. **37**, 93–98 (1999). doi:10.1007/BF02513272
55. Ryan, D.B., Frye, G.E., Townsend, G., Berry, D.R., Mesa-G, S., Gates, N.A., Sellers, E.W.: Predictive spelling with a P300-based brain–computer interface: Increasing the rate of communication. Int. J. Hum. Comput. Interact. **27**, 69–84 (2011). DOI 10.1080/10447318.2011.535754
56. Schalk, G., Wolpaw, J.R., McFarland, D.J., Pfurtscheller, G.: EEG-based communication: presence of an error potential. Clin. Neurophysiol. **111**, 2138–2144 (2000)
57. Schlögl, A., Kronegg, J., Huggins, J.E., Mason, S.G.: Evaluation criteria for BCI research. In: Toward brain–computer interfacing. MIT Press (2007)
58. Shannon, C.E., Weaver, W.: A mathematical theory of communication. University of Illinois Press (1964)
59. Singh, A.K., Phillips, S.: Hierarchical control of false discovery rate for phase locking measures of EEG synchrony. NeuroImage **50**(1), 40–47 (2010). DOI 10.1016/j.neuroimage.2009.12.030, http://dx.doi.org/10.1016/j.neuroimage.2009.12.030
60. Sitaram, R., Zhang, H., Guan, C., Thulasidas, M., Hoshi, Y., Ishikawa, A., Shimizu, K., Birbaumer, N.: Temporal classification of multichannel near-infrared spectroscopy signals of motor imagery for developing a brain–computer interface. NeuroImage **34**, 1416–1427 (2007)
61. Townsend, G., LaPallo, B.K., Boulay, C.B., Krusienski, D.J., Frye, G.E., Hauser, C.K., Schwartz, N.E., Vaughan, T.M., Wolpaw, J.R., Sellers, E.W.: A novel P300-based brain–computer interface stimulus presentation paradigm: Moving beyond rows and columns. Clin. Neurophysiol. **121**, 1109–1120 (2010)
62. Vidaurre, C., Blankertz, B.: Towards a cure for BCI illiteracy. Brain Topogr. **23**, 194–198 (2010). DOI 10.1007/s10548-009-0121-6
63. Volosyak, I.: SSVEP-based Bremen-BCI interface – boosting information transfer rates. J. Neural Eng. **8**, 036,020 (2011). DOI 10.1088/1741-2560/8/3/036020
64. Wolpaw, J.R.: Brain-computer interfaces as new brain output pathways. J. Physiol. **579**, 623–619 (2007). DOI 10.1113/jphysiol.2006.125948
65. Wolpaw, J.R., Flotzinger, D., Pfurtscheller, G., McFarland, D.J.: Timing of EEG-based cursor control. J. Clin. Neurophysiol. **14**(6), 529–538 (1997)
66. Wolpaw, J.R., Birbaumer, N., Heetderks, W.J., McFarland, D.J., Peckham, P.H., Schalk, G., Donchin, E., Quatrano, L.A., Robinson, C.J., Vaughan, T.M.: Brain-computer interface technology: a review of the first international meeting. IEEE Trans. Rehabil. Eng. **8**, 164–173 (2000). DOI 10.1109/TRE.2000.847807
67. Wolpaw, J.R., Birbaumer, N., McFarland, D.J., Pfurtscheller, G., Vaughan, T.M.: Brain-computer interfaces for communication and control. Clin. Neurophysiol. **113**, 767–791 (2002). DOI 10.1016/S1388-2457(02)00057-3

Chapter 18
Principles of Hybrid Brain–Computer Interfaces

Gernot R. Müller-Putz, Robert Leeb, José d.R. Millán, Petar Horki, Alex Kreilinger, Günther Bauernfeind, Brendan Z. Allison, Clemens Brunner, and Reinhold Scherer

18.1 Introduction

Persons having severe disabilities for various reasons can use a wide range of assistive devices (ADs) for managing their daily needs as well as using them for communication and entertainment purposes. The set of ADs ranges from simple switches connected to a remote controller to complex sensors (e.g., mouth mouse) attached to a computer and to eye tracking systems. All of these systems work very well after being adjusted individually for each person. However, there are still situations where these systems do not work properly, e.g., in the case of fatigue of residual muscles. In such a case, a Brain–Computer Interface (BCI) might be a good option, using brain signals (most likely the electroencephalogram, EEG) for control without the need for movement.

BCIs are systems that establish a direct connection between the human brain and a computer [48], thus providing an additional communication channel. For individuals suffering from severe palsy caused by muscle dystrophy, amyotrophic lateral sclerosis (ALS), or brain stem stroke, such a BCI constitutes a possible way to communicate with the environment [5, 21, 34]. BCIs can also be used to control neuroprostheses in patients suffering from a high spinal cord injury (SCI), for example by using functional electrical stimulation (FES) for grasp restoration [28].

G. Müller-Putz (✉) · G. Bauernfeind · C. Brunner · P. Horki · A. Kreilinger · B.Z. Allison · R. Scherer
Institute for Knowledge Discovery, BCI-Lab, Graz University of Technology, Graz, Austria
e-mail: gernot.mueller@tugraz.at; g.bauernfeind@tugraz.at; clemens.brunner@tugraz.at; petar.horki@tugraz.at; alex.kreilinger@tugraz.at; allison@tugraz.at; reinhold.scherer@tugraz.at

R. Leeb · J.d.R. Millán
Chair in Non-Invasive Brain-Machine Interface, École Polytechnique Fédérale de Lausanne, Lausanne, Switzerland
e-mail: robert.leeb@epfl.ch; jose.millan@epfl.ch

B. Z. Allison et al. (eds.), *Towards Practical Brain-Computer Interfaces*, Biological and Medical Physics, Biomedical Engineering, DOI 10.1007/978-3-642-29746-5_18,
© Springer-Verlag Berlin Heidelberg 2012

Various types of BCI exist, either based on evoked activities like evoked potentials (e.g., P300, [10]), slow cortical potentials [4], steady-state evoked potentials [15,29], or based on induced activities resulting in changes of oscillatory components like the event-related (de)synchronization (ERD/ERS, [38]). Besides EEG- based BCIs, BCIs based on metabolic measures also exist, where practical relevant are near-infrared spectroscopy based BCIs (e.g., [8]).

The aim of this chapter is to introduce a new type of a BCI—the so-called hybrid BCI (hBCI). A hybrid BCI is assembled by a collection of systems that work together to provide a communication pathway between the human brain and a computer (machine). To be undoubtedly accepted as a BCI, the hybrid system must include at least one component that fulfills the basic BCI criteria: (a) it must provide volitional control, (b) it must rely on brain signals, (c) it must provide feedback, and (d) it must work online [42].

This means, the BCI should be available if the user wishes to extend the types of inputs available to an assistive technology system, but the user can also choose not to use the BCI at all. Here it is of importance that the BCI itself is running, which means online EEG analysis is performed all the time. The hBCI might decide on the one hand which input channel(s) offer the most reliable signal(s) and switch between input channels to improve information transfer rate, usability, or other factors, or on the other hand fuse various input channels.

Coming to the ultimate general idea of the hybrid BCI, in recent papers already subtypes of BCI have been shown and their functionality demonstrated. Different steps of the development of hBCI are demonstrated in subsections, beginning from the specialized to the very general hBCI:

- hBCI based on two different EEG-based BCIs
- hBCI based on EEG-based BCI and a non-EEG based BCI
- hBCI based on EEG-based BCI and another biosignal
- hBCI based on EEG-based BCI and EEG-based monitoring
- hBCI based on EEG-based BCI and other signals
- Outlook: hBCI based on EEG-based BCI and EEG-based monitoring and other biosignals.

18.2 hBCI Based on Two Different EEG-Based BCIs

18.2.1 BCIs Based on ERD and Evoked Potentials

Some hybrid BCIs combine a BCI with another BCI. Such hybrid BCIs are called "pure" hybrid BCIs. For example, a few different groups have developed hybrid BCIs that combine the P300 with other measures. Panicker et al. [37] introduced a P300 BCI in which some of the display oscillated to elicit SSVEPs. If the system did not detect SSVEP activity, then it assumed that the user was not paying attention to the BCI, and thus did not produce an output. Hence, the SSVEP activity served as a passive "brain switch." This system is only a hybrid BCI if the definition of BCI is expanded to include passive BCIs [35, 50], a terminological matter we do not pursue within this chapter.

While Panicker et al. combined a P300 BCI with an SSVEP system, Li et al. [23] instead combined a P300 BCI with a BCI based on imagined movement. This approach also differed in the overall goal, which was to move a cursor in two dimensions rather than directly spell. To move the cursor vertically, subjects focused on a particular target box; different boxes flashed that contained the word "up," "down," or "stop." The resulting P300s could move the cursor in steps. Subjects could control the horizontal position by imagining left or right hand movement.

Jin et al. [20] combined a P300 BCI with a new type of BCI based on motion visual evoked potentials (mVEPs) [16, 18, 25]. This study compared three conditions: a "P300" condition in which stimuli flashed (like a conventional BCI); an "mVEP" condition in which stimuli moved (like the new mVEP BCI) and a hybrid condition in which stimuli flashed and moved. This new hybrid condition yielded significant improvements in accuracy and information transfer rate over the other two conditions without making subjects feel tired or overwhelmed. The authors noted that further manipulations to the stimulus and task parameters could yield further improvements.

Su et al. [47] combined P300 and motor imagery activity to navigate in virtual environments using a hybrid BCI. They implemented a sequential protocol, where motor imagery of left and right hands controlled movement to the left or right, and P300 activity controlled virtual objects in a discrete way. The authors showed that users performed well with the hybrid approach, there was no performance different as compared to each single approach alone.

Another research effort combines SSVEP and ERD activity. Allison et al. [1] showed that subjects could produce both SSVEP and ERD activity within the same trial, and Brunner et al. [6] explored improved signal processing approaches with the same data. For example, these studies optimized SSVEP feature extraction and assessed the influence of ERD activity on SSVEP activity (and vice versa). These publications laid the foundations for the first online BCI that combined ERD and SSVEP activities [7]. Subjects could move a cursor in one dimension using both ERD and SSVEP measures, which could provide additional information to improve classification. While these authors did not find a significant performance increase in the hybrid condition, they could show in a follow up study that more advanced classifiers can indeed improve the hybrid condition over the simple ones. This approach was later adapted to a two dimensional BCI, in which vertical movement was controlled by imagined movement and horizontal movement by SSVEPs [2]. These studies also employed questionnaires to assess subjective factors, which found that hybrid BCIs were not substantially more difficult or annoying.

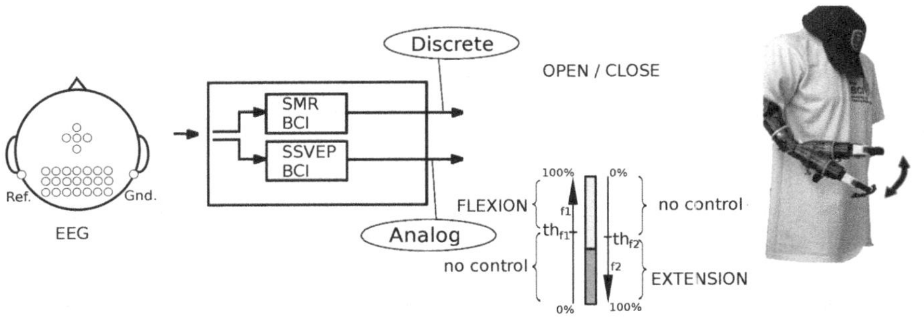

Fig. 18.1 Combined MI- and SSVEP-BCI system for control of a 2-DoF artificial upper limb. The beta rebound after brisk feet MI is used to control the grasp function, and an SSVEP-BCI the elbow function

18.2.2 Combined Motor Imagery and SSVEP Based BCI Control of a 2 DoF Artificial Upper Limb

SSVEP and ERS measures were combined in a different kind of BCI system that allowed independent simultaneous operation of these BCIs [19]. This combined system built upon previous research on restoration of hand and elbow control in spinal cord injured patients: in [40] the lateral grasp was restored in a spinal cord injured patient by sequentially switching between grasp phases by imagining foot movements; in [27, 36] healthy participants used SSVEP to control a prosthesis and an orthosis, respectively. Based on these results, the next logical step was to combine MI and SSVEP-BCIs for the independent control of the grasp and elbow functions. To this end, a control method was investigated where the MI-BCI controlled the grasp function and the SSVEP-BCI the elbow function of an artificial upper limb with 2 degrees-of-freedom (DoF). Since the SSVEP-BCIs require little or no training, a similar MI pattern was desirable in order to allow for a fast and practical set-up. Such a pattern, i.e. strong and stable without any subject training, is the post-movement beta rebound [31,41].

The combined MI- and SSVEP-BCI system is shown in Fig. 18.1. The grasp function could be toggled between opened and closed state by imagining brisk feet movement. The elbow could be gradually moved from full extension to full flexion by using the two SSVEP classes for flexion or extension, respectively. Such a hybrid design allowed the two BCIs to be operated independently with two different purposes that serve the common goal of controlling a 2 DoF artificial arm.

The combined BCI control, where subjects performed a predefined sequence of movements, is examplified in Fig. 18.2. Most of the MI-BCI commands occurred in the first 10 s following the experimenter indications (see Fig. 18.2b), with the histogram of gripper activations approximating a decaying exponential. The online SSVEP-BCI control is examplified in Fig. 18.2a showing elbow movement trajectories and the corresponding control tasks during a single run for one subject. Generally, the subjects were able to move the elbow to the desired position, but had

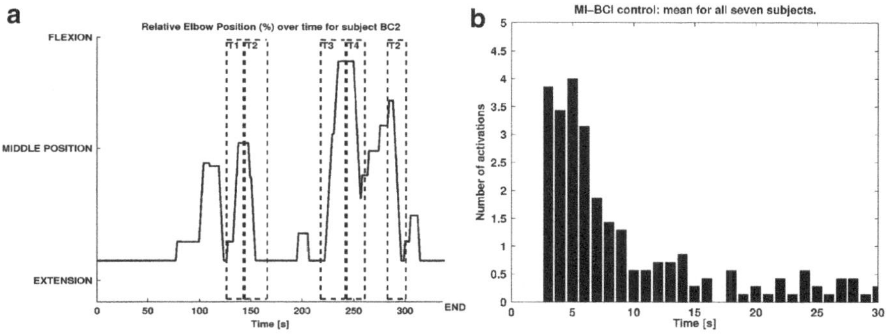

Fig. 18.2 (**a**) Shown here are the elbow movement trajectory and the corresponding control tasks during a single run for one of the subjects. The SSVEP tasks were the following: move the elbow from extended to middle position (T1); move the elbow from middle to extended position (T2); move the elbow from extended to flexed position (T3); move the elbow from extended to middle position (T4). (**b**) Shown here is the average number of gripper activations across time, for all seven subjects, as they perform the motor imagery task

difficulties sustaining the reached position due to false activations. This was also confirmed by the subjective measures, assessed through a questionnaire, which also showed a slight preference for the SSVEP control (elbow). Future work will thus focus on improving the performance of the system during the non-control periods, and on developing a fully self-paced BCI system, with the final goal of controlling hand and elbow neuroprosthesis.

18.3 hBCI Based on EEG-Based BCI and a Non-EEG Based BCI

The variety of brain signals (electrical, magnetical, metabolical) have different signal characteristics and can be used therefore for distinct functions. In this section one example of a hBCI consisting of near-infrared spectroscopy (NIRS) and EEG-based SSVEP BCI is presented here.

Self-activation is an important factor for BCI systems to become more practical and user-friendly devices [46]. This means, for a higher independency in daily use, the user should be able to switch on or off the BCI system autonomously.

In a initial study [42] we investigated the realization of an asynchronous hybrid BCI, by combining NIRS with SSVEP. Therefore, we used a one channel NIRS-system developed by our group [3] to turn on and off a 4-step electrical hand orthosis [24]. NIRS is a functional brain imaging method and allows, similar to functional magnetic resonance imaging (fMRI), to study hemodynamic changes during cortical activation. NIRS has been used to measure hemodynamic responses (changes of oxy- and deoxyhemoglobin (oxy-Hb, deoxy-Hb)) to cognitive, visual,

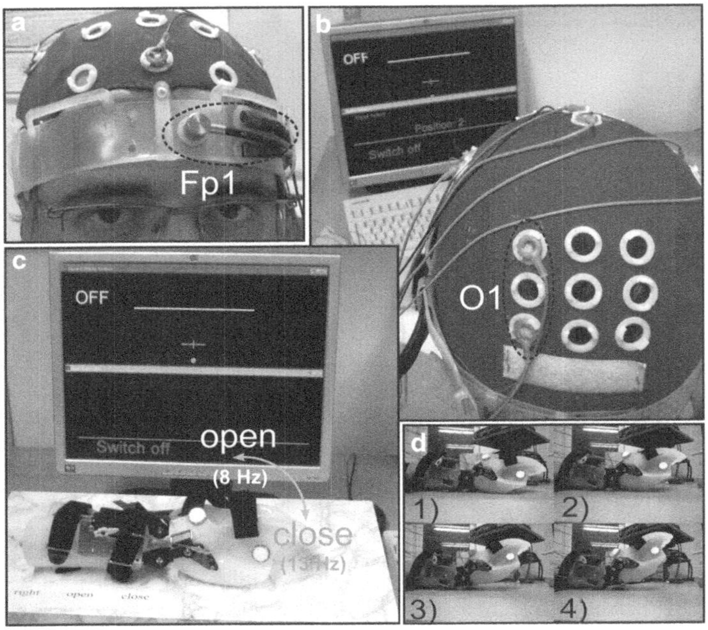

Fig. 18.3 **(a)** One channel NIRS placement over Fp1. **(b)** EEG recording to measure SSVEPs (electrode position O1). **(c)** Hybrid BCI system: Orthosis in front of the presentation screen. To open the orthosis, focused attention on an 8 Hz LED was necessary, and to close it on a 13 Hz LED. The upper part of the presentation screen displays the varying concentration change with a ball. The bold horizontal line indicates the on/off toggle switch threshold. In the lower part, the current status of the SSVEP orthosis control (on/off) and the detected command were shown. **(d)** Stepwise SSVEP orthosis control. Modified from [42]

visuomotor and motor tasks (e.g., [17, 43, 49]). One healthy subject, familiar with NIRS recording but naive using SSVEP, performed 4 runs with the hybrid system. In each run, the subject had to open and close (one activation block contained positions 0-1-2-3-2-1-0) the orthosis (Fig. 18.3d) three times, each at self-paced intervals, with 60 s breaks between the blocks (resting periods). To open the orthosis the subject had to focus on an 8 Hz flickering LED. To close it, the subject had to pay attention on a 13 Hz flickering LED (Fig. 18.3c). Only if the whole open/close sequence was finished, the resting period was initiated.

To measure SSVEPs the EEG was recorded bipolarly from electrodes placed over the occipital cortex (electrode position O1, 2.5 cm inter-electrode distance, ground Fz, Fig. 18.3b). Prior to the first block, the subject had to self-initiate the SSVEP orthosis control using the NIRS system (brain switch). To this end, the NIRS measurement was split up into 8 s periods. A pre-waiting period was included prior to the first segment, which started at second 18. Within the periods, the relative oxy-Hb concentration change (measured over position Fp1) was used as a visual feedback (green ball). The measured concentration change, referred to a 4 s baseline

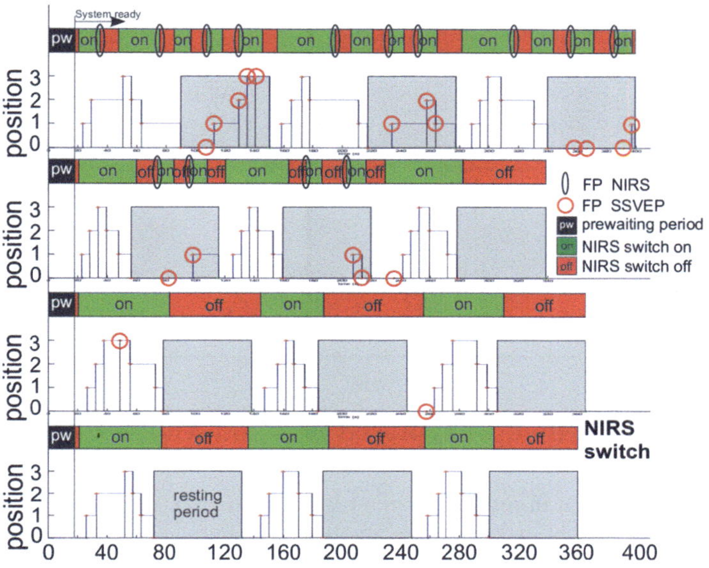

Fig. 18.4 Timing course of the 4 runs, *green areas* indicate an activated SSVEP control. The *gray areas* indicate resting periods. *Red circles* indicate FPs in SSVEP control, *black ellipsis* in NIRS. The *black areas* at the beginning of each run indicate a pre-waiting period. Modified from [42]

interval (mean concentration of the last 4 s prior the period), represents the ball position on screen. If the change exceeded a subject-specific threshold, indicated as a yellow bar on the screen (Fig. 18.3c), an on/off (off/on) state switch of the orthosis control was triggered. The threshold for the first run was selected from previous NIRS measurements of the subject and adapted after the first run to minimize false positives (FP). After each switch, no other switch command was accepted for a period of 8 s (refractory period, black screen). During the resting periods and after the last activation block, the subject was instructed to switch off the SSVEP orthosis control system by using the brain switch again to avoid FP SSVEP activations.

During the first two runs FPs were detected in the activation as well as in the resting period (NIRS and SSVEP, Fig. 18.4). In the third run the subject displayed a perfect performance with the NIRS switch and only two FP detections occurred during the SSVEP orthosis control. In the last run, the subject displayed a perfect performance with 100 % accuracy, meaning no FP in the NIRS and SSVEP control, respectively. Table 18.1 summarizes these results.

These preliminary results provide evidence that the combination of NIRS and SSVEP within a hybrid BCI system may be a viable control interface. However, using NIRS as a "brain switch" is just only one possible approach for a hBCI. In a recent publication Fazli et al. [13] investigated whether NIRS can be used also to enhance the performance of a BCI based on sensory motor rhythm. They found that in a multimodal setup the simultaneous use of NIRS and EEG features can significantly improve their classification accuracy.

Table 18.1 TP and FP detections in self-paced orthosis and NIRS control. The parameters are given for the activation as well as the resting period

	Activation period				Resting period			Time		
	NIRS		SSVEP		NIRS		SSVEP	Act. period	rest period	total period
Run	TP_a	FP_a	TP_a (min^{-1})	FP_a (min^{-1})	TP_r	FP_r	FP_a (min^{-1})	(s)	(s)	(s)
1	7	4	5.4	0.0	9	6	4.0	201.6	180.0	381.6
2	3	0	7.7	0.4	7	4	1.3	140.6	180.0	320.6
3	3	0	6.4	0.7	3	0	0.0	167.8	180.0	347.8
4	3	0	6.6	0.0	3	0	0.0	162.6	180.0	342.6
Mean	4.0	1.00	6.5	0.3	5.5	2.5	1.3	168.1	180.0	348.1
SD	2.0	2.00	1.0	0.4	3.0	3.0	1.9	25.2	0.0	25.2

18.4 hBCI Based on EEG-Based BCI and Another Biosignal

In the definition given in introduction, a hBCI can also exist of other signals than brain signals, as long brain signals are involved. The human body produces a series of other biosignals which can be controlled by the user. One example is to use an eyetracking device for cursor control but an EEG-based BCI for the target selection, recently shown by Zander et al. [51].

In this section two studies using either heart rate changes or electromyographical patterns as additional input source for the hBCI.

18.4.1 Heart Rate Changes to Power On/Off an SSVEP-BCI

The heart has a constant intrinsic rhythm with a period of about 1 s, which is modulated especially by respiration, blood pressure waves and "central commands." This means that central processes, such as, for example, motor preparation, mental simulation, stimulus anticipation and translation, can result in a heart rate (HR) response. If such a centrally induced HR response can be detected in the ongoing electrocardiogram (ECG) signal, then the HR can be used to encode messages and thus act as additional communication channel.

In an initial feasibility study to explore this prospect, we used brisk inspiration to modulate the HR [46]. The HR-triggered switch could turn the SSVEP-operated prosthetic hand on and off. We recorded the ECG and computed the HR. Changes of the HR measured in beat-to-beat intervals (RRI) were computed and used to initiate the SSVEP-BCI control. An on/off event was generated each time the relative change (dRRI), induced by brisk inspiration and exceeded the subject-specific threshold. The dRRI with the highest true positive rates during the cue-guided inspiration, and the lowest false positive detections during the remaining tasks, were selected through receiver operating analysis and used as basis for the online experiments. Four light emitting diodes were affixed on the hand prosthesis, each flickering at a different frequency between 6.3 and 17.3 Hz (stimulation frequency).

The EEG was recorded bipolarly from EEG electrodes placed 2.5 cm anterior and posterior to electrode position O2. The harmonic sum decision algorithm [30] was used for the SSVEP classification. The flickering light source with the highest harmonic sum within a given time period triggered the prosthetic hand movement. A typical selection time period of about 1.5 s was estimated empirically for each subject. The online experiment used to evaluate the performance of the HR-switch lasted about 30 min. Subjects were verbally instructed to turn on the SSVEP-BCI, perform a pre-defined motion sequence with the prosthetic hand and then turn the BCI off. The motion sequence to be performed was:

- O: open the hand
- L: rotate the hand 90° to the left
- R: rotate the hand 90° to the right
- C: close the hand
- R: rotate the hand 90° to the right
- O: open the hand
- C: close the hand, and
- L: rotate 90° left, back to the original position.

The whole sequence had to be performed four times within 30 min. The start time of each sequence was randomly chosen by the experimenter, who talked to the subjects between the motor sequences. Subjects succeeded in switching on and off the BCI by brisk inspiration and operating the SSVEP-actuated hand prosthesis. Eight true positive HR switches were required to turn the BCI on and off for the four movement trials. The average number of false positive RRI detections was 2.9. The average number of erroneous (true negative) RRI detections was 4.9. The average selection speed for one out of the four SSVEP classes was about 9.5 s (6.3 commands per minute). On average, one SSVEP detection per minute was erroneous. These results, based on ten able-bodied subjects, suggest that transient HR changes, induced by brisk inspiration, are feasible signals in a hybrid BCI.

18.4.2 Fusion of Brain and Muscular Activities

Practical BCIs for disabled people should allow them to exploit *all* their remaining functionalities as control possibilities so that they can use the *currently best* available ones at any time. Sometimes these people have residual activity of their muscles, most likely in the morning when they are not exhausted. Therefore, in our hybrid BCI framework we can combine electroencephalographic and electromyographic (EMG) activities, whereby both channels are fused to produce a more robust control signal (see Fig. 18.5a). Indeed, subjects were able to achieve a good control of their hybrid BCI independently of their level of muscular fatigue and the fused condition yielded a more stable performance compared to the single modalities [22].

Twelve healthy subjects participated in synchronous BCI recordings, whereby repetitive left and right hand motor execution (depending on a visual cue) was carried out over a period of 5 s (resulting in 60 trials per class). The recorded

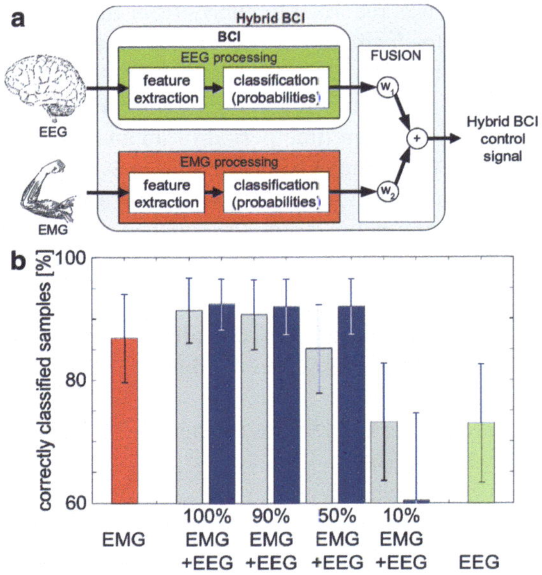

Fig. 18.5 (**a**) Fusion principle of muscular and brain activities. (**b**) Performance result over the six conditions (mean ± SD of correctly classified samples over the task period). The *outer bars* represent the single modalities (EMG: *leftmost/red*; EEG: *rightmost/green*). The *middle bars* correspond to the fused modalities with different levels remaining EMG amplitude (100 %–10 %). For each of these conditions we provide two performances according to the fusion modality: simple fusion (*left/light grey*) and Bayesian fusion (*right/dark blue*)

brain and muscular activities were separately processed and classified. (a) Four EMG channels were recorded over the flexor and extensor of the left and right forearm. The prehensile EMG activities were rectified and averaged (0.3 s) to get the envelopes. The resulting features were subject-specific thresholded, normalized and classified based on maximum distance. (b) The brain activity was acquired via 16 EEG channels over the motor cortex. From the Laplacian filtered EEG the power spectral density was calculated and the selected features were classified with a Gaussian classifier [14, 26]. The evidence about the executed task was temporary accumulated (exponential smoothing), provided the confidence was above a rejection threshold. (c) Finally the two classifier probabilities were fused together in order to generate one control signal. In this work we explored two classifier fusion techniques. In the first approach the fusion weights were equally balanced between the two classifiers, while in the second one we adopted a naïve Bayesian fusion approach [44].

The performances of either one modality alone (EEG or EMG) or the fusion of both were compared based on the correctly classified samples over the task period (0–5 s after the cue). Furthermore, to simulate fatigue of exhausted muscles, the amplitudes of the EMG channel were degraded over the run time (attenuation from 10 % up to 100 %) [9], so that the EEG activity became more and more important in the fusion. Importantly, however, the same classifier weights for EEG and EMG and the same fusion rules were kept over all conditions. This simulates the realistic situation of a patient who becomes more and more fatigued over the day.

Figure 18.5b shows that the subjects could achieve a good control of their hybrid BCI independently of their level of muscular fatigue. Furthermore, although EMG alone yields good performance, it is outperformed by the hybrid fusion of EEG and EMG. Remarkably, thanks to the fusion, increasing muscular fatigue led to a moderate and graceful degradation of performance. Such a system allows a very reliable control and a smooth handover, even though the subjects is getting more and more exhausted or fatigued during the day. In more detail, the Bayesian fusion outperformed the simple one, except in the case of 90 % attenuation [22]. The reason is that the assumption of stable input patterns while setting up the Bayesian confusion matrices were violated and the performance dropped.

In summary, the experiment demonstrated the benefits of a hybrid BCI: (a) Multimodal fusion techniques allow the combination of brain control with other residual motor control signals and thereby achieve better and more reliable performances. (b) Increasing muscular fatigue led only to a moderate and graceful degradation of performance compared to the non-fatigued case. (c) The Bayesian fusion approach led to a very constant behavior over a wide range of muscular fatigue, compared to the steadily decreasing performance in case of the simple fusion (see also [22]).

In our future work we will adapt dynamically the way of weighting the contribution of the single modalities. These weights reflect the reliability of the channels, or the confidence/certainty the system has on its outputs. Generally these weights can be estimated from supervision signals such as cognitive mental states (e.g., fatigue, error potentials) and physiological parameters (e.g., muscular fatigue). Another source to derive the weights is to analyze the performance of the individual channels in achieving the task (e.g., stability over time, influence of noise . . .).

Finally, patients with progressive loss of muscular activity (as in muscular dystrophy, amyotrophic lateral sclerosis and spinal muscular atrophies) could benefit from such a hybrid BCI with dynamic fusion. For example, during early hybrid BCI training the user could still exploit her/his residual motor functions, while with increasing long-term use of the assistive product the transition between the hybrid assistive device and pure BCI (when muscular activity is too weak to operate them) would be smooth.

18.5 hBCI Based on EEG-Based BCI and EEG-Based Monitoring

18.5.1 Simultaneous Usage of Motor Imagery and Error Potential

Like many other interaction modality based on physiological signals Brain–Computer Interfaces based on motor imagery are unfortunately prone to errors in the recognition of subject's intent. In contrast to the other interaction modalities, a unique feature of the "brain channel" is that it conveys both information, from which we can derive mental control commands to operate as well as information about cognitive signals like the awareness of erroneous responses [45]. Therefore, an elegant approach to improve the accuracy of BCIs consists of a verification procedure directly based on the presence of error-related potentials (ErrP) in the EEG recorded right after the occurrence of an error [11].

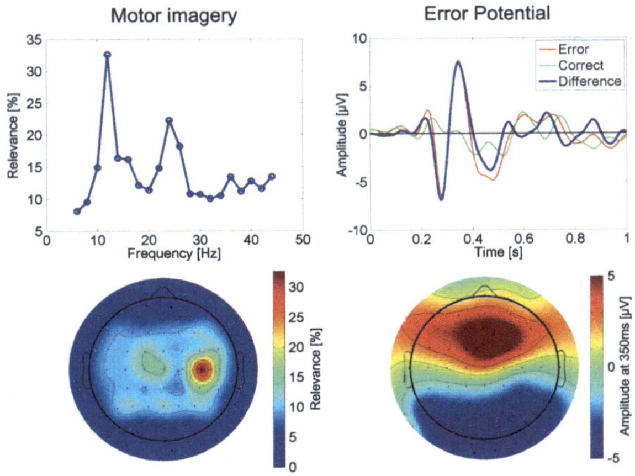

Fig. 18.6 (*Left column*) Features relevance for motor imagery classification (one subject): Discriminant Power of frequencies (*top left*) and of electrodes (*bottom left*). (*Right column*) Error Potential detection (one subject): Grand averages (*top right*) of error trials, correct trials and the difference between them (channel Cz). Scalp potential topography (*bottom right*) at the peak occurring 350 ms after the feedback presentation

Such a simultaneously detection of erroneous responses of the interface and classification of motor imagery at the level of single trials in a real-time BCI system was presented in [12]. Two subjects had the task to bring a squared cursor to targets located three steps away. Left and right movements of the cursor were achieved via MI, analyzed over the last second. After the response of the BCI (i.e., a step bringing the cursor closer to or farther away from the target), a 400 ms window was used to detect the presence of an ErrP. If an ErrP was detected the last erroneous step was cancelled. Figure 18.6 shows the used features of both BCIs, the discriminant power of the frequencies and channels in case of the MI-BCI and the time course and topographic average in case of the ErrP-BCI. The analysis showed that the BCI error rate without the integration of ErrP detection was around 32% for both subjects. However, when integrating ErrP detection, the averaged online error rate dropped to 7%, which would yield an increase of the bit rate above 200 %. For more details see [12].

These results confirm that it is possible to simultaneously control a brain-controlled device (via motor imagery) as well as to extract the error-related potentials of this interaction and combined the outcome of both. The combined (hybrid) BCI approach improves the quality of the brain–computer interaction, although neither of these two input channels is perfect.

18.6 hBCI Based on EEG-Based BCI and Other Signals

18.6.1 Combination of an EEG-Based BCI and a Joystick

Persons with remaining muscle functions can use these muscles directly to control assistive devices. The derived control signal is not depending on biosignals like EMG, which have to be translated into control commands, but on the functionality of the muscles themselves. This functionality can suffer from tremors, spasms, and fatigue, especially when used over a long time period. To compensate these issues BCI, as an alternative which is not relying on muscular activity, could take over control in case of a reduced functionality. Fatigue of muscles may occur faster than when using EMG because muscles have to be contracted strong enough to cause movement, whereas EMG signals can be detected even with a weaker activation. However, the use of remaining muscle functions has the benefit of a more natural sense and delivers immediate feedback. A possible scenario deals with the combination of manual control via a joystick (JS) and BCI. Here, a switch to BCI can be used to restore control over an assistive system as soon as the JS is not working any more due to fatigue of the muscles. To study this combination, ten healthy subjects were asked to control a car game with JS and BCI. In the game the goal was to collect coins and avoid barriers. Both input signals were monitored and a fusion system could switch between them in case of bad signals.

A possible scenario deals with the combination of manual control via a joystick (JS) and BCI. Here, a switch to BCI can be used to restore control over an assistive system as soon as the JS is not working any more due to fatigue of the muscles, which can happen after a long time of usage. To study this combination, ten healthy subjects were asked to control a car game with JS and BCI. In the game the goal was to collect coins and avoid barriers. Both input signals were monitored and a fusion system could switch between them in case of bad signals.

EEG for BCI mode was recorded with six electrodes over C3, Cz, and C4. The data was sampled with 512 Hz and filtered between 0.5 and 30 Hz. The JS was controlled manually but was affected with randomly occurring artificial spasms and tremors and was deteriorated increasingly over time with weakness, resulting in a reduction of the range of movement. The task used for BCI was based on motor imagery [33]. A classifier was trained [39] to distinguish between two classes: MI of the right hand versus both feet.

Online, both signals were monitored individually with four quality measures. These measures weighted the currently active control mode and adapted the specific quality rating accordingly. A quality rating below 20 % induced a switch to the other mode, provided that the quality of the other signal was above 50 %. A measure decreased the quality when active but could recover otherwise. BCI measures monitored noise, instability and invariability of the classifier, and bias. JS measures monitored shaking, low amplitude, invariability, and also bias. Noise and shaking had the strongest impact on the quality, reducing it with 10 %/s. All the measures can be seen in Table 18.2.

After setting up the BCI classifier, subjects performed 2 runs with only JS and then 6 runs with JS + BCI with the car game. One run consisted of 40 trials during which six coins appeared on one of the street sides, accompanied by six barriers on the opposite side. Subjects were asked to collect the coins while avoiding the barriers. A coin collection increased the score +1, whereas a barrier decreased it by −1. When using only JS, the weakness reached its maximum (no more reaching of coins possible) after 30 trials. In JS + BCI mode this value was reached already

Table 18.2 Quality measures for both control modes, BCI and JS.The measures either decrease the quality (100 %—numbers in the second and fifth column) when they are currently detected but also recover over time otherwise (third and sixth column)

BCI			JS		
	QL↑↓ $\frac{\%}{s}$			QL↑↓ $\frac{\%}{s}$	
Measures	↓	↑	Measures	↓	↑
EMG noise	10	−3	Shaking	10	−2
Instability	5	−1	Low Amplitude	2	−4
Invariability	1	−4	Invariability	1	−4
Bias	∝Bias	∝Bias	Bias	∝Bias	∝Bias

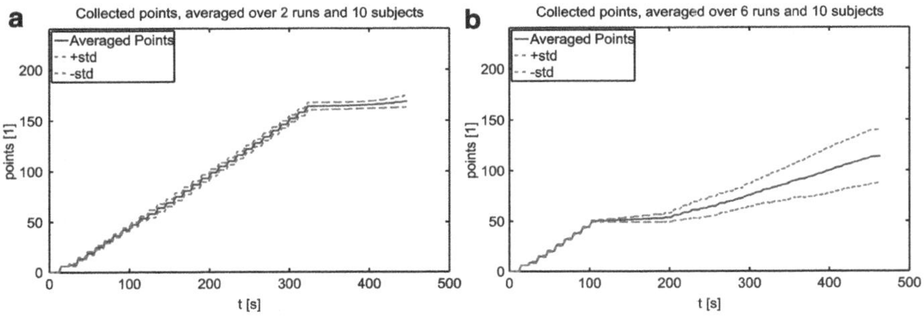

Fig. 18.7 The *left figure* shows the averaged score over 2 runs from 10 subjects during "JS only" mode. As soon as the maximum weakness was reached after 30 trials, collecting coins was only possible with forced overshooting. The *right figure* shows the averaged collection of points during six averaged runs over all 40 trials. The first points were always collected with the JS which was weakened within the first ten trials up to a point when no more collection was possible. After this stagnation, the fusion initiated the first switch to BCI and would continue to monitor both input qualities and decide which control method was best at the moment. (**a**) JS only, (**b**) JS + BCI

after ten trials. In JS + BCI mode each run started in JS mode. However, since the JS signal was deteriorated, a switch to BCI was imminent after the first ten trials. After that, subjects could either stay in BCI mode or, in case of a bad BCI quality, return to JS mode again. Switching back to a control mode was possible since measures could recover individually, as seen in Table 18.2, and generally by 1 %/s if the other control mode was currently active.

The measures affecting the quality were called long term quality measures. Short term quality measures were applied additionally. These measures were used only to give immediate feedback about strong impairments like heavy noise or shaking. In case of these effects, control was inhibited totally and subjects were not able to move the car any more, it was fixed to the middle of the street.

The effect of the fusion was that low quality signals were discarded soon in favor of the other signal. Subjects were still able to collect coins after the JS signal was too weak to reach any coins. Figure 18.7 shows the scores for all 10 subjects. The maximum number of points per run was 240. There was also a trend indicating that good BCI performers tended to stay in BCI mode for a longer time, $r = 0.6$ ($p = 0.09$), when removing one outlier with a strong classifier bias.

For patients, problems like fatigue and other factors which deteriorate the functionality of assistive devices are highly anticipated. Therefore, the introduced fusion of two signals and switching between them according to the current quality might become very useful in the near future. One major drawback, however, is the need for a very specific adaptation of the measures. These measures need to be carefully adjusted to provide a meaningful quality prediction of the monitored signals. As an example, one of the subjects had a heavily biased classifier output in BCI mode which resulted in a bad BCI performance. However, the measure weighting the bias was not set high enough to reduce the quality of BCI. Still, if these measures are adapted carefully, this kind of fusion offers a strong boost in functionality and can also be expanded to deal with more than two signals, or to allow more complex fusion rules, like a combination of inputs [22] depending on the individual qualities of the involved signals.

18.7 Outlook: hBCI Based on EEG-Based BCI and EEG-Based Monitoring and Other Biosignals

The idea of having a hybrid solution is not entirely new. In this chapter and in a recent work by Pfurtscheller et al. [42], an overview of existing hybrid BCIs is given. However, they all combine a BCI with another BCI or a BCI with another biosignal or a BCI with another signal.

Having an assistive technology system, which consists of a BCI, such a system must be able to reliably work most of the time during a day. Therefore, also monitoring (see Sect. 18.5.1) as well as adaptive classifiers have to be introduced in such a hBCI system. In Fig. 18.8 such a system is shown. In addition to the EEG-based BCI, there are other input and control signals shown. These include other biosignals as well as signals from manual controls such as from ADs (e.g., mouth mouse, push buttons …). Furthermore, mental monitoring gives insights about, e.g., the tiredness of the patient. The "fusion" generates a new control signal out of all inputs. Besides a quality check (e.g., artifact detection), those signals will be weighted and fused to a control signal, or the most reliable one will be chosen. In the so-called "shared control," sensor signals from the application (neuroprosthesis, software, assistive robot) will also be included and used to generate an accurate final control signal (see Chaps. 6 and 9).

All of these parts have been demonstrated already in several studies (many of them shown in this chapter), but in the near future one complete system has to be created (more details see [32]).

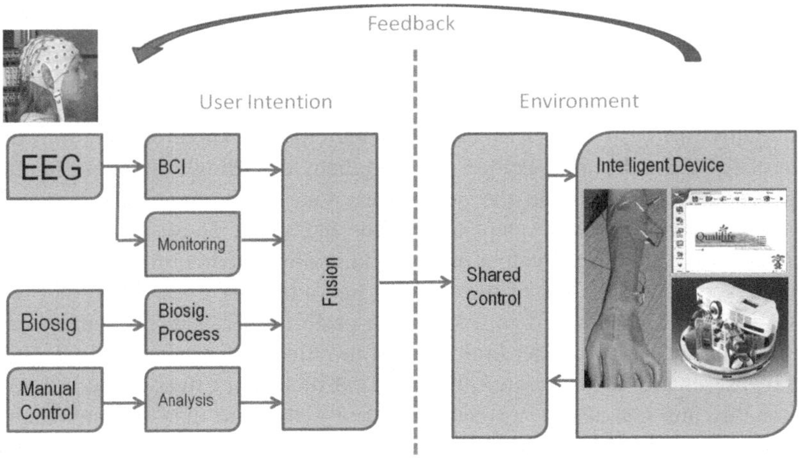

Fig. 18.8 General principle of an hBCI

18.8 Conclusion and Future Work

In conclusion, even though hybrid BCIs are a fairly new research endeavor, many different kinds of hybrid BCIs have been introduced and validated. Hybrid BCIs may use a variety of different input signals, from different sources that are measured in different ways, and may combine these signals to accomplish a variety of goals. Combining BCIs with physiological measures of alertness or errors, could lead to more user-friendly interfaces that adapt according to the user's state.

Hybrid BCIs could benefit users in three general ways. First, hybrid BCIs can extend the capabilities of current BCIs by "pushing the envelope" such as by allowing users to control more dimensions of movement or send otherwise unavailable command combinations. Second, hybrid BCIs can make human–computer interaction more intuitive and adaptive. BCIs could provide new options to a user who is tired or just made a mistake, or could turn themselves off if a user is not interested. Third, hybrid BCIs can help make modern BCIs and ADs practical for a wider variety of users. For example, hybrid BCIs can reduce errors, possibly enough to enable control in a previously illiterate user, or provide a communication option when other channels are unavailable due to fatigue or other factors. These are all significant benefits, and we expect that hybrid BCI research will remain a promising research direction. Eventually, most BCIs in real world settings will be hybrid BCIs.

Acknowledgements This work is partly supported by the European ICT programme projects TOBI: Tools for Brain–Computer Interaction (FP7-224631) and fBNCI: Future Directions for Brain/Neuronal Computer Interaction (FP7-248320). Also, parts were supported by the "Land Steiermark" (project A3-22.N-13/2009-8) and the NeuroCenterStyria. This paper only reflects the authors' views and funding agencies are not liable for any use that may be made of the information contained herein.

References

1. Allison, B.Z., Brunner, C., Kaiser, V., Müller-Putz, G.R., Neuper, C., Pfurtscheller, G.: Toward a hybrid brain–computer interface based on imagined movement and visual attention. J Neural Eng. **7**, 026,007 (2010). DOI 10.1088/1741-2560/7/2/026007
2. Allison, B.Z., Brunner, C., Altstätter, C., Wagner, I., Grissmann, S., Neuper, C.: A hybrid ERD/SSVEP BCI for continuous simultaneous two dimensional cursor control. J. Neurosci. Methods, in press. DOI 10.1016/j.jneumeth.2012.06.022
3. Bauernfeind, G., Leeb, R., Wriessnegger, S., Pfurtscheller, G.: Development, set-up and first results of a one-channel near-infrared spectroscopy system. Biomed. Tech. (Berl.) **53**, 36–43 (2008). DOI 10.1515/BMT.2008.005
4. Birbaumer, N.: Operant control of slow cortical potentials: a tool in the investigation of the potentials' meaning and its relation to attentional dysfunction. In: Elbert, T., Rockstroh, B., Lutzenberger, W., Birbaumer, N. (eds.) Self-regulation of the brain and behaviour, pp. 227–239. Springer, New York (1984)
5. Birbaumer, N., Ghanayim, N., Hinterberger, T., Iversen, I., Kotchoubey, B., Kübler, A., Perelmouter, J., Taub, E., Flor, H.: A spelling device for the paralysed. Nature **398**, 297–298 (1999). DOI 10.1038/18581
6. Brunner, C., Allison, B.Z., Krusienski, D.J., Kaiser, V., Müller-Putz, G.R., Pfurtscheller, G., Neuper, C.: Improved signal processing approaches in an offline simulation of a hybrid brain–computer interface. J. Neurosci. Methods **188**, 165–173 (2010). DOI 10.1016/j.jneumeth.2010.02.002
7. Brunner, C., Allison, B.Z., Altstätter, C., Neuper, C.: A comparison of three brain–computer interfaces based on event-related desynchronization, steady state visual evoked potentials, or a hybrid approach using both signals. J. Neural Eng. **8**, 025,010 (2011). DOI 10.1088/1741-2560/8/2/025010
8. Coyle, S., Ward, T., Markham, C., McDarby, G.: On the suitability of near-infrared (NIR) systems for next-generation brain–computer interfaces. Physiol. Meas. **25**, 815–822 (2004)
9. Dimitrova, N.A., Dimitrov, G.V.: Interpretation of EMG changes with fatigue: facts, pitfalls, and fallacies. J. Electromyogr. Kinesiol. **13**(1), 13–36 (2003). DOI 10.1016/S1050-6411(02)00083-4, http://www.sciencedirect.com/science/article/B6T89-47DPS3S-1/2/b0fc595d474b4418ed7ded7fa4ecc746
10. Donchin, E., Spencer, K.M., Wijesinghe, R.: The mental prosthesis: assessing the speed of a P300-based brain–computer interface. IEEE Trans. Neural Syst. Rehabil. Eng. **8**, 174–179 (2000). DOI 10.1109/86.847808
11. Ferrez, P.W., Millán, J.: Error-related EEG potentials generated during simulated brain–computer interaction. IEEE Trans. Biomed. Eng. **55**, 923–929 (2008a). DOI 10.1109/TBME.2007.908083
12. Ferrez, P.W., Millán, J.: Simultaneous real-time detection of motor imagery and error-related potentials for improved BCI accuracy. In: Proceedings of 4th International Brain–Computer Interface Workshop and Training Course, Graz, Austria (2008b)
13. Fazli, S., Mehnert, J., Steinbrink, J., Curio, G., Villringer, A., Müller, K.R., Blankertz, B.: Enhanced performance by a Hybrid NIRS-EEG Brain Computer Interface. Neuroimage **59**(1), 519–529 (2011).
14. Galán, F., Nuttin, M., Lew, E., Ferrez, P.W., Vanacker, G., Philips, J., Millán, J.D.R.: A brain-actuated wheelchair: Asynchronous and non-invasive brain–computer interfaces for continuous control of robots. Clin. Neurophysiol. **119**(9), 2159–2169 (2008). http://dx.doi.org/10.1016/j.clinph.2008.06.001

15. Gao, X., Xu, D., Cheng, M., Gao, S.: A BCI-based environmental controller for the motion-disabled. IEEE Trans. Neural Syst. Rehabil. Eng. **11**, 137–140 (2003). DOI 10.1109/TNSRE. 2003.814449
16. Guo, F., Hong, B., Gao, X., Gao, S.: A brain–computer interface using motion-onset visual evoked potential. J. Neural Eng. **5**, 477–485 (2008). DOI 10.1088/1741-2560/5/4/011
17. Herrmann, M.J., Ehlis, A.C., Wagener, A., Jacob, C.P., Fallgatter, A.J.: Near-infrared optical topography to assess activation of the parietal cortex during a visuo-spatial task. Neuropsychologia **43**, 1713–1720 (2005)
18. Hong, B., Guo, F., Liu, T., Gao, X., Gao, S.: N200-speller using motion-onset visual response. Clin. Neurophysiol. **120**, 1658–1666 (2009). DOI 10.1016/j.clinph.2009.06.026
19. Horki, P., Solis-Escalante, T., Neuper, C., Müller-Putz, G.: Combined motor imagery and SSVEP based BCI control of a 2 DoF artificial upper limb. Med. Biol. Eng. Comput. Vol. **49**(5), 567–577 (2011). DOI 10.1007/s11517-011-0750-2
20. Jin, J., Allison, B.Z., Wang, X., Neuper, C.: A combined brain-computer interface based on P300 potentials and motion-onset visual evoked potentials. J. Neurosci. Methods **205**(2), 265–276 (2012)
21. Kübler, A., Furdea, A., Halder, S., Hammer, E.M., Nijboer, F., Kotchoubey, B.: A brain–computer interface controlled auditory event-related potential (P300) spelling system for locked-in patients. Ann. N Y Acad. Sci. **1157**, 90–100 (2009). DOI 10.1111/j.1749-6632.2008. 04122.x
22. Leeb, R., Sagha, H., Chavarriaga, R., Millán, J.: A hybrid brain–computer interface based on the fusion of electroencephalographic and electromyographic activities. J. Neural Eng. **8**(2), 025,011 (2011)
23. Li, Y., Long, J., Yu, T., Yu, Z., Wang, C., Zhang, H., Guan, C.: An EEG-based BCI system for 2-D cursor control by combining mu/beta rhythm and P300 potential. IEEE Trans. Biomed. Eng. **57**, 2495–2505 (2010). DOI 10.1109/TBME.2010.2055564
24. Linortner, P., Ortner, R., Müller-Putz, G.R., Neuper, C., Pfurtscheller, G.: Self-paced control of a hand orthosis using SSVEP-based BCI. In: Proceedings of the13th International Conference on Human–Computer Interaction (2009)
25. Liu, T., Goldberg, L., Gao, S., Hong, B.: An online brain–computer interface using non-flashing visual evoked potentials. J. Neural Eng. **7**, 036,003 (2010). DOI 10.1088/1741-2560/ 7/3/036003
26. Millán, J., Ferrez, P.W., Galán, F., Lew, E., Chavarriaga, R.: Non-invasive brain-machine interaction. Intern. J. Pattern Recognit. Artif. Intell. **22**(5), 959–972 (2008)
27. Müller-Putz, G.R., Pfurtscheller, G.: Control of an electrical prosthesis with an SSVEP-based BCI. IEEE Trans. Biomed. Eng. **55**, 361–364 (2008). DOI 10.1109/TBME.2007.897815
28. Müller-Putz, G.R., Scherer, R., Pfurtscheller, G., Rupp, R.: EEG-based neuroprosthesis control: a step towards clinical practice. Neurosci. Lett. **382**, 169–174 (2005)
29. Müller-Putz, G.R., Scherer, R., Neuper, C., Pfurtscheller, G.: Steady-state somatosensory evoked potentials: suitable brain signals for brain–computer interfaces? IEEE Trans. Neural Syst. Rehabil. Eng. **14**, 30–37 (2006). DOI 10.1109/TNSRE.2005.863842
30. Müller-Putz, G.R., Eder, E., Wriessnegger, S.C., Pfurtscheller, G.: Comparison of DFT and lock-in amplifier features and search for optimal electrode positions in SSVEP-based BCI. J. Neurosci. Meth. **168**, 174–181 (2008). DOI 10.1016/j.jneumeth.2007.09.024
31. Müller-Putz, G.R., Kaiser, V., Solis-Escalante, T., Pfurtscheller, G.: Fast set-up asynchronous brain-switch based on detection of foot motor imagery in 1-channel EEG. Med. Biol. Eng. Comput. **48**, 229–233 (2010). DOI 10.1007/s11517-009n-0572-7
32. Müller-Putz, G.R., Breitwieser, C., Cincotti, F., Leeb, R., Schreuder, M., Leotta, F., Tavella, M., Bianchi, L., Kreilinger, A., Ramsay, A., Rohm, M., Sagebaum, M., Tonin, L., Neuper, C., del R Millán, J.: Tools for brain–computer interaction: a general concept for a hybrid BCI (hBCI). Frontiers in Neuroinformatics **5**, 30 (2011, in revision)
33. Neuper, C., Wörtz, M., Pfurtscheller, G.: ERD/ERS patterns reflecting sensorimotor activation and deactivation. In: Neuper, C., Klimesch, W. (eds.) Event-Related Dynamics of Brain Oscillations, Progress in Brain Research, vol. 159, chap. 14, pp. 211–222. Elsevier, Netherlands (2006). DOI 10.1016/S0079-6123(06)59014-4

34. Nijboer, F., Furdea, A., Gunst, I., Mellinger, J., McFarland, D.J., Birbaumer, N., Kübler, A.: An auditory brain–computer interface (BCI). Neurosci. Lett. **167**, 43–50 (2008)
35. Nijholt, A.: BCI for games: a "state of the art" survey. In: Stevens, S., Saldamarco, S. (eds.) Entertainment Computing – ICEC 2008, pp. 225–228. Springer, Berlin/Heidelberg (2009). DOI 10.1007/978-3-540-89222-9_29
36. Ortner, R., Allison, B., Korisek, G., Gaggl, G., Pfurtscheller, G.: An SSVEP BCI to control a hand orthosis for persons with tetraplegia. IEEE. Trans. Neural Syst. Rehabil. Eng. **19**(1), 1–5 (2011)
37. Panicker, R.C., Puthusserypady, S., Sun, Y.: An asynchronous P300 BCI with SSVEP-based control state detection. IEEE Trans. Biomed. Eng. **58**, 1781–1788 (2011). DOI 10.1109/TBME. 2011.2116018
38. Pfurtscheller, G., Lopes da Silva F.H.: Event-related desynchronization (ERD) and event-related synchronization (ERS). In: Electroencephalography: basic principles, clinical applications and related fields. Williams & Wilkins, Philadelphia (2005)
39. Pfurtscheller, G., Neuper, C.: Motor imagery and direct brain–computer communication. Proc. IEEE **89**, 1123–1134 (2001). DOI 10.1109/5.939829
40. Pfurtscheller, G., Wörtz, M., Supp, G., da Silva, F.H.L.: Early onset of post-movement beta electroencephalogram synchronization in the supplementary motor area during self-paced finger movement in man. Neurosci. Lett. **339**, 111–114 (2003)
41. Pfurtscheller, G., Leeb, R., Friedman, D., Slater, M.: Centrally controlled heart rate changes during mental practice in immersive virtual environment: a case study with a tetraplegic. Int. J. Psychophysiol. **68**, 1–5 (2008). DOI 10.1016/j.ijpsycho.2007.11.003
42. Pfurtscheller, G., Allison, B.Z., Brunner, C., Bauernfeind, G., Solis-Escalante, T., Scherer, R., Zander, T.O., Müller-Putz, G., Neuper, C., Birbaumer, N.: The hybrid BCI. Front. Neurosci. **4**, 30 (2010a). DOI 10.3389/fnpro.2010.00003
43. Pfurtscheller, G., Bauernfeind, G., Wriessnegger, S.C., Neuper, C.: Focal frontal (de)oxyhemoglobin responses during simple arithmetic. Int. J. Psychophysiol. **76**, 186–192 (2010b). DOI 10.1016/j.ijpsycho.2010.03.013
44. Rogova, G.L., Nimier, V.: Reliability in information fusion: literature survey. In: Proc. of. the 7th Intl. Conference on Informatin Fusion, Stockholm. pp. 1158–1165 (2004)
45. Schalk, G., Wolpaw, J.R., McFarland, D.J., Pfurtscheller, G.: EEG-based communication: Presence of an error potential. Clin. Neurophysiol. **111**(12), 2138–2144 (2000)
46. Scherer, R., Müller-Putz, G.R., Pfurtscheller, G.: Self-initiation of EEG-based brain–computer communication using the heart rate response. J. Neural Eng. **4**, L23–L29 (2007). DOI 10.1088/ 1741-2560/4/4/L01
47. Su, Y., Qi, Y., Luo, J.X., Wu, B., Yang, F., Li, Y., Zhuang, Y.T., Zheng, X.X., Chen, W.D.: A hybrid brain–computer interface control strategy in a virtual environment. J. Zhejiang Univ. Sci. C. **12**, 351–361 (2011). DOI 10.1631/jzus.C1000208
48. Wolpaw, J.R., Birbaumer, N., McFarland, D.J., Pfurtscheller, G., Vaughan, T.M.: Brain-computer interfaces for communication and control. Clin. Neurophysiol. **113**, 767–791 (2002). DOI 10.1016/S1388-2457n(02)00057-3
49. Wriessnegger, S.C., Kurzmann, J., Neuper, C.: Spatio-temporal differences in brain oxygenation between movement execution and imagery: a multichannel near-infrared spectroscopy study. Int. J. Psychophysiol. **67**, 54–63 (2008). DOI 10.1016/j.ijpsycho.2007.10.004
50. Zander, T.O., Kothe, C.: Towards passive brain–computer interfaces: applying brain–computer interface technology to human-machine systems in general. J. Neural Eng. **8**, 025,005 (2011). DOI 10.1088/1741-2560/8/2/025005
51. Zander, T.O., Gaertner, M., Kothe, C., Vilimek, R.: Combining eye gaze input with a brain–computer interface for touchless human–computer interaction. Int. J. Hum. Comput. Interaction **27**, 38–51 (2011)

Chapter 19
Non-visual and Multisensory BCI Systems: Present and Future

Isabella C. Wagner, Ian Daly, and Aleksander Väljamäe

19.1 Introduction

During the past decade, brain–computer interfaces (BCIs) have developed rapidly, both in terms of their application and the technologies themselves. However, most of these interfaces rely on the visual modality for providing users with control and feedback signals. Only a few research groups have been studying non-visual BCIs, primarily based on auditory and, rarely, on somatosensory signals.

For severely disabled patients with poor vision, non-visual BCI approaches may be the only option. For example, Jacobs and colleagues [31] showed that vision deterioration is an inevitable aspect of the later stages of amyotrophic lateral sclerosis (ALS), a common target user-group for BCIs (for example, see [39, 62, 71]). Gradually decreasing, or even complete loss of eye-movement control prevents the use of common BCI technologies that rely on visual displays and spatial vision [46]. Similarly, many potential BCI users can have cortical or subcortical lesions, which may lead to neuropsychological conditions such as hemineglect or agnosia that make it difficult or even impossible to focus attention on visual stimuli. For non-visually impaired BCI users, there are strong neurophysiological reasons to use multisensory BCIs. In the last two decades multisensory research has clearly demonstrated that human perception and cognition is largely multisensory [19] which may have important implications for future BCI systems development.

The shift from the traditional unisensory view on brain sensory processing towards a multisensory one can have a strong impact on a number of different applications. The benefits may extend from joint processing of brain signals from different multisensory and unisensory modalities to an amodal, multisensory oriented design of information and communication technology (ICT) applications.

I.C. Wagner · I. Daly · A. Väljamäe (✉)
Institute for Knowledge Discovery, BCI-Lab, Graz University of Technology, Graz, Austria
e-mail: i.wagner@donders.ru.nl; ian.daly@tugraz.at; aleksander.valjamae@gmail.com

B. Z. Allison et al. (eds.), *Towards Practical Brain-Computer Interfaces*, Biological and Medical Physics, Biomedical Engineering, DOI 10.1007/978-3-642-29746-5_19,
© Springer-Verlag Berlin Heidelberg 2012

In such a design, the modality-specific properties of our perception can define the necessary quality of each sensory modality when providing cues in a multisensory display [73]. For example, the auditory system allows for the presentation of omnidirectional sound cues outside of the visual field. Therefore, the categories of space, time, single events or desired affective responses may serve as primary amodal parameters when designing multisensory displays. Such a multisensory design then may offer new opportunities for the perceptual optimization of BCI displays while reducing the users' sensory load.

The aim of this chapter is two-fold. First, we will review non-visual and multimodal BCI systems that have been reported so far. We will concentrate on EEG-based BCI systems, mentioning other brain imaging studies based on fMRI, NIRS or ECoG where necessary. We make use of four categories of noninvasive BCI paradigms in this review [77]: (1) P300 evoked potentials, (2) steady-state evoked potentials, (3) slow cortical potentials, and (4) sensorimotor rhythms and other brain activity related to mental tasks. The first part of this chapter reviews non-visual BCIs according to this categorization. Secondly, we outline possible directions for future research and promising sensory combinations that future multisensory BCIs could utilize. It is important to stress that the increasing attention paid to hybrid BCIs (see Chap. 18 by Müller-Putz et al. in this volume) should not underestimate the importance of multisensory hybrid BCIs where different sensory modalities can be linked to provide different control and feedback paradigms.

19.2 P300 Based BCI Systems

19.2.1 The "P300" Matrix Speller

Farwell and Donchin [16] were among the first to incorporate event-related potentials (ERPs) into the design of a BCI. A character matrix that contained the letters of the alphabet was designed, and rows' and columns' relative brightness was increased in a random manner. The subjects' task was to attend to a specific character, constituting a rare event within a series of frequent stimulus presentations—a concept that is referred to as the *oddball paradigm* [54]. With this approach, it is possible to elicit the P300 response [70]: a positive deflection around 300 ms after stimulus onset that provides information about the discrimination of targets and non-targets and thus can be utilized to determine a users' intent during the operation of the BCI spelling device.

Based on this first method of P300 detection, further studies tried to enhance the spellers' success by increasing the number of sensory modalities employed, primarily through the use of sound-based cues. Furdea et al. assigned numbers to the different rows and columns of the matrix, spoken out by a voice in sequential fashion [18]. Despite lower accuracies when utilizing the auditory modality for stimulation, Kübler et al. [39] then showed that this paradigm is also feasible for application in a home user's environment. Four disabled subjects suffering from ALS performed either a visual or auditory spelling task identical to [18]. For communication using the auditory speller, performance was relatively low-ranging from 25 % to 58.3 %, whereas the visual speller lead to accuracies exceeding 70 %.

In another study Klobassa et al. [36] augmented the visual P300 speller matrix with environmental sounds. In their application, six sounds (e.g., a bell or chord) were each associated in turn with one of six columns and subsequently one of six rows thus splitting the process of character selection into two successive steps. Subjects participated in eleven sessions each and received either auditory cues, or a combination of audio–visual cues. Individual subjects reached mean accuracies of up to 95 % when auditory and visual modalities were combined and performance rates of up to 77 % when stimuli were presented only auditorily. This finding highlights the advantage of modality combination, which leads to improved performance for individuals. Recently, Höhne et al. [29] expanded the speller matrix with sounds of diverse pitch (high/medium/low) and direction (left/middle/right) ascribed to a 3×3 character matrix and there by combined aspects of auditory sound presentation that emerged as being useful in previous studies [18, 23, 36, 59]. On average the ten participants were able to select the correct stimulus with 78 % accuracy.

A system that used visual and auditory stimuli within a matrix has been tested by Belitski and colleagues [3]. Again, numbers were spoken out loud and presented from different spatial locations arranged frontally around the subjects' heads. Instead of ascribing certain sound cues for rows and columns, the matrix was flipped by 90° after a first selection of the row containing the particular target was made. Results showed that multimodal audio–visual stimulus presentation combined with the matrix rotation lead to significantly better performance (above 80 %) than solely visual presentation (around 77 %) with or without matrix rotation, or solely auditory presentation (around 65 % including matrix rotation). The combination of different modalities led to stronger P300 amplitudes, improving the discrimination between targets and non-targets. These findings are in-line with results from the multisensory research presented in [60], where audio–visual stimuli led to behavioral performance and neurophysiological activity enhancement, compared to unimodal conditions.

19.2.2 Moving Beyond the "Matrix": Other Oddball Paradigms

Using a set of characters arranged in a matrix as a starting point for designing a BCI that relies on auditory stimuli may not be ideal; row and column selections require two distinct steps, and thus the time required to select a character and the complexity are both increased. Several groups studied different methods for presenting auditory control commands. The studies reviewed mark an important step towards improved auditory BCIs, showing that feasible results can be obtained by "moving beyond the matrix."

In one study, a simplified four-choice "Yes/No/Pass/End" paradigm was used [61]. While this paradigm continued to use visual presentations, a voice enunciated various choices randomly. When testing this design on ALS patients, Sellers and Donchin demonstrated that users were able to articulate their choices successfully by focusing on either a specific visual or auditory cue, as well as attending to both stimuli in combination. No substantial differences between healthy subjects and subjects suffering from ALS were reported in terms of accuracy. This clearly demonstrates the benefit of auditory BCIs for ALS patients, and other patient groups with visual impairment.

In another set of studies by Guo and colleagues, the oddball paradigm was used with eight spoken digits [21,22,30]. Subjects were asked to focus on a desired number that was either presented diotically (i.e., both ears heard the same sound material simultaneously), laterally [21], or vocalized by a male or female voice [22]. For quantitative discrimination between targets and non-targets, ERPs such as the N200 and the late positive complex (LPC) were used—indicating that not only P300 responses can provide useful information about directed auditory attention.

A number of the studies described used other aspects of auditory stimulus presentation, depending on either the presentation of different auditory streams, spatially arranged stimulus material, or auditory cues that vary in terms of stimulus intensity (modulated in terms of their loudness or pitch). According to Bregman, subjects should be able to consciously divide simultaneously presented auditory streams that differ in regard to certain characteristics (e.g., frequencies) and attend to either one of them, based upon the principle of *auditory stream segregation* [5].

Hill et al. utilized this auditory stream approach. Subjects listened to two distinct sequences of beeps concurrently presented to the left and right ear [24]. By attending to the target beep in one of these streams they were able to make a binary decision. Although classification rates showed a high variation between users, results were very promising and produced offline accuracies of up to 97 %. This approach was used in a later study by Kanoh et al. [32] who used a minimalistic setup with only two electrode sites and achieved equivalent results.

More recently, Schreuder et al. [59] used spatial cues in an oddball task. First, neurophysiological experiments were performed using a ring of eight speakers surrounding a listener's head. The task was to attend to a specific target sound location. For later BCI experiments, only five frontal speakers were used, reducing the difficulty of the task. Results demonstrated clear P300 as well as N100 and N200 ERP components, which were strongest above frontal and temporal brain areas. The effects of different inter-stimulus interval (ISI) sizes on BCI performance were investigated. Binary classification results were compared and accuracies over 90 % were obtained, one subject even reached 100 % correctly classified trials when an ISI of 175 ms was applied. These impressive results encourage the inclusion of spatial cues in auditory BCI paradigms in the future.

While it is still not clear what the best auditory parameters are for a P300 based BCI, recent work by Wagner [75] utilized a four-choice oddball paradigm with two possible targets for elicitation of ERPs. Stimulus material was presented in a diotic and dichotic manner, with the main goal being to enhance binary classification of

ERPs through the inclusion of lateralized information in response classification. Offline analysis revealed average performance scores around 81.8 %, two subjects even reached a 100 % correct-classification rate.

Finally, Halder et al. [23] directly compared sound pitch, loudness, and spatial location (dichotic task) paradigms. A three-stimulus oddball paradigm with two possible targets was used where three conditions were compared. Each task produced classification and communication transfer rates that could feasibly support real-world communication applications. Choosing the optimum task for each user led to a mean accuracy of 78.5 %, with the pitch task giving the best over-all results across subjects.

19.2.3 Tactile P300 Based BCIs

As compared to visual and auditory systems, tactile BCIs still remain relatively uncommon. Recently, Brouwer and van Erp [6] proposed a BCI system that relied on P300 responses resulting from attention to different sites of tactile stimulation, produced by tactors placed around the waist area. The effects of different numbers of stimuli (2, 4 and 6 tactors) and varying stimulus timing on classification accuracy were tested. Online classification was greater than chance (i.e., 16.67 %) and resulted in 58 % correctly classified trials when using six different tactors, whereas the inclusion of only two tactors led to 73 % accuracy (with chance equalling 50 %). In addition, optimal stimulus onset asynchrony (SOA) values were similar to visual P300 BCIs. The follow-up study directly compared the tactile paradigm with visual and visual-tactile stimulation [7]. Encouragingly, the bi-modal stimulation produced stronger EEG response amplitudes, demonstrating the potential of multisensory BCIs that include tactile stimulation.

19.3 BCIs Based on Steady-State Evoked Responses

19.3.1 Auditory Steady-State Responses

Besides the elicitation of transient responses, such as ERPs, another type of BCI control is offered by *steady-state evoked potentials* (SSEPs) that are evoked by a repetitive external stimulus and can be modulated by user attention. SSEPs related to auditory stimulation are referred to as *auditory steady-state responses* (ASSRs) and can be evoked by either click trains, short tone bursts and amplitude or square-wave modulated sinusoidal tones or noise [66]. In the case of amplitude modulation frequencies in the range of 10 to 100 Hz, a frequency of ~40 Hz has been shown to provide maximum bandpower at the modulation frequency and its harmonics [57].

While commonly used in visual BCIs [1, 34, 76], SSEPs only recently gained attention as a potential control paradigm for auditory BCIs. Lopez et al. [40] used two amplitude modulated (AM) tones presented to both ears simultaneously. Stimuli consisted of 1 kHz and 2.5 kHz carriers and had modulating frequencies of 38 Hz (left ear) and 42 Hz (right ear). Results demonstrated that attention modulated the spectral density at the AM frequency of the carrier tones. This initial study was followed by Kim et al. [35] where the first auditory ASSR-based BCI was tested. Task instructions and stimulus material were similar to [40]. Subjects reached maximum accuracies ranging from 80 % to 92 %, located significantly above a chance level of 50 % for unbiased binary classification.

Desain and colleagues explore another technique for the modulation of steady-state signals, utilizing a more widespread spectrum for labeling called *noise or frequency tagging* [14, 15]. In [15], the modulating envelopes that watermarked the target sounds consisted of pseudo-random noise sequences that permitted the later decoding of attended stimuli. Two tasks were assessed: (a) the Serial Selective Attention task, or (b) the Parallel Selective Attention task. In task (a) the tones were presented as an oddball sequence and subjects counted the target stimuli, while during task (b) stimuli were presented simultaneously, thereby probing the phenomenon of selective attention. For noise tags, classification rates went up to 89 %, proving the feasibility of this concept as a method for expanding the number of classes in future ASSR-based BCI systems.

However, in another study Hill et al. [25] attempted to detect shifts of attention during a dichotic listening task. In this very simple task, users were able to make reliable binary choices by focusing their attention and thereby modulating the ERPs elicited by rapid, regular auditory stimuli. In addition to ERPs, the stimuli simultaneously elicited strong ASSRs at two frequencies close to 40 Hz. Offline analysis showed that while the N200 component of the ERP (and to some extent the P300) was modulated successfully by the users' attention, such control was less apparent with respect to ASSRs. These results support the comparison of the feasibility of both features in further developments of novel auditory BCIs.

To conclude, ASSRs may certainly be useful for the design of auditory BCIs in the future, although their feasibility and neurophysiological aspects are not fully explored yet and need further elucidation.

19.3.2 Tactile Steady-State Responses

Both auditory and tactile BCIs may circumvent some of the difficulties that arise when using the visual modality for communication systems currently available to severely paralyzed subjects. Tactile stimuli that are presented to the subject none-transiently produce *steady-state somatosensory potentials* (SSSEPs). An initial study by Müller-Putz and colleagues [48] explored the nature of the SSSEP signal, testing various stimulation frequencies between 17 and 31 Hz. Ten subjects received mechanical stimulation of left and right index fingers, and bandpower was computed in eight 2 Hz width bands from 16–18 Hz up to 30–32 Hz at contralateral electrode sites C3 and C4. Maximum bandpower was found at 27 Hz, indicating a feasible rate of stimulation for later SSSEP-based BCI experiments.

In a later study, stimulation patterns with subject-specific frequencies were applied to both index fingers via transducers [49]. Subjects were asked to focus attention on one of their index fingers, indicated by a visual cue, and to count the changes in tactile stimuli at the desired index finger. Online performances of four subjects ranged between 70 % and 80 % (offline accuracies between 84 % and 88 %). This work showed that it is possible to implement a SSSEP-based BCI. The SSSEP amplitude was stable and constant, and subjects could indeed modulate activity to produce robust changes on a single trial basis.

19.4 Controlling BCIs with Slow Cortical Potentials

Another possibility for controlling a BCI system is the self-regulation of *slow cortical potentials* (SCPs). This approach requires training supported through feedback and positive reinforcement. BCI systems based on SCP have been actively studied by Niels Birbaumer's group and are sometimes referred to as *thought translation devices* (TTDs) [4].

Sonification efficiencies of SCPs have been compared with visual and audio–visual feedback during three training sessions in [27, 28]. Negative SCP was mapped to upward cursor movement (or a higher sound pitch) while positive shifts moved it downwards (which resulted in a lower sound pitch). During the third session, results showed that visual presentation and feedback led to higher average accuracies (67 %), compared to auditory (59 %) or combined auditory-visual conditions (57 %). Another study on SCP control used a similar methodology on three different subject groups that received either visual, auditory or audio–visual feedback [53]. Results showed that most subjects who received visual cues reached at least 60 % correct responses (11/19 subjects), whereas with auditory stimulation and feedback or a combination of audio–visual presentation a smaller proportion (8/20 and 5/20 subjects, respectively) of the sample reached the same accuracy. Even fewer achieved a 70 % correct classification rate, a benchmark that is critical for free spelling as stated by the authors.

In conclusion, this set of studies showed that multimodal feedback was not beneficial for SCP self-regulation, and that subject performance was comparable to visual stimuli-only conditions. Several methodological issues, e.g., the slow 16 Hz feedback refresh rate used for pitch changes or the fact that only three training sessions were completed could have influenced these results. Thus, more refined studies are needed to verify whether such findings are specific to SCP self-regulation, or to a particular type of multisensory feedback.

19.5 Sensorimotor Rhythms and Different Mental Tasks

Several strategies for BCI control involve a different type of mental activity—kinaesthetic or visual imagination of movement, auditory imagination of music, and speech. Some of these tasks are non-visual, and some have been used in combination with non-visual feedback.

19.5.1 Sonification of Motor Imagery

Many BCI systems are based on monitoring mu and beta rhythm activity over the motor cortical areas while subjects imagine movement. For example, imagination of a left hand movement leads to an *event-related desynchronization* (ERD; [51]) above contralateral electrode positions. An experiment with auditory real-time feedback in response to mu-rhythm activity has been reported in [27]. There, imagined left or right hand movement, and the corresponding mu-rhythm dynamics were sonified using a corresponding left or right loudspeaker. Classification accuracies were found in the range of 60 % and higher, after a relatively short period of training of about 200 trials.

Nijboer et al. [52] conducted experiments comparing success rates for visual and auditory feedback in motor imagery based BCIs. In the visual condition, a vertical cursor movement on the screen was controlled with motor imagery and feedback was given by target flashes. In the sound condition, the amplitude of the signal was sonified by environmental sounds which varied in volume according to the degree of *sensorimotor rhythm* (SMR) (de-)synchronization. A negativity, or desynchronization related to motor imagery was expressed by bongo sounds, and synchronization by harp sounds. Instructions and feedback about success or failure were given through a spoken voice. Within visual trials subjects reached on average higher accuracies (74 %) than in the auditory condition (56 %). After the third training session, however, performance did not differ significantly between visual and auditory conditions, indicating that more training might be needed when stimulus and feedback material is presented auditorily.

19.5.2 Somatosensory Feedback for Motor Imagery

In a motor imagery study done by Chatterjee et al. [10] participants received a combination of visual and tactile stimulation. Feedback consisted of visual information about the position of a horizontal bar relative to the two levels, as well as the varying intensity of vibrotactile stimulation applied via tactors placed on the subject's arm. When training was conducted with visual and vibrotactile feedback simultaneously, only vibrotactile feedback was used during the online testing. Additionally, the influence of congruent vibrotactile stimulation was studied. Tactors were placed ipsilaterally or contralaterally relative to the imagined left or right hand movement. Subjects reached an average accuracy of 56 %, with the highest performance being 72 %. Importantly, the performance was influenced by the tactors' position: subjects were more successful if tactors were placed ipsilateral to the imagined movement. This bias towards congruency between lateralized imagery and vibrotactile feedback provides valuable knowledge for further studies which include tactile information in BCI control.

Cincotti et al. [11] completed several studies investigating the feasibility of vibrotactile feedback in BCIs. Tactors were placed on neck or shoulder positions. A preliminary study showed that users were able to classify tactile stimulation according to its location and intensity. During the subsequent BCI experiment, subjects achieved accuracies between 56 % and 80 % with no differences found between performance rates obtained from training phases with solely visual or vibrotactile feedback, indicating that those feedback modes are comparable. Similar classification results were obtained in two other experiments when using this visual/tactile BCI system during navigation in virtual reality (VR). The results also demonstrated the benefits of vibrotactile feedback when performing complex visual tasks.

Powered exoskeletons provide another type of somatesensory feedback in motor imagery based BCI systems, typically in the rehabilitation setting. In a study by Gomez–Rodriguez et al. [20] healthy volunteers imagined extensions and flexions of their right forearm. Their arm was attached to a robot arm that could move according to BCI commands. The conditions of receiving and not receiving such additional somatosensory feedback were compared in training and testing phases. The results showed that additional feedback facilitated a higher success rate for decoding motor imagery.

19.5.3 BCIs Based Upon Imagination of Music and Rhythmization

The ideas of sonifying neuro- and peripheral physiological activity has been extensively explored by media artists since the early 1970s [56]. Recently introduced *brain–computer musical interfaces* (BCMIs) allow subjects to modulate or create music through modulation of brain signals, either by evaluating ongoing EEG activity [43], or by eliciting SSVEP responses for cursor-based selection [44]. However, the majority of such systems mainly transform brain activity into music and allow little conscious control over their output. A BCI system that relies on the imagery of music, also referred to as *audiation*, was proposed in [37]. In pilot experiments the subjects imagined inner tones and some evidence was found for a possible differentiation between imagined material. In [9, 12, 17] different mental tasks including music imagery have been compared to test the feasibility of this approach for future BCI systems. In particular, [17] instructed subjects to imagine a familiar tune and to focus on the melody rather than spoken language, in addition to performing other mental tasks (e.g., mental subtraction, word association, motor imagery, mental rotation and navigation). Although the best classification performance was obtained for the other mental tasks, music imagery could also be used for discrimination of brain activity related to different mental strategies.

While non-invasive BCI paradigms based on imagined music are still in a very early stage, the rhythm perception studies by Vlek and colleagues give further support to auditory imagery-based BCIs. Studies in [38, 74] rely on the cognitive mechanism of subjective rhythmization (also referred to as the *clock illusion*). In such a paradigm, identical auditory pulses at an isochronous rate are perceived as having different musically accented patterns (e.g., as a march: *one*-two; a waltz: *one*-two-three; or a common 4-beat rhythm: *one*-two-three-four [74]). A person can freely choose the different accents in a steady metronome pattern, and such activity can be decoded and used for BCI control. The study results show that accented and non-accented beats obtained from a single trial basis could be successively distinguished. An offline study [74] showed that it is possible to decode subjective accents from single-trial EEG.

19.5.4 BCIs Based Upon Speech

An idea for improving the intuitiveness of BCI operation is to base control upon the imagination of speech. In such a paradigm the user would simply be asked to imagine speaking a control command in order to enact control. For example, to operate a wheel chair to go left they might imagine speaking the word "left." More interestingly, such a paradigm could theoretically be used to make a highly intuitive and fast BCI speller.

Classifying imagined speech has been shown to be possible using functional magnetic resonance imaging (fMRI). For example, in [42] fMRI is used to classify which of three syllables is covertly spoken (imagined) by five subjects with accuracies between 63 % and 78 % with the specific aim of developing an imagined-speech based BCI. It is also possible to classify the speech via implanted cortical electrodes. Ongoing work reported in [8] classifies the silent production of phonemes by a locked-in patient recorded via an implanted electrode array. However, for the majority of users, BCIs generally require cheaper neuroimaging techniques then fMRI which are more practical for subjects to use on a day-to-day basis, and do not carry the risks associated with long term implants. The most common neuroimaging technique used with BCIs is the EEG. Results achieved thus far when attempting to identify imagined speech from the EEG have not been successful.

In [67, 68] and [69] results are presented which claim to accurately identify imagined speech from the EEG with accuracies of up to 97 %. However as shown in [13] and independently in [55] these high accuracies may be accounted for by faults in the methodology and artifacts. Thus, the ability to identify imagined speech from the EEG remains an idea that is yet to be proven.

19.5.5 Conceptual BCIs

An alternative approach that could be adopted is to base control of a BCI on the identification of cognitive processes relating to specific concepts. Thus, for example, control of the BCI users' television could be achieved by thinking about the concept of a television or a care giver could be called by thinking about the concept of a caregiver.

In [45], for example, fMRI blood-oxygen-level-dependent (BOLD) signals related to the meanings of nouns are classified. Intriguingly, in [63] EEG is recorded while concepts related to the different semantic categories "animals" and "tools" are presented via different modalities. Concepts are presented via their spoken names, their visual representations and their written names. Binary classification accuracies of up to 89 % were achieved. [50] has also shown a similar result, again using EEG: different concepts are presented to a subject in the form of a series of nouns describing objects pertaining to that concept. For example, for representing the concept "tools" the words "hammer", or "saw" might be presented to the user. By applying data mining techniques, features are identified that allow classification of the EEG within the correct semantic category with an accuracy of 98%.

This suggests that semantic categories may be identified from the EEG and that a BCI based upon these semantic categories could be a feasible possibility. Such a system could help improve the intuitiveness of a BCI design. This comes with the restriction that each unique control command must relate to a different semantic category. Thus, two commands could be "hand" and (category "body part") to move a robotic hand, and "television" (category "objects in the home") to turn on the television. However, the commands "hand" and "foot" used perhaps to move a robotic hand or robotic foot, are less likely to be differentiable.

19.6 New Directions for Multisensory BCI Research

In this section we attempt to summarize the current efforts related to the development of multisensory BCIs. We have followed a classification scheme that is similar to that of the first part of this chapter: (1) P300 evoked potentials, (2) steady-state evoked potentials, (3) slow cortical potentials, and (4) sensorimotor rhythms and other brain activity related to mental tasks. To discuss the multisensory combinations, we have created Fig. 19.1 with a grid corresponding to this classification, that contains visual (rows) and auditory (columns) dimensions.

The subsections below describe the rows of the grid in Fig. 19.1 and discuss potentially interesting further directions of multisensory BCI systems.

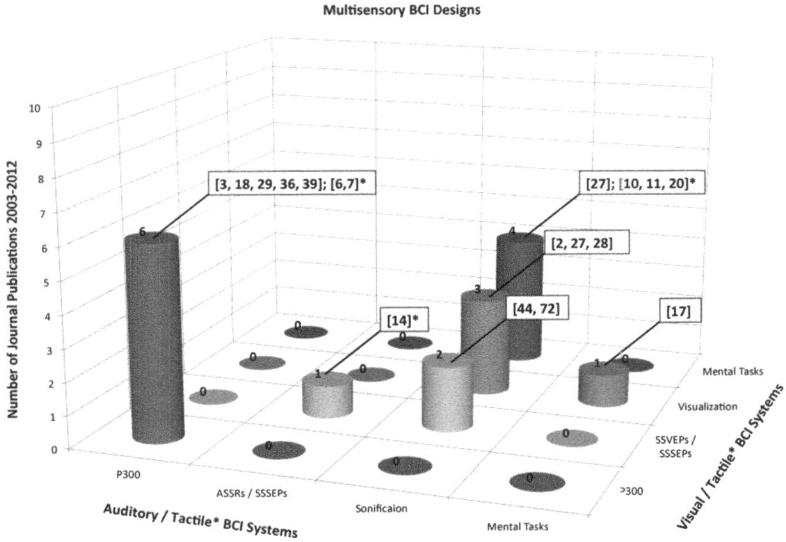

Fig. 19.1 The cells of the grid contain the references to the work that cover these particular multisensory combinations. Since somatosensory developments are rather sparse, we add this modality both to visual and auditory parts of the grid, and mark it with *asterisks* to refer to auditory-tactile or visual-tactile BCI systems. All cells, whether containing some work, or empty as of current state-of-the-art, are discussed in the following sections. The empty cells may be seen as potential directions of multisensory BCI research and are described in this section and in the discussion section

19.6.1 Combining Visual P300 BCIs with Other Modalities

The visually evoked P300-based paradigms currently represent the most well developed BCI applications where a character matrix is used for spelling or other control mechanisms, e.g., home operations. Hence, it is natural that these efforts incorporate sound [18, 36, 39]. Some of these studies (e.g., [39]) also used visual information to help users to learn the non-visual interface during the training phase of a purely auditory interface. It should be noted that these initial efforts utilized auditory spatial cues and the framework of auditory BCIs developed in [59] may greatly enhance the speed of such future audio–visual spellers as has been shown in the studies by [3, 29]. Apart from the behavioral benefits of such multisensory BCI systems, this can also lead to enhanced EEG feature sets, creating better BCI system performance. This has been shown in visual-tactile P300-based BCI systems in [7], and in audio-visual systems [3]. Further research on these BCI applications can benefit from work in the sensory substitution field, especially in auditory-based vision substitution systems [72].

ERP-based BCIs rely largely on the P300, and contributions from other *transient visual evoked potentials* (t-VEP) components such as the N100 and N200 have often been overlooked, e.g., [2]. Recent work introduced a new paradigm in which stimuli move instead of flash [30], eliciting *motion VEPs* (mVEPs) [64]. Such new mVEP based BCI paradigms can also benefit from a multisensory design since many studies show that motion processing is largely multisensory [65].

19.6.2 Combining Visual SSVEP BCIs with Other Modalities

BCI studies based on SSVEP signals are rapidly growing. However, there have been almost no efforts to use these control signals in combination with auditory or tactile stimuli either for control or feedback. To the best of the authors knowledge, there is only one study that describes two different sonification modes of EEG signals for a SSVEP feedback quality [58]. Unfortunately, no experimental evidence was provided as to whether such feedback improved BCI system performance. Another study that used musical feedback of SSVEP signals with ALS patient was reported in [44] and this concept may prove to be a viable tool for refining such hybrid multisensory BCI systems.

Similarly to multisensory P300-based systems, one can expect emerging systems that will combine SSEP signals evoked by different modalities. The initial study in [14] mentions a possible combination of auditory and tactile frequency tagged stimuli. However, no results or follow up studies have been presented. Combinations of different sensory cues may enhance SSEP amplitudes in such systems either directly, or indirectly. For example, emotional arousal has been shown to modulate SSVEP amplitudes [33], and sound can serve as an easy emotional booster of artificial visual or tactile stimuli.

19.6.3 Combining Visual Feedback with Other Modalities

Several studies provided audio-visual feedback for SCP-based BCI systems, but showed no enhancement compared to visual only training [27, 28]. Further studies may clarify the reason for inefficiency of this multisensory feedback. In addition, visual feedback can be provided while performing auditory imagery tasks, as was done recently in [17]. Again, future studies with such systems using multisensory feedback might prove to be more efficient than unisensory feedback. In addition, non-specific visualisation and sonification techniques of brain activity, such as described in [26], can lead to new paradigms for BCI control.

19.6.4 Mental Tasks and Multisensory Feedback

A number of studies on motor imagery have been conducted using multisensory feedback to enhance BCI system performance. In [27] spatial sound was used for marking left or right hand imagery. Several studies used tactile, visual or kinesthetic (passive hand movement using a robotic arm) feedback to enhance the ERD-based BCI systems [10, 11, 20]. Promising results of such initial systems will lead to the expansion and refinement of multisensory BCIs in this sector.

19.7 Conclusion

Multisensory BCI systems are emerging. The first prototypes already show an enhancement in performance on both user and system levels compared to purely vision based systems. In these studies, researchers mostly explored P300-based spelling and motor imagery-based systems where the use of non-visual information remained minimal. Unfortunately, the research progress in multisensory BCIs has been inequitably slow when compared to their potential benefits. Building upon the work from other fields such as auditory and multisensory displays, sensory substitution, and synaesthesia research, these initial systems can certainly be improved in the near future. Other emerging topics of research are auditory imagery and rhythmization, and non-visual SSEPs. The combination of control and feedback systems based on different sensory modalities may lead to the appearance of many new hybrid BCIs, as highlighted by the empty cells in Fig. 19.1. An important additional topic is the use of ecological stimuli which also bear emotional characteristics. Recent studies show that emotional processing can enhance both BCI control signals and motivation of the user (see [47], for a recent review).

Within multisensory BCI developments, systems that do not use visual cues for BCI control and feedback are of special importance. These auditory, tactile, and auditory-tactile systems are essential for users whose visual system has deteriorated due to progressive illness such as in ALS patients or visually impaired users. It should be noted that more multisensory studies with patients should reveal the potential of non-visual stimuli for BCI systems. For example, recent work showed functional deficits in secondary/higher-order sensory processing brain areas in ALS patients [41]. However, non-visual BCIs could also be useful for persons with "situational disability" which prevents them from using visual BCIs safely or effectively. Importantly, work on such non-visual developments provides important know-how for future multisensory BCI systems.

Indeed, the results reviewed in this chapter show that different sensory modalities can complement each other in the design of control or feedback signals. For example, the development of an omnidirectional spatial audio-based BCI in [59] or a mental rhythmization-based system in [74] explores the features of the auditory system that surpasses our visual capabilities. Hence, we can envisage the development of new hybrid BCI systems where multisensory design will define necessary quality of each sensory modality when providing cues in a multisensory display [73]. Neuroergonomic studies must guide the research of such perceptual optimization of BCI displays. The proper multisensory design, e.g., congruent sensory combinations, should lead to a reduction in the BCI users' sensory load, which is especially important for patient groups.

Acknowledgements This work was supported by Support action "FutureBNCI," Project number ICT-2010-248320.

References

1. Allison, B., McFarland, D., Schalk, G., Zheng, S., Jackson, M., Wolpaw, J.: Towards an independent brain–computer interface using steady state visual evoked potentials. Clin. Neurophysiol. **119**(2), 399–408 (2008)
2. Allison, B.Z., Pineda, J.A.: ERPs evoked by different matrix sizes: implications for a brain computer interface (BCI) system. IEEE Trans. Neural Syst. Rehabil. Eng. **11**, 110–113 (2003). DOI 10.1109/TNSRE.2003.814448
3. Belitski, A., Farquhar, J., Desain, P.: P300 audio–visual speller. J. Neural Eng. **8**(2), 025,022 (2011). DOI 10.1088/1741-2560/8/2/025022, http://dx.doi.org/10.1088/1741-2560/8/2/025022
4. Birbaumer, N., Hinterberger, T., Kübler, A., Neumann, N.: The thought-translation device (TTD): neurobehavioral mechanisms and clinical outcome. IEEE Trans. Neural Syst. Rehabil. Eng. **11**(2), 120–123 (2003). DOI 10.1109/TNSRE.2003.814439, http://dx.doi.org/10.1109/TNSRE.2003.814439
5. Bregman, A.: Auditory scene analysis: Hearing in complex environments. In: McAdams, S., Bigand, E. (eds.) Thinking in sound: the cognitive psychology of human audition, pp. 10–36. Oxford University Press, Oxford (1993)
6. Brouwer, A.M., van Erp, J.B.: A tactile P300 brain–computer interface. Front. Neurosci. **4**, 19 (2010). DOI 10.3389/fnins.2010.00019
7. Brouwer, A.M., van Erp, J.B.F., Aloise, F., Cincotti, F.: Tactile, visual and bimodal P300s: Could bimodal P300s boost BCI performance? SRX Neuroscience, Article ID:967027
8. Brumberg, J.S., Wright, E.J., Andreasen, D.S., Guenther, F.H., Kennedy, P.R.: Classification of intended phoneme production from chronic intracortical microelectrode recordings in speech-motor cortex. Front. Neurosci. **5**, 65 (2011). DOI 10.3389/fnins.2011.00065, http://dx.doi.org/10.3389/fnins.2011.00065
9. Cabrera, A., Dremstrup, K.: Auditory and spatial navigation imagery in brain–computer interface using optimized wavelets. J. Neurosci. Methods **174**(1), 135–146 (2008)
10. Chatterjee, A., Aggarwal, V., Ramos, A., Acharya, S., Thakor, N.: A brain–computer interface with vibrotactile biofeedback for haptic information. J. Neuroeng. Rehabil. **4**(1), 40 (2007)
11. Cincotti, F., Kauhanen, L., Aloise, F., Palomäki, T., Caporusso, N., Jylänki, P., Mattia, D., Babiloni, F., Vanacker, G., Nuttin, M., et al.: Vibrotactile feedback for brain–computer interface operation. Comput. Intell. Neurosci. **2007**:48937 (2007)
12. Curran, E., Sykacek, P., Stokes, M., Roberts, S., Penny, W., Johnsrude, I., Owen, A.: Cognitive tasks for driving a brain–computer interfacing system: a pilot study. IEEE Trans. Neural Syst. Rehabil. Eng. **12**(1), 48–54 (2004)
13. Daly, I., Nasuto, S., Warwick, K.: Towards natural human computer interaction in BCI. In: AISB 2008 Convention Communication, Interaction and Social Intelligence, vol 1, p. 26 (2008)
14. Desain, P., Hupse, A., Kallenberg, M., de Kruif, B., Schaefer, R.: Brain–computer interfacing using selective attention and frequency-tagged stimuli. In: Proceedings of the 3rd International Brain–Computer Interface Workshop & Training Course, Graz, Austria, pp. 98–99 (2006)
15. Farquhar, J., Blankespoor, J., Vlek, R., Desain, P.: Towards a noise-tagging auditory BCI-paradigm. In: Proceedings of the 4th International BCI Workshop and Training Course, Graz, Austria, pp. 50–55 (2008)

16. Farwell, L.A., Donchin, E.: Talking off the top of your head: toward a mental prosthesis utilizing event-related brain potentials. Electroencephalogr. Clin. Neurophysiol. **70**(6), 510–523 (1988)
17. Friedrich, E., Scherer, R., Neuper, C.: The effect of distinct mental strategies on classification performance for brain–computer interfaces. International J. Psychophysiol. (2012)
18. Furdea, A., Halder, S., Krusienski, D., Bross, D., Nijboer, F., Birbaumer, N., Kübler, A.: An auditory oddball (P300) spelling system for brain–computer interfaces. Psychophysiology **46**(3), 617–625 (2009). DOI 10.1111/j.1469-8986.2008.00783.x
19. Ghazanfar, A., Schroeder, C.: Is neocortex essentially multisensory? Trends in Cognitive Sciences **10**(6), 278–285 (2006)
20. Gomez-Rodriguez, M., Peters, J., Hill, J., Schölkopf, B., Gharabaghi, A., Grosse-Wentrup, M.: Closing the sensorimotor loop: Haptic feedback facilitates decoding of arm movement imagery. In: Systems Man and Cybernetics (SMC), 2010 IEEE International Conference on IEEE, pp. 121–126 (2010)
21. Guo, J., Hong, B., Guo, F., Gao, X., Gao, S.: An auditory BCI using voluntary mental response. In: Neural Engineering, 2009. NER'09. 4th International IEEE/EMBS Conference on IEEE, pp. 455–458 (2009)
22. Guo, J., Gao, S., Hong, B.: An auditory brain–computer interface using active mental response. IEEE Trans. Neural Syst. Rehabil. Eng. **18**(3), 230–235 (2010)
23. Halder, S., Rea, M., Andreoni, R., Nijboer, F., Hammer, E.M., Kleih, S.C., Birbaumer, N., Kübler, A.: An auditory oddball brain–computer interface for binary choices. Clin. Neurophysiol. **121**(4), 516–523 (2010). DOI 10.1016/j.clinph.2009.11.087, http://dx.doi.org/10.1016/j.clinph.2009.11.087
24. Hill, N., Lal, T., Bierig, K., Birbaumer, N., Schölkopf, B.: An auditory paradigm for brain–computer interfaces. Adv. Neural Inf. Process. Syst. **17**, 569–76 (2005)
25. Hill, N.J., Schölkopf, B.: An online brain-computer interface based on shifting attention to concurrent streams of auditory stimuli. J Neural Eng. **9**(2):026011 (2012)
26. Hinterberger, T.: The sensorium: a multimodal neurofeedback environment. Adv. Hum. Comput. Interact. **2011**, 3 (2011)
27. Hinterberger, T., Hill, J., Birbaumer, N.: An auditory brain–computer communication device. In: Biomedical Circuits and Systems, 2004 IEEE International Workshop on IEEE, pp. S3–6 (2004a)
28. Hinterberger, T., Neumann, N., Pham, M., Kübler, A., Grether, A., Hofmayer, N., Wilhelm, B., Flor, H., Birbaumer, N.: A multimodal brain-based feedback and communication system. Exp. Brain Res. **154**, 521–526 (2004b). DOI 10.1007/s00221-003-1690-3
29. Höhne, J., Schreuder, M., Blankertz, B., Tangermann, M.: Frontiers: A novel 9-class auditory ERP paradigm driving a predictive text entry system. Front. Neuroprosthetics **5**:99 (2011)
30. Hong, B., Lou, B., Guo, J., Gao, S.: Adaptive active auditory brain computer interface. In: Engineering in Medicine and Biology Society, 2009. EMBC 2009. Annual International Conference of the IEEE, IEEE, pp. 4531–4534 (2009)
31. Jacobs, L., Bozian, D., Heffner, R., Barron, S.: An eye movement disorder in amyotrophic lateral sclerosis. Neurology **31**(10), 1282–1287 (1981)
32. Kanoh, S., Miyamoto, K., Yoshinobu, T.: A brain–computer interface (BCI) system based on auditory stream segregation. Conf Proc IEEE Eng Med Biol Soc., 2008:642–645 (2008)
33. Keil, A., Gruber, T., Müller, M., Moratti, S., Stolarova, M., Bradley, M., Lang, P.: Early modulation of visual perception by emotional arousal: evidence from steady-state visual evoked brain potentials. Cogn. Affect. Behav. Neurosci. **3**(3), 195–206 (2003)
34. Kelly, S., Lalor, E., Finucane, C., McDarby, G., Reilly, R.: Visual spatial attention control in an independent brain–computer interface. IEEE Trans. Biomed. Eng. **52**(9), 1588–1596 (2005)
35. Kim, D.W., Hwang, H.J., Lim, J.H., Lee, Y.H., Jung, K.Y., Im, C.H.: Classification of selective attention to auditory stimuli: toward vision-free brain–computer interfacing. J. Neurosci. Methods **197**(1), 180–185 (2011). DOI 10.1016/j.jneumeth.2011.02.007, http://dx.doi.org/10.1016/j.jneumeth.2011.02.007

36. Klobassa, D.S., Vaughan, T.M., Brunner, P., Schwartz, N.E., Wolpaw, J.R., Neuper, C., Sellers, E.W.: Toward a high-throughput auditory P300-based brain–computer interface. Clin. Neurophysiol. **120**(7), 1252–1261 (2009). DOI 10.1016/j.clinph.2009.04.019, http://dx.doi.org/10.1016/j.clinph.2009.04.019
37. Klonowski, W., Duch, W., Perovic, A., Jovanovic, A.: Some computational aspects of the brain computer interfaces based on inner music. Comput. Intell. Neurosci. 2009:950403 (2009)
38. de Kruif, B., Schaefer, R., Desain, P.: Classification of imagined beats for use in a brain computer interface. Conf Proc IEEE Eng Med Biol Soc., 2007:678–681 (2007)
39. Kübler, A., Furdea, A., Halder, S., Hammer, E., Nijboer, F., Kotchoubey, B.: A brain–computer interface controlled auditory event-related potential (P300) spelling system for locked-in patients. Ann. N. Y. Acad. Sci. **1157**, 90–100 (2009). DOI 10.1111/j.1749-6632.2008.04122.x
40. Lopez, M., Pomares, H., Pelayo, F., Urquiza, J., Perez, J.: Evidences of cognitive effects over auditory steady-state responses by means of artificial neural networks and its use in brain–computer interfaces. Neurocomputing **72**(16-18), 3617–3623 (2009)
41. Lulé D., Diekmann, V., Müller, H., Kassubek, J., Ludolph, A., Birbaumer, N.: Neuroimaging of multimodal sensory stimulation in amyotrophic lateral sclerosis. J. Neurol. Neurosurg. Psychiatry **81**(8), 899 (2010)
42. McCorry, D.: Using statistical classification algorithms to decode covert speech states with functional magnetic resonance imaging. PhD thesis, George Mason University (2010)
43. Miranda, E.: Brain–computer music interface for composition and performance. Int. J. Disabil. Hum. Dev. **5**(2), 119 (2006)
44. Miranda, E., Magee, W., Wilson, J., Eaton, J., Palaniappan, R.: Brain–computer music interfacing (BCMI): From basic research to the real world of special needs. Music Med. 3:134–140 (2011)
45. Mitchell, T., Shinkareva, S., Carlson, A., Chang, K.M., Malave, V., Mason, R., Just, M.: Predicting human brain activity associated with the meanings of nouns. Science **320**(5880), 1191–1195 (2008). DOI 10.1126/science.1152876, http://dx.doi.org/10.1126/science.1152876
46. Mitsumoto, H., Przedborski, S., Gordon, P. (eds.): Amyotrophic Lateral Sclerosis. Taylor & Francis Group: New York, NY (2006)
47. Molina, G., Tsoneva, T., Nijholt, A.: Emotional brain–computer interfaces. In: Affective Computing and Intelligent Interaction and Workshops, 2009. ACII 2009. 3rd International Conference on IEEE, pp. 1–9 (2009)
48. Müller-Putz, G., Neuper, C., Pfurtscheller, G.: Resonance-like frequencies of sensorimotor areas evoked by repetitive tactile stimulation. Biomed. Tech. (Berl.) **46**, 186–190 (2001)
49. Müller-Putz, G., Scherer, R., Neuper, C., Pfurtscheller, G.: Steady-state somatosensory evoked potentials: suitable brain signals for brain–computer interfaces? IEEE Trans. Neural Syst. Rehabil. Eng. **14**(1), 30–37 (2006)
50. Murphy, B., Poesio, M., Bovolo, F., Bruzzone, L., Dalponte, M., Lakany, H.: EEG decoding of semantic category reveals distributed representations for single concepts. Brain Lang. **117**(1), 12–22 (2011). DOI 10.1016/j.bandl.2010.09.013, http://dx.doi.org/10.1016/j.bandl.2010.09.013
51. Neuper, C., Pfurtscheller, G.: Event-related dynamics of cortical rhythms: frequency-specific features and functional correlates. Int. J. Psychophysiol. **43**(1), 41–58 (2001)
52. Nijboer, F., Furdea, A., Gunst, I., Mellinger, J., McFarland, D., Birbaumer, N., Kübler, A.: An auditory brain–computer interface (BCI). J. Neurosci. methods **167**(1), 43–50 (2008)
53. Pham, M., Hinterberger, T., Neumann, N., Kübler, A., Hofmayer, N., Grether, A., Wilhelm, B., Vatine, J., Birbaumer, N.: An auditory brain–computer interface based on the self-regulation of slow cortical potentials. Neurorehabil. Neural Repair **19**(3), 206 (2005)
54. Polich, J.: Updating P300: an integrative theory of P3a and P3b. Clin. Neurophysiol. **118**(10), 2128–2148 (2007). DOI 10.1016/j.clinph.2007.04.019, http://dx.doi.org/10.1016/j.clinph.2007.04.019
55. Porbadnigk, A., Wester, M., Calliess, J.P., Schultz, T.: EEG-based speech recognition – impact of temporal effects. In: Proceedings of the International Conference on Bio-inspired Systems and Signal Processing (2009)

56. Rosenboom, D.: Extended musical interface with the human nervous system. Leonardo Monograph Series International Society for the Arts, Sciences and Technology (ISAST) 1 (1997)
57. Roß B., Borgmann, C., Draganova, R., Roberts, L., Pantev, C.: A high-precision magnetoencephalographic study of human auditory steady-state responses to amplitude-modulated tones. J. Acoust. Soc. Am. **108**, 679 (2000)
58. Rutkowski, T., Vialatte, F., Cichocki, A., Mandic, D., Barros, A.: Auditory feedback for brain computer interface management–an EEG data sonification approach. In: Knowledge-Based Intelligent Information and Engineering Systems, pp. 1232–1239. Springer-Verlag: Berling Heidelberg (2006)
59. Schreuder, M., Blankertz, B., Tangermann, M.: A new auditory multi-class brain–computer interface paradigm: spatial hearing as an informative cue. PLoS One **5**, e9813 (2010). DOI 10.1371/journal.pone.0009813
60. Schröger, E., Widmann, A.: Speeded responses to audiovisual signal changes result from bimodal integration. Psychophysiology **35**(6), 755–759 (1998). DOI 10.1111/1469-8986.3560755, http://dx.doi.org/10.1111/1469-8986.3560755
61. Sellers, E., Donchin, E.: A P300-based brain–computer interface: initial tests by ALS patients. Clin. Neurophysiol. **117**(3), 538–548 (2006). DOI 10.1016/j.clinph.2005.06.027, http://dx.doi.org/10.1016/j.clinph.2005.06.027
62. Sellers, E., Kübler, A., Donchin, E.: Brain-computer interface research at the University of South Florida Cognitive Psychophysiology Laboratory: the P300 speller. IEEE Trans. Neural Syst. Rehabil. Eng. **14**, 221–224 (2006). DOI 10.1109/TNSRE.2006.875580
63. Simanova, I., van Gerven, M., Oostenveld, R., Hagoort, P.: Identifying object categories from event-related EEG: toward decoding of conceptual representations. PLoS One **5**(12), e14465 (2010). DOI 10.1371/journal.pone.0014465, http://dx.doi.org/10.1371/journal.pone.0014465
64. Skrandies, W., Jedynak, A., Kleiser, R.: Scalp distribution components of brain activity evoked by visual motion stimuli. Exp. Brain Res. **122**(1), 62–70 (1998)
65. Soto-Faraco, S., Väljamäe, A.: Multisensory interactions during motion perception: From basic principles to media applications. Taylor & Francis Group: New York, NY (2011)
66. Stapells, D., Herdman, A., Small, S., Dimitrijevic, A., Hatton, J.: Current status of the auditory steady-state responses for estimating an infant's audiogram. A sound foundation through early amplification, pp. 43–59 (2004)
67. Suppes, P., Han, B., Lu, Z.L.: Brain wave recognition of words. Proc. Natl. Acad. Sci. USA **94**(26), 14,965–14,969 (1997)
68. Suppes, P., Han, B., Lu, Z.L.: Brain-wave recognition of sentences. Proc. Natl. Acad. Sci. USA **95**(26), 15,861–15,866 (1998)
69. Suppes, P., Han, B., Epelboim, J., Lu, Z.: Invariance between subjects of brain wave representations of language. Proc. Natl. Acad. Sci . **96**(22), 12,953 (1999)
70. Sutton, S., Braren, M., Zubin, J., John, E.: Evoked-potential correlates of stimulus uncertainty. Science **150**(700), 1187–1188 (1965)
71. Townsend, G., LaPallo, B., Boulay, C., Krusienski, D., Frye, G., Hauser, C., Schwartz, N., Vaughan, T., Wolpaw, J., Sellers, E.: A novel P300-based brain–computer interface stimulus presentation paradigm: moving beyond rows and columns. Clin. Neurophysiol. **121**, 1109–1120 (2010). DOI 10.1016/j.clinph.2010.01.030
72. Väljamäe, A., Kleiner, M.: Spatial sound in auditory vision substitution systems. In: Audio Engineering Society Convention, pp. 120 (2006). http://www.aes.org/e-lib/browse.cfm?elib=13599
73. Väljamäe, A., Tajadura-Jimenez, A., Larsson, P., Västfjäll, D., Kleiner, M.: Handheld experiences: Using audio to enhance the illusion of self-motion. IEEE MultiMedia, pp. 68–75 (2008)
74. Vlek, R., Schaefer, R., Gielen, C., Farquhar, J., Desain, P.: Sequenced subjective accents for brain–computer interfaces. J. Neural Eng. **8**(3), 036,002 (2011). DOI 10.1088/1741-2560/8/3/036002, http://dx.doi.org/10.1088/1741-2560/8/3/036002

75. Wagner, I.: An auditory brain–computer interface for binary choices using event-related potentials and lateralized hemispheric brain activity: Tests with healthy controls. Master Thesis, University of Graz, Graz, Austria (2011)
76. Wang, Y., Gao, X., Hong, B., Jia, C., Gao, S.: Brain–computer interfaces based on visual evoked potentials. IEEE Eng. Med. Biol. Mag. **27**(5), 64–71 (2008)
77. Wolpaw, J., Birbaumer, N., McFarland, D., Pfurtscheller, G., Vaughan, T.: Brain-computer interfaces for communication and control. Clin. Neurophysiol. **113**, 767–791 (2002). DOI 10.1016/S1388-2457(02)00057-3

Chapter 20
Characterizing Control of Brain–Computer Interfaces with BioGauges

Adriane B. Randolph, Melody M. Moore Jackson, and Steven G. Mason

20.1 Introduction

A recent review article [30] wrote that: "The central tenet of a [brain–computer interface] BCI is the capability to distinguish different patterns of brain activity, each being associated to a particular intention or mental task." Hence, one of the *a priori* decisions of any BCI design is which mental activities (and hence corresponding brain activity patterns) are available to the user. This issue has long been a challenge in BCI research, and many articles have also discussed matching the right BCI to the right user [1, 16, 31, 34, 46].

Recent research trends concerning "BCI illiteracy" in the community have provided additional information that must be considered when deciding the right BCI for each user. This means that some users cannot use a BCI, which has long been documented [6,19,32,49], but only recently explored parametrically [9,41,46]. Additionally, new results show that people who cannot use one type of BCI (such as a BCI based on imagined movement) could use a different type (such as a BCI based on visual attention) [9, 46].

However, research efforts such as these have three major drawbacks. First, they typically only compare a few different types of BCIs (often two). Second, they rarely assess other novel assistive technologies (ATs), such as devices based on eye-tracking or electrodermal activity (EDA) (a.k.a. galvanic skin response), that

A.B. Randolph (✉)
Kennesaw State University, Information Systems, Kennesaw, GA 30144, USA
e-mail: arandolph@kennesaw.edu

M. Moore Jackson
Georgia Institute of Technology, College of Computing, Atlanta, GA 30332 USA
e-mail: melody@cc.gatech.edu

S.G. Mason
Left Coast Biometrics Ltd., 4871 Sophia Street, Vancouver BC, Canada V5V 3W5
e-mail: smason.van@gmail.com

B. Z. Allison et al. (eds.), *Towards Practical Brain-Computer Interfaces*, Biological and Medical Physics, Biomedical Engineering, DOI 10.1007/978-3-642-29746-5_20,
© Springer-Verlag Berlin Heidelberg 2012

some users may want to consider as alternatives to a BCI, or in combination with a BCI as a hybrid system [9, 22, 37]. Third, they often only explore one factor such as information throughput, whereas many other factors may also affect a decision about which BCI or other assistive technology to use [31].

Our primary goals in this article are to review some of the challenges and issues in choosing the right BCI or other novel AT for each user, and discuss some solutions. We focus heavily on the BioGauges approach, which has been developed over several years. We conclude that the BioGauges system could provide a solid framework for comparing different interface systems, and propose some additional future directions.

20.2 Key Factors for BCI Use

A recent article [2] summarized the many factors that might influence the adoption of BCIs and related technologies and is shown in Fig. 20.1. Ideally, a potential BCI user should be provided with as much information as possible about each BCI (or other system) that might meet his or her needs. Because this is a distant goal, it is important to determine which factors are most relevant today and focus on them. Recent work [14, 51] assessed some of these factors by asking actual or potential end-users from communities with severely disabled people what matters most. Critical factors included bandwidth, reliability, cosmesis, and different aspects of support—patients strongly preferred systems that did not create a strong dependence on other people for help to set up the electrode cap for recording brain signals, to customize the BCIs for each user, or to wash the hair of the electrode gel used.

The decision about the match with a BCI system is complicated by illiteracy—users might choose a system that seems to work well, only to learn that it will not work for them. There are many reasons why some types of mental activities might be unfeasible, including:

1. Different activities might yield better accuracy, faster communication, or more selections [1, 5, 49].
2. In the most extreme example of poor accuracy—a fairly small percentage of healthy subjects cannot produce discriminable EEG patterns while performing some mental tasks used in BCIs [4, 6, 9, 20]. This phenomenon is exacerbated in some patient groups for various reasons, such as damage to different brain areas, blindness or other visual difficulties, attentional disorders, uncontrollable movements, or difficulty pursuing goal-directed activity [18, 19, 35].
3. Users may find some tasks more fun, less distracting, or easier to learn, perform, change, multitask, or sustain. Some types of visual stimuli may be annoying or fatiguing to some users, such as flickering boxes or LEDs [3, 9].
4. Some activities may be better suited to some goals. For example, P300 BCIs tend to be used in tasks to directly select one of several targets, without any intermediate feedback [12, 15, 26, 31]; whereas event-related desynchronization

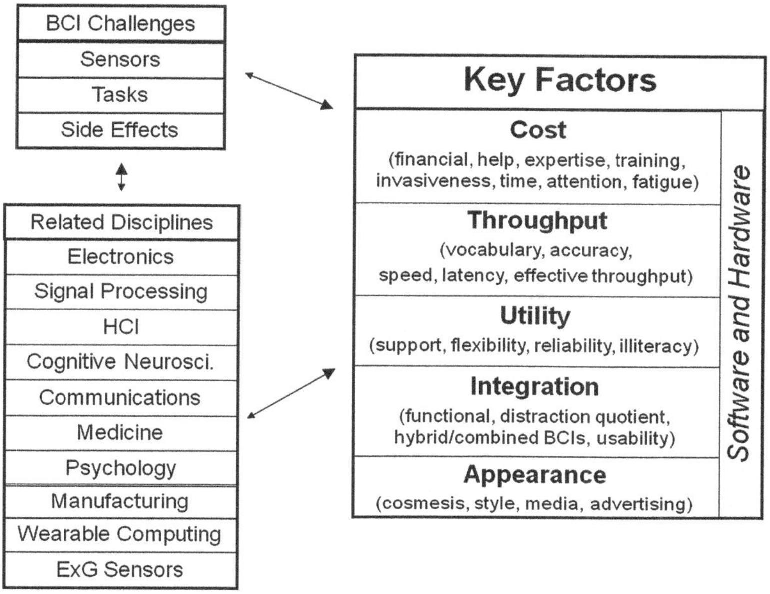

Fig. 20.1 Factors affecting the adoption of BCIs and related technologies. Used with permission [2]

(ERD) BCIs tend to be used to control cursor movement in one or more dimensions [8, 29, 39, 47].

5. Some EEG signals may be more robust in difficult environmental conditions, such as bad lighting, background noise, or distractions [3, 31, 47].
6. Since different mental activities produce signals that are stronger over different areas, some types of head-mounted devices may be incompatible with some BCIs. A hair beret or gamer headset might have electrodes over central but not occipital sites, whereas a headband might have the opposite sites available [2, 23, 24].
7. Some mental tasks may be incompatible with other tasks the user wishes to concurrently perform, such as watching a movie, playing a game, talking to a friend, performing other movements, or using another BCI. Given the recent enthusiasm for using BCIs to play games, it is especially important to explore ways to seamlessly integrate any visual stimuli that a BCI requires within the game environment [2, 21, 25, 36].
8. Some communication systems may be especially prone to the Midas Touch problem [32], in which users send unintended commands. Ideally, a BCI should support asynchronous operation with a standby mode, and only function when the user so desires.

Interestingly, modern BCIs tend to rely on electrodes over the top and back of the head—central, parietal, and occipital sites. Frontal and temporal activity is used less often—not because these regions have been ignored, but because studies typically show they provide little or no useful information for control purposes in most BCI systems [17,38,40]. Future research efforts might identify new tasks for BCI control that produce activity with different scalp distributions.

Once the appropriate type of BCI has been preliminarily selected, the designer must consider many other factors, including:

- Numerous differences in the nature and scalp distribution of the relevant EEG activity, and hence many aspects of the best filtering and signal processing
- The software and hardware
- The way different tasks map to different goals
- Ensuring an effective support infrastructure so the user can get help when needed
- The timing and nature of the interactions between the user and system, such as synchronous versus asynchronous operation.

These factors are largely relevant for selection of novel AT, as well. This decision then affects many things from the user's perspective alone, including:

- Different characteristics of the stimuli necessary to elicit relevant brain activity
- The location and number of electrodes
- The nature and extent of errors
- Performance fluctuations that may occur in different environments, usage sessions, mental states, and concurrent tasks
- Many aspects of training.

In summary, finding the appropriate BCI or novel AT for a particular user is very complicated. Many factors may influence this decision, and some trial and error may be necessary. Software tools such as BioGauges could greatly reduce the time, cost, and subjectivity of this decision. We next review our work with the BioGauges project.

20.3 Characterizing BCI Systems

Novel AT systems follow a similar architecture as that used in BCIs and here will be included in the definition of BCI system components with the understanding that input comes from some peripheral physiology influenced by brain processes and not directly from the brain. At the heart of each BCI or novel AT system is a *transducer*. Similar to the general definition of a transducer, which is "a device that receives a signal in the form of one type of energy and converts it to a signal in another form" [48], we define a *BCI transducer* as a device or system that translates electrophysiological or metabolic signals, such as human brain signals, into control signals. The control signals result from a complex combination of factors including the signal detection technology, the individual user's abilities and state, and environmental influences on the user. Compared to more traditional devices that are based on direct physical movement, the BCI transducers that record electrophysiological and metabolic signals often have high error rates and low information transfer rates, or bandwidth [32]. Therefore, particularly for people with disabilities, determining which BCI transducer will provide the best results for a particular individual is a difficult process.

Further complicating this process is the myriad of techniques under study for BCIs for control purposes. The output and performance of the various approaches are reported in many different ways that are difficult to compare objectively. Typically, results are reported in bit rates, error rates, or performance on a specific control task [31]. The parameters and design of control interfaces make an enormous difference in the significance and meaning of the results. For example, performance on a simple cursor-based binary selection task is affected by variables such as whether the selection space is bounded; whether both alternatives are always available and thereby allowing a person to create a false positive where s/he may indicate "yes" when "no" is intended; and Fitts' Law factors [10, 13] such as the size of icons that might be selected and distance travelled. Therefore, evaluating the suitability of a BCI transducer for the capabilities of a specific individual or objectively comparing the potential of multiple BCI transducers is very challenging.

20.3.1 BioGauges and Controllability

The goal of the BioGauges project is to provide a method to objectively compare the outputs of BCI transducers by reducing control tasks to their simplest, or atomic, levels. It represents a first step away from the current trial-and-error testing for matching individuals to BCIs and may help form a basis for future determinations to take place offline. *BioGauges* are very simple control interfaces that directly measure and record users' electrophysiological and metabolic outputs for the basic components of interaction. BioGauges can be used to determine the range, spatial accuracy, temporal accuracy, and granularity of control for a specific user and a particular transducer configuration; this constitutes a person's ability to control a particular transducer, or his/her *controllability*. Such controllability information could then be used to help choose a device with which a user achieves his/her best performance, to better configure a BCI system for a user, or to more objectively assess the potential of a BCI technology for control.

20.3.2 Transducer Categories

The design of appropriate BioGauges to measure BCI transducer performance depends heavily on the category of the signals output by the transducer and the desired state of user control. In general, the three transducer categories are: discrete, continuous, and spatial reference [27,28]. A user's control states may be categorized as no-control and intentional-control. These categories are described in more detail as follows.

Discrete transducers output a series of discrete states, as in a switch. There may be any number of states but typically discrete transducers produce two states (such as flipping a light switch) or a momentary-on state (such as pressing a button). BioGauges for discrete transducers include measurements for temporal accuracy to a predictable event (such as pressing a button every 2 s), response rate to an unpredictable event (such as pressing a button when a randomly-timed stimulus appears), repetition rate (what is the fastest rate that the transducer can be reactivated, such as in calibrating a mouse double-click), and the ability to hold-and-release (for transducers that support sustained activation). An example of a discrete BCI transducer is the Neil Squire Society's Low-Frequency Asynchronous Signal Detector (LF-ASD) [7] which is a single-trial switch that detects the difference between active and idle states of voluntary motor-related potentials in the brain.

Continuous transducers output a constant stream of values potentially varying in amplitude within a specified range, such as moving a mouse across a computer screen. BioGauges for continuous transducers include measurements for the range of output that can be attained (highest and lowest values), ability to attain a particular value or range of values (spatial granularity of control), and the ability to attain-and-hold a value or value range. Continuous BioGauges also can include temporal measurements such as time-to-attain values or value ranges, and repetition rate for attaining values. Two examples of continuous BCI transducers are the Wadsworth mu-based transducer [50] and the Georgia State University (GSU)/Georgia Institute of Technology (GT) BrainLab transducer based on EDA [33,42]. The Wadsworth mu-based transducer directly inputs users' scalp-recorded fluctuations in the 8–12 Hz mu rhythm, a signal from the area of the brain responsible for real and imagined movement. The GSU/GT BrainLab EDA-based transducer takes as input changes in electrical conductivity of the skin from sensors placed on the fingers and serves as an indirect measure of brain activity.

Spatial reference transducers output a particular location in a 2-D or 3-D space, such as a touch-screen. BioGauges for spatial reference transducers include measurements for granularity of selection (what is the smallest possible difference between locations), accuracy of selection, and repetition rate. An example of an EEG-based spatial reference transducer is Donchin's P300 matrix [11] where the users are presented with a grid of characters or icons from which they may make a selection by attending to the desired item, which generates a detectable response in the brain called the P300; the result is the equivalent of the user pointing to the desired item.

One important measurement with all BCI transducers is the difference between the transducer output during intentional-control and no-control states of a user. In the *no-control state*, the user is not trying to operate the BCI device. The user may be performing another task (such as reading a website page) referred to as active no-control, or performing no task at all (such as staring at a constant image) termed passive no-control. In the *intentional-control state*, the user is trying to operate

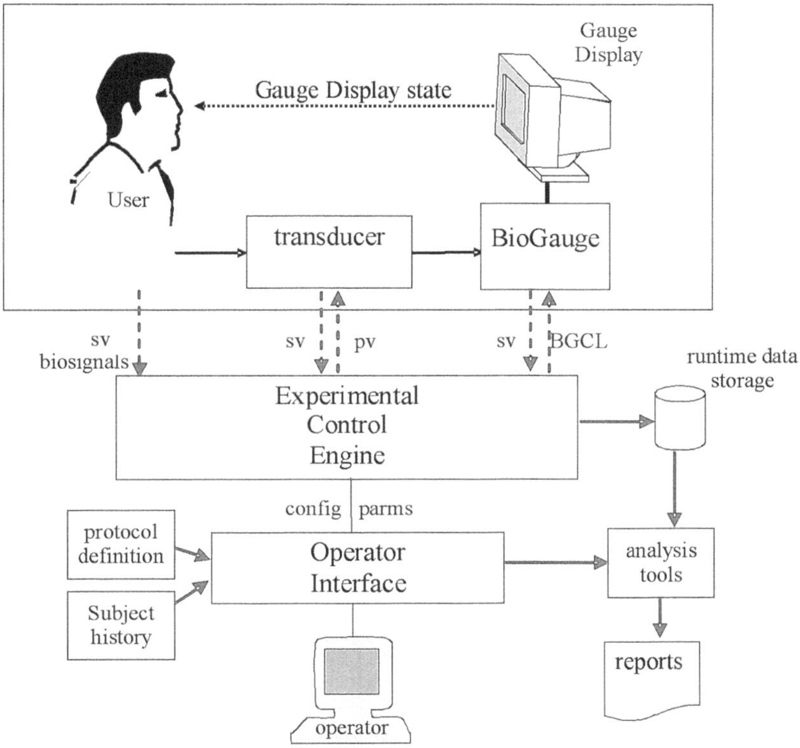

Fig. 20.2 BioGauges system architecture illustrating component connectivity and transmittance of system values in the form of state vectors (sv), parameter vectors (pv), and BioGauge Control Language (BGCL)

the BCI transducer. The difference between user outputs on intentional-control and no-control states can significantly affect the usability of a BCI transducer for real-world applications such as communication.

20.3.3 The BioGauges Experimental System

As shown in Fig. 20.2, the BioGauges toolset is implemented in a configurable architecture that consists of an experimental control engine, a set of BioGauges, and a BCI transducer. This toolset has been iteratively developed and tested over the past five years with over fifty users ranging from able-bodied to completely locked-in. The following describes the toolset components in more detail.

The Experimental Control Engine receives configuration parameters from an Operator Interface. It then sends commands, in the form of BioGauge Control

Table 20.1 Attributes of the BioGauge control language

Keyword	Definition
GaugeID	Code for BioGauge type being run.
DesiredGMode	Indicates one of five modes in which the gauge may run: 0) Break, 1) Ping Test, 2) Data Test, 3) Operating, and 4) Playback.
TID	Transducer ID. Code for the name of the transducer being used.
TrialEndMessage	Message to show the operator at end of a run.
ReportRate	State vector report rate.
ITIDistribution	Inter-trial interval distribution. Allows for randomization of time period between trials.
RateDistribution	Distribution of times it takes for the indicator to appear and reach the target line over which the gauge randomly selects.
PreTAW	Pre-Target Acquisition Window for indicator to hit target. This gives the number of seconds before target entry that indicator activation is detected.
PostTAW	Post-Target Acquisition Window for indicator to hit target. This gives the time after target exit that indicator activation is detected.
FeedbackOn	Allows subject to receive feedback upon an activation or not.
TotalNumTrials	Number of automatically repeated trials in a protocol.
FeedbackDuration	Amount of time that onscreen feedback is generated after an activation.
TrialStartMessage	Message to show the operator before the start of a run.
ImageIndex	Index number of image to display for a no-control BioGauge.
RSDuration	Response Stimulus Duration. Time to display a reaction stimulus onscreen for discrete BioGauges.
TimeOutPeriod	Maximum time allowed: (1) for possible activation before timeout in discrete BioGauges, (2) to achieve the target in continuous BioGauges, and (3) the image display time in no-control BioGauges.
HoldTime	Targeted time to hold the indicator within the target window.
IndicatorStartPoint	Percentage offset in the control space for the indicator's starting point.
WSDuration	Warning Stimulus Duration. Time to display the warning stimulus on the subject's screen in preparation for starting a run.
PWPDistribution	Post Warning Period Distribution. Used to randomly select the time period between the warning stimulus and the next event.
TargetGenMap	Structure defining target generation parameters.
DisplayOrientation	Orientation display of indicator movement.
Task	Name of protocol for operator to identify a BGCL file.

Language (BGCL) to the BioGauge currently being used in the test. BioGauge Control Language represents the parameters that need to be set in a BioGauge to affect a particular behavior or experimental design. BioGauge Control Language is input as a delimited character stream that the BioGauge parses at initialization time. BioGauge Control Language specifications consist of case-insensitive keyword attributes and string values.

Table 20.1 provides the keyword attributes used within BGCL and their definitions, explaining the capabilities of each BioGauge.

After processing inputs from the BCI transducer for which it listens over a TCP/IP connection, the BioGauge reports its state back to the Experimental Control Engine as a state vector (SV), which records the system state according

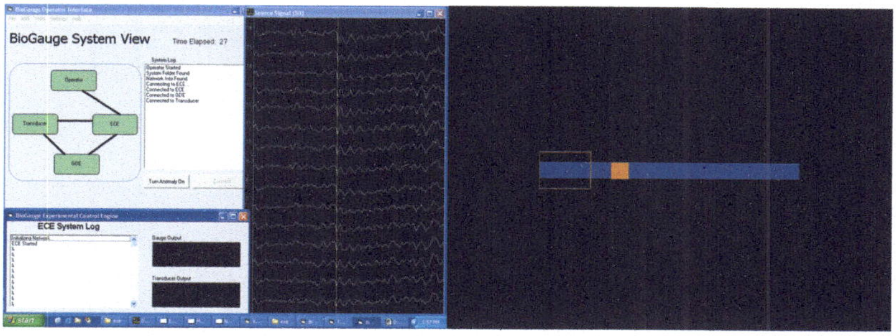

Fig. 20.3 The BioGauges system display showing the operator's view of the Operator Interface and EEG signals being recorded

to parameters of the specified experimental protocol or parameter vectors (PVs). There are agent tasks that act as translators between system components to ensure compatibility. The separation of the system components provides flexibility, as transducers and BioGauges can be easily reconfigured or swapped out without extensive programming.

When the BioGauges toolset is first launched, it verifies that it can see all system components and that it is receiving values from the transducer as illustrated in Fig. 20.3. The human operator must then calibrate the system manually. For one minute, the operator observes the numeric values assigned to the transducer output while asking the subject to generate mental imagery related to movement and relaxation as requested during the screening procedure. The operator records the highest and lowest values generated by the transducer within the Operator Interface. From these values, the BioGauges toolset calculates a midpoint value to be used during the experimental tasks. This value does not change for the duration of the session as the system does not dynamically self-calibrate. At any point, if the BioGauges toolset components stop sending values or if it no longer receives values from the transducer it registers an error and stops.

20.3.4 Analysis Methods

The BioGauges toolset incorporates an analysis component intended to express the characterization of a BCI transducer in graphical as well as numeric form. In addition to displaying and processing raw data from experimental sessions, the analysis toolset provides researchers the capability to visualize transducer output trends in charts termed *Mason–Moore Maps* (M–M Maps). For example, Fig. 20.4 contains an M–M Map for data resulting from an experimental session employing a Relative-Attain BioGauge with a starting point in the center of the path bar and two potential targets. The three dimensional M–M map shows the start position on the x-axis, the target position on the y-axis, and the time to move the indicator on the z-axis. The

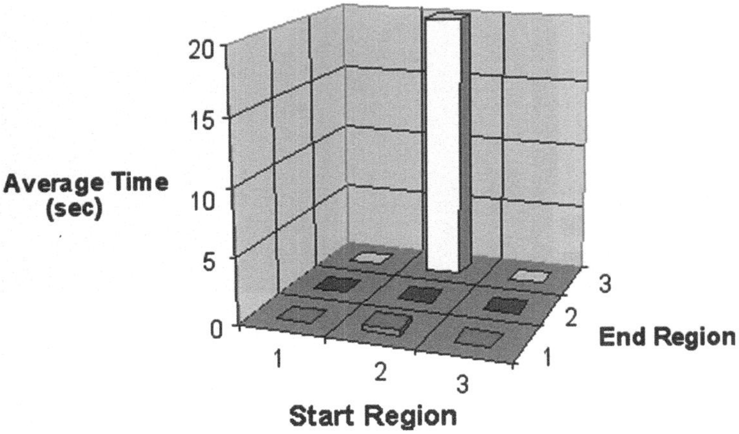

Fig. 20.4 Sample M–M Map for a temporal performance measurement BioGauge

path bar was divided into three regions where region 2 was the starting point in the center of the path bar and regions 1 and 3 were regions on the far ends of the path bar. Visual inspection of the M–M map in Fig. 20.4 allows a researcher to quickly see that for this transducer-user combination, from a start position in region 2 it took a small amount of time to reach targets in the first region and the system generally timed out while subjects attempted to reach targets in the third region.

The BioGauges toolset currently implements a large variety of M–M Maps for discrete and continuous transducers. Examples of maps for discrete transducers include time between false positives, distribution of time accuracy, repetition accuracy, and others. Continuous transducer maps include time to attain target as shown in Fig. 20.4, distance travelled in a time interval as a percentage of the screen size, and hold-time stability mappings. In addition, the BioGauges toolset assesses the percentage success for achieving a prescribed task.

20.3.5 Validation

The BioGauges toolset and methodology has been demonstrated with discrete and various continuous transducers including those based on mu, functional near-infrared (fNIR), and EDA. BioGauges have in some cases shown how individuals have better literacy with one type of BCI transducer over another [45, 46]. Further, BioGauges have been tested with both able-bodied participants and those with varying stages of paralysis due to amyotrophic lateral sclerosis (ALS).

A discrete, EEG-based transducer was tested with five able-bodied participants [43, 44]. With the LF-ASD, subjects achieved upwards of 73% accuracy for reaction time, 96% for temporal accuracy, and 82% with repeated accuracy. More extensive tests have been run with continuous transducers.

In a study with six able-bodied participants using a continuous, EDA-based transducer [43], BioGauges showed that it is difficult for participants to hold their EDA signal at heightened arbitrary levels for prolonged periods of time or to hold plateaus of excitement. Participants reported mental exhaustion after tasks that required higher excitement levels, and one participant complained of a headache that developed during the test sessions. The study showed that it is possible for untrained participants to exhibit stability and control at upwards of 87% accuracy when using EDA but that there are some challenges. In another study comparing mu control to EDA control [45], ten able-bodied participants evidenced little variation in their control with either transducer because adaptive algorithms were not utilized. Further, due to the low strength of the raw mu values, all participants were able to obtain targets in one direction but not the other.

Lastly, in a study with 33 able-bodied participants and five participants with ALS comparing fNIR control to EDA control [46], people were able to exhibit some level of control of both the fNIR and EDA technologies. Seventy-four percent of all participants and 60% of participants with ALS were able to achieve greater than chance results with fNIR. For EDA, 60% of all participants and 40% of participants with ALS were able to achieve greater than chance results. There were individuals who were unable to generate a response with the EDA device at all during calibration due to their self-reported inability to become very "sweaty."

20.4 Summary and Future Work

Here, we have introduced a method and toolset to characterize the output of BCI transducers. We hope that by presenting the idea of BioGauges that we will lay the foundation for a continuing dialogue in the research community about BCIs and their use by varying individuals. We hope such conversations will lead to improved communication among various research groups, methods developed to more accurately and objectively measure performance of BCIs, and better facilitated reporting and comparison of study results. Furthermore, there is a great deal of opportunity to study the profound implications of understanding BCI literacy with a particular transducer on the design of control interfaces. Our goal is to eventually be able to systematically match an individual to the appropriate and optimum BCI for his or her needs. Researchers interested in collaborating to utilize the BioGauges methodology and toolset in their work are welcomed to contact the first author.

Acknowledgements We would like to thank Dr. Brendan Allison for his shared expertise on characterizing control of BCIs and his encouragement of this chapter. We would also like to thank our research sponsor, the National Science Foundation, CISE/IIS for their support on this project. Lastly, we would like to thank members of the GSU/GT BrainLab for their expertise in implementing the BioGauges toolset.

References

1. Allison, B.Z. (ed.): The I of BCIs: Next generation interfaces for brain–computer interface systems that adapt to individual users. Human–Computer Interaction: Novel Interaction Methods and Techniques. Springer, Berlin, Heidelberg (2009)
2. Allison, B.Z.: Toward ubiquitous BCIs. In: Graimann, B., Allison, B.Z., and Pfurtscheller, G. (eds.) Brain–computer interfaces: Revolutionizing Human–Computer Interaction, pp. 357–387. Springer, Berlin, Heidelberg (2010)
3. Allison, B.Z., Lüth. T., et al.: BCI demographics: How many (and what kinds of) people can use an SSVEP BCI? IEEE Trans. Neural. Syst. Rehabil. Eng. **18**(2), 107–116 (2010)
4. Allison, B.Z., Neuper, C.: Could anyone use a BCI? in Brain–Computer Interfaces: Applying Our Minds to Human-Computer Interaction, Human-Computer Interaction Series In: Tan, D.S., Nijholt, A. (eds.) pp. 35–54, Springer Verlag, London (2010)
5. Bin, G., Gao, X., et al.: An online multi-channel SSVEP-based braincomputer interface using a canonical correlation analysis method. J. Neural Eng. **6**(4) (2009)
6. Birbaumer, N., Cohen, L.: Brain–computer interfaces: communication and restoration of movement in paralysis. J. Physiol. **579**, 621–636 (2007)
7. Birch, G.E., Mason, S.G.: Brain–computer interface research at the Neil Squire Foundation. IEEE Trans. Rehab. Eng. **8**(2), 193–195 (2000)
8. Blankertz, B., Sannelli. C., et al.: Neurophysiological predictor of SMR-based BCI performance. NeuroImage **51**(4), 1303–1309 (2010)
9. Brunner, C., Allison, B.Z., et al.: A comparison of three brain–computer interfaces based on event-related desynchronization, steady state visual evoked potentials, or a hybrid approach using both signals. J. Neural Eng. **8**(2), 025010 (2011)
10. Card, S.K., English. W.K., et al.: Evaluation of mouse, rate-controlled isometric joystick, step keys and text keys for text selection on a CRT. Ergonomics **21**(8), 601–613 (1978)
11. Donchin, E., Spencer, K.M., et al.: The mental prosthesis: Assessing the speed of a P300-based brain–computer interface. IEEE Trans. Neural Syst. Rehabil. Eng. **8**(2), 174–179 (2000)
12. Farwell, L.A., Donchin, E.: Talking off the top of your head: Toward a mental prothesis utilizing event-related brain potentials. Electroencephalogr. Clin. Neurophysiol. **70**(6), 510–523 (1988)
13. Fitts, P.M.: The information capacity of the human motor system in controlling the amplitude of movement. J. Exp. Psychol. **47**, 381–391 (1954)
14. Huggins, J.E., Wren, P.A., et al.: What would brain–computer interface users want? Opinions and priorities of potential users with amyotrophic lateral sclerosis. Amyotroph. Lateral Scler. **12**(5), 1–8 (2011)
15. Jin, J., Allison, B.Z., et al.: An adaptive P300 based control system. J. Neural Eng. **8**(3), 036006 (2011)
16. Kennedy, P.R., Adams, K.D.: A decision tree for brain–computer interface devices. IEEE Trans. Neural Syst. Rehabil. Eng. **11**(2), 148–150 (2003)
17. Krusienski, D., Sellers, E., et al.: Toward enhanced P300 speller performance. J. Neurosci. Methods **167**(1), 15–21 (2008)
18. Kübler, A., Birbaumer, N.: Brain–computer interfaces and communication in paralysis: Extinction of goal directed thinking in completely paralysed patients? Clin. Neurophysiol. **119**(11), 2658–2666 (2008)
19. Kübler, A., Kotchoubey, B., et al.: Brain–computer communication: unlocking the locked-in. Psychol. Bull. **127**, 358–375 (2001)
20. Kübler, A., Müller, K.R.: An Introduction to Brain Computer Interfacing. in Toward Brain–Computer Interfacing, Neural Information Processing Series, G. Dornhege, Millan, J.d.R., Hinterberger, T., McFarland, D.J., Muller, K.R., (eds.), MA: MIT Press, Cambridge, pp. 1–25 (2007)

21. Lalor, E., Kelly, S.P., et al.: Brain–computer interface based on the steady-state VEP for immersive gaming control. 2nd International Brain–Computer Interface Workshop Training Course, Graz, vol. 49, pp. 63–64 (2004)
22. Li, Y., Long, J., et al.: An EEG-based BCI system for 2-D cursor control by combining Mu/Beta rhythm and P300 potential. IEEE Trans. Biomed. Eng. **57**(10), 2495–2505 (2010)
23. Lin, Y.-P., Wang, C.-H., et al.: EEG-based emotion recognition in music listening. IEEE Trans. Biomed. Eng. **57**(7), 1798–1806 (2010)
24. Luo, A., Sullivan, T.J.: A user-friendly SSVEP-based brain–computer interface using a time-domain classifier. J. Neural Eng. **7**(2), 026010 (2010)
25. Martinez, P., Bakardjian, H., Cichocki, A.: Fully online multicommand brain–computer interface with visual neurofeedback using SSVEP paradigm. Computational Intelligence and Neuroscience **9** (2007). DOI 10.1155/2007/94561
26. Mason, S.G., Bashashati, A., et al.: A comprehensive survey of brain interface technology designs. Ann. Biomed. Eng. **35**(2), 137–169 (2007)
27. Mason, S.G., Moore Jackson, M.M., et al.: A general framework for characterizing studies of brain interface technology. Ann. Biomed. Eng. **33**(11), 1653–1670 (2005)
28. Mason, S.G., Moore, M.M., et al.: Designing pointing devices using brain–computer interface technology. First International IEEE EMBS Conference on Neural Engineering, Capri Island, Italy (2003)
29. McFarland, D.J., Sarnacki, W.A., et al.: Electroencephalographic (EEG) control of three-dimensional movement. J. Neural Eng. **7**(3), 036007 (2010)
30. Millan, E.Z., Furlong, T.M., et al.: Accumbens shell-hypothalamus interactions mediate extinction of alcohol seeking. J. Neurosci. **30**, 4626–4635 (2010)
31. Moore Jackson, M.M., Mason, S.M., et al.: Analyzing trends in brain interface technology: A method to compare studies. Ann. Biomed. Eng. **34**(5), 859–877 (2006)
32. Moore, M.M.: Real-world applications for brain–computer interface technology. IEEE Trans. Neural Syst. Rehabil. Eng. **11**(2), 162–165 (2003)
33. Moore, M.M., Dua, U.: A galvanic skin response interface for people with severe motor disabilities. 6th International ACM SIGACCESS Conference on Computers and Accessibility (ASSETS), Atlanta, GA (2004)
34. Neumann, N., Kübler, A.: Training locked-in patients: a challenge for the use of brain–computer interfaces. IEEE Trans. Neural Syst. Rehabil. Eng. **11**, 169–172 (2003)
35. Nijboer, F., Broermann, U.: Brain–computer interfaces for communication and control in locked-in patients. In: Graimann, B., Pfurtscheller, G., Allison, B.Z. (eds.) Brain–Computer Interfaces – revolutionizing Human–Computer Interaction, pp. 185–201. Springer, Berlin (2010)
36. Nijholt, A., Tan, D.S., et al.: Brain–computer interfacing for intelligent systems. IEEE Intell. Syst. **23**(3), 72–79 (2008)
37. Pfurtscheller, G., Allison, B.Z., et al.: The hybrid BCI. Front. Neurosci. **4**(42) (2010)
38. Pfurtscheller, G., Flotzinger, D., et al.: EEG-based brain computer interface (BCI). Search for optimal electrode positions and frequency components. Med. Prog. Technol. **21**(3), 111–121 (1995–1996)
39. Pfurtscheller, G., Neuper, C.: Dynamics of sensorimotor oscillations in a motor task. Brain–Computer Interfaces: In: Graimann, B., Pfurtscheller, G., Allison, B.Z. (eds.) Revolutionizing Human–Computer Interaction, pp. 47–64. Springer, Berlin Heidelberg (2010)
40. Pregenzer, M., Pfurtscheller, G., et al.: Selection of electrode positions for an EEG-based Brain Computer Interface. Biomed. Tech. (Berl.) **39**(10), 264–269 (1994)
41. Randolph, A.B., Jackson, M.M., et al.: Individual characteristics and their effect on predicting Mu rhythm modulation. Int. J. Hum. Comput. Interact. **27**(1), 1–14 (2011)
42. Randolph, A.B., McCampbell, L.A., et al.: Methodology for characterizing biometric interface systems. Neuroscience 2005: The 35th Annual Meeting of the Society for Neuroscience, Washington, DC (2005a)
43. Randolph, A.B., McCampbell, L.A., et al.: Controllability of galvanic skin response. 11th International Conference on Human–Computer Interaction (HCII), Las Vegas, NV (2005b)

44. Randolph, A.B., Moore Jackson, M.M., et al.: BioGauges for characterizing biometric interface systems. 3rd International Meeting of Brain–Computer Interface Technology, Rensselaerville, NY (2005)
45. Randolph, A.B., Moore Jackson, M.M., et al.: BioGauges: Toward more objective evaluation of biometrically-based interfaces. 6th Annual Workshop on HCI Research in MIS, Montreal, Canada, Association for Information Systems (AIS) (2007)
46. Randolph, A.B., Moore Jackson, M.M.: Assessing fit of nontraditional assistive technologies. ACM Trans. Access. Comput. **2**(4), 1–31 (2010)
47. Scherer, R., Lee, F., et al.: Towards self-paced brain–computer communication: Navigation through virtual worlds. IEEE Trans. Biomed. Eng. **55**(2), 675–682 (2008)
48. Transducer. Random House Webster's Unabridged Dictionary, Random House Reference: 2256 (2005)
49. Wolpaw, J.R., Birbaumer, N., et al.: Brain–computer interfaces for communication and control. Clin. Neurophysiol. **113**(6), 767–791 (2002)
50. Wolpaw, J.R., McFarland, D.J., et al.: An EEG-based brain–computer interface for cursor control. Electroencephalogr. Clin. Neurophysiol. **78**(3), 252–259 (1991)
51. Zickler, C., Di Donna, V., et al.: BCI applications for people with disabilities: Defining user needs and user requirements. In: Emiliani, P.L., Burzagli, L., Como, A., Gabbanini F., Salimen, A.-L. (eds.) Assistive Technology from Adapted Equipment to Inclusive Environments, pp. 185–189. IOS Press, Amsterdam (2009)

Index

B. Z. Allison et al. (eds.), *Towards Practical Brain-Computer Interfaces*, Biological and Medical Physics, Biomedical Engineering, DOI 10.1007/978-3-642-29746-5,
© Springer-Verlag Berlin Heidelberg 2012